Neural Networks, Fuzzy Systems and Other Computational Intelligence Techniques for Advanced Process Control

Neural Networks, Fuzzy Systems and Other Computational Intelligence Techniques for Advanced Process Control

Editors

Jie Zhang
Meihong Wang

Basel • Beijing • Wuhan • Barcelona • Belgrade • Novi Sad • Cluj • Manchester

Editors

Jie Zhang
Newcastle University
Newcastle upon Tyne
UK

Meihong Wang
University of Sheffield
Sheffield
UK

Editorial Office
MDPI AG
Grosspeteranlage 5
4052 Basel, Switzerland

This is a reprint of articles from the Special Issue published online in the open access journal *Processes* (ISSN 2227-9717) (available at: https://www.mdpi.com/journal/processes/special_issues/advanced_process_control).

For citation purposes, cite each article independently as indicated on the article page online and as indicated below:

Lastname, A.A.; Lastname, B.B. Article Title. *Journal Name* **Year**, *Volume Number*, Page Range.

ISBN 978-3-7258-1653-8 (Hbk)
ISBN 978-3-7258-1654-5 (PDF)
doi.org/10.3390/books978-3-7258-1654-5

Contents

About the Editors

Jie Zhang

Dr Jie Zhang received his PhD in Control Engineering from City University, London, in 1991. He has been with the School of Chemical Engineering and Advanced Materials, Newcastle University, UK, since 1991 and is a Reader in Process Systems Engineering. His research interests are in process system engineering, including process modeling, batch process control, process monitoring, and computational intelligence. He has published over 400 papers in international journals, books, and conference proceedings. He is on the Editorial Boards of several journals, including *Neurocomputing, Network: Computation in Neural Systems, Processes*, and *PLoS ONE*.

Meihong Wang

Professor Meihong Wang leads the Process and Energy Systems Engineering Research Group at the University of Sheffield, UK, with expertise in modeling, optimization, and control for power plants, carbon capture utilization and storage, energy storage, and bio-energy. He joined the Centre for Process Systems Engineering at Imperial College London and UCL in 1999. He worked at Alstom Power, Cranfield University, and the University of Hull from 2004 to 2016. He has published over 200 papers. He has been an investigator in over 20 research projects worth over £20 million from UK Research Councils and the European Union. One of them was awarded the Ludwig Mond Prize 2014 by IMechE. He was the joint winner of the Nigeria Prize for Science 2019 (the most important prize in the African Continent). A research project on intensified carbon capture led by Prof. Wang was runner-up in the IChemE Global Award 2019 (Energy Category). He was Siemens PSE MBI Prize Winner 2022.

processes

MDPI

Editorial

Special Issue: Neural Networks, Fuzzy Systems and Other Computational Intelligence Techniques for Advanced Process Control

Jie Zhang [1,*] and Meihong Wang [2]

[1] School of Engineering, Merz Court, Newcastle University, Newcastle upon Tyne NE1 7RU, UK
[2] Department of Chemical and Biological Engineering, The University of Sheffield, Mappin Street, Sheffield S1 3JD, UK; meihong.wang@sheffield.ac.uk
*   Correspondence: jie.zhang@newcastle.ac.uk; Tel.: +44-191-2087240

Citation: Zhang, J.; Wang, M. Special Issue: Neural Networks, Fuzzy Systems and Other Computational Intelligence Techniques for Advanced Process Control. *Processes* **2023**, *11*, 2278. https://doi.org/10.3390/pr11082278

Received: 17 July 2023
Accepted: 26 July 2023
Published: 28 July 2023

Computational intelligence (CI) techniques have developed very fast over the past two decades, with many new methods emerging. Novel machine learning techniques, such as deep learning, convolutional neural networks, deep belief networks, long short-term memory networks, and reinforcement learning, have been successfully applied to solve many complicated problems ranging from image processing to natural language processing. These novel CI techniques have also been applied to process systems engineering areas, with many successful applications reported, such as in the data-driven modelling of nonlinear processes, inferential estimation and soft sensors, intelligent process monitoring, and process optimisation. This Special Issue (https://www.mdpi.com/journal/processes/special_issues/advanced_process_control accessed on 17 July 2023) includes 17 papers on CI techniques applied to the areas related to advanced process control.

Liu et al. [1] present a reference-model-based neural network (NN) control method for a multi-input multi-output (MIMO) temperature system. A reference model is introduced to provide the teaching signal for the NN controller. The control inputs for the MIMO system are given by the sum of the outputs of the conventional integral-proportional-derivative (I-PD) controller and the outputs of the neural network controller. It is shown that the proposed NN control method can not only improve the transient response of the system, but also realize temperature uniformity in the MIMO temperature system. The proposed control system is demonstrated on simulations in the MATLAB/SIMULINK environment and a Digital-Signal-Processor-based experimental platform.

Hung [2] presents a memetic particle swarm optimization (MPSO) algorithm combined with a noise variance estimator to address the issue of performance decay in a direction of arrival (DOA) estimator under a non-uniform noise and low signal-to-noise ratio (SNR) environment. The proposed MPSO incorporates the re-estimation of noise variance and iterative local search algorithms into the particle swarm optimization (PSO) algorithm, resulting in higher efficiency and a reduction in non-uniform noise effects under a low SNR. In the proposed algorithm, PSO is initially utilized to evaluate the signal DOA using a subspace maximum-likelihood (SML) method. Then, the best position of the swarm to estimate the noise variance is determined and the iterative local search algorithm is built to reduce the non-uniform noise effect. The proposed method uses the SML criterion to rebuild the noise variance for the iterative local search algorithm to reduce non-uniform noise effects. The proposed method is demonstrated by simulation.

Xue et al. [3] propose a quality integrated fuzzy inference system (QFIS) to quantify the deviations of the operating variables and the product quality from their target values in order to overcome the measurement delay of the product quality and to estimate the reliability of the operation status, as well as the product quality, to enhance the performance of the safety monitoring system. A quality-weighted multivariate inverted normal loss function is proposed to quantify the deviation of the product quality from the target value in order to overcome the measurement delay. Vital process safety variables are identified according to the expert knowledge. Then, the quality loss and the vital variables are used as inputs for an elaborate fuzzy inference system to estimate the process reliability of the fluorochemical engineering processes. By integrating the abundant expert knowledge and a data-driven quality prediction model to design the fuzzy rules of QFIS, the operation reliability can be enhanced and the product quality can also be monitored on-line. The proposed method is applied to a real fluorochemical engineering process located in East China and the benchmark Tennessee Eastman process.

Zhang et al. [4] propose an improved finite control set model predictive torque control strategy for induction motor (IM) control. The proposed strategy is based on a novel fuzzy adaptive speed controller and an adaptive weighting factor for the tuning strategy to reduce the speed, torque, and flux ripples caused by different factors. Both simulation and hardware-in-loop tests are conducted on a 1.1 kW IM drive to verify the proposed ripple reduction algorithms.

Aguilar-López et al. [5] propose a two-input two-output control strategy for an exothermic continuous chemical reactor. The reactor temperature is regulated by a standard proportional-integral controller. An optimal controller is activated to increase the reactor productivity in terms of the mass of the product. The optimal control strategy is based on a Euler–Lagrange framework, where the Lagrangian is based on the model equations of the reactor and the optimal controller is coupled with an uncertainty estimator to infer the unknown terms required by the proposed controller. The proposed method is demonstrated on a simulated continuous stirred tank reactor with a Van de Vusse chemical reaction.

Almarashi et al. [6] study the group acceptance sampling plan in the case where (i) the lifetime of the items follows the Marshall–Olkin Kumaraswamy exponential distribution and (ii) a large number of items, considered as a group, can be tested at the same time. When the consumer's risk and the test termination period are defined, the key design parameters can be extracted. The minimum ratios of the true average life to the specified average life are calculated. The proposed technique is explained using real-world data on the breaking stress of carbon fibres.

Zhang et al. [7] propose double-layer back propagation neural networks for the learning of PID control parameters. One network is used to fit the relationship among the working parameters, the control parameters, and the control performance. Another network is used to fit the relationship between the working condition parameters and the selected control parameters, and to realize the adaptive adjustment of the PID control parameters according to the working condition parameters. The effectiveness of the proposed control method was verified by a simulation and experiment on the hydraulic drive unit of a legged robot.

Yang et al. [8] propose a self-organizing radial basis function neural network (RBFNN) based on network sensitivity to improve the generalization performance for nonlinear process modelling. In the proposed approach, a self-organizing structure optimization strategy is designed based on the sensitivity measurement to adjust the structure and parameters of RBFNN. The convergence of the proposed RBFNN-GP is analysed. The proposed method is applied to two numerical case studies and a membrane bio-reactor in a wastewater treatment plant.

Zhai et al. [9] propose an adaptive depth-wise separable dilated convolution and multigrained cascade forest (ADSD-gcForest) fault diagnosis model for fault diagnosis in bearings. The multiscale convolution, combined with the convolutional attention mechanism, concentrates on effectively extracting fault information under strong noise, and the Meta-Activate or Not (Meta-ACON) activation function is integrated to adaptively optimize the model structure according to the characteristics of input samples. Then, gcForest, as the classifier, outputs the final diagnosis result. The proposed method is applied to bearings failure diagnoses under various noise and load conditions.

Wang et al. [10] present a new fault detection scheme using the mutual *k*-nearest neighbour (MkNN) method to solve the problem of pseudo neighbour caused by outliers or large noises in the dataset. In the proposed method, the distance statistics for process monitoring are calculated using the MkNN rule instead of kNN, so that the influence of outliers in the training data is eliminated. The effectiveness of the proposed method is demonstrated through numerical examples and the benchmark Tennessee Eastman process.

Wu et al. [11] present a parameter identification method based on the hybrid genetic algorithm for the control system of double-fed induction generator converters. A strategy of "individual identification, elite retention, and overall identification" is proposed in the improved genetic algorithm, which adopts the generation gap value and immune strategy. The proposed parameter identification method is applied to a wind farm in North China for maximum power point tracking, constant speed, and the constant power operation conditions of the wind turbine.

Chen et al. [12] present an event-triggered H_∞ asynchronous filtering for Markov jump nonlinear systems with varying delay and unknown probabilities. The devised filter is mode dependent and asynchronous compared with the original system, which is represented by a hidden Markov model. Both the probability information involved in the original system and the filter are assumed to be only partly available. Under this framework, via employing the Lyapunov–Krasovskii functional and matrix inequality transformation techniques, a sufficient condition is given and the filter is further devised to ensure that the resulting filtering error dynamic system is stochastically stable, with a desired H_∞ disturbance attenuation performance. The proposed filter design method is demonstrated through a numerical example.

Muhsin and Zhang [13] present the multi-objective optimization of a crude oil hydrotreating (HDT) process with a crude atmospheric distillation unit using data-driven models based on bootstrap-aggregated neural networks. The HDT of the whole crude oil has economic benefit compared to the conventional HDT of individual oil products. Reliable data-driven models for this process are developed using bootstrap-aggregated neural networks to overcome the difficulty in developing accurate mechanistic models and the computational burden of utilizing such models in optimization. Reliable optimal process operating conditions are derived by solving a multi-objective optimization problem, incorporating the minimization of the widths of model prediction confidence bounds as additional objectives. The multi-objective optimization problem is solved using the goal-attainment method. The proposed method is demonstrated on the HDT of crude oil, with a crude distillation unit simulated using Aspen HYSYS.

Gao et al. [14] propose using the extended Kalman filter algorithm and backpropagation neural network to build a state of charge (SOC) estimation model of the electric vehicle battery (E-cell) to improve the estimation accuracy. Three working conditions, constant current discharge, pulse discharge, and urban dynamometer driving schedule, were considered. The enhanced estimation and tracking of the SOC of the E-cell can provide a data reference for vehicle battery management, and is of great significance for improving the battery performance and energy utilization in electric vehicles.

Berard et al. [15] present a range of rate of error change–fuzzy logic controller designs to demonstrate the tunability of the controller for different haemorrhage scenarios. Five different controller setups are configured with different membership functions to create more- and less-aggressive controller designs. It is shown that the proposed controllers are well-suited for haemorrhagic shock resuscitation and can be tuned to meet the response rates set by clinical practice guidelines for this application.

Wang et al. [16] propose a traffic light timing optimization method based on a double duelling deep Q-network, MaxPressure, and self-organizing traffic lights, which control traffic flows by dynamically adjusting the duration of traffic lights in a cycle, and the phase can be switched depending on the rules set in advance and the pressure of the lane. In the proposed method, each intersection corresponds to an agent, and the road entering the intersection is divided into grids, with each grid storing the speed and position of a car, thus forming the vehicle information matrix and acting as the state of the agent. Experimental results show that the proposed method has superior performance in light and heavy traffic flow scenarios, and can reduce the waiting time and travel time of vehicles and improve the traffic efficiency of an intersection.

Ang et al. [17] propose a modified particle swarm optimization (PSO) variant with two-level learning phases to train an artificial neural network (ANN) for image classification. A multi-swarm approach and a social learning scheme are designed in the primary learning phase to enhance the population diversity and the solution quality, respectively. Two modified search operators with different search characteristics are incorporated into the secondary learning phase to improve the algorithm's robustness in handling various optimization problems. The proposed algorithm is used to train ANN, by optimizing its weights, biases, and the selection of activation function for the given classification dataset. It is shown that ANN models trained by the proposed algorithm outperform those trained by existing PSO variants in terms of classification accuracy.

The above papers in this Special Issue demonstrate that CI techniques can significantly improve the performance of process control systems. As more advanced CI techniques have emerged in recent years, more CI-based advanced process control techniques will be reported in the near future.

Funding: This research received no external funding.

Conflicts of Interest: The authors declare no conflict of interest.

References

1. Liu, Y.; Xu, S.; Hashimoto, S.; Kawaguchi, T. A reference-model-based neural network control method for multi-input multi-output temperature control system. *Processes* **2020**, *8*, 1365. [CrossRef]
2. Hung, J.C. DOA estimation in non-uniform noise based on subspace maximum likelihood using MPSO. *Processes* **2020**, *8*, 1429. [CrossRef]
3. Xue, F.; Li, X.; Zhou, K.; Ge, X.; Deng, W.; Chen, X.; Song, K. A Quality integrated fuzzy inference system for the reliability estimating of fluorochemical engineering processes. *Processes* **2021**, *9*, 292. [CrossRef]
4. Zhang, Z.; Wei, H.; Zhang, W.; Jiang, J. Ripple attenuation for induction motor finite control set model predictive torque control using novel fuzzy adaptive techniques. *Processes* **2021**, *9*, 710. [CrossRef]
5. Aguilar-López, R.; Mata-Machuca, J.L.; Godinez-Cantillo, V. A TITO control strategy to increase productivity in uncertain exothermic continuous chemical reactors. *Processes* **2021**, *9*, 873. [CrossRef]
6. Almarashi, A.M.; Khan, K.; Chesneau, C.; Jamal, F. Group acceptance sampling plan using Marshall–Olkin Kumaraswamy exponential (MOKw-E) distribution. *Processes* **2021**, *9*, 1066. [CrossRef]
7. Zhang, M.-L.; Zhang, Y.-J.; He, X.-L.; Gao, Z.-J. Adaptive PID control and its application based on a double-layer BP neural network. *Processes* **2021**, *9*, 1475. [CrossRef]
8. Yang, Y.; Wang, P.; Gao, X. A novel radial basis function neural network with high generalization performance for nonlinear process modelling. *Processes* **2022**, *10*, 140. [CrossRef]
9. Zhai, S.; Wang, Z.; Gao, D. Bearing fault diagnosis based on a novel adaptive ADSD-gcForest model. *Processes* **2022**, *10*, 209. [CrossRef]
10. Wang, J.; Zhou, Z.; Li, Z.; Du, S. A novel fault detection scheme based on mutual k-nearest neighbor method: Application on the industrial processes with outliers. *Processes* **2022**, *10*, 497. [CrossRef]
11. Wu, L.; Liu, H.; Zhang, J.; Liu, C.; Sun, Y.; Li, Z.; Li, J. Identification of control parameters for converters of doubly fed wind turbines based on hybrid genetic algorithm. *Processes* **2022**, *10*, 567. [CrossRef]
12. Chen, H.; Liu, R.; Xia, W.; Li, Z. Event-triggered filtering for delayed Markov jump nonlinear systems with unknown probabilities. *Processes* **2022**, *10*, 769. [CrossRef]
13. Muhsin, W.; Zhang, J. Multi-objective optimization of a crude oil hydrotreating process with a crude distillation unit based on bootstrap aggregated neural network models. *Processes* **2022**, *10*, 1438. [CrossRef]
14. Gao, Y.; Ji, W.; Zhao, X. SOC Estimation of E-cell combining BP neural network and EKF algorithm. *Processes* **2022**, *10*, 1721. [CrossRef]
15. Berard, D.; Vega, S.J.; Avital, G.; Snider, E.J. Dual input fuzzy logic controllers for closed loop hemorrhagic shock resuscitation. *Processes* **2022**, *10*, 2301. [CrossRef]

16. Wang, B.; He, Z.; Sheng, J.; Chen, Y. Deep reinforcement learning for traffic light timing optimization. *Processes* **2022**, *10*, 2458. [CrossRef]

17. Ang, K.M.; Chow, C.E.; El-Kenawy, E.-S.M.; Abdelhamid, A.A.; Ibrahim, A.; Karim, F.K.; Khafaga, D.S.; Tiang, S.S.; Lim, W.H. A modified particle swarm optimization algorithm for optimizing artificial neural network in classification tasks. *Processes* **2022**, *10*, 2579. [CrossRef]

Article

A Modified Particle Swarm Optimization Algorithm for Optimizing Artificial Neural Network in Classification Tasks

Koon Meng Ang [1], Cher En Chow [1], El-Sayed M. El-Kenawy [2,*], Abdelaziz A. Abdelhamid [3], Abdelhameed Ibrahim [4], Faten Khalid Karim [5], Doaa Sami Khafaga [5], Sew Sun Tiang [1,*] and Wei Hong Lim [1,*]

[1] Faculty of Engineering, Technology and Built Environment, UCSI University, Kuala Lumpur 56000, Malaysia
[2] Department of Communications and Electronics, Delta Higher Institute of Engineering and Technology, Mansoura 35111, Egypt
[3] Department of Computer Science, Faculty of Computer and Information Sciences, Ain Shams University, Cairo 11566, Egypt
[4] Computer Engineering and Control Systems Department, Faculty of Engineering, Mansoura University, Mansoura 35516, Egypt
[5] Department of Computer Sciences, College of Computer and Information Sciences, Princess Nourah bint Abdulrahman University, P.O. Box 84428, Riyadh 11671, Saudi Arabia
* Correspondence: skenawy@ieee.org (E.-S.M.E.-K.); tiangss@ucsiuniversity.edu.my (S.S.T.); limwh@ucsiuniversity.edu.my (W.H.L.)

Citation: Ang, K.M.; Chow, C.E.; El-Kenawy, E.-S.M.; Abdelhamid, A.A.; Ibrahim, A.; Karim, F.K.; Khafaga, D.S.; Tiang, S.S.; Lim, W.H. A Modified Particle Swarm Optimization Algorithm for Optimizing Artificial Neural Network in Classification Tasks. *Processes* **2022**, *10*, 2579. https:// doi.org/10.3390/pr10122579

Academic Editors: Jie Zhang and Meihong Wang

Received: 25 October 2022
Accepted: 1 December 2022
Published: 3 December 2022

Publisher's Note: MDPI stays neutral with regard to jurisdictional claims in published maps and institutional affiliations.

Abstract: Artificial neural networks (ANNs) have achieved great success in performing machine learning tasks, including classification, regression, prediction, image processing, image recognition, etc., due to their outstanding training, learning, and organizing of data. Conventionally, a gradient-based algorithm known as backpropagation (BP) is frequently used to train the parameters' value of ANN. However, this method has inherent drawbacks of slow convergence speed, sensitivity to initial solutions, and high tendency to be trapped into local optima. This paper proposes a modified particle swarm optimization (PSO) variant with two-level learning phases to train ANN for image classification. A multi-swarm approach and a social learning scheme are designed into the primary learning phase to enhance the population diversity and the solution quality, respectively. Two modified search operators with different search characteristics are incorporated into the secondary learning phase to improve the algorithm's robustness in handling various optimization problems. Finally, the proposed algorithm is formulated as a training algorithm of ANN to optimize its neuron weights, biases, and selection of activation function based on the given classification dataset. The ANN model trained by the proposed algorithm is reported to outperform those trained by existing PSO variants in terms of classification accuracy when solving the majority of selected datasets, suggesting its potential applications in challenging real-world problems, such as intelligent condition monitoring of complex industrial systems.

Keywords: particle swarm optimization; artificial neural network; training algorithm; machine learning; two-level learning phases

1. Introduction

The human nervous system inspires an artificial neural network (ANN), and many artificial neurons act as interconnected processing elements in ANN to emulate the cerebral cortex of brain structure. In contrast to other conventional machine learning methods, ANNs demonstrate outstanding capabilities in generalizing, organizing, and learning the nonlinear data that are commonly observed in real-world scenarios [1]. Due to these appealing features, ANNs have been widely implemented to solve various real-world application, including intelligent condition monitoring [2], speech recognition [3], fault diagnosis [4], pothole classification [5], etc. Generally, ANN structure can be separated into three parts: the input layer, the hidden layer, and the output layer. Information flows in a single direction within an ANN model, i.e., from the input layer to the hidden layer, followed by the output layer [6]. Each neuron of the layers is incorporated with an activation function to convert the summed weighted input received by each neuron into the nonlinear output. This nonlinear characteristic serves as the cornerstone for ANN to have competitive performance in tackling various challenging machine learning tasks, such as the learning of complex data, functions approximation, and predictions [7].

Before deploying an ANN model to solve machine learning tasks, a training process is performed to determine the optimal combinations of weight and bias values for all neurons that can achieve minimum error. A gradient-based algorithm known as the backpropagation (BP) method is conventionally used to train ANN models [7]. At the beginning stage of ANN training, the initial weight and bias values of neurons are randomly generated, and the actual output values of the network are determined. The error signals between the desired and actual outputs are then calculated and backpropagated to the network to adjust the weight and bias values of each neuron. Despite its popularity, some drawbacks were reported when using the classical BP method to train ANN models, especially when solving complex nonlinear problems. For instance, the BP method has a high tendency to be trapped in local optima and fail to reach global optimum when dealing with solution regions with complex characteristics. The performance of the ANN model when solving complex problems can be significantly degraded due to suboptimal weight and bias values assigned to all neurons [7]. The convergence characteristic of the classical BP method in ANN training is also sensitive to the initial values of weight, bias, and network parameters (e g , activation function). There might be scenarios where the classical BP method produces poor initial weight and bias values for ANN during the training process, leading to compromised network performances [7]. These limitations of classical BP methods have motivated researchers to seek robust alternatives that can address ANN training problems more efficiently and effectively.

In recent years, metaheuristic search algorithms (MSAs) have emerged as promising solutions to tackle complex optimization problems, such as those reported in [8–13]. The excellent global search ability and stochastic characteristic of these MSAs can also be harnessed for training ANN models to solve classification or regression problems. Particle swarm optimization (PSO) is one of the most popular MSAs, and it is motivated by the collective behavior of bird flocks in locating food sources [14]. Each PSO particle can memorize its previous best searching experiences, and this unique feature distinguishes PSO from other MSAs [15]. The search trajectory of each PSO particle is adjusted based on its previous best experience and the best experience achieved by the entire population during the optimization process. Since the inception of the original PSO in 1995, many new variants have been proposed to solve various real-world optimization problems. Despite its appealing features, such as simplicity in implementation and fast convergence speed, the capability of original PSO to solve complex optimization problems such as ANN training remain questionable due to its high tendency to suffer from premature convergence issues when solving complex problems with high-dimensional search space. Therefore, more robust search mechanisms must be incorporated into the original PSO to balance the algorithm's exploration and exploitation searches.

1.1. Research Motivations

A PSO variant known as particle swarm optimization without velocity (PSOWV) was proposed in [16] by discarding the velocity component of each particle during the search process. PSOWV has demonstrated more competitive performance than the conventional PSO by solving simple benchmark problems with better accuracy in fewer iteration numbers. Despite its promising performance, PSOWV still suffers from some inherent drawbacks that might restrict its feasibility to solve more challenging optimization problems, such as the training of ANN models for classification or regression tasks.

Similar to most MSAs, PSOWV employs a conventional initialization scheme to randomly generate the initial population of particles during the search process. This conventional initialization scheme produces the initial position of each particle using uniform distribution without considering the characteristics or fitness landscapes of given optimization problems [17]. When particles are mistakenly initialized in local optima, it may lead to premature convergence issues and poor solution accuracy. In contrary, the convergence speed of the algorithm can be significantly deteriorated if particles are initialized at solution regions far away from the global optimum. It is more desirable to have a robust initialization scheme that can generate the initial position of each particle more systematically to ensure the initial population has a better solution quality.

When carefully inspecting the search operator of PSOWV, the new position of each particle is directly affected by the directional information of historical best positions, i.e., the personal best position of the particle itself and the global best position of the population. While these historical best positions are beneficial to accelerate the convergence speed of PSOWV at the early stage of optimization, they are less frequently updated at the latter stages [18]. When both personal and global best positions are overemphasized during the search process of PSOWV, the entrapment of these historical best positions into local optima at the early stage of the search process could misguide the remaining particles converging towards the inferior solution regions and lead to premature convergence. To prevent these undesirable scenarios, it is necessary for PSOWV to be incorporated with a more robust diversity preservation scheme when dealing with complex problems. The directional information offered by other non-historical best positions should be leveraged during the search process to prevent the potential negative impacts brought by personal and global best positions.

Finally, it is observed that PSOWV has limited ability to achieve proper trade-off between exploration and exploitation searches because only one search operator is used to perform the search process. Hence, PSOWV can only perform well in certain types of optimization problems (i.e., unimodal problem), and it exhibits poor optimization performance in other problem categories. When solving optimization problems with different complexity levels as governed by the characteristics of fitness landscapes, it is crucial for an algorithm to intelligently regulate its exploration and exploitation strength to locate the global optima [19]. The incorporation of multiple search operators with different levels of exploration and exploitation strengths can be envisioned as an alternative to enhance the robustness of the algorithm to solve different complex optimization problems competitively.

1.2. Research Contributions

To address the shortcomings of PSOWV in solving complex optimization problems such as the ANN model training, an enhanced variant known as multi-swarm-based particle swarm optimization with two-level learning phases (MSPSOTLP) is proposed. Apart from optimizing the weight and bias values of ANN models during the training process, the proposed MSPSOTLP can also determine the optimal activation function of an ANN model to solve the given classification problem. The main modifications introduced into MSPSOTLP to highlight its research contributions are summarized as follows:

1. A modified initialization scheme is incorporated into MSPSOTLP to generate an initial population with better robustness and diversity by leveraging the benefits of the chaotic system (CS) and oppositional-based learning (OBL).
2. In the primary learning phase of MSPSOTLP, both the multi-swarm concept and social learning concept are incorporated to promote rapid convergence of the population towards the optimal regions by enabling particles to learn from other superior population members while preserving the diversity level of population.
3. In the secondary learning phase of MSPSOTLP, two modified search operators with different characteristics are designed for each particle to perform searching with different levels of exploration and exploitation strengths, hence enabling the proposed algorithm to solve different types of optimization problems more competitively.

4. The performance of MSPSOTLP in solving global optimization problems is investigated using CEC 2014 benchmark functions. The classification performances of ANN models trained using MSPSOTLP are also evaluated with 16 datasets selected from UCI Machine Learning Repository. The proposed MSPSOTLP is proven more competitive than its peer algorithms at solving the benchmark functions and ANN training.

The remaining sections of this article are organized as follows. Related works of this study are explained in Section 2. The detailed search mechanisms of MSPSOTLP are described in Section 3. Extensive simulation studies performed to investigate the performance of MSPSOTLP in solving global optimization problems and its capability to train ANN models in solving classification problems are presented in Section 4. Finally, Section 5 concludes the research findings and future works.

2. Related Works

2.1. Particle Swarm Optimization (PSO)

PSO is inspired by the collective behavior of bird flocking or fish schooling to search for food sources [14]. Suppose that N is the population size and D is the dimensional size of an optimization problem. Each PSO particle is a potential solution to solve an optimization problem represented with a velocity vector of $V_n = [V_{n,1}, \ldots, V_{n,d}, \ldots, V_{n,D}]$ and a position vector of $X_n = [X_{n,1}, \ldots, X_{n,d}, \ldots, X_{n,D}]$, where $n = 1, \ldots, N$ and $d = 1, \ldots, D$ refer to the population index and dimension index, respectively. Unlike other MSAs, each PSO particle can memorize its previous best experience and the best experience achieved by the population, denoted as personal best position of $X_n^{Pbest} = \left[X_{n,1}^{Pbest}, \ldots, X_{n,d}^{Pbest}, \ldots, X_{n,D}^{Pbest} \right]$ and global best position of $G^{best} = \left[G_1^{best}, \ldots, G_d^{best}, \ldots, G_D^{best} \right]$, respectively. During the t-th iteration of the search process, the new velocity $V_{n,d}^{t+1}$ of each n-th particle in any d-th dimension is adjusted based on the corresponding dimensional components of the personal best position $X_{n,d}^{Pbest,t}$ (i.e., self-cognitive component) and global best position $G_d^{best,t}$ (i.e., social component) as follows:

$$V_{n,d}^{t+1} = \omega V_{n,d}^t + c_1 r_1 \left(X_{n,d}^{Pbest,t} - X_{n,d}^t \right) + c_2 r_2 \left(G_d^{best,t} - X_{n,d}^t \right) \tag{1}$$

where ω is an inertia weight; c_1 and c_2 are acceleration coefficients; $r_1, r_2 \in [0,1]$ are two random numbers generated from a uniform distribution. Referring to $V_{n,d}^{t+1}$, the new position of each n-th particle in every d-th dimension is updated as:

$$X_{n,d}^{t+1} = V_{n,d}^{t+1} + X_{n,d}^t \tag{2}$$

The fitness of every n-th particle with updated position is evaluated as $f\left(X_n^{t+1} \right)$ and compared with those of the personal best position and global best position denoted as $f\left(X_n^{Pbest,t} \right)$ and $f\left(G^{best,t} \right)$, respectively. Both of $X_n^{Pbest,t}$ and $G^{best,t}$ will be updated as X_n^{t+1} if the latter solution is superior. The search process of PSO using Equations (1) and (2) is iterated until the predefined termination criteria are satisfied.

2.2. Particle Swarm Optimization without Velocity (PSOWV)

PSOWV is a velocity-discarded version of PSO proposed in [16], aiming to enhance the ability of PSO to locate the global optimum of a given problem with a lesser iteration number. During the search process, the d-th dimension of position for every n-th PSOWV particle (i.e., $X_{n,d}^{t+1}$) can be updated based on the random linear combination between its personal best position (i.e., $X_{n,d}^{Pbest,t}$) and the global best position (i.e., $G_d^{best,t}$) of the population in the same dimensional component as follows:

$$X_{n,d}^{t+1} = c_1 r_1 X_{n,d}^{Pbest,t} + c_2 r_2 G_d^{best,t} \tag{3}$$

The fitness evaluation and procedure to update historically best positions (i.e., $X_n^{Pbest,t}$ and $G^{best,t}$) of PSOWV are similar to those of the original PSO. Although PSOWV can solve the simple benchmark functions with a faster convergence speed than the original PSO, its feasibility to solve more challenging real-world optimization problems, such as training ANN models, remains unexplored. Furthermore, preliminary studies also revealed the high tendency of PSOWV to suffer premature convergence issues because its search behavior is governed by a single search operator that highly relies on the directional information provided by historically best positions.

2.3. PSO Variants

Despite appealing characteristics, such as rapid convergence speed and simplistic implementation, the original PSO tends to suffer from rapid loss of population diversity and premature convergence issues when solving more complex optimization problems. Proper balancing of exploration and exploitation searches is considered a fundamental cornerstone for MSAs such as PSO to solve different optimization problems competitively. Therefore, various modification schemes were proposed by researchers over the years to address the demerits of the original PSO.

Parameter adaptation is a popular enhancement strategy of PSO by determining the proper combination of control parameters that can govern its search trajectories. The control parameters to be adjusted include inertia weight, constriction factor, acceleration coefficients, and a variety of these parameters. Some notable PSO variants that were proposed with parameter adaptation approaches are reported in [20–23]. Neighborhood structure modification is another promising technique of PSO because it governs the broadcast rate of information between population members. Particularly, the fully connected population topology tends to be more exploitative, whereas the partially connected one has stronger exploitation strengths. PSO variants reported in [19,24–28] can achieve proper tradeoffs between the exploration and exploitation searches by varying their population topology with time or based on current search environments. Apart from modifying the neighborhood structure, it is also feasible to introduce single or multiple modified learning strategies into PSO for achieving performance enhancements such as those proposed in [29–36]. To alleviate potential negative impacts brought by the historically best solution (e.g., personal and global best positions), novel exemplars might be constructed by these modified learning strategies from the non-fittest solutions to guide the search process of particles more effectively while preserving the population diversity. Finally, the PSO's robustness in solving high-complexity real-world problems can be enhanced using the hybridization approach, such as those reported in [37–40]. Different hybridization frameworks [41] can be designed to leverage the strengths of search operators incorporated in other MSAs for compensating the drawbacks of PSO in solving certain problem classes. The strengths and limitations of selected PSO variants are analyzed and summarized in Table 1.

Table 1. Strengths and limitations of selected PSO variants.

Reference	Year	Strengths	Limitations
[20]	2019	• Acceleration coefficients and inertia weight of each particle were adjusted adaptively based on current search environment.	• High computation costs incurred by the novel method used to estimate the search environment in each iteration.
[21]	2019	• Inertia weights of particles were adjusted based on their personal best fitness. • Mutation scheme was performed on stagnated particles to preserve swarm diversity.	• Both adaptive inertia weight and mutation scheme were highly relying on the directional information of global best position.

Table 1. *Cont.*

Reference	Year	Strengths	Limitations
[22]	2019	• Periodic trigonometric functions were used to adjust the inertia weight and acceleration coefficient of particle.	• Limited ability to adjust exploration and exploitation searches with single search operator. • Search process only relied on personal and global best positions.
[23]	2021	• Gaussian white noise with different intensity was used to adjust the acceleration coefficients of particles adaptively. • Wider exploration search.	• Limited ability to adjust exploration and exploitation searches with single search operator. • Search process only relied on personal and global best positions.
[24]	2022	• Neighborhood structure of each particle was gradually increased from ring topology to fully connected topology.	• Limited flexibility to regulate exploration and exploitation searches of algorithm because the swarm diversity level is increased monotonically.
[19]	2018	• Neighborhood structure of each particle can be adaptively maintained, decreased, increased or shuffled by referring to the search track record of population.	• Expensive computation cost used to adaptively adjust the neighborhood structure of each particle. • Laborious works to fine tune the newly introduced parameters.
[25]	2018	• Flexible neighborhood structure concept was introduced to achieve proper tradeoff between exploration and exploitation searches.	• Expensive computation cost used to adaptively adjust the neighborhood structure of each particle. • Relatively poor performances when dealing with unimodal problems.
[26]	2020	• A reinforcement learning concept (i.e., Q-learning) was used to select the optimal topology of particle.	• Expensive computation cost due to the involvement of Q-learning and computation of swarm diversity. • Search process only relied on personal and global best positions.
[27]	2020	• Multiple good quality subswarms were constructed based on correlations between group sequences. • A dynamic regrouping strategy was introduced to promote information sharing between different subswarms and accelerate their convergence speed.	• Overemphasized on the influences of historically best positions to guide the search process of subswarms. • Complex population division scheme. • Laborious works to fine tune the newly introduced parameters.
[28]	2020	• Benefits of holonic organization in multiagent system were leveraged to achieve proper tradeoff between exploration and exploitation searches.	• Relatively slow convergence speed when locating the global optima of unimodal problems.
[29]	2017	• Multiple subswarms were constructed based on the fitness levels of particles. • Directional information of non-fittest particles was used to guide the search process.	• Relatively poor performances when dealing with low and medium scale optimization problems.
[30]	2020	• Exemplar used to guide the general swarm was derived from the mean positions of elitist swarm.	• Limited enhancement of swarm diversity because mean position used to guide the population search was shared by all particles.
[31]	2020	• Positions of other particles were updated using mainstream and stochastic learning strategies. • Global worst position was handled using terminal replacement mechanism.	• Limited enhancement of swarm diversity because mean position used to guide the population search was shared by all particles.

Table 1. *Cont.*

Reference	Year	Strengths	Limitations
[32]	2020	• Forgetting ability was introduced to maintain the population diversity of algorithm.	• Neglected the potential benefits brought by non-fittest particles to guide the search process.
[33]	2019	• Dimensional learning strategy and comprehensive learning strategy were introduced to achieve proper tradeoff between exploration and exploitation searches.	• High fitness evaluation numbers were consumed by dimensional learning strategy when generating the exemplar.
[34]	2019	• Oppositional-based learning and convex combination concepts were used to generate exemplar for the fittest particle.	• Strong dependency on historically best positions used to guide the non-fittest particles.
[35]	2020	• Optimal guide creation module was designed to generate a global exemplar based on two nearest neighbors of global best position.	• Expensive computation cost due to the construction of global exemplar in every iteration.
[36]	2021	• Construction of main swarm and hover swarm as diversity maintenance scheme. • Construction of unique exemplar for main swarm and hover swarm,	• Expensive computation cost due to the binary population division scheme and construction of unique exemplar in every iteration.
[37]	2019	• Crossover and mutation of genetic algorithm were used to enhance the exploitation and exploration searches of PSO, respectively.	• Huge memory consumption to store the individual solutions that offered significant performance gains.
[38]	2019	• Grey wolf optimizer was used to update the positions of some particles to enhance exploration.	• Increasing execution time due to sequential cascading of grey wolf optimizer and PSO.
[39]	2020	• Butterfly optimization algorithm was hybridized with PSO to improve the exploration ability.	• Neglected the potential benefits brought by non-fittest particles to guide the search process.
[40]	2022	• Differential evolution was hybridized with PSO to achieve better balancing of exploration and exploitation searches of algorithm	• Increasing execution time due to sequential cascading of differential evolution and PSO.

2.4. Application of MSAs in Training ANN Models

Given the drawbacks of the conventional BP method, numerous MSAs were designed as promising alternatives to train ANN models with more robust network performances [42]. A two-layer PSO (PSO-PSO) algorithm was proposed in [43] to train a multilayer perceptron (MLP) neural network by interleaving two PSO algorithms. The first layer of PSO was used to optimize the number of perceptrons in the network architecture, whereas the second layer of PSO optimized the weight and bias values based on the network architecture obtained by the first layer. Nevertheless, simulation studies revealed that the ANN models optimized by PSO-PSO did not significantly outperform their peers when solving the classification benchmark datasets. A hybridized PSO and gravitational search algorithm (PSOGSA) was proposed [44] to optimize the weights and biases of the ANN model by leveraging the unique searching behaviors of PSO and GSA. When training the ANN model, PSO was incorporated to alleviate the inherent drawbacks of GSA, i.e., low convergence rate and high tendency to be trapped in local optima. A hybrid improved opposition-based PSO with a backpropagation algorithm (IOPSO-BPA) [45] was introduced to optimize the weight values of ANNs. Both oppositional-based learning and mutation schemes were incorporated as diversity preservation schemes of IOPSO-BPA.

Furthermore, the concepts of time-varying parameters were introduced to improve the convergence characteristic of the algorithm in optimizing the weights of ANN models. Kandasamy and Rajendran [46] proposed a hybrid algorithm to overcome the drawbacks of the conventional BP methods overfitting and entrapment in local optima when training ANN models. PSO was firstly employed to search for the optimal combination of trainable weights of the ANN model by minimizing the classification error function. Subsequently, the steepest descent method was applied to fine tune the near-optimal weight values encoded in the global best position to further improve classification accuracy. In [47], a self-adaptive and strategy-based PSO (SPS-PSO) was designed to solve the large-scale feature selection and ANN training problems for large-scale datasets. Five search operators with different characteristics were introduced in SPS-PSO, and an adaptive mechanism was used to assign a suitable operator for each particle to ensure it can perform searching with balanced exploration and exploitation strengths. A comprehensive adaptive PSO (ACPSO) was designed in [48] to tackle the denoising issue of ultrasound images by optimizing the functional-link neural network (FLNN). The velocity of the ACPSO was only dependent on the global best position, and its controlling parameters were adaptively adjusted based on the personal and global best positions. To mitigate the flooding issue, an ANN model was trained using PSO in [49] to formulate river flow modelling based on the weather and meteorological data. It was revealed that the river flow strongly correlates with selected variables, such as temperature, evaporation, and rainfall. In [50], hybrid intelligent modeling was developed using PSO and ANN (PSO-ANN) to predict the soil matric sanction, i.e., a useful metric to indicate the soil shear strength in addressing sudden landslides issue, with improved prediction accuracy.

The genetic algorithm (GA) [51] is another popular MSA used to train ANN models. A hybrid model of ANN and GA (ANN-GA) was proposed in [52] to optimize the weights and biases of the ANN model used for modeling the slump of ready-mix concrete. The initial weights and biases of ANN-GA were determined using GA and fine-tuned with the BP algorithm. ANN-GA was reported to outperform its peers by leveraging the benefits of GA and BP to promote its global and local search abilities, respectively. In [53], GABPNN was proposed by integrating GA into a BP-trained ANN to optimize the thickness of blow-molded polypropylene bellows. In GABPNN, the BP algorithm was first applied to train the weights of the ANN model using lesser learning samples, and these weights were further evolved in the feasible solution regions using GA. Contrary to ANN-GA, GA promoted the local search behavior of GABPNN by adopting an elitist strategy and a simulated annealing algorithm. GABPNN demonstrated its effectiveness and competitive performance in solving blow molding problems. A hybrid MLP ANN and GA approach (MLPANN-GA) was introduced in [54] to predict sludge bulking for water treatment plants. GA was incorporated to train the weights, activation functions, and thresholds of the MLPANN model. Simulation studies reported that the incorporation of GA to train MLPANN can increase the accuracy of the ANN model in estimating the sludge volume index.

In addition to GA, teaching-learning-based optimization (TLBO) [55] is another popular MSA widely used for ANN optimization. A TLBO-based ANN (TLBOANN) was proposed in [56] to estimate the energy consumption in Turkey. TLBO was applied to search for the optimal weights and biases of the ANN model by minimizing the error function based on the input data provided, i.e., gross domestic product, population, and import and export data in Turkey. An improved hybrid TLBO and ANN (iTLBO-ANN) was proposed in [57] to solve real-world building energy consumption forecasting problems. Three modifications, known as feedback stage, accuracy factor, and worst solution elimination, were introduced to improve the performance of TLBO in optimizing the weights and thresholds of the ANN model within a shorter time. The iTLBO-ANN outperformed other ANN models trained by GA and PSO by predicting building energy consumption with better accuracy and computational speed. Another TLBO-optimized ANN [58] was proposed to improve the performance in predicting the axial capacity of pile foundations. TLBO was used to train the weights of the ANN model by minimizing the mean square error produced when predicting the ultimate capacity of both driven and drilled shaft piles embedded in uncemented soils. The ANN model trained by TLBO outperformed that trained by BP by producing better variance accounted for and determination coefficient. A new TLBO variant known as TLBO-MLPs was used to train ANN model for data classification [59]. Additional mechanisms were introduced in both the teacher and learner phases of TLBO-MLPs to achieve realistic emulation of classroom teaching and learning that can lead to performance gain in solving complex optimization problems [60].

3. Proposed Methodology

3.1. Formulation of ANN Training as an Optimization Problem

An ANN model to be optimized in this study consists of a three-layer structure with P input neurons, Q hidden neurons, and R output neurons in input, hidden, and output layers, respectively, as illustrated in Figure 1. The neurons in each layer are considered a set of processing elements that are connected by weights with other layers. Suppose that I_p refers to the value of p-th neuron at the input layer, H_q represents the value of q-th neuron at the hidden layer, and O_r is the r-th neuron at the output layer, where $p = 1, \ldots, P, q = 1, \ldots, Q$, and $r = 1, \ldots, R$. Denote $W_{p,q}^H$ as the connection weight between I_p and H_q; $W_{q,r}^O$ as the connection weight between H_q and O_r; B_q^H and B_r^O as the biases of H_q and O_r, respectively. The values of each q-th hidden neuron H_q and r-th output neuron O_r can be produced by computing the sum of input weight with the presence of biases, followed by the non-linearization process of this weighted summation with an activation function of $\Phi(\cdot)$ expressed as follows:

$$H_q = \Phi\left(\sum_{p=1}^{P} W_{p,q}^H I_p + B_q^H\right) \tag{4}$$

$$O_r = \Phi\left(\sum_{q=1}^{Q} W_{q,r}^O I_q + B_r^O\right) \tag{5}$$

Contrary to the majority of existing ANN training algorithms that only focus on searching optimal weights and biases, this study also considers the optimal selection of the activation function used to solve a given classification task. The decision variables to be optimized by the proposed MSPSOTLP when training the ANN model include: (a) weights $W_{p,q}^H, W_{q,r}^O \in [-1, 1]$, (b) biases $B_q^H, B_r^O \in [-1, 1]$, and (c) index $K = \{1, 2, 3, 4, 5\}$ that refer to the candidate activation functions, where Binary Step, Sigmoid [7], Hyperbolic Tangent (Tanh) [7], Inverse Tangent (ATan), and Rectified Linear Unit (ReLU) [61] functions are assigned with indices of 1, 2, 3, 4, and 5, respectively. The mathematical formulation of the candidate activation functions considered in this study is presented in Table 2.

Table 2. The mathematical formulation of the considered activation functions.

Activation Functions	Mathematical Formulation
Binary Step	$\Phi(X) = \begin{cases} 0, & \text{for } X < 0 \\ 1, & \text{for } X \geq 0 \end{cases}$
Sigmoid	$\Phi(X) = \dfrac{X}{1 + e^{-X}}$
Hyperbolic Tangent (Tanh)	$\Phi(X) = \left(\dfrac{e^X - e^{-X}}{e^X + e^{-X}}\right)$
Inverse Tangent (ATan)	$\Phi(X) = \tan^{-1}(X)$
Rectified Linear Unit (ReLU)	$\Phi(X) = \begin{cases} X, & \text{for } X \geq 0 \\ 0, & \text{for } X < 0 \end{cases}$

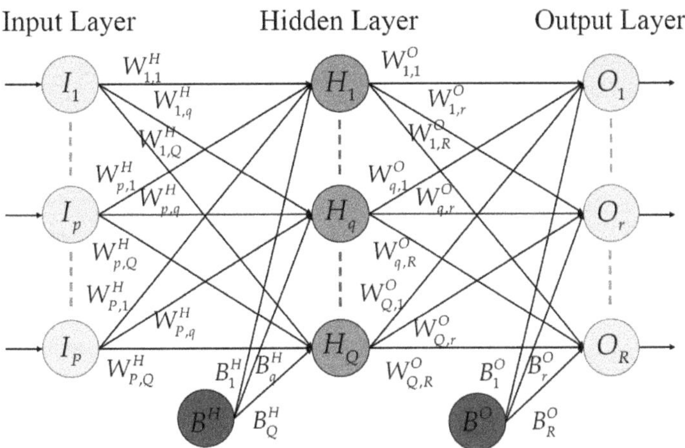

Figure 1. The network architecture of an ANN model with three-layer structure.

The decision variables to be optimized by the proposed MSPSOTLP in ANN training are encoded into the position vector X of each particle as follows:

$$X = \left[W_{1,1}^H, \ldots, W_{p,q}^H, \ldots, W_{P,Q}^H, W_{1,1}^O, \ldots, W_{q,r}^O, \ldots, W_{Q,R}^O, B_1^H, \ldots, B_q^H, \ldots, B_Q^H, B_1^O, \ldots, B_r^O, \ldots, B_R^O, K\right] \tag{6}$$

Referring to Equation (6), the ANN training considered in the current study can be formulated as an optimization problem with a dimensional size of D, where

$$D = PQ + QR + Q + R + 1 \tag{7}$$

In this study, the fitness of each MSPSOTLP particle when it is used to train the ANN model can be evaluated by measuring the mean square error between the predicted and expected output values. Given a dataset with a total of G data samples, the predicted output of g-th data sample produced by an ANN model trained by MSPSOTLP and its corresponding expected outcome stored in the dataset are indicated as $\Im_g^{pred}$ and $\Im_g^{exp}$, respectively, where $g = 1, \ldots, G$. The mean square error $\varepsilon(X)$ produced by an ANN model constructed using the decision variables stored in the position vector X of the particle is calculated as the fitness value $f(X)$ as follows:

$$f(X) = \varepsilon(X) = \frac{1}{G} \sum_{g=1}^{G} \left[\Im_g^{pred}(X) - \Im_g^{exp}\right]^2 \tag{8}$$

Based on Equation (8), the ANN training is considered a minimization problem because it is more desirable to produce an ANN model with a minor error, implying its high classification accuracy in solving a given dataset.

3.2. Proposed MSPSOTLP Algorithm

In this study, a new PSO variant known as MSPSOTLP is proposed to solve challenging optimization problems, including the ANN training problem formulated in Equation (8), with improved performances. For the latter problem, the proposed MSPSOTLP is used to optimize the weights, biases, and selection of activation functions of the ANN model when solving the given datasets. The essential modifications introduced to enhance the search performance of MSPSOTLP are described in the following subsections.

3.2.1. Modified Initialization Scheme of MSPSOTLP

Population initialization is considered a crucial process to develop robust MSAs because the quality of initial candidate solutions can influence the algorithm's convergence rate and searching accuracy [17]. Most PSO variants employed random initialization to generate the initial population without considering any meaningful information about the search environment [17]. The stochastic behavior of the random initialization scheme might produce particles at inferior solution regions at the beginning stage of optimization. This undesirable scenario can prevent the algorithm's convergence towards the global optimum, thus compromising the algorithm's overall performance.

In this study, a modified initialization scheme incorporated with the chaotic system (CS) and oppositional-based learning (OBL), namely the CSOBL initialization scheme, is designed for the proposed MSPSOTLP to overcome the drawbacks of the conventional initialization scheme. Unlike a random system that demonstrates completely unpredictable behaviors, CS is considered a more powerful initialization scheme that can produce an initial swarm with better diversity by leveraging its ergodicity and non-repetition natures. Denote ϑ_0 as the initial condition of a chaotic variable that is randomly generated in each independent run. ϑ_z refers to the value of the chaotic variable in z-th iteration with $z = 1, \ldots, Z$, where Z represents the maximum sequence number. Given the bifurcation coefficient of $\mu = \pi$, the chaotic sequence is updated using a chaotic sine map [62] as:

$$\vartheta_{z+1} = \sin(\mu\vartheta_z), \quad \text{where } z = 1, \ldots, Z \tag{9}$$

Let X_d^U and X_d^L be the upper and lower limits of the decision variable in each d-th dimension, respectively, where $1 = 1, \ldots, D$. Given the chaotic variable ϑ_Z produced in the final iteration of Z, the d-th dimension of each n-th chaotic swarm member $X_{n,d}^{CS}$ can be initialized as:

$$X_{n,d}^{CS} = X_d^L + \vartheta_Z \left(X_d^U - X_d^L \right) \tag{10}$$

Referring to Equation (10), a chaotic population with a swarm size of N can be produced and represented as a population set of $\mathbf{P^{CS}} = \left[X_1^{CS}, \ldots, X_n^{CS}, \ldots, X_N^{CS} \right]$.

Despite the benefits of the chaotic map in enhancing population diversity, these chaotic swarm members can be initialized in solution regions far away from the global optimum, leading to the low convergence rate of the algorithm. To overcome this drawback, a solution set opposite with $\mathbf{P^{CS}}$ is generated by leveraging the OBL concept [63]. For every d-th dimension of n-th chaotic swarm member represented as $X_{n,d}^{CS}$, the corresponding opposite swarm member $X_{n,d}^{OL}$ is calculated using OBL strategy [17,64] as follows:

$$X_{n,d}^{OL} = X_d^L + X_d^U - X_{n,d}^{CS} \tag{11}$$

Similarly, an opposite population with a swarm size of N can be generated using Equation (11) and represented as another population set of $\mathbf{P^{OL}} = \left[X_1^{OL}, \ldots, X_n^{OL}, \ldots, X_N^{OL} \right]$.

To produce an initial population with better fitness and wider coverage in the solution space, both of $\mathbf{P^{CS}}$ and $\mathbf{P^{OL}}$ are merged as a combined population set of $\mathbf{P^{MRG}}$ with a swarm size of $2N$ as follows:

$$\mathbf{P^{MRG}} = \mathbf{P^{CS}} \cup \mathbf{P^{OL}} \tag{12}$$

Subsequently, the fitness of all solution members in $\mathbf{P^{MRG}}$ are evaluated, and a sorting operator of $\Psi(\cdot)$ is then applied to rearrange these solution members from the best to worst based on their fitness to produce a sorted population set of $\mathbf{P^{Sort}}$, where

$$\mathbf{P^{Sort}} = \Psi\left(\mathbf{P^{MRG}}\right) \tag{13}$$

Finally, a truncation operator $\Gamma(\cdot)$ is applied to select the top N solution members of on $\mathbf{P^{Sort}}$ to construct the initial population of MSPSOTLP, i.e., $\mathbf{P} = [X_1, \ldots, X_n, \ldots, X_N]$, where

$$\mathbf{P} = \Gamma\left(\mathbf{P^{Sort}}\right) \tag{14}$$

The pseudocode used to describe the CSOBL initialization scheme of MSPSOTLP is presented in Algorithm 1.

Algorithm 1. Pseudocode of CSOBL initialization scheme for MSPSOTLP

Input: N, D, X^U, X^L, Z
01: Initialize $\mathbf{P^{CS}} \leftarrow \varnothing$ and $\mathbf{P^{OL}} \leftarrow \varnothing$;
02: **for** each n-th particle **do**
03:   **for** each d-th dimension **do**
04:     Randomly generate initial chaotic variable $\vartheta_0 \in [0, 1]$;
05:     Initialize the chaotic sequence as $z = 1$;
06:     **while** z is smaller than Z **do**
07:      Update chaotic variable ϑ_{z+1} using Equation (9);
08:     **end while**
09:     Compute $X^{CS}_{n,d}$ using Equation (10);
10:     Compute $X^{OL}_{n,d}$ using Equation (11);
11:   **end for**
12:   $\mathbf{P^{CS}} \leftarrow \mathbf{P^{CS}} \cup X^{CS}_n$;/* *Store new chaotic swarm member* */
13:   $\mathbf{P^{OL}} \leftarrow \mathbf{P^{OL}} \cup X^{OL}_n$;/* *Store new opposite swarm member* */
14: **end for**
15: Construct $\mathbf{P^{MRG}}$ using Equation (12);
16: Evaluate the fitness of all solution members in $\mathbf{P^{MRG}}$;
17: Sort all solution members of $\mathbf{P^{MRG}}$ from the best to worst using Equation (13);
18: Produce the initial population $\mathbf{P}$ using Equation (14);
Output: $\mathbf{P} = [X_1, \ldots, X_n, \ldots, X_N]$

3.2.2. Primary Learning Phase of MSPSOTLP

Most PSO variants, including PSOWV, rely on the global best position to adjust the search trajectories of particles during the optimization process without considering the useful information of other non-fittest particles in the population. Although the directional information of the global best position might be useful to solve simple unimodal problems, it might not necessarily be the best option to handle complex problems with multiple numbers of local optima due to the possibility of the global best position being trapped at the local optima in an earlier stage of optimization. Without a proper diversity preservation scheme, other population members tend to be attracted by misleading information about the global best position and converge towards the inferior region, leading to premature convergence and poor optimization results.

To address the aforementioned issues, several modifications are incorporated into the primary learning phase of MSPSOTLP to achieve a proper balancing of exploration and exploitation searches. The multiswarm concept is first introduced as a diversity preservation scheme at the beginning stage of the primary learning phase by randomly dividing the main population of $\mathbf{P} = [X_1, \ldots, X_n, \ldots, X_N]$ into S subswarms. Each s-th subswarm is denoted by $\mathbf{P^{sub}_s} = \left[X^{sub}_1, \ldots, X^{sub}_n, \ldots, X^{sub}_{N^{sub}} \right]$ consists of $N^{sub} = N/S$ particles, where $s = 1, \ldots, S$. To produce each s-th subswarm $\mathbf{P^{sub}_s}$ from the main population $\mathbf{P}$, a reference point of $\mathcal{R}_s = \left[X^{ref}_{s,1}, \ldots, X^{ref}_{s,d}, \ldots, X^{ref}_{s,D} \right]$ is randomly generated in search space. The normalized Euclidean distance between the reference point $\mathcal{R}_s$ and personal best position of each n-th particle, i.e., $X^{Pbest}_n = \left[X^{Pbest}_{n,1}, \ldots, X^{Pbest}_{n,d}, \ldots, X^{Pbest}_{n,D} \right]$ are measured as $\Xi(s,n)$, i.e.,

$$\Xi(s,n) = \sqrt{\sum_{d=1}^{D} \left(\frac{X^{ref}_{s,d} - X^{Pbest}_{n,d}}{X^U_d - X^L_d} \right)^2} \tag{15}$$

Referring to the $\Xi(s,n)$ values computed for all N particles, the N^{sub} particles with the nearest $\Xi(s,n)$ distances from $\mathcal{R}_s$ are identified as the members of s-th subswarm and stored in $\mathbf{P}_s^{sub}$ before they are discarded from the main population $\mathbf{P}$. From Algorithm 2, the reference-point-based population division scheme used to generate the multiswarm for the primary learning phase of MSPSOTLP is repeated until all S subswarms are generated.

Algorithm 2. Pseudocode of reference-point-based population division scheme used to generate the multiswarm for the primary learning phase of MSPSOTLP

Input: $\mathbf{P}$, N, S, D, N^{sub}, X^U, X^L
01: Initialize $s \leftarrow 1$;
02: **while** main population $\mathbf{P}$ is not empty **do**
03:        Randomly generate $\mathcal{R}_s = [X_{s,1}^{ref}, \ldots, X_{s,d}^{ref}, \ldots, X_{s,D}^{ref}]$ in search space;
04:        **for** each n-th particle **do**
05:              Calculate $\Xi(s, n)$ using Equation (15);
06:        **end for**
07:        Select N^{sub} particles with the nearest $\Xi(s, n)$ from $\mathcal{R}_s$ to construct $\mathbf{P}_s^{sub}$;
08:        Eliminate the members of $\mathbf{P}_s^{sub}$ from $\mathbf{P}$;
09:        $s \leftarrow s + 1$; /* *Update the index of subswarm*/
10: **end while**
Output: $\mathbf{P}_s^{sub} = [X_1^{sub}, \ldots, X_n^{sub}, \ldots, X_{N^{sub}}^{sub}]$ where $s = 1, \ldots, S$

Define $f\left(X_n^{Pbest,s}\right)$ as the personal best fitness of each n-th particle stored in the s-th subswarm $\mathbf{P}_s^{sub}$, where $n = 1, \ldots, N^{sub}$ and $s = 1, \ldots, S$. All N^{sub} particles stored in each s-th subswarm $\mathbf{P}_s^{sub}$ are then sorted from the worst to best based on their personal best fitness values, as shown in Figure 2. Accordingly, any k-th particle stored in the sorted $\mathbf{P}_s^{sub}$ is considered to have better or equally good personal best fitness than that of n-th particle if the condition of $n \leq k \leq N^{sub}$ is satisfied. Referring to Figure 2, it is notable that the first particle stored in $\mathbf{P}_s^{sub}$ has the worst personal best fitness, whereas the final particle stored in $\mathbf{P}_s^{sub}$ has the most competitive personal best fitness after the sorting process. Therefore, the personal best position of the final particle is also considered as the subswarm best position of $\mathbf{P}_s^{sub}$ represented as $X_s^{Sbest} = X_{N^{sub}}^{Pbest,s}$ for $s = 1, \ldots, S$.

For each s-th sorted subswarm, define $\Omega_n^s = \{X_k^{Pbest,s} | k \in [n, N^{sub}]\}$ as a set variable used to store the personal best position of all solution members that are superior to that of n-th solution member for $n = 1, \ldots, N^{sub} - 1$. Notably, the set variable $\Omega_{N^{sub}}^s$ is not constructed for the final solution member because none of the solution members stored in $\mathbf{P}_s^{sub}$ has better personal best fitness than $X_{N^{sub}}^{Pbest,s}$ or X_s^{Sbest}. Referring to the solution members stored in each Ω_n^s, a unique mean position denoted as $X_n^{mean,s}$ is then specifically constructed to guide the search process of every n-th solution member in the s-th subswarm, where:

$$X_n^{mean,s} = \frac{1}{N^{sub} - n + 1} \left(\sum_{k=n}^{N^{sub}} X_k^{Pbest,s} \right) \tag{16}$$

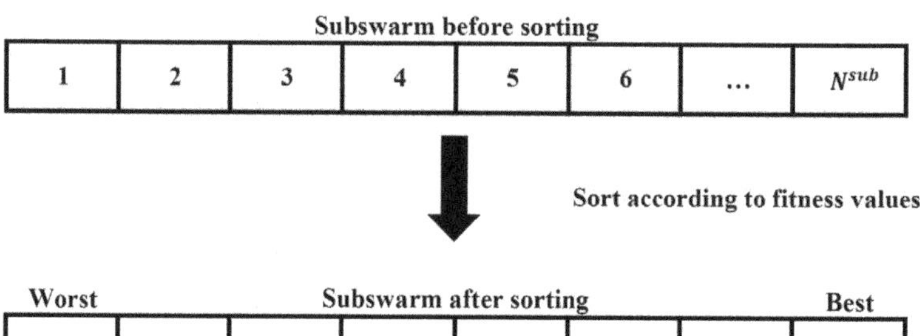

Figure 2. Graphical illustration of sorting all particles stored in each subswarm based on their personal best fitness values.

Apart from $X_n^{mean,s}$, a social exemplar $X_n^{SOC,s} = [X_{n,1}^{SOC,s}, \ldots, X_{n,d}^{SOC,s}, \ldots, X_{n,D}^{SOC,s}]$ that plays crucial roles to adjust the search trajectory of each n-th particle stored in $\mathbf{P}_s^{sub}$ is also formulated. In contrary to global best position, the social exemplar constructed for each n-th particle is unique and it can guide the search process with better diversity by fully utilizing the promising directional information of other particles stored in Ω_n^s with better personal best fitness values. Specifically, each d-th dimension of n-th social learning exemplar for s-th subswarm, i.e., $X_{n,d}^{SOC,s}$, can be contributed by the same dimensional component of any randomly selected solution members of Ω_n^s. The procedures used to construct the social exemplar $X_n^{SOC,s}$ for each n-th particle stored in $\mathbf{P}_s^{sub}$ are described in Algorithm 3, where α refers to a random integer generated between the indices of n and N^{sub}.

Algorithm 3. Pseudocode used to generate the social exemplar for each non-fittest solution member in each subswarm

---

Input: D, N^{sub}, n, Ω_n^s
01:    **for** each d-th dimension **do**
02:        Randomly generate an integer α between indices of n and N^{sub};
03:        Extract the associated component of $X_{\alpha,d}^{Pbest,s}$ from Ω_n^s;
04:        $X_{n,d}^{SOC,s} \leftarrow X_{\alpha,d}^{Pbest,s}$;
05:    **end for**
Output: $X_n^{SOC,s}$

Given the subswarm best position X_s^{Sbest}, mean position $X_n^{mean,s}$ and social exemplar $X_n^{SOC,s}$, the new position X_n^s of each n-th non-fittest solution member stored in the s-th subswarm, where $n = 1, \ldots, N^{sub} - 1$ and $s = 1, \ldots, S$, is updated as follows:

$$X_n^s = X_n^s + c_1 r_1 \left(X_n^{SOC,s} - X_n^s \right) + c_2 r_2 \left(X_s^{Sbest} - X_n^s \right) + c_3 r_3 \left(X_n^{mean,s} - X_n^s \right) \tag{17}$$

where c_1, c_2 and c_3 represent the acceleration coefficients; $r_1, r_2, r_3 \in [0,1]$ are random numbers generated from uniform distributions. Referring to Equation (17), the directional information contributed by $X_n^{SOC,s}$ and $X_n^{mean,s}$ are unique for each n-th non-fittest solution member of $\mathbf{P}_s^{sub}$ because the better solution members are stored in every set variable Ω_n^s are different for $n = 1, \ldots, N^{sub} - 1$. The social learning concept incorporated in Equation (17) also ensures that only the useful information brought by better-performing solutions is used to guide the search process of each n-th particle to accelerate the algorithm's convergence rate. Furthermore, this learning strategy does not consider the global best position in updating the new position of each n-th particle; therefore, it has better robustness against premature convergence issues.

On the other hand, different approaches are proposed to generate the mean position and social exemplar used for guiding the search process of the final particle stored in every n-th subswarm because none of the solution members of $\mathbf{P}_s^{sub}$ can have better personal best fitness than that of $X_{N^{sub}}^{Pbest,s}$. Define $\Omega_{N^{sub}}^s$ as a set variable used to store the subswarm best position of any b-th subswarm $\mathbf{P}_b^{sub}$ if $f\left(X_b^{Sbest} \right)$ is better than $f\left(X_s^{Sbest} \right)$, i.e.,

$$\Omega_{N^{sub}}^s = \left\{ X_b^{Sbest} \,\middle|\, f\left(X_b^{Sbest} \right) \text{ is better than } \left(X_s^{Sbest} \right), \ b \in \mathbf{B}^s \right\} \tag{18}$$

where $\mathbf{B}^s$ is a set containing the indices of all subswarms that have better subswarm best fitness than that of s-th subswarm; $|\mathbf{B}^s|$ refers to the size of set $\mathbf{B}^s$ in the range of 0 to $S - 1$. Obviously, $|\mathbf{B}^s| = 0$ is the subswarm best position of s-th subswarm is the same as the global best position G^{best} of population, therefore the empty sets of $\mathbf{B}^s = \Omega_{N^{sub}}^s = \varnothing$ are obtained under this circumstance.

The subswarm consists of G^{best}, the unique mean position $X^{mean,s}_{Nsub}$ used to guide the search process of the final particle X^s_{Nsub} stored in each s-th subswarm based on Ω^s_{Nsub} are calculated as follows:

$$X^{mean,s}_{Nsub} = \frac{1}{|\mathbf{B}^s|} \left(\sum_{b \in \mathbf{B}^s, |\mathbf{B}^s| \neq 0} X^{Sbest}_b \right) \tag{19}$$

Similarly, a social exemplar of $X^{SOC,s}_{Nsub} = \left[X^{SOC,s}_{Nsub,1}, \cdots, X^{SOC,s}_{Nsub,d}, \cdots, X^{SOC,s}_{Nsub,D} \right]$ is also derived from adjusting the search trajectory of the final particle X^s_{Nsub} stored in each s-th subswarm except for the one consisting of G^{best}. As shown in Algorithm 4, each d-th dimension of the social exemplar is assigned to the final particle of s-th subswarm $\mathbf{P}^{sub}_s$, i.e., $X^{SOC,s}_{Nsub,d}$ is contributed by the same dimensional component of any subswarm best position X^{Sbest}_b randomly selected from Ω^s_{Nsub}, where b refers to a subswarm index randomly selected from $\mathbf{B}^s$.

Algorithm 4. Social Exemplar Scheme for the Best Particle in Each Subswarm

---

Input: $D, \Omega^s_{Nsub}, \mathbf{B}^s, N^{sub}$
01: **for** each d-th dimension **do**
02:     Randomly generated a subswarm index of $b \in \mathbf{B}^s$;
03:     Extract the corresponding component of $X^{Sbest}_{b,d}$ from Ω^s_{Nsub};
04:     $X^{SOC,s}_{Nsub,d} \leftarrow X^{Sbest}_{b,d}$;
05: **end for**
Output: $X^{SOC,s}_{Nsub} = \left[X^{SOC,s}_{Nsub,1}, \cdots, X^{SOC,s}_{Nsub,d}, \cdots, X^{SOC,s}_{Nsub,D} \right]$

Except for the subswarm consisting of G^{best} with $\mathbf{B}^s = \Omega^s_{Nsub} = \varnothing$, the position X^s_{Nsub} of final particle stored in each s-th subswarm can be updated as follows:

$$\begin{aligned} X^s_{Nsub} = X^s_{Nsub} + c_1 r_4 \left(X^{SOC,s}_{Nsub} - X^s_{Nsub} \right) + c_2 r_5 \left(G^{best} - X^s_{Nsub} \right) \\ + c_3 r_6 \left(X^{mean,s}_{Nsub} - X^s_{Nsub} \right) \end{aligned} \tag{20}$$

where $r_4, r_5, r_6 \in [0, 1]$ are random numbers generated from a uniform distribution. Similarly, the directional information provided by $X^{SOC,s}_{Nsub}$ and $X^{mean,s}_{Nsub}$ are unique for each final particle X^s_{Nsub} in each s-th subswarm $\mathbf{P}^{sub}_s$ because the solution members stored in each Ω^s_{Nsub} set are different. Contrary to Equation (17), the social learning concept introduced in Equation (20) allows each subswarm to converge towards the promising solution regions without experiencing rapid loss of population diversity by facilitating the information exchanges between different subswarms. In addition, the employment of G^{best} in Equation (20) is expected to improve the convergence rate of MSPSOTLP.

The overall procedures of the primary learning phase proposed for MSPSOTLP is described in Algorithm 5. For each new position X^s_n obtained from Equations (17) or (20), boundary checking is first performed to ensure all decision variables' upper and lower limits are not violated. The fitness value corresponding to the updated X^s_n for each particle in s-th subswarm is then evaluated as $f(X^s_n)$ and compared with those of its personal best position and global best position denoted as $f(X^{Pbest,s}_n)$ and $f(G^{best})$, respectively. Both $X^{Pbest,s}_n$ and G^{best} are replaced by the updated X^s_n if the latter solution is proven to be superior to the latter two solutions.

Algorithm 5. Pseudocode of primary learning phase for MSPSOTLP

Input: P, N, S, D, N^{sub}, X^U, X^L, G^{best}
01: Divide main population into multiple subswarm using **Algorithm 2**;
02: Sort the solution members of $\mathbf{P}_s^{sub}$ from the worst to best based on their personal best fitness values and create Ω_n^s for each s-th subswarm;
03: Identify the subswarm best position X_s^{Sbest} of each s-th subswarm and create Ω_{Nsub}^s for the final particle in all S subswarms using Equation (18);
04: Determine $\mathbf{B}^s$ and $|\mathbf{B}^s|$ for the final particle of all S subswarms based on Ω_{Nsub}^s;
05: **for** each s-th subswarm **do**
06:     **for** each n-th particle **do**
07:         **if** $n \neq N^{sub}$ **then**
08:             Calculate $X_n^{mean,s}$ based on Ω_n^s using Equation (16);
09:             Generate $X_n^{SOC,s}$ based on Ω_n^s using **Algorithm 3**;
10:             Update X_n^s using Equation (17);
11:         **else if** $n = N^{sub}$ **then**
12:             **if** $\mathbf{B}^s \neq \varnothing$ and $\Omega_{Nsub}^s \neq \varnothing$ **then**
13:                 Calculate $X_{Nsub}^{mean,s}$ based on Ω_{Nsub}^s using Equation (19);
14:                 Generate $X_{Nsub}^{SOC,s}$ based on Ω_{Nsub}^s using **Algorithm 4**;
15:                 Update X_{Nsub}^s using Equation (20);
16:             **end if**
17:         **end if**
18:         Evaluate $f(X_n^s)$ of the updated X_n^s;
19:         **if** $f(X_n^s)$ is better than $f(X_n^{Pbest,s})$ **then**
20:             $X_n^{Pbest,s} \leftarrow X_n^s$, $f(X_n^{Pbest,s}) \leftarrow f(X_n^s)$;
21:             **if** $f(X_n^s)$ is better than $f(G^{best})$ **then**
22:                 $G^{best} \leftarrow X_n^s$, $f(G^{best}) \leftarrow f(X_n^s)$;
23:             **end if**
24:         **end if**
25:     **end for**
26: **end for**
Output: X_n^s, $f(X_n^s)$, $X_n^{Pbest,s}$, $f(X_n^{Pbest,s})$, G^{best} and $f(G^{best})$

3.2.3. Secondary Learning Phase of MSPSOTLP

Substantial studies [19,65] reported that most PSO variants employed single search operators that can only solve specific optimization problems with good results, but fail to perform well in the remaining problems due to the limited variations of exploration and exploitation strengths. For some challenging optimization problems, the fitness landscapes contained in different subregions of search space can be significantly different. Therefore, the particles need to adjust their exploration and exploitation strengths dynamically when searching in different regions of the solution space to locate global optimum and deliver good optimization results.

Motivated by these findings, a secondary phase is designed as an alternative framework of MSPSOTLP, where two search operators with different search characteristics are incorporated to guide the search process of particles with varying levels of exploration and exploitation strengths. Unlike the primary learning phase, both search operators assigned in the secondary learning phase aim to further refine those already found promising regions by searching around the personal best positions of all MSPSOTLP particles. Before initiating the secondary learning phase, all S subswarms constructed during the primary learning phase, i.e., $\mathbf{P}_s^{sub}$ for $s = 1, \ldots, S$, are merged to form the main population $\mathbf{P}$ with N particles as shown below:

$$\mathbf{P} = \mathbf{P}_1^{sub} \cup \ldots \cup \mathbf{P}_s^{sub} \cup \ldots \cup \mathbf{P}_S^{sub} \tag{21}$$

To assign a search operator for each n-th particle during the secondary learning phase of MSPSOTLP, a randomly selected particle with a population index of e is randomly generated, where $e \in [1, N]$ and $e \neq n$. Define X_e^{Pbest} as the personal best position of this randomly selected e-th particle and its personal best fitness is evaluated as $f(X_e^{Pbest})$. If the e-th particle has better personal best fitness than that of X_n^{Pbest}, the new personal best position of latter particle can be updated as $X_n^{Pbest,new}$, where

$$X_n^{Pbest,new} = X_n^{Pbest} + r_7(X_e^{Pbest} - X_n^{Pbest}), \text{ if } f(X_e^{Pbest}) \text{ is better than } f(X_n^{Pbest}). \tag{22}$$

where $r_7 \in [0, 1]$ are random numbers generated from a uniform distribution. The search operator of Equation (22) can attract of n-th particle towards the promising solution regions covered by the e-th peer particle, hence it behaves more exploratively.

For the case, if e-th particle has more inferior personal best fitness than that of n-th particle, the former solution is discarded. Another four distinct particles with population indices of w, x, y and z are randomly selected instead, where $w, x, y, z \in [1, N]$ and $w \neq x \neq y \neq z$. Denote $X_{w,d}^{Pbest}$, $X_{x,d}^{Pbest}$, $X_{y,d}^{Pbest}$ and $X_{z,d}^{Pbest}$ as the d-th dimension of the personal best position for the w-th, x-th, y-th and z-th particles, respectively. Let G_d^{best} be the same d-th dimension of the global best position. For every n-th particle, each of the d-th dimension of its new personal best position can be calculated as:

$$X_{n,d}^{Pbest,new} = \begin{cases} G_d^{best} + \tau_1(X_{w,d}^{Pbest} - X_{x,d}^{Pbest}) + \tau_2(X_{y,d}^{Pbest} - X_{z,d}^{Pbest}), & \text{if } r_8 > 0.5 \\ X_{n,d}^{Pbest} & , \text{ otherwise} \end{cases} \tag{23}$$

where $\tau_1, \tau_2, r_8 \in [0, 1]$ are random numbers generated from a uniform distribution. From Equation (23), there is a probability for each $X_{n,d}^{Pbest,new}$ to inherit its original information from $X_{n,d}^{Pbest}$ or to perform searching around the nearby region of G_d^{best} with small perturbations based on the information of $X_{w,d}^{Pbest}$, $X_{x,d}^{Pbest}$, $X_{y,d}^{Pbest}$ and $X_{z,d}^{Pbest}$. Hence, the search operator of Equation (23) is considered more exploitative than that of Equation (22). The procedures used to implement the secondary learning phase of MSPSOTLP is shown in Algorithm 6. For each new $X_n^{Pbest,new}$ obtained from Equations (22) or (23), boundary checking is performed. For each n-th particle, the fitness value of its updated $X_n^{Pbest,new}$ is obtained as $f(X_n^{Pbest,new})$ and compared with $f(X_n^{Pbest})$ and $f(G^{best})$. If $X_n^{Pbest,new}$ is better than X_n^{Pbest} and G^{best}, the latter two solutions are replaced by the former one.

3.2.4. Overall Framework of MSPSOTLP

The overall framework of MSPSOTLP is described in Algorithm 7. The initial population of MSPSOTLP is first generated using the CSOBL initialization scheme, where the chaotic map and oppositional-based learning concepts are incorporated to enhance the quality and coverage of the initial solutions in the search space, respectively. The primary learning phase is performed to update the positions of particles by leveraging the benefits of the multiswarm and social learning concepts. The secondary learning phase is then executed to fine-tune the personal best position of each particle based on two newly proposed search operators with different search characteristics. These iterative search processes are repeated until the termination criterion of $\gamma > \Gamma^{max}$ is satisfied, where γ is a counter used to record the fitness evaluation number consumed by MSPSOTLP and Γ^{max} is a predefined maximum fitness evaluation number. At the end of the optimization process, G^{best} is returned as the best-optimized result found by the proposed algorithm. If MSPSOTLP is used to train the ANN model, then G^{best} can be decoded as the optimal combination of weights, biases, and activation functions used by an ANN model to solve a given dataset.

Algorithm 6. Pseudocode of secondary learning phase for MSPSOTLP

Input: $\mathbf{P}_s^{sub}$ for $s = 1, \ldots, S$, G_d^{best}, $f(G^{best})$
01:    Reconstruct main population $\mathbf{P}$ using Equation (21);
02:    **for** each n-th particle **do**
03:        Randomly select e-th particle from $\mathbf{P}$ with X_e^{Pbest} and $f(X_e^{Pbest})$;
04:        **if** $f(X_e^{Pbest})$ is better than $f(X_n^{Pbest})$ **then**
05:            Calculate $X_n^{Pbest,new}$ using Equation (22);
06:        **else if** $f(X_e^{Pbest})$ is not better than $f(X_n^{Pbest})$ **then**
07:            **for** each d-th dimension **do**
09:                Calculate $X_{n,d}^{Pbest,new}$ using Equation (23);
10:            **end for**
11:        **end if**
12:        Evaluate $f(X_n^{Pbest,new})$;
13:        **if** $f(X_n^{Pbest,new})$ is better than $f(X_n^{Pbest})$ **then**
14:            $X_n^{Pbest} \leftarrow X_n^{Pbest,new}$, $f(X_n^{Pbest}) \leftarrow f(X_n^{Pbest,new})$;
15:            **if** $f(X_n^{Pbest,new})$ is better than $f(G^{best})$ **then**
16:                $G^{best} \leftarrow X_n^{Pbest}$, $f(G^{best}) \leftarrow f(X_n^{Pbest})$;
17:            **end if**
18:        **end if**
19:    **end for**
Output: X_n^{Pbest}, $f(X_n^{Pbest})$, G^{best}, $f(G^{best})$

Algorithm 7 (Main Algorithm). MSPSOTLP

Input: $N, D, , X^U, X^L, \Gamma^{max}$
01:    Initialize G^{best} as an empty vector and $f(G^{best}) \leftarrow \infty$;
02:    Initialize $\gamma \leftarrow 0$;
03:    Generate the initial population $\mathbf{P}$ using **Algorithm 1**;
04:    $\gamma \leftarrow \gamma + 2N$;
05:    **for** each n-th particle do
06:        $X_n^{Pbest} \leftarrow X_n$, $f(X_n^{Pbest}) \leftarrow f(X_n)$;
07:        **If** $f(X_n)$ is better than $f(G^{best})$ **then**
08:            $G^{best} \leftarrow X_n$, $f(G^{best}) \leftarrow f(X_n)$;
09:        **end if**
10:    **end for**
11:    **while** $\gamma \leq \Gamma^{max}$ **do**
12:        Perform the primary learning phase using **Algorithm 5**;
13:        $\gamma \leftarrow \gamma + N$;
14:        Perform the secondary learning phase using **Algorithm 6**;
15:        $\gamma \leftarrow \gamma + N$;
16:    **end while**
Output: G^{best}, $f(G^{best})$

---

4. Performance Analysis of MSPSOTLP

The performance of the proposed MSPSOTLP in solving various types of challenging optimization problems is investigated and compared with well-established PSO variants. This includes the performance of MSPSOTLP in solving CEC 2014 benchmark functions, followed by its performance in optimizing the weights, biases, and activation functions of ANN models for solving classification problems.

4.1. Performance Evaluation of MSPSOTLP in Solving Global Optimization Problems

4.1.1. CEC 2014 Benchmark Functions

The optimization performance of the proposed MSPSOTLP is evaluated using 30 benchmark functions of CEC 2014 [66]. As described in Table 3, the benchmark functions with different fitness landscape characteristics can be classified into four categories, known as (i) unimodal functions (F1–F3), (ii) simple multimodal functions (F4–F16), (iii) hybrid functions (F17–F22) and (iv) composition functions (F23–F30). For all benchmark functions in CEC 2014 with D dimensions, the search range X of each decision variable is constrained between -100 to 100. Furthermore, the fitness value of theoretical global optimum $f(X^*)$ of each function is presented in Table 3.

Table 3. CEC 2014 benchmark functions and its fitness value of theoretical global optimum.

Categories	No.	Function Name	$f(X^*)$
Unimodal	F1	Rotated High Conditioned Elliptic Function	100
	F2	Rotated Bent Cigar Function	200
	F3	Rotated Discus Function	300
Simple Multimodal	F4	Shifted and Rotated Rosenbrock's Function	400
	F5	Shifted and Rotated Ackley's Function	500
	F6	Shifted and Rotated Weierstrass Function	600
	F7	Shifted and Rotated Griewank's Function	700
	F8	Shifted Rastrigin's Function	800
	F9	Shifted and Rotated Rastrigin's Function	900
	F10	Shifted Schewefel's Function	1000
	F11	Shifted and Rotated Schwefel's Function	1100
	F12	Shifted and Rotated Katsuura Function	1200
	F13	Shifted and Rotated HappyCat Function	1300
	F14	Shifted and Rotated HGBat Function	1400
	F15	Shifted and Rotated Expanded Griewank's plus Rosenbrock's Function	1500
	F16	Shifted and Rotated Expanded Schaffer's F6 Function	1600
Hybrid	F17	Hybrid Function1	1700
	F18	Hybrid Function 2	1800
	F19	Hybrid Function 3	1900
	F20	Hybrid Function 4	2000
	F21	Hybrid Function 5	2100
	F22	Hybrid Function 6	2200
Composition	F23	Composition Function 1	2300
	F24	Composition Function 2	2400
	F25	Composition Function 3	2500
	F26	Composition Function 4	2600
	F27	Composition Function 5	2700
	F28	Composition Function 6	2800
	F29	Composition Function 7	2900
	F30	Composition Function 8	3000

4.1.2. Performance Metrics for Solving Benchmark Functions

Performances of all compared algorithms are measured using the mean fitness F_{mean} and standard deviation SD. Specifically, F_{mean} is the mean error between the fitness of the best solution produced by an algorithm and the actual global optimum of a benchmark function in multiple runs. At the same time, the SD value measures the consistency of an algorithm in solving a given problem. Smaller F_{mean} and SD imply the capability of an algorithm to solve a function with better accuracy and consistency, respectively.

A set of non-parametric statistical analysis procedures [67,68] are applied to analyze the performance of the proposed MSPSOTLP and its competitors from a statistical point of view. The Wilcoxon signed rank test [68] is applied to conduct a pairwise comparison between the proposed MSPSOTLP and each competitor at a significance level of $\alpha = 0.05$. The results generated by the Wilcoxon signed rank test are presented in terms of R^+, R^-, p, and h values. R^+ and R^- summarize the sum of ranks where MSPSOTLP outperforms or underperforms its peer algorithm, respectively. The p-value indicates the minimum significance level required to identify the performance deviations between the two algorithms. If the p-value is smaller than $\alpha = 0.05$, the better result achieved by an algorithm is considered statistically significant. Based on the obtained p-value and predefined α, the corresponding h value is concluded to be significantly better (i.e., h = "+"), statistically insignificant (i.e., h = "="), or significantly worse (i.e., h = "-").

Multiple comparisons among the proposed MSPSOTLP and its competitors is also conducted using the Friedman test [67]. The Friedman test first produces the average ranking of each algorithm. The p-value obtained from the Friedman test measures the global differences among all compared algorithms at a significance level of $\alpha = 0.05$. If significant global differences are observed, three post-hoc analyses [67], known as Bonferroni-Dunn, Holm, and Hochberg, are performed to analyze the substantial differences among all compared algorithms based on the adjusted p-values (APVs).

4.1.3. Parameter Settings for Solving Benchmark Functions

The performance of the proposed MSPSOTLP in solving CEC 2014 benchmark functions is compared with seven well-established PSO variants. The selected PSO variants include the conventional PSO (PSO) [14], PSO without velocity (PSOWV) [16], unconstrained version of multi-swarm PSO without velocity (MPSOWV) [15], competitive swarm optimizer (CSO) [69], social learning PSO (SLPSO) [18], hybridized PSO with gravitational search algorithm (PSOGSA) [44], and accelerated PSO (APSO) [70].

The parameter settings of all of the compared algorithms are set with the recommended values in their respective literature and presented in Table 4. All compared algorithms are configured with a population size of $N = 100$ to solve each benchmark function at $D = 30$ for 30 independent times. The maximum fitness evaluation numbers of all algorithms are set as $\Gamma^{max} = 10,000 \times D$. All compared algorithms are simulated using Matlab 2019b on a personal computer with Intel$^{®}$ Core i7-7500 CPU @ 2.70 GHz.

Table 4. Parameter settings of all compared algorithms.

Algorithms	Parameter Settings
PSO	Inertia weight $\omega : 0.9 \rightarrow 0.2$, acceleration coefficients $c_1 = c_2 = 2.05$
PSOWV	$c_1 = 1.00,\ c_2 = 1.70$
MPSOWV	Subpopulation size $N_s^{sub} = 10$, where $s - 1, \ldots, 10$, acceleration coefficients $c_1 = c_2 = c_2 = 4.1/3$
CSO	Parameter control the influence of mean position $\varphi \in [0, 0.3]$
SLPSO	Exponential component to adjust learning probability $\tilde{\alpha} = 0.5$, parameter to control social influence factor $\tilde{\beta} = 0.01$
PSOGSA	Initial gravitational constant $G_0 = 1$, descending coefficient of gravitational constant $\hat{\alpha} = 20$
APSO	Additional acceleration coefficient $A = 0.3$, $\omega = 1$, $c_1 = c_2 = 2.05$
MSPSOTLP	$N_s^{sub} = 10$, where $s = 1, \ldots, 10$, $c_1 = c_2 = c_2 = 4.1/3$

4.1.4. Performance Comparison in Solving CEC 2014 Benchmark Functions

The simulation results of F_{mean} and SD produced by the proposed MSPSOTLP and other selected PSO variants in solving the CEC 2014 benchmark functions are presented in Table 5. The algorithms' best and second-best Fmean obtained are indicated in boldface and underlined, respectively. Moreover, the performance comparison between the proposed MSPSOTLP and selected PSO variants is summarized in #BMF and w/t/l. Specifically, #BMF indicates the number of best F_{mean} obtained by an algorithm in solving all function, while w/t/l reports that the proposed MSPSOTLP has better performance in w function, similar performance in t function, and worse performance in l function as compared to the particular compared algorithm.

Table 5. F_{mean} and SD values produced by the proposed MSPSOTLP and other PSO variants in solving CEC 2014 benchmark functions.

Func.	Criteria	MSPSOTLP	PSO	PSOWV	CSO	MPSOWV	SLPSO	PSOGSA	APSO
F1	F_{mean}	$\mathbf{5.61 \times 10^3}$	6.69×10^6	4.70×10^8	3.12×10^5	1.61×10^8	4.88×10^5	2.47×10^5	1.25×10^8
	SD	2.57×10^3	8.99×10^6	3.79×10^8	1.32×10^5	7.38×10^7	2.86×10^5	7.24×10^4	2.02×10^7
F2	F_{mean}	$\mathbf{0.00 \times 10^0}$	7.27×10^1	6.94×10^{10}	6.84×10^3	2.12×10^{10}	1.57×10^4	1.25×10^4	3.01×10^7
	SD	0.00×10^0	2.90×10^2	1.56×10^{10}	5.63×10^3	2.74×10^9	1.10×10^4	6.97×10^3	5.81×10^7
F3	F_{mean}	$\mathbf{2.18 \times 10^{11}}$	4.87×10^1	3.18×10^5	7.85×10^3	8.20×10^4	7.26×10^3	6.93×10^3	1.97×10^5
	SD	2.52×10^{-11}	6.61×10^1	5.70×10^4	5.62×10^3	2.44×10^4	5.50×10^3	1.33×10^3	3.56×10^4
F4	F_{mean}	1.44×10^{-2}	1.79×10^2	8.66×10^3	5.71×10^1	1.39×10^3	2.56×10^1	4.49×10^1	2.71×10^2
	SD	1.23×10^{-2}	5.12×10^1	3.26×10^3	2.31×10^1	3.56×10^2	3.04×10^1	2.91×10^1	3.29×10^1
F5	F_{mean}	2.01×10^1	2.09×10^1	2.09×10^1	2.10×10^1	2.09×10^1	2.09×10^1	$\mathbf{2.00 \times 10^1}$	$\mathbf{2.00 \times 10^1}$
	SD	5.32×10^{-2}	8.52×10^{-2}	9.48×10^{-2}	3.65×10^{-2}	3.01×10^{-2}	6.49×10^{-2}	2.99×10^{-4}	3.26×10^{-2}
F6	F_{mean}	3.25×10^0	2.12×10^1	3.83×10^1	$\mathbf{4.08 \times 10^{-1}}$	3.38×10^1	5.43×10^{-1}	1.80×10^1	2.45×10^1
	SD	1.02×10^0	5.40×10^0	2.41×10^0	6.73×10^{-1}	9.46×10^{-1}	6.38×10^{-1}	5.22×10^0	3.50×10^0
F7	F_{mean}	$\mathbf{0.00 \times 10^0}$	2.62×10^{-2}	5.61×10^2	1.48×10^{-3}	1.62×10^2	1.97×10^{-3}	1.23×10^{-2}	1.25×10^0
	SD	0.00×10^0	2.21×10^{-2}	2.05×10^2	3.12×10^{-3}	1.44×10^1	4.41×10^{-3}	7.38×10^{-3}	6.16×10^{-1}
F8	F_{mean}	1.15×10^1	1.15×10^2	3.87×10^2	$\mathbf{8.66 \times 10^0}$	2.68×10^2	1.37×10^1	1.70×10^2	1.37×10^2
	SD	1.57×10^0	2.51×10^1	3.23×10^1	1.88×10^0	2.21×10^1	3.25×10^0	9.24×10^0	2.75×10^1
F9	F_{mean}	1.20×10^1	1.39×10^2	4.85×10^2	9.65×10^0	3.29×10^2	1.65×10^1	1.91×10^2	1.98×10^2
	SD	3.16×10^0	1.42×10^1	3.67×10^1	3.78×10^0	1.25×10^1	5.52×10^0	7.06×10^0	2.18×10^1
F10	F_{mean}	$\mathbf{1.52 \times 10^2}$	2.63×10^3	7.66×10^3	1.78×10^2	6.22×10^3	5.47×10^2	3.89×10^3	3.71×10^3
	SD	7.91×10^1	4.52×10^2	4.67×10^2	1.36×10^2	2.45×10^2	2.60×10^2	3.20×10^2	5.04×10^2
F11	F_{mean}	2.42×10^3	3.76×10^3	7.56×10^3	$\mathbf{2.85 \times 10^2}$	7.24×10^3	7.21×10^2	4.36×10^3	4.16×10^3
	SD	3.73×10^2	6.08×10^2	4.57×10^2	2.28×10^2	4.44×10^2	2.29×10^2	1.82×10^2	8.01×10^2
F12	F_{mean}	1.62×10^0	1.74×10^0	2.70×10^0	2.39×10^0	2.65×10^0	2.50×10^0	$\mathbf{2.07 \times 10^{-1}}$	5.61×10^{-1}
	SD	2.07×10^{-1}	3.84×10^{-1}	1.22×10^{-1}	3.10×10^{-1}	1.63×10^{-1}	3.68×10^{-1}	1.66×10^{-1}	2.68×10^{-1}
F13	F_{mean}	$\mathbf{1.29 \times 10^{-1}}$	4.82×10^{-1}	6.12×10^0	1.34×10^{-1}	3.09×10^0	1.97×10^{-1}	5.62×10^{-1}	5.92×10^{-1}
	SD	7.49×10^{-3}	1.19×10^{-1}	5.58×10^{-1}	1.61×10^{-2}	2.33×10^{-1}	2.41×10^{-2}	7.52×10^{-2}	1.17×10^{-1}
F14	F_{mean}	$\mathbf{1.93 \times 10^{-1}}$	2.90×10^{-1}	1.46×10^2	3.94×10^{-1}	4.48×10^1	4.42×10^{-1}	2.82×10^{-1}	2.57×10^{-1}
	SD	8.61×10^{-3}	9.55×10^{-2}	6.54×10^1	4.61×10^{-2}	7.08×10^0	7.32×10^{-2}	2.23×10^{-2}	5.35×10^{-2}
F15	F_{mean}	$\mathbf{2.30 \times 10^0}$	1.10×10^0	3.58×10^6	3.14×10^0	9.90×10^4	5.07×10^0	8.15×10^0	3.16×10^1
	SD	2.08×10^{-1}	3.82×10^0	7.86×10^5	4.29×10^{-1}	2.26×10^4	4.93×10^0	4.94×10^0	2.06×10^1
F16	F_{mean}	$\mathbf{7.60 \times 10^0}$	1.16×10^1	1.34×10^1	1.08×10^1	1.30×10^1	1.21×10^1	1.25×10^1	1.28×10^1
	SD	3.86×10^{-1}	5.88×10^{-1}	2.05×10^{-1}	4.73×10^{-1}	1.38×10^{-1}	1.79×10^{-1}	2.12×10^{-1}	6.66×10^{-1}
F17	F_{mean}	$\mathbf{2.10 \times 10^3}$	2.86×10^5	1.77×10^7	1.49×10^5	6.01×10^6	1.05×10^5	1.86×10^4	2.05×10^6
	SD	5.36×10^2	3.25×10^5	1.08×10^7	6.66×10^4	1.12×10^6	5.42×10^4	5.76×10^3	2.37×10^6
F18	F_{mean}	$\mathbf{1.22 \times 10^2}$	3.79×10^3	5.88×10^8	2.17×10^3	2.58×10^8	8.42×10^2	3.84×10^2	1.41×10^5
	SD	6.67×10^1	4.62×10^3	2.87×10^8	3.17×10^3	8.05×10^7	4.97×10^2	1.85×10^2	3.12×10^5
F19	F_{mean}	$\mathbf{4.08 \times 10^0}$	1.24×10^1	3.64×10^2	5.30×10^0	1.41×10^2	6.13×10^0	1.59×10^1	2.93×10^1
	SD	2.68×10^{-1}	3.21×10^0	1.51×10^2	6.33×10^{-1}	2.38×10^1	5.50×10^{-1}	2.89×10^0	2.74×10^1
F20	F_{mean}	$\mathbf{7.96 \times 10^1}$	3.74×10^2	2.78×10^5	1.26×10^4	2.90×10^4	2.48×10^4	2.84×10^3	1.02×10^5
	SD	8.58×10^0	2.38×10^2	1.88×10^5	9.04×10^3	9.09×10^3	1.36×10^4	2.73×10^2	4.65×10^4
F21	F_{mean}	$\mathbf{8.02 \times 10^2}$	6.12×10^4	6.03×10^6	6.92×10^4	1.56×10^6	8.18×10^4	1.18×10^4	1.16×10^6
	SD	2.07×10^2	6.19×10^4	6.96×10^6	3.80×10^4	5.59×10^5	3.84×10^4	1.29×10^3	9.69×10^5
F22	F_{mean}	$\mathbf{3.52 \times 10^1}$	6.77×10^2	1.08×10^3	1.28×10^2	9.76×10^2	1.51×10^2	7.94×10^2	6.23×10^2
	SD	1.30×10^1	2.68×10^2	2.42×10^2	5.43×10^1	1.96×10^2	1.32×10^1	1.59×10^2	173×10^2
F23	F_{mean}	3.15×10^2	3.16×10^2	3.15×10^2	3.15×10^2	4.17×10^2	3.15×10^2	$\mathbf{2.00 \times 10^2}$	3.62×10^2
	SD	3.20×10^{-13}	3.52×10^{-1}	8.04×10^1	9.16×10^{-12}	1.26×10^1	4.99×10^{-13}	1.06×10^{-7}	1.51×10^1
F24	F_{mean}	2.24×10^2	2.30×10^2	4.66×10^2	2.26×10^2	3.20×10^2	2.33×10^2	$\mathbf{2.01 \times 10^2}$	2.50×10^2
	SD	1.50×10^{-1}	6.64×10^0	3.90×10^1	4.74×10^0	1.13×10^1	7.32×10^0	1.09×10^{-1}	2.84×10^0
F25	F_{mean}	$\mathbf{2.00 \times 10^2}$	2.13×10^2	2.59×10^2	2.06×10^2	2.25×10^2	2.06×10^2	$\mathbf{2.00 \times 10^2}$	2.22×10^2
	SD	0.00×10^0	4.81×10^0	2.21×10^1	1.55×10^0	5.87×10^0	2.72×10^0	8.94×10^{-10}	2.99×10^0
F26	F_{mean}	$\mathbf{1.00 \times 10^2}$	1.70×10^2	1.06×10^2	$\mathbf{1.00 \times 10^2}$	1.03×10^2	1.40×10^2	$\mathbf{1.00 \times 10^2}$	1.83×10^2
	SD	2.32×10^{-2}	4.81×10^1	1.57×10^0	1.83×10^{-2}	5.44×10^{-1}	5.47×10^1	1.35×10^{-1}	4.59×10^1
F27	F_{mean}	3.94×10^2	8.49×10^2	1.28×10^3	3.55×10^2	1.14×10^3	3.69×10^2	$\mathbf{2.00 \times 10^2}$	8.02×10^2
	SD	2.91×10^1	3.58×10^2	3.92×10^1	5.77×10^1	6.81×10^1	2.77×10^1	5.10×10^{-9}	2.03×10^2
F28	F_{mean}	8.55×10^2	4.04×10^3	1.82×10^3	8.57×10^2	1.36×10^3	9.77×10^2	$\mathbf{2.00 \times 10^2}$	2.50×10^3
	SD	1.80×10^1	9.59×10^2	5.12×10^2	4.42×10^1	3.69×10^2	1.07×10^2	2.07×10^{-8}	5.80×10^2
F29	F_{mean}	8.60×10^2	1.43×10^3	1.04×10^7	1.63×10^3	2.15×10^6	1.55×10^3	$\mathbf{2.02 \times 10^2}$	1.23×10^7
	SD	5.36×10^1	7.53×10^2	9.14×10^6	4.77×10^2	4.02×10^6	4.57×10^2	2.00×10^{-1}	1.88×10^7
F30	F_{mean}	1.72×10^3	3.56×10^3	3.79×10^5	2.69×10^3	4.04×10^4	3.17×10^3	$\mathbf{2.00 \times 10^2}$	9.88×10^4
	SD	5.39×10^2	1.93×10^3	339×10^5	1.06×10^3	8.14×10^3	$3,17 \times 10^3$	3.62×10^{-3}	3.25×10^4
	#BMF	18	0	0	5	0	0	10	1
	w/t/l	-	29/1/0	30/0/0	22/3/5	30/0/0	26/1/3	19/2/9	28/0/2

For unimodal functions (i.e., F1–F3), the proposed MSPSOTLP has the most dominating search performance by producing the best F_{mean} values to solve these three functions. MSPSOTLP is also the only PSO variant to locate the global and near-global optimum solutions of F2 and F3, respectively. Apart from MSPSOTLP, both PSO and PSOGSA are considered to have relatively better search performance than the rest of the algorithms in solving unimodal functions by producing one second-best F_{mean}. Meanwhile, PSOWV, MPSOWV and APSO have inferior search performance by producing F_{mean} values that are generally larger than those of the other algorithms.

For simple multimodal functions (i.e., F4-F16), the proposed MSPSOTLP has the most competitive performance in solving these 13 functions with the seven best F_{mean} (i.e., F4, F7, F10, F13, and F14) and three second-best F_{mean} (i.e., F5, F8, and F9). MSPSOTLP is also the only algorithm that successfully locates the global optimum of function F7. Contrary to unimodal functions, CSO and SLPSO have exhibited relatively better performance in solving several simple multimodal functions, such as F6, F8, F9, and F11. Although PSO and PSOGSA have good performance in solving unimodal functions, their performances in solving F4, F6, F7, F8, F9, F10, F11, F13, F15, and F16 are relatively inferior. The performance degradations of PSO and PSOGSA reflect the limitation of both algorithms in tackling optimization problems with multiple local optima. Meanwhile, APSO shows relatively better search performance at solving F5, F12, and F14 than PSOWV and MPSOWV.

The excellent optimization performance of the proposed MSPSOTLP is also demonstrated in the hybrid function category (i.e., F17–F22) by solving all six functions with the best F_{mean} values. PSOGSA follows this, producing three second-best F_{mean} values for F17, F18, and F21. However, the performance of PSOGSA in solving hybrid functions is not consistent, as shown by its relatively inferior results in F19 and F22. A similar scenario is observed from PSO, CSO, and SLPSO, producing mediocre performance in solving most hybrid functions. Specifically, PSO can solve F20 and F21 with relatively good performance, but it performs poorly in F17, F18, F19, and F22. Meanwhile, CSO is reported to solve F19 and F22 with competitive performance, but delivers poor results for F17, F18, F20, and F21. On the other hand, APSO, PSOWV, and MPSOWV are reported to have inferior search performance in solving all hybrid functions with higher complexity levels.

In the category of composition function (i.e., F23–F30), PSOGSA shows its competitive performance in dealing with these more complex functions, followed by the proposed MSP-SOTLP. PSAGSA can solve all eight composite functions with the best F_{mean} values, while MSPSOTLP produces two best F_{mean} (i.e., F25 and F26) and five second-best F_{mean} (i.e., F23, F24, F28, F29, and F30). Although PSOGSA performs better than MSPSOTLP when solving composite functions, PSOGSA performs more inferiorly than MSPSOTLP in three other problem categories. CSO produces one best F_{mean} (i.e., F26) and three second-best F_{mean} (i.e., F23, F25, and F27), implying its competitive performance in solving composite functions. Meanwhile, SLPSO is observed to have relatively good performance in solving F23 and F25, but mediocre performance in the remaining composite functions.

Overall, the proposed MSPSOTLP has demonstrated the best search accuracy among all compared PSO variants by producing 18 best F_{mean} values in solving 30 functions, implying that the search mechanisms incorporated are sufficiently robust to handle optimization problems with different levels of complexity as compared to most of its peer algorithms. This is followed by PSOGSA and CSO, which are reported to have 10 and 5 best F_{mean} values, respectively. On the other hand, PSOWV is identified as the worst algorithm by producing 26 worst F_{mean} values in solving 30 CEC 2014, implying its limitations in solving the benchmark functions with simple fitness landscapes.

4.1.5. Non-Parametric Statistical Analyses

Based on the reported F_{mean} values, Wilcoxon signed rank test [68] is applied to perform a pairwise comparison between the proposed MSPSOTLP and the selected PSO variants. The results in terms of R^+, R^-, p, and h values, are presented in Table 6. Accordingly, MSPSOTLP performs significantly better than all other PSO variants at a significance level of $\alpha = 0.05$ as indicated by the h value of "+". Notably, the proposed MSPSOTLP completely dominates PSO, PSOWV, and PSOWV to solve CEC 2014 benchmark functions based on the promising R^+, R^-, p and h values reported.

Table 6. Pairwise comparisons by Wilcoxon signed rank test between MSPSOTLP and each peer algorithm.

MSPSOTLP vs.	R^+	R^-	p Value	h Value
PSO	465.0	0.0	2.00×10^{-6}	+
PSOWV	465.0	0.0	2.00×10^{-6}	+
CSO	382.5	82.5	1.91×10^{-3}	+
MPSOWV	465.0	0.0	2.00×10^{-6}	+
SLPSO	390.0	45.0	1.76×10^{-4}	+
PSOGSA	347.5	117.5	1.75×10^{-2}	+
APSO	459.0	6.0	3.00×10^{-6}	+

The Friedman test [67] is further conducted for multiple comparisons between the proposed MSPSOTLP and selected PSO variants based on their F_{mean} values. The results, in terms of average ranking, chi-square statistics, and p-value, are reported in Table 7. The Friedman test reports that MSPSOTLP has the best performance by scoring an average rank of 1.6833, followed by CSPSO, PSOGSA, SLPSO, PSO, APSO, MPSOWV, and PSOWV with average ranks of 3.0500, 3.0667, 3.7667, 4.4167, 5.6500, 6.6500, and 7.7167, respectively. The p-value determined by the Friedman test through chi-square statistics is smaller than the predefined significance level of $\alpha = 0.05$. Therefore, significant global performance deviations among all compared algorithms are observed.

Table 7. Average ranking and p-value produced by Friedman test.

Algorithm	Ranking	Chi-Square Statistics	p value
MSPSOTLP	1.6833		
PSOGSA	3.0667		
CSO	3.0500		
PSO	4.4167	144.636111	0.00×10^0
SLPSO	3.7667		
APSO	5.6500		
MPSOWV	6.6500		
PSOWV	7.7167		

Given the global performance difference observed from the Friedman test, three post-hoc statistical analyses [67], known as Bonferroni-Dunn, Holm, and Hochberg, are utilized to identify other concrete performance differences between the proposed MSPSOTLP and different PSO variants. The results, in terms of z values, unadjusted p values, and adjusted p values (APVs), produced by three procedures are reported in Table 8. All post-hoc procedures confirm the significant performance enhancement of MSPSOTLP against PSOWV, MPSOWV, APSO, SLPSO, and PSO at $\alpha = 0.05$. The Hochberg procedure has higher sensitivity to detect the significant performance difference between MSPSOTLP, CSO, and PSAGSA. Notably, the Holm procedure can detect the significant performance improvement of MSPSOTLP against CSO and PSOGSA if the threshold level is adjusted to $\alpha = 0.10$.

4.1.6. Performance Analysis of Proposed Improvement Strategies

A further performance analysis is conducted in this subsection to investigate the contribution brought by each improvement strategy introduced into MSPSOTLP, i.e., modified initialization scheme (i.e., chaotic map and oppositional based learning), primary learning phase (i.e., multiswarm concept and construction of social exemplar), and secondary learning phase (i.e., two search operators with different search characteristic). The original PSOWV is chosen as the baseline method to be compared in this subsection. Another three variants of MSPSOTLP, i.e., MSPSOTLP-1, MSPSOTLP-2 and MSPSOTLP-3, are also introduced to analyze the performance gains brought by the modified initialization scheme, primary learning phase, and secondary learning phase, respectively. Particularly, MSPSOTLP-1 refers to the PSOWV enhanced with the CSOBL initialization scheme. Meanwhile, MSPSOTLP-2 refers to PSOWV enhanced with the CSOBL initialization scheme and primary learning phase. Finally, MSPSOTLP-3 refers to PSOWV enhanced with the CSOBL initialization scheme and secondary learning phase, where its primary phase is replaced with the original search operator of PSOWV. The performance gain achieved by each MSPSOTLP variants against the original PSOWV when solving every benchmark function is measured as ΔG as follow:

$$\Delta G = \frac{F_{mean}(MSPSOTLP\ variant) - F_{mean}(PSOWV)}{|F_{mean}(PSOWV)|} \times 100\% \tag{24}$$

Table 8. Adjusted p values produced by each algorithm through three post-hoc analysis procedures.

MSPSOTLP vs.	z	Unadjusted p	Bonferroni-Dunn p	Holm p	Hochberg p
PSOWV	9.54×10^0	0.00×10^0	0.00×10^0	0.00×10^0	0.00×10^0
MPSOWV	7.85×10^0	0.00×10^0	0.00×10^0	0.00×10^0	0.00×10^0
APSO	6.27×10^0	0.00×10^0	0.00×10^0	0.00×10^0	0.00×10^0
SLPSO	4.32×10^0	1.50×10^{-5}	1.08×10^{-4}	6.20×10^{-3}	6.20×10^{-5}
PSO	3.29×10^0	9.88×10^{-4}	6.91×10^{-3}	2.96×10^{-3}	2.96×10^{-3}
CSO	2.19×10^0	2.87×10^{-2}	2.01×10^{-1}	5.75×10^{-2}	3.07×10^{-2}
PSOGSA	2.16×10^0	3.07×10^{-2}	2.15×10^{-1}	5.75×10^{-2}	3.07×10^{-2}

Referring to Equation (24), it is evident that a positive value of ΔG can be obtained if a particular MSPSOTLP variant can solve the benchmark functions with better F_{mean} value than that of PSOWV and vice versa.

The simulation results in terms of F_{mean} and ΔG obtained by PSOWV and all MSPSOTLP variants when solving all CEC 2014 benchmark functions are presented in Table 9. Accordingly, all MSPSOTLP variants have successfully solved the majority of CEC 2014 benchmark functions with different degrees of performance gains. MPSOTLP-1 is observed to outperform PSOWV in the majority of CEC 2014 benchmark functions except for F1, F5, F6, F11, F12, F16, F17, F21, F29, and F30. Although the CSOBL can produce an initial population with better solution quality in terms of fitness and diversity that can lead to performance gain of algorithm, it is not sufficient to solve the complex problem because CSOBL is only executed once during the search process. Some interesting findings can be observed when both variants of MSPSOTLP-2 and MSPSOTLP-3 are used to solve CEC 2014 benchmark functions. Particularly, MSPSOTLP-2 can perform better than MSPSOTLP-3 when solving the unimodal (i.e., F1 to F3), simple multimodal (i.e., F4 to F16) and hybrid (i.e., F17 to F22) functions. Meanwhile, MSPSOTLP-3 is revealed to be more competitive than MSPSOTLP-2 in dealing with the most complex composition function. The performance differences between MSPSOTLP-2 and MSPSOTLP03 can be justified based on their inherent search mechanisms. For MSPSOTLP-2, the primary learning phase is incorporated with multiswarm and social learning concepts used to accelerate the convergence characteristic of the algorithm without compromising its population diversity. When dealing with optimization functions with less complex fitness landscapes (i.e., unimodal, simple multimodal, and hybrid functions), the benefits brought by both the multiswarm and social learning concepts can still suppress the potential negative drawbacks of the historically best position (i.e., global best position in this case). When the fitness landscapes of the optimization problems are increased further, such as those in composition functions, the numbers of local optima in the solution space have increased exponentially. Under this circumstance, the diversity maintenance scheme introduced in the primary learning phase of MSPSOTLP-2 is not sufficient to curb the high tendency of historically best positions to be trapped in these local optima. On the other hand, the secondary learning phase of

MSPSOTLP-3 can leverage the useful directional information of other non-fittest solutions to perform searching with greater exploration strengths; thus, it has a higher chance to escape from the inferior regions of the solution space. Finally, the complete MSPSOTLP has exhibited the best performance when solving all CEC 2014 benchmark functions with 26 best and 4 s-best F_{mean} values. The simulation results reported in Table 9 have verified that each improvement strategy incorporated into MSPSOTLP indeed has different contributions to enhancing the search performance of the proposed algorithm.

Table 9. Simulation results of F_{mean} and ΔP obtained by PSOWV and all MSPSOTLP variants when solving CEC 2014 benchmark functions.

Func	F_{mean} (ΔG)				
	PSOWV	**MSPSOTLP-1**	**MSPSOTLP-2**	**MSPSOTLP-3**	**MSPSOTLP**
F1	4.70×10^8 ($-$)	8.08×10^8 (-71.93%)	8.42×10^6 (98.21%)	1.37×10^8 (70.83%)	$\mathbf{5.61 \times 10^3}$ (100.00%)
F2	6.94×10^{10} ($-$)	1.47×10^{10} (78.79%)	2.70×10^7 (99.96%)	4.95×10^9 (92.87%)	$\mathbf{0.00 \times 10^0}$ (100.00%)
F3	3.18×10^5 ($-$)	8.56×10^4 (73.09%)	1.10×10^4 (96.53%)	4.09×10^4 (87.13%)	$\mathbf{2.18 \times 10^{-11}}$ (100.00%)
F4	8.66×10^3 ($-$)	6.80×10^3 (21.52%)	1.48×10^2 (98.29%)	2.43×10^2 (97.20%)	$\mathbf{1.44 \times 10^{-2}}$ (100.00%)
F5	2.09×10^1 ($-$)	2.10×10^1 (-0.60%)	2.09×10^1 (0.04%)	2.08×10^1 (0.43%)	$\mathbf{2.01 \times 10^1}$ (3.83%)
F6	3.83×10^1 ($-$)	4.34×10^1 (-13.26%)	2.45×10^1 (35.92%)	2.78×10^1 (27.39%)	$\mathbf{3.25 \times 10^0}$ (91.51%)
F7	5.61×10^2 ($-$)	2.23×10^2 (60.17%)	1.19×10^0 (99.79%)	3.68×10^0 (99.34%)	$\mathbf{0.00 \times 10^0}$ (100.00%)
F8	3.87×10^2 ($-$)	2.49×10^2 (35.67%)	4.13×10^1 (89.34%)	1.53×10^2 (60.36%)	$\mathbf{1.15 \times 10^1}$ (97.03%)
F9	4.85×10^2 ($-$)	3.16×10^2 (34.90%)	9.96×10^1 (79.47%)	1.67×10^2 (65.50%)	$\mathbf{1.20 \times 10^1}$ (97.53%)
F10	7.66×10^3 ($-$)	6.85×10^3 (10.55%)	2.12×10^3 (72.36%)	4.47×10^3 (41.70%)	$\mathbf{1.52 \times 10^2}$ (98.02%)
F11	7.56×10^3 ($-$)	8.54×10^3 (-12.92%)	3.60×10^3 (52.35%)	4.82×10^3 (36.24%)	$\mathbf{2.42 \times 10^3}$ (67.99%)
F12	2.70×10^0 ($-$)	3.61×10^0 (-33.67%)	2.15×10^0 (20.26%)	2.15×10^0 (20.48%)	$\mathbf{1.62 \times 10^0}$ (40.00%)
F13	6.12×10^0 ($-$)	5.14×10^0 (16.06%)	2.64×10^{-1} (95.68%)	7.40×10^{-1} (87.91%)	$\mathbf{1.29 \times 10^{-1}}$ (97.89%)
F14	1.46×10^2 ($-$)	1.40×10^2 (3.96%)	9.63×10^{-1} (99.34%)	2.98×10^{-1} (99.80%)	$\mathbf{1.93 \times 10^{-1}}$ (99.87%)
F15	3.58×10^6 ($-$)	3.64×10^4 (98.98%)	1.89×10^1 (100.00%)	9.12×10^1 (100.00%)	$\mathbf{2.30 \times 10^0}$ (100.00%)
F16	1.34×10^1 ($-$)	1.37×10^1 (-2.00%)	1.31×10^1 (2.56%)	1.36×10^1 (-1.63%)	$\mathbf{7.60 \times 10^0}$ (43.28%)
F17	1.77×10^7 ($-$)	5.56×10^7 (-100.00%)	8.27×10^5 (95.33%)	6.08×10^5 (95.56%)	$\mathbf{2.10 \times 10^3}$ (99.99%)
F18	5.88×10^8 ($-$)	4.07×10^8 (30.77%)	2.59×10^4 (100.00%)	7.51×10^5 (99.87%)	$\mathbf{1.22 \times 10^2}$ (100.00%)
F19	3.64×10^2 ($-$)	1.44×10^2 (60.49%)	1.06×10^1 (97.09%)	2.26×10^1 (93.80%)	$\mathbf{4.08 \times 10^0}$ (98.88%)
F20	2.78×10^5 ($-$)	2.54×10^5 (8.77%)	1.98×10^4 (92.88%)	5.06×10^3 (98.18%)	$\mathbf{7.96 \times 10^1}$ (99.97%)
F21	6.03×10^6 ($-$)	1.22×10^7 (-100.00%)	1.21×10^5 (97.99%)	7.86×10^5 (86.97%)	$\mathbf{8.02 \times 10^2}$ (99.99%)
F22	1.08×10^3 ($-$)	8.96×10^2 (17.07%)	3.15×10^2 (70.81%)	4.25×10^2 (60.65%)	$\mathbf{3.52 \times 10^1}$ (96.74%)
F23	6.74×10^2 ($-$)	$\mathbf{2.00 \times 10^2}$ (70.33%)	3.15×10^2 (53.21%)	$\mathbf{2.00 \times 10^2}$ (70.33%)	3.15×10^2 (53.26%)
F24	4.66×10^2 ($-$)	$\mathbf{2.00 \times 10^2}$ (57.08%)	2.12×10^2 (54.45%)	$\mathbf{2.00 \times 10^2}$ (57.08%)	2.24×10^2 (51.93%)
F25	2.59×10^2 ($-$)	$\mathbf{2.00 \times 10^2}$ (22.78%)	2.07×10^2 (19.92%)	$\mathbf{2.00 \times 10^2}$ (22.78%)	$\mathbf{2.00 \times 10^2}$ (22.78%)
F26	1.06×10^2 ($-$)	1.04×10^2 (1.80%)	1.01×10^2 (5.19%)	1.01×10^2 (5.13%)	$\mathbf{1.00 \times 10^2}$ (5.66%)
F27	1.28×10^3 ($-$)	$\mathbf{2.00 \times 10^2}$ (84.38%)	7.58×10^2 (40.76%)	$\mathbf{2.00 \times 10^2}$ (84.38%)	3.94×10^2 (69.22%)
F28	1.82×10^3 ($-$)	$\mathbf{2.00 \times 10^2}$ (89.01%)	8.98×10^2 (50.65%)	$\mathbf{2.00 \times 10^2}$ (89.01%)	8.55×10^2 (53.02%)
F29	1.04×10^7 ($-$)	3.21×10^7 (-100.00%)	1.47×10^3 (99.99%)	5.95×10^4 (99.43%)	$\mathbf{8.60 \times 10^2}$ (99.99%)
F30	3.79×10^5 ($-$)	8.38×10^5 (-100.00%)	2.54×10^3 (99.33%)	9.93×10^4 (73.80%)	$\mathbf{1.72 \times 10^3}$ (99.55%)

4.2. Performance Evaluation of MSPSOTLP in Training ANN Models

4.2.1. Classification Datasets for Training ANN Models

Apart from general optimization performance, the capability of the proposed MSP-SOTLP in training ANN models for data classification tasks is also evaluated using sixteen standard datasets extracted from the University of California Irvine (UCI) machine learning repository [71]. The sixteen datasets selected for performance evaluation include Iris, Liver Disorder, Blood Transfusion, Statlog Heart, Hepatitis, Wine, Breast Cancer, Seeds, Australian Credit Approval, Haberman's Survival, New Thyroid, Glass, Balance Scale, Dermatology, Landsat and Bank Note. The properties of each dataset are summarized in Table 10. Each selected dataset is separated into two parts, known as 70% of training samples and 30% of testing samples. Specifically, the training samples are used by MSPSOTLP and other PSO variants to optimize the parameters of ANN models (i.e., weights, biases, and activation function). In contrast, the testing samples are used to evaluate the generalization performance of ANN models trained by all compared algorithms.

Table 10. Properties of datasets selected for ANN model training.

No.	Dataset	# Attributes	# Classes	# Samples
DS1	Iris	4	3	150
DS2	Liver Disorder	6	2	345
DS3	Blood Transfusion	4	2	748
DS4	Statlog Heart	13	2	270
DS5	Hepatitis	19	2	80
DS6	Wine	13	3	178
DS7	Breast Cancer	9	2	277
DS8	Seeds	7	3	210
DS9	Australian Credit Approval	14	2	690
DS10	Haberman's Survival	3	2	306
DS11	New Thyroid	5	3	215
DS12	Glass	9	6	214
DS13	Balance Scale	4	3	625
DS14	Dermatology	34	6	338
DS15	Landsat	36	6	4435
DS16	Bank Note	4	2	1372

4.2.2. Performance Metrics for ANN Training

Classification accuracy $\mathbb{R}^C$ is a popular performance metric used to measure the classification performance of an ANN model. Suppose that $\widetilde{R}$ refers to the number of correctly classified data samples by the ANN model, while R represents the total number of data samples in each dataset. The $\mathbb{R}^C$ value is calculated as follow:

$$\mathbb{R}^C = \frac{\widetilde{R}}{R} \times 100\% \tag{25}$$

An ANN model with a larger value of $\mathbb{R}^C$ is more desirable because it can produce better results in terms of classification accuracy. In addition, the ANN model produces a larger $\mathbb{R}^C$ value when solving the testing samples s also considered to have better generalization performance due to its excellent capability to accurately classify unseen data in different classes with its understanding of the existing data. Furthermore, the standard deviation SD values are also recorded to observe the consistency of ANN models trained by compared PSO variants in solving classification datasets.

4.2.3. Parameter Settings for ANN Training

Similar to the global optimization, the performance of the ANN model trained by the proposed MSPSOTLP in solving sixteen datasets extracted from the UCI machine learning repository is compared with the seven PSO variants reported in Section 4.1.3. The parameters of all compared algorithms are configured as reported in Table 3 as recommended in their original literature. The same values of $N = 100$ and $\Gamma^{max} = 10,000 \times D$ are also configured for all compared PSO variants in solving ANN training problems, where the value of D is calculated based on Equation (7). The ANN model to be trained in this study is constructed by an input layer, a hidden layer, and an output layer. The number of input and output neurons of each ANN model are configured based on the number of attributes and classes of a given dataset, respectively, as presented in Table 8. Meanwhile, the number of neurons in the hidden layer is 15. Similarly, all compared algorithms in training ANN model are simulated using Matlab 2019b on a personal computer with Intel ® Core i7-7500 CPU @ 2.70GHz.

4.2.4. Performance Comparison in Training ANN Models

The $\mathbb{R}^C$ and SD values produced by the ANN models optimized by all compared PSO variants when classifying the training and testing samples are presented in Tables 11 and 12, respectively. Similarly, the best and second best $\mathbb{R}^C$ values produced by the compared methods in each dataset are indicated in boldface and underlined, respectively. The comparative studies between the ANN models trained by MSPSOTLP and other PSO variants are summarized in $\#B\mathbb{R}^C$ and $w/t/l$ as similar to those in Table 5. Specifically, $\#B\mathbb{R}^C$ records the number of best $\mathbb{R}^C$ values produced by each algorithm in training the ANN models for 16 datasets extracted from UCI machine learning repository.

Table 11. The $\mathbb{R}^C$ and SD values produced by ANN models trained by the proposed MSPSOTLP and other PSO variants when solving the training samples.

Dataset	Criteria	MSPSOTLP	PSO	PSOWV	CSO	MPSOWV	SLPSO	PSOGSA	APSO
DS1	$\mathbb{R}^C$	**99.58**	94.58	67.42	79.08	<u>98.67</u>	77.00	90.92	19.83
	SD	4.39×10^{-1}	2.25×10^0	1.11×10^1	6.30×10^0	3.73×10^{-1}	2.20×10^0	9.84×10^0	4.01×10^1
DS2	$\mathbb{R}^C$	72.72	**73.44**	52.68	58.55	69.60	57.03	<u>73.26</u>	58.59
	SD	2.45×10^{-1}	5.37×10^{-1}	5.30×10^0	1.05×10^0	9.12×10^{-1}	2.67×10^0	2.09×10^{-1}	2.43×10^0
DS3	$\mathbb{R}^C$	<u>80.59</u>	80.55	78.63	78.93	79.45	78.93	**81.29**	78.90
	SD	5.29×10^{-2}	4.83×10^{-1}	2.70×10^0	0.00×10^0	3.34×10^{-1}	0.00×10^0	1.67×10^{-1}	0.00×10^0
DS4	$\mathbb{R}^C$	**96.20**	<u>94.07</u>	69.82	80.56	91.90	77.69	92.18	52.45
	SD	8.11×10^{-1}	2.09×10^0	7.87×10^0	2.12×10^0	5.35×10^{-1}	2.41×10^0	1.17×10^0	1.86×10^1
DS5	$\mathbb{R}^C$	**96.41**	94.53	62.19	75.94	<u>95.00</u>	80.31	92.81	60.94
	SD	1.81×10^0	1.80×10^0	9.55×10^0	0.00×10^0	1.80×10^0	3.61×10^0	9.02×10^{-1}	1.18×10^1
DS6	$\mathbb{R}^C$	**100.00**	94.09	59.58	82.32	<u>99.79</u>	66.41	99.51	28.80
	SD	0.00×10^0	1.08×10^0	2.22×10^1	3.33×10^0	4.07×10^{-1}	4.30×10^0	4.07×10^{-1}	4.64×10^1
DS7	$\mathbb{R}^C$	<u>82.48</u>	**83.29**	67.34	74.32	79.69	74.73	79.55	69.73
	SD	3.82×10^{-1}	6.88×10^{-1}	2.31×10^0	9.38×10^{-1}	4.50×10^{-1}	6.88×10^{-1}	2.74×10^0	4.82×10^0
DS8	$\mathbb{R}^C$	**97.50**	87.62	60.30	75.77	<u>96.37</u>	77.20	93.10	69.05
	SD	1.04×10^0	7.35×10^0	2.03×10^1	2.81×10^0	3.44×10^{-1}	4.81×10^0	9.09×10^{-1}	4.70×10^1
DS9	$\mathbb{R}^C$	**89.67**	<u>89.40</u>	73.55	82.63	88.97	83.97	89.10	88.46
	SD	2.96×10^{-1}	1.41×10^0	7.60×10^0	1.73×10^0	4.56×10^{-1}	2.20×10^0	3.14×10^{-1}	1.56×10^1
DS10	$\mathbb{R}^C$	75.14	<u>75.43</u>	70.94	72.33	75.14	72.04	**75.71**	72.20
	SD	2.32×10^{-1}	2.36×10^{-1}	3.89×10^0	8.16×10^{-1}	2.36×10^{-1}	4.08×10^{-1}	1.08×10^0	0.00×10^0
DS11	$\mathbb{R}^C$	**98.55**	93.20	80.35	92.85	<u>95.87</u>	85.23	94.48	71.28
	SD	1.29×10^0	1.01×10^1	5.53×10^0	3.20×10^0	1.34×10^0	6.40×10^0	6.71×10^{-1}	1.11×10^1
DS12	$\mathbb{R}^C$	**54.39**	45.73	13.22	15.97	41.17	13.74	<u>49.06</u>	45.61
	SD	1.04×10^1	1.32×10^1	1.14×10^1	4.39×10^0	2.06×10^1	2.05×10^0	1.88×10^0	1.49×10^1
DS13	$\mathbb{R}^C$	**91.18**	86.96	65.06	80.78	89.42	81.46	<u>89.82</u>	77.00
	SD	1.50×10^0	2.31×10^0	4.63×10^0	1.93×10^0	1.15×10^{-1}	4.05×10^0	7.02×10^{-1}	2.70×10^1
DS14	$\mathbb{R}^C$	**29.02**	<u>28.50</u>	24.34	27.20	**29.02**	24.86	27.94	21.33
	SD	9.90×10^{-3}	3.03×10^0	2.29×10^0	2.13×10^0	4.35×10^{-15}	6.06×10^{-1}	3.03×10^0	6.06×10^{-1}
DS15	$\mathbb{R}^C$	**76.63**	71.16	33.35	52.30	<u>75.57</u>	35.53	69.50	33.74
	SD	3.67×10^{-1}	1.30×10^0	3.48×10^1	2.20×10^0	8.15×10^0	1.10×10^{-1}	8.03×10^0	2.55×10^1
DS16	$\mathbb{R}^C$	**100.00**	98.48	73.96	96.07	99.46	92.71	<u>99.68</u>	63.37
	SD	0.00×10^0	1.90×10^{-1}	6.01×10^0	6.59×10^{-1}	0.00×10^0	8.71×10^0	1.58×10^{-1}	1.46×10^1
	$\#B\mathbb{R}^C$	12	2	0	0	1	0	2	0
	$w/t/l$	-	13/0/3	16/0/0	16/0/0	14/2/0	16/0/0	13/0/3	16/0/0

According to Table 11, ANN models trained by the proposed MSPSOTLP are reported to have the best performance for being able to solve 12 out of 16 sets of training samples with the best $\mathbb{R}^C$ values. Specifically, MSPSOTLP emerges as the best training algorithm for ANN models in dealing with datasets of Iris, Statlog Hearts, Hepatitis, Wine, Seeds, Australian Credit Approval, New Thyroid, Glass, Balance Scale, Dermatology, Landsat, and Bank Note. It is also noteworthy that the proposed MSPSOTLP is the only algorithm that has successfully trained ANN models with 100% of $\mathbb{R}^C$ values to classify the training sets of Wine and Bank Note. The ANN models trained by PSO and PSAGSA can occasionally deliver good performances by producing two best $\mathbb{R}^C$ values in solving training samples. Although the $\mathbb{R}^C$ values produced by ANN models trained by the proposed MSPSOSLP in classifying training samples of Liver Disorder, Blood Transfusion, Breast Cancer, and Haberman's Survival are slightly lower than PSO and PSOGSA, the performance differences between these algorithms are marginal and not more than 1%. On the other hand, the ANN models trained by MSPSOTLP are reported to have more notable performance differences than PSO and PSOGSA in terms of $\mathbb{R}^C$ values when classifying the training samples of Iris, Statlog Heart, Hepatitis, Seeds, New Thyroid, Glass, and Landsat. The ANN models trained by PSOWV are reported to have the most inferior performance by producing eight worst and seven second-worst $\mathbb{R}^C$ values when solving all 16 classification datasets except for Landsat.

Table 12. The $\mathbb{R}^C$ and SD values produced by ANN models trained by the proposed MSPSOTLP and other PSO variants when solving the testing samples.

Dataset	Criteria	MSPSOTLP	PSO	PSOWV	CSO	MPSOWV	SLPSO	PSOGSA	APSO
DS1	$\mathbb{R}^C$	**100.00**	<u>99.67</u>	56.67	88.33	97.33	92.33	87.67	16.67
	SD	0.00×10^0	1.49×10^0	3.16×10^1	2.23×10^1	0.00×10^0	0.00×10^0	3.35×10^1	4.88×10^1
DS2	$\mathbb{R}^C$	<u>50.00</u>	47.68	35.80	40.73	47.83	37.10	49.28	**55.22**
	SD	7.64×10^{-1}	6.48×10^{-1}	1.51×10^1	7.67×10^0	1.45×10^0	1.35×10^1	5.23×10^0	3.89×10^0
DS3	$\mathbb{R}^C$	58.53	58.53	60.67	63.80	**65.33**	<u>65.47</u>	49.73	63.13
	SD	2.81×10^{-1}	1.19×10^1	1.27×10^1	2.40×10^0	4.37×10^0	7.70×10^{-1}	4.23×10^0	2.31×10^0
DS4	$\mathbb{R}^C$	**80.37**	75.93	61.30	75.00	<u>77.22</u>	72.78	73.15	42.04
	SD	1.56×10^0	3.85×10^0	2.38×10^1	2.14×10^0	1.07×10^0	3.21×10^0	3.85×10^0	1.57×10^1
DS5	$\mathbb{R}^C$	**72.50**	60.00	51.88	55.63	<u>60.63</u>	60.63	53.13	46.88
	SD	3.23×10^0	3.61×10^0	7.22×10^0	1.30×10^1	9.55×10^0	1.08×10^1	0.00×10^0	1.44×10^1
DS6	$\mathbb{R}^C$	**90.28**	76.94	34.44	61.11	82.50	42.78	<u>83.06</u>	16.11
	SD	1.46×10^0	3.21×10^0	1.58×10^1	8.93×10^0	2.78×10^0	1.79×10^1	6.99×10^0	4.01×10^1
DS7	$\mathbb{R}^C$	**73.82**	<u>72.55</u>	59.46	65.09	70.73	66.36	66.91	69.27
	SD	9.39×10^{-1}	1.05×10^0	7.57×10^0	3.15×10^0	1.05×10^0	4.81×10^0	2.78×10^0	4.58×10^0
DS8	$\mathbb{R}^C$	**100.00**	78.33	30.24	32.06	<u>96.19</u>	40.24	74.29	54.76
	SD	7.53×10^{-1}	5.22×10^1	3.23×10^1	5.23×10^1	3.71×10^1	2.75×10^0	4.76×10^0	4.81×10^1
DS9	$\mathbb{R}^C$	84.57	82.39	67.75	76.38	82.54	**87.31**	<u>87.05</u>	86.86
	SD	6.87×10^{-1}	2.54×10^0	0.00×10^0	3.16×10^0	2.09×10^0	2.09×10^0	3.83×10^0	1.51×10^1
DS10	$\mathbb{R}^C$	**78.69**	78.03	68.85	73.12	<u>78.53</u>	76.39	67.38	78.53
	SD	0.00×10^0	0.00×10^0	4.41×10^1	1.64×10^0	0.00×10^0	5.91×10^0	1.46×10^1	1.89×10^0
DS11	$\mathbb{R}^C$	**84.19**	<u>67.21</u>	38.84	48.37	54.19	42.79	47.67	40.47
	SD	1.05×10^1	4.65×10^0	3.55×10^0	4.65×10^0	1.28×10^1	8.70×10^{-15}	8.06×10^0	4.03×10^0
DS12	$\mathbb{R}^C$	**46.51**	<u>44.19</u>	8.14	14.19	29.07	15.58	29.07	25.58
	SD	6.39×10^0	5.85×10^0	2.01×10^1	5.37×10^0	6.15×10^0	7.48×10^0	9.30×10^0	7.48×10^0
DS13	$\mathbb{R}^C$	**88.72**	85.84	54.80	82.08	<u>87.36</u>	80.24	86.88	80.80
	SD	1.22×10^0	2.01×10^0	8.09×10^0	3.23×10^0	1.67×10^0	7.26×10^0	4.62×10^{-1}	3.80×10^1
DS14	$\mathbb{R}^C$	<u>65.83</u>	**65.97**	57.78	62.64	65.14	59.17	62.78	54.17
	SD	9.71×10^{-1}	4.88×10^0	7.13×10^0	6.56×10^0	1.39×10^0	4.01×10^0	2.41×10^0	0.00×10^0
DS15	$\mathbb{R}^C$	**80.47**	72.66	12.22	25.78	<u>78.19</u>	10.67	63.84	21.76
	SD	1.49×10^0	1.14×10^1	9.85×10^0	2.22×10^0	3.03×10^1	7.81×10^0	3.06×10^1	3.41×10^1
DS16	$\mathbb{R}^C$	**92.15**	<u>90.18</u>	33.98	4.9	81.35	55.99	86.28	31.68
	SD	3.30×10^0	2.64×10^0	0×10^0	3.39×10^0	1.48×10^1	3.45×10^0	1.16×10^1	3.16×10^1
	$\#B\mathbb{R}^C$	12	1	0	0	1	1	0	1
	$w/t/l$	-	14/1/1	15/0/1	15/0/1	15/0/1	14/0/2	15/0/1	13/0/3

In addition to training samples, 30% of datasets are extracted as testing samples to evaluate the generalization performances of ANN models optimized by all compared PSO variants. According to Table 12, ANN models trained by the proposed MSPSOTLP are reported to have the best generalization performance for being able to produce 12 best and 2 s-best $\mathbb{R}^C$ values when classifying the testing samples of 16 selected datasets. Specifically, ANN models trained by MSPSOTLP successfully produce the best $\mathbb{R}^C$ in solving the testing samples of Iris, Statlog Heart, Hepatitis, Wine, Breast Cancer, Seeds, Haberman's Survival, New Thyroid, Glass, Balance Scale, Landsat, and Bank Note. Moreover, MSPSOTLP is also reported to be the only algorithm that can train ANN models to solve the testing samples of Iris and Seeds with 100% of $\mathbb{R}^C$. On the other hand, the ANN models trained by PSO, MPSOWV, SLPSO, and APSO are reported to produce the best $\mathbb{R}^C$ values when classifying the testing samples of Dermatology, Blood Transfusion, Australian Credit Approval, and Liver Disorder, respectively. Although ANN models trained by MSPSOTLP are observed to produce relatively inferior $\mathbb{R}^C$ values in solving the testing samples of these four datasets, some performance differences are insignificant. On the contrary, ANN models trained by MSPSOTLP produce much better $\mathbb{R}^C$ values than PSO, MPSOWV, SLPSO, and APSO in solving the testing samples of certain datasets, such as Hepatitis, Wine, Seeds, New Thyroid, etc. The ANN models trained by PSOWV perform the worst by producing seven lowest $\mathbb{R}^C$ values (i.e., testing samples of Liver Disorder, Breast Cancer, Seeds, Australian Credit Approval, New Thyroid, Glass, and Balance Scale) and eight second-worst $\mathbb{R}^C$ values (i.e., testing samples of Iris, Statlog Heart, Hepatitis, Wine, Haberman's Survival, Dermatology, Landsat, and Bank Note).

4.2.5. Non-Parametric Statistical Analyses

Similar to those reported in Section 4.1.5, the Wilcoxon signed rank test [68] is applied to perform a pairwise comparison between the proposed MSPSOTLP and the selected PSO variants based on the reported $\mathbb{R}^C$ values. The R^+, R^-, p, and h values produced by ANN models trained by the compared algorithms in classifying the training and testing samples are presented in Tables 13 and 14, respectively. From Table 13, ANN models trained by the proposed MSPSOTLP have significantly better performance than those of other PSO variants at a significance level of $\alpha = 0.05$, as indicated by the h value of "+". Notably, the ANN model optimized by the proposed MSPSOTLP can completely dominate those of PSOWV, CSO, SLPSO, and APSO when solving the training samples of selected datasets. Similarly, ANN models trained by MSPSOTLP are observed to perform significantly better than other PSO variants in solving all testing samples of datasets chosen, as indicated by the h values of "+" in Table 14. These pairwise comparison results imply the excellent generalization ability of ANN model trained by MSPSOTLP due to its ability to handle unseen data of testing samples effectively.

Table 13. Wilcoxon signed rank test as a pairwise comparison between the ANN models optimized by MSPSOTLP and each PSO variant when classifying the training samples.

MSPSOTLP vs.	R^+	R^-	p Value	h Value
PSO	122.0	14.0	3.36×10^{-3}	+
PSOWV	136.0	0.0	3.05×10^{-5}	+
CSO	136.0	0.0	3.05×10^{-5}	+
MPSOWV	119.0	1.0	1.22×10^{-4}	+
SLPSO	136.0	0.0	3.05×10^{-5}	+
PSOGSA	122.0	14.0	3.36×10^{-3}	+
APSO	136.0	0.0	3.05×10^{-5}	+

Table 14. Wilcoxon signed rank test as pairwise comparison between the ANN models optimized by MSPSOTLP and each PSO variant when classifying the testing samples.

MSPSOTLP vs.	R^+	R^-	p Value	h Value
PSO	134.0	2.0	9.16×10^{-5}	+
PSOWV	135.0	1.0	6.10×10^{-5}	+
CSO	134.0	2.0	9.16×10^{-5}	+
MPSOWV	125.0	11.0	1.68×10^{-3}	+
SLPSO	130.0	6.0	4.27×10^{-4}	+
PSOGSA	133.0	3.0	1.53×10^{-4}	+
APSO	125.0	11.0	1.68×10^{-3}	+

Apart from pairwise comparison, multiple comparisons among the ANN models trained by all compared algorithms are also conducted by Friedman Test [67]. The average ranking values produced by the ANN models trained by all compared algorithms in solving training and testing samples are reported in Tables 15 and 16, respectively. Table 15 shows that the ANN models trained by MSPSOTLP score the best average ranking when classifying the training datasets, followed by PSOGSA, PSO, MPSOWV, CSO, SLPSO,

APSO, and PSOWV. Although Table 7 reports that MPSOWV has a relatively poor ranking in solving CEC 2014 benchmark functions, it performs relatively well in training ANN models in solving training samples. Conversely, CSO does not perform well in training the ANN model despite producing relatively competitive performance in solving CEC 2014 benchmark functions. Table 16 reveals that the ANN model trained by MSPSOTLP has the best average ranking value in classifying the testing samples, followed by MPSOWV, PSO, PSOGSA, SLPSO, CSO, APSO, and PSOWV. Similarly, MPSOWV shows its competitive performance in training the ANN model, despite having inferior performance in solving CEC 2014 benchmark functions. Although the ANN model trained by PSOGSA has the second-best average ranking in solving training samples, as reported in Table 15, it does not perform well in solving testing samples, as reported in Table 16, implying the tendency of PSAGSA to produce the ANN models that suffer with overfitting issues and have poor generalization performance to handle unseen data.

Table 15. The average ranking values of ANN models are trained by all compared algorithms in solving training samples.

Algorithm	Ranking	Chi-Square Statistics	p Value
MSPSOTLP	1.4688		
PSOGSA	2.8125		
PSO	2.8750		
MPSOWV	2.9062	94.875000	0.00×10^0
CPSO	5.5312		
SLPSO	5.9062		
APSO	6.9375		
PSOWV	7.5625		

Table 16. The average ranking values of ANN models are trained by all compared algorithms in solving testing samples.

Algorithm	Ranking	Chi-Square Statistics	p value
MSPSOTLP	1.6250		
MPSOWV	2.9688		
PSO	3.3750		
PSOGSA	4.4688	59.864583	0.00×10^0
SLPSO	5.2188		
CSO	5.3125		
APSO	5.7188		
PSOWV	7.3125		

Referring to the p values reported in Tables 15 and 16, significant global performance differences among the ANN models trained by all compared algorithms to solve training and testing samples are observed at a significance level of $\alpha = 0.05$. The concrete differences between the ANN models trained by MSPSOTLP and other PSO variants in classifying the training and testing samples are further analyzed using the Bonferroni-Dunn, Holm, and Hochberg procedures, as shown in Tables 17 and 18, respectively. According to the APVs for solving training samples, as reported in Table 17, all post-hoc procedures confirm the significant performance improvement of ANN models trained by MSPSOTLP against those of PSOWV, APSO, SLPSO, and CPSO at $\alpha = 0.05$. On the other hand, all post-hoc procedures can detect the significant performance improvement of ANN models trained by MSPSOTLP against those of PSOWV, APSO, CSO, SLPSO, and PSOGSA in solving testing samples, as indicated by the APVs values in Table 18.

Table 17. Adjusted p values produced by trained ANN models in solving training samples through three post-hoc analysis procedures.

MSPSOTLP vs.	z	Unadjusted p	Bonferroni-Dunn p	Holm p	Hochberg p
PSOWV	7.04×10^0	0.00×10^0	0.00×10^0	0.00×10^0	0.00×10^0
APSO	6.31×10^0	0.00×10^0	0.00×10^0	0.00×10^0	0.00×10^0
SLPSO	5.12×10^0	0.00×10^0	2.00×10^{-6}	1.00×10^{-5}	1.00×10^{-5}
CSO	4.69×10^0	3.00×10^{-6}	1.70×10^{-5}	1.10×10^{-5}	1.10×10^{-5}
MPSOWV	1.66×10^0	9.69×10^{-2}	6.79×10^{-1}	2.91×10^{-1}	1.21×10^{-1}
PSO	1.62×10^0	1.04×10^{-1}	7.31×10^{-1}	2.91×10^{-1}	1.21×10^{-1}
PSOGSA	1.55×10^0	1.21×10^{-1}	8.45×10^{-1}	2.91×10^{-1}	1.21×10^{-1}

Table 18. Adjusted p values produced by trained ANN models in solving testing samples through three post-hoc analysis procedures.

MSPSOTLP vs.	z	Unadjusted p	Bonferroni-Dunn p	Holm p	Hochberg p
PSOWV	6.57×10^0	0.00×10^0	0.00×10^0	0.00×10^0	0.00×10^0
APSO	4.73×10^0	0.00×10^0	1.60×10^{-5}	1.40×10^{-5}	1.40×10^{-5}
CSO	4.26×10^0	2.10×10^{-5}	1.44×10^{-4}	1.03×10^{-4}	1.03×10^{-4}
SLPSO	4.15×10^0	3.30×10^{-5}	2.33×10^{-4}	1.33×10^{-4}	1.33×10^{-4}
PSOGSA	3.28×10^0	1.03×10^{-3}	7.17×10^{-3}	3.07×10^{-3}	3.07×10^{-3}
PSO	2.02×10^0	4.33×10^{-2}	3.03×10^{-1}	8.66×10^{-1}	8.66×10^{-1}
MPSOWV	1.55×10^0	1.21×10^{-1}	8.45×10^{-1}	1.21×10^{-1}	1.21×10^{-1}

5. Conclusions

This study proposes a new PSO variant known as multi-swarm-based particle swarm optimization with two-level learning phases (MSPSOTLP) to address the potential drawbacks of PSOWV. Three significant modifications are introduced into the proposed MSPSOTLP to ensure that proper balancing of the exploration and exploitation searches of the algorithm can be achieved in handling more challenging optimization problems, including the training process of the ANN model. A new population initialization scheme, the CSOBL initialization scheme, is incorporated to replace the conventional random initialization scheme in generating initial solutions with better diversity and broader coverage in the solution space. Both multiswarm and social learning concepts are incorporated into the primary learning phase of MSPSOTLP to guide the search process of particles more effectively without losing population diversity by leveraging the directional information contributed by other non-fittest particles in the population. Additionally, a secondary learning phase is introduced with the adoption of two search operators with different levels of exploration and exploitation strengths, aiming to address the limitations of a single search operator adopted by many existing PSO variants. Extensive simulation studies report that the proposed MSPSOTLP outperforms the selected seven PSO variants in solving benchmark problems from CEC 2014 by producing 18 best mean fitness values out of 30 functions. Moreover, the training process of the ANN model is also formulated as an optimization problem, where the objective is to produce the optimal values of weights and biases and the selection of activation functions. The proposed MSPSOTLP is reported to have the best overall performance in training ANN models to solve classification datasets extracted from the UCI machine learning repository.

While MSPSOTLP has demonstrated competitive performances to solve the CEC 2014 benchmark functions and train ANN model for classifying UCI machine learning datasets, the proposed work still has room for improvement in terms of its search mechanisms and potential real-world applications. First, the main population of MSPSOTLP is divided by the reference-point-based population division scheme into a predefined number of subswarms during the primary learning phase. It is nontrivial to determine the optimal subswarm numbers for optimization problems with different types of fitness landscapes.

Second, the solution update process of MSPSOTLP is performed by comparing the fitness values of current and new particles. It is noteworthy that such a greedy selection scheme tends to suppress the survival of novel particles that might have temporary poor performances at the earlier stage of search process, but can contribute for long-term success if given sufficient iteration numbers. Third, the performance of the ANN optimized by MSPSOTLP is currently evaluated using the datasets obtained from a public database. Despite exhibiting promising classification accuracy in most selected datasets, the feasibility of the proposed method to solve more challenging real-world classification and regression problems remains unexplored. Some potential future works are then suggested to address these aforementioned limitations. First, the population division scheme of MSPSOTLBP can be further enhanced such that the optimal subswarm numbers can be determined adaptively based on the types of fitness landscapes encountered by the population. Second, other criteria, such as the fitness improvement rate and population diversity, should be considered by MSPSOTLP during the solution update process to preserve the novel particles that can bring long-term success for the algorithm. Finally, it is worth it to investigate the feasibility of ANN optimized by MSPSOTLP to address challenging issues encountered in the intelligent condition monitoring of complex industrial systems [2], such as the remaining useful life prediction of gear pumps [72], the time series prognosis of fuel cells [73], and predictive maintenance of renewable energy systems [74].

Author Contributions: Conceptualization, W.H.L., S.S.T. and E.-S.M.E.-K.; methodology, K.M.A., W.H.L. and E.-S.M.E.-K.; software, K.M.A., C.E.C., A.A.A. and A.I.; validation, K.M.A., C.E.C., F.K.K. and D.S.K.; formal analysis, K.M.A., S.S.T. and W.H.L.; investigation, K.M.A., W.H.L. and E.-S.M.E.-K.; resources, E.-S.M.E.-K., C.E.C., F.K.K. and D.S.K.; data curation, K.M.A., C.E.C., A.A.A., A.I. and F.K.K.; writing—original draft preparation, K.M.A., C.E.C., A.A.A., A.I. and F.K.K.; writing—review and editing, W.H.L., S.S.T. and E.-S.M.E.-K.; visualization, K.M.A., C.E.C. and D.S.K.; supervision, W.H.L., S.S.T. and E.-S.M.E.-K.; project administration, W.H.L. and E.-S.M.E.-K.; funding acquisition, E.-S.M.E.-K. and D.S.K. All authors have read and agreed to the published version of the manuscript.

Funding: Princess Nourah bint Abdulrahman University Researchers Supporting Project number (PNURSP2022R300), Princess Nourah bint Abdulrahman University, Riyadh, Saudi Arabia.

Data Availability Statement: The data will be provided upon reasonable request.

Acknowledgments: Princess Nourah bint Abdulrahman University Researchers Supporting Project number (PNURSP2022R300), Princess Nourah bint Abdulrahman University, Riyadh, Saudi Arabia.

Conflicts of Interest: The authors declare no conflict of interest.

References

1. Khashei, M.; Bijari, M. An artificial neural network (p, d, q) model for timeseries forecasting. *Expert Syst. Appl.* **2010**, *37*, 479–489. [CrossRef]
2. Berghout, T.; Benbouzid, M. A Systematic Guide for Predicting Remaining Useful Life with Machine Learning. *Electronics* **2022**, *11*, 1125. [CrossRef]
3. Abdelhamid, A.A.; El-Kenawy, E.-S.M.; Alotaibi, B.; Amer, G.M.; Abdelkader, M.Y.; Ibrahim, A.; Eid, M.M. Robust Speech Emotion Recognition Using CNN+ LSTM Based on Stochastic Fractal Search Optimization Algorithm. *IEEE Access* **2022**, *10*, 49265–49284. [CrossRef]
4. El-kenawy, E.-S.M.; Albalawi, F.; Ward, S.A.; Ghoneim, S.S.; Eid, M.M.; Abdelhamid, A.A.; Bailek, N.; Ibrahim, A. Feature selection and classification of transformer faults based on novel meta-heuristic algorithm. *Mathematics* **2022**, *10*, 3144. [CrossRef]
5. Alhussan, A.A.; Khafaga, D.S.; El-Kenawy, E.-S.M.; Ibrahim, A.; Eid, M.M.; Abdelhamid, A.A. Pothole and Plain Road Classification Using Adaptive Mutation Dipper Throated Optimization and Transfer Learning for Self Driving Cars. *IEEE Access* **2022**, *10*, 84188–84211. [CrossRef]
6. Wu, H.; Zhou, Y.; Luo, Q.; Basset, M.A. Training feedforward neural networks using symbiotic organisms search algorithm. *Comput. Intell. Neurosci.* **2016**, *2016*, 9063065. [CrossRef] [PubMed]
7. Feng, J.; Lu, S. Performance analysis of various activation functions in artificial neural networks. *J. Phys. Conf. Ser.* **2019**, *1237*, 022030. [CrossRef]
8. Abu-Shams, M.; Ramadan, S.; Al-Dahidi, S.; Abdallah, A. Scheduling Large-Size Identical Parallel Machines with Single Server Using a Novel Heuristic-Guided Genetic Algorithm (DAS/GA) Approach. *Processes* **2022**, *10*, 2071. [CrossRef]

9. Sharma, A.; Khan, R.A.; Sharma, A.; Kashyap, D.; Rajput, S. A Novel Opposition-Based Arithmetic Optimization Algorithm for Parameter Extraction of PEM Fuel Cell. *Electronics* **2021**, *10*, 2834. [CrossRef]
10. Singh, A.; Sharma, A.; Rajput, S.; Mondal, A.K.; Bose, A.; Ram, M. Parameter Extraction of Solar Module Using the Sooty Tern Optimization Algorithm. *Electronics* **2022**, *11*, 564. [CrossRef]
11. El-Kenawy, E.-S.M.; Mirjalili, S.; Abdelhamid, A.A.; Ibrahim, A.; Khodadadi, N.; Eid, M.M. Meta-heuristic optimization and keystroke dynamics for authentication of smartphone users. *Mathematics* **2022**, *10*, 2912. [CrossRef]
12. Khafaga, D.S.; Alhussan, A.A.; El-Kenawy, E.-S.M.; Ibrahim, A.; Eid, M.M.; Abdelhamid, A.A. Solving Optimization Problems of Metamaterial and Double T-Shape Antennas Using Advanced Meta-Heuristics Algorithms. *IEEE Access* **2022**, *10*, 74449–74471. [CrossRef]
13. El-Kenawy, E.-S.M.; Mirjalili, S.; Alassery, F.; Zhang, Y.-D.; Eid, M.M.; El-Mashad, S.Y.; Aloyaydi, B.A.; Ibrahim, A.; Abdelhamid, A.A. Novel Meta-Heuristic Algorithm for Feature Selection, Unconstrained Functions and Engineering Problems. *IEEE Access* **2022**, *10*, 40536–40555. [CrossRef]
14. Kennedy, J.; Eberhart, R. Particle swarm optimization. In Proceedings of the ICNN'95—International Conference on Neural Networks, Perth, Australia, 27 November–1 December 1995; Volume 1944, pp. 1942–1948.
15. Ang, K.M.; Lim, W.H.; Isa, N.A.M.; Tiang, S.S.; Wong, C.H. A constrained multi-swarm particle swarm optimization without velocity for constrained optimization problems. *Expert Syst. Appl.* **2020**, *140*, 112882. [CrossRef]
16. El-Sherbiny, M.M. Particle swarm inspired optimization algorithm without velocity equation. *Egypt. Inform. J.* **2011**, *12*, 1–8. [CrossRef]
17. Tian, D.; Zhao, X.; Shi, Z. DMPSO: Diversity-guided multi-mutation particle swarm optimizer. *IEEE Access* **2019**, *7*, 124008–124025. [CrossRef]
18. Cheng, R.; Jin, Y. A social learning particle swarm optimization algorithm for scalable optimization. *Inf. Sci.* **2015**, *291*, 43–60. [CrossRef]
19. Lim, W.H.; Isa, N.A.M.; Tiang, S.S.; Tan, T.H.; Natarajan, E.; Wong, C.H.; Tang, J.R. A self-adaptive topologically connected-based particle swarm optimization. *IEEE Access* **2018**, *6*, 65347–65366. [CrossRef]
20. Isiet, M.; Gadala, M. Self-adapting control parameters in particle swarm optimization. *Appl. Soft Comput.* **2019**, *83*, 105653. [CrossRef]
21. Li, M.; Chen, H.; Wang, X.; Zhong, N.; Lu, S. An improved particle swarm optimization algorithm with adaptive inertia weights. *Int. J. Inf. Technol. Decis. Mak.* **2019**, *18*, 833–866. [CrossRef]
22. Ghasemi, M.; Akbari, E.; Rahimnejad, A.; Razavi, S.E.; Ghavidel, S.; Li, L. Phasor particle swarm optimization: A simple and efficient variant of PSO. *Soft Comput.* **2019**, *23*, 9701–9718. [CrossRef]
23. Liu, W.; Wang, Z.; Zeng, N.; Yuan, Y.; Alsaadi, F.E.; Liu, X. A novel randomised particle swarm optimizer. *Int. J. Mach. Learn. Cybern.* **2021**, *12*, 529–540. [CrossRef]
24. Ang, K.M.; Juhari, M.R.M.; Cheng, W.-L.; Lim, W.H.; Tiang, S.S.; Wong, C.H.; Rahman, H.; Pan, L. New Particle Swarm Optimization Variant with Modified Neighborhood Structure. In Proceedings of the 2022 International Conference on Artificial Life and Robotics (ICAROB2022), Oita, Japan, 20–23 January 2022. [CrossRef]
25. Wu, D.; Jiang, N.; Du, W.; Tang, K.; Cao, X. Particle swarm optimization with moving particles on scale-free networks. *IEEE Trans. Netw. Sci. Eng.* **2018**, *7*, 497–506. [CrossRef]
26. Xu, Y.; Pi, D. A reinforcement learning-based communication topology in particle swarm optimization. *Neural Comput. Appl.* **2020**, *32*, 10007–10032. [CrossRef]
27. Chen, K.; Xue, B.; Zhang, M.; Zhou, F. Novel chaotic grouping particle swarm optimization with a dynamic regrouping strategy for solving numerical optimization tasks. *Knowl. Based Syst.* **2020**, *194*, 105568. [CrossRef]
28. Roshanzamir, M.; Balafar, M.A.; Razavi, S.N. A new hierarchical multi group particle swarm optimization with different task allocations inspired by holonic multi agent systems. *Expert Syst. Appl.* **2020**, *149*, 113292. [CrossRef]
29. Yang, Q.; Chen, W.-N.; Da Deng, J.; Li, Y.; Gu, T.; Zhang, J. A level-based learning swarm optimizer for large-scale optimization. *IEEE Trans. Evol. Comput.* **2017**, *22*, 578–594. [CrossRef]
30. Li, W.; Meng, X.; Huang, Y.; Fu, Z.-H. Multipopulation cooperative particle swarm optimization with a mixed mutation strategy. *Inf. Sci.* **2020**, *529*, 179–196. [CrossRef]
31. Liu, H.; Zhang, X.-W.; Tu, L.-P. A modified particle swarm optimization using adaptive strategy. *Expert Syst. Appl.* **2020**, *152*, 113353. [CrossRef]
32. Xia, X.; Gui, L.; He, G.; Wei, B.; Zhang, Y.; Yu, F.; Wu, H.; Zhan, Z.-H. An expanded particle swarm optimization based on multi-exemplar and forgetting ability. *Inf. Sci.* **2020**, *508*, 105–120. [CrossRef]
33. Xu, G.; Cui, Q.; Shi, X.; Ge, H.; Zhan, Z.-H.; Lee, H.P.; Liang, Y.; Tai, R.; Wu, C. Particle swarm optimization based on dimensional learning strategy. *Swarm Evol. Comput.* **2019**, *45*, 33–51. [CrossRef]
34. Wang, C.; Song, W. A modified particle swarm optimization algorithm based on velocity updating mechanism. *Ain Shams Eng. J.* **2019**, *10*, 847–866. [CrossRef]
35. Karim, A.A.; Isa, N.A.M.; Lim, W.H. Modified particle swarm optimization with effective guides. *IEEE Access* **2020**, *8*, 188699–188725. [CrossRef]
36. Karim, A.A.; Isa, N.A.M.; Lim, W.H. Hovering Swarm Particle Swarm Optimization. *IEEE Access* **2021**, *9*, 115719–115749. [CrossRef]

37. Wei, B.; Zhang, W.; Xia, X.; Zhang, Y.; Yu, F.; Zhu, Z. Efficient feature selection algorithm based on particle swarm optimization with learning memory. *IEEE Access* **2019**, *7*, 166066–166078. [CrossRef]

38. Şenel, F.A.; Gökçe, F.; Yüksel, A.S.; Yiğit, T. A novel hybrid PSO–GWO algorithm for optimization problems. *Eng. Comput.* **2019**, *35*, 1359–1373. [CrossRef]

39. Zhang, M.; Long, D.; Qin, T.; Yang, J. A chaotic hybrid butterfly optimization algorithm with particle swarm optimization for high-dimensional optimization problems. *Symmetry* **2020**, *12*, 1800. [CrossRef]

40. Ang, K.M.; Juhari, M.R.M.; Lim, W.H.; Tiang, S.S.; Ang, C.K.; Hussin, E.E.; Pan, L.; Chong, T.H. New Hybridization Algorithm of Differential Evolution and Particle Swarm Optimization for Efficient Feature Selection. In Proceedings of the 2022 International Conference on Artificial Life and Robotics (ICAROB2022), Oita, Japan, 20–23 January 2022; Volume 27, p. 5. [CrossRef]

41. Grosan, C.; Abraham, A. Hybrid evolutionary algorithms: Methodologies, architectures, and reviews. In *Hybrid Evolutionary Algorithms*; Springer: Berlin/Heidelberg, Germany, 2007; pp. 1–17.

42. Abdolrasol, M.G.; Hussain, S.S.; Ustun, T.S.; Sarker, M.R.; Hannan, M.A.; Mohamed, R.; Ali, J.A.; Mekhilef, S.; Milad, A. Artificial neural networks based optimization techniques: A review. *Electronics* **2021**, *10*, 2689. [CrossRef]

43. Carvalho, M.; Ludermir, T.B. Particle swarm optimization of neural network architectures andweights. In Proceedings of the 7th International Conference on Hybrid Intelligent Systems (HIS 2007), Kaiserslautern, Germany, 17–19 September 2007; pp. 336–339.

44. Mirjalili, S.; Hashim, S.Z.M.; Sardroudi, H.M. Training feedforward neural networks using hybrid particle swarm optimization and gravitational search algorithm. *Appl. Math. Comput.* **2012**, *218*, 11125–11137. [CrossRef]

45. Yaghini, M.; Khoshraftar, M.M.; Fallahi, M. A hybrid algorithm for artificial neural network training. *Eng. Appl. Artif. Intell.* **2013**, *26*, 293–301. [CrossRef]

46. Kandasamy, T.; Rajendran, R. Hybrid algorithm with variants for feed forward neural network. *Int. Arab J. Inf. Technol.* **2018**, *15*, 240–245.

47. Xue, Y.; Tang, T.; Liu, A.X. Large-scale feedforward neural network optimization by a self-adaptive strategy and parameter based particle swarm optimization. *IEEE Access* **2019**, *7*, 52473–52483. [CrossRef]

48. Kumar, M.; Mishra, S.K.; Joseph, J.; Jangir, S.K.; Goyal, D. Adaptive comprehensive particle swarm optimisation-based functional-link neural network filtre model for denoising ultrasound images. *IET Image Process.* **2021**, *15*, 1232–1246. [CrossRef]

49. Hayder, G.; Solihin, M.I.; Mustafa, H.M. Modelling of river flow using particle swarm optimized cascade-forward neural networks: A case study of Kelantan River in Malaysia. *Appl. Sci.* **2020**, *10*, 8670. [CrossRef]

50. Davar, S.; Nobahar, M.; Khan, M.S.; Amini, F. The Development of PSO-ANN and BOA-ANN Models for Predicting Matric Suction in Expansive Clay Soil. *Mathematics* **2022**, *10*, 2825. [CrossRef]

51. Melanie, M. *An Introduction to Genetic Algorithms*; Massachusetts Institute of Technology: Cambridge, MA, USA, 1996; p. 158.

52. Chandwani, V.; Agrawal, V.; Nagar, R. Modeling slump of ready mix concrete using genetic algorithms assisted training of Artificial Neural Networks. *Expert Syst. Appl.* **2015**, *42*, 885–893. [CrossRef]

53. Huang, H.-X.; Li, J.-C.; Xiao, C.-L. A proposed iteration optimization approach integrating backpropagation neural network with genetic algorithm. *Expert Syst. Appl.* **2015**, *42*, 146–155. [CrossRef]

54. Bagheri, M.; Mirbagheri, S.A.; Bagheri, Z.; Kamarkhani, A.M. Modeling and optimization of activated sludge bulking for a real wastewater treatment plant using hybrid artificial neural networks-genetic algorithm approach. *Process Saf. Environ. Prot.* **2015**, *95*, 12–25. [CrossRef]

55. Rao, R.V.; Savsani, V.J.; Vakharia, D.P. Teaching–learning-based optimization: A novel method for constrained mechanical design optimization problems. *Comput.-Aided Des.* **2011**, *43*, 303–315. [CrossRef]

56. Uzlu, E.; Kankal, M.; Akpınar, A.; Dede, T. Estimates of energy consumption in Turkey using neural networks with the teaching–learning-based optimization algorithm. *Energy* **2014**, *75*, 295–303. [CrossRef]

57. Li, K.; Xie, X.; Xue, W.; Dai, X.; Chen, X.; Yang, X. A hybrid teaching-learning artificial neural network for building electrical energy consumption prediction. *Energy Build.* **2018**, *174*, 323–334. [CrossRef]

58. Benali, A.; Hachama, M.; Bounif, A.; Nechnech, A.; Karray, M. A TLBO-optimized artificial neural network for modeling axial capacity of pile foundations. *Eng. Comput.* **2021**, *37*, 675–684. [CrossRef]

59. Ang, K.M.; Lim, W.H.; Tiang, S.S.; Ang, C.K.; Natarajan, E.; Ahamed Khan, M. Optimal Training of Feedforward Neural Networks Using Teaching-Learning-Based Optimization with Modified Learning Phases. In Proceedings of the 12th National Technical Seminar on Unmanned System Technology 2020; Springer: Singapore, 2022; pp. 867–887.

60. Chong, O.T.; Lim, W.H.; Isa, N.A.M.; Ang, K.M.; Tiang, S.S.; Ang, C.K. A Teaching-Learning-Based Optimization with Modified Learning Phases for Continuous Optimization. In *Science and Information Conference*; Springer: Cham, Switzerland, 2020; pp. 103–124.

61. Lin, G.; Shen, W. Research on convolutional neural network based on improved Relu piecewise activation function. *Procedia Comput. Sci.* **2018**, *131*, 977–984. [CrossRef]

62. Gao, W.; Liu, S.; Huang, L. A global best artificial bee colony algorithm for global optimization. *J. Comput. Appl. Math.* **2012**, *236*, 2741–2753. [CrossRef]

63. Tizhoosh, H.R. Opposition-based learning: A new scheme for machine intelligence. In Proceedings of the International Conference on Computational Intelligence for Modelling, Control and Automation and International Conference on Intelligent Agents, Web Technologies and Internet Commerce (CIMCA-IAWTIC'06), Vienna, Austria, 28–30 November 2005; pp. 695–701.

64. Dong, N.; Wu, C.-H.; Ip, W.-H.; Chen, Z.-Q.; Chan, C.-Y.; Yung, K.-L. An opposition-based chaotic GA/PSO hybrid algorithm and its application in circle detection. *Comput. Math. Appl.* **2012**, *64*, 1886–1902. [CrossRef]

65. Vrugt, J.A.; Robinson, B.A.; Hyman, J.M. Self-adaptive multimethod search for global optimization in real-parameter spaces. *IEEE Trans. Evol. Comput.* **2008**, *13*, 243–259. [CrossRef]

66. Liang, J.J.; Qu, B.Y.; Suganthan, P.N. Problem Definitions and Evaluation Criteria for the CEC 2014 Special Session and Competition on Single Objective Real-Parameter Numerical Optimization. In *Technical Report 201311, Computational Intelligence Laboratory, Zhengzhou University, Zhengzhou China and Technical Report*; Nanyang Technological University: Singapore, 2013.

67. Derrac, J.; García, S.; Molina, D.; Herrera, F. A practical tutorial on the use of nonparametric statistical tests as a methodology for comparing evolutionary and swarm intelligence algorithms. *Swarm Evol. Comput.* **2011**, *1*, 3–18. [CrossRef]

68. García, S.; Molina, D.; Lozano, M.; Herrera, F. A study on the use of non-parametric tests for analyzing the evolutionary algorithms' behaviour: A case study on the CEC'2005 special session on real parameter optimization. *J. Heuristics* **2009**, *15*, 617–644. [CrossRef]

69. Cheng, R.; Jin, Y. A competitive swarm optimizer for large scale optimization. *IEEE Trans. Cybern.* **2014**, *45*, 191–204. [CrossRef] [PubMed]

70. Yang, X.-S.; Deb, S.; Fong, S. Accelerated particle swarm optimization and support vector machine for business optimization and applications. In *International Conference on Networked Digital Technologies*; Springer: Berlin/Heidelberg, Germany, 2011; pp. 53–66.

71. Lichman, M. UCI Machine Learning Repository. 2013. Available online: https://archive.ics.uci.edu/ml/index.php (accessed on 3 June 2022).

72. Zhang, P.; Jiang, W.; Shi, X.; Zhang, S. Remaining Useful Life Prediction of Gear Pump Based on Deep Sparse Autoencoders and Multilayer Bidirectional Long ands Short Term Memory Network. *Processes* **2022**, *10*, 2500. [CrossRef]

73. Wang, P.; Liu, H.; Hou, M.; Zheng, L.; Yang, Y.; Geng, J.; Song, W.; Shao, Z. Estimating the Remaining Useful Life of Proton Exchange Membrane Fuel Cells under Variable Loading Conditions Online. *Processes* **2021**, *9*, 1459. [CrossRef]

74. Benbouzid, M.; Berghout, T.; Sarma, N.; Djurović, S.; Wu, Y.; Ma, X. Intelligent Condition Monitoring of Wind Power Systems: State of the Art Review. *Energies* **2021**, *14*, 5967. [CrossRef]

Article

Deep Reinforcement Learning for Traffic Light Timing Optimization

Bin Wang, Zhengkun He, Jinfang Sheng * and Yu Chen

School of Computer Science and Engineering, Central South University, Changsha 410083, China
* Correspondence: jfsheng@csu.edu.cn

Abstract: Existing inflexible and ineffective traffic light control at a key intersection can often lead to traffic congestion due to the complexity of traffic dynamics, how to find the optimal traffic light timing strategy is a significant challenge. This paper proposes a traffic light timing optimization method based on double dueling deep Q-network, MaxPressure, and Self-organizing traffic lights (SOTL), namely EP-D3QN, which controls traffic flows by dynamically adjusting the duration of traffic lights in a cycle, whether the phase is switched based on the rules we set in advance and the pressure of the lane. In EP-D3QN, each intersection corresponds to an agent, and the road entering the intersection is divided into grids, each grid stores the speed and position of a car, thus forming the vehicle information matrix, and as the state of the agent. The action of the agent is a set of traffic light phase in a signal cycle, which has four values. The effective duration of the traffic lights is 0–60 s, and the traffic light phases switching depends on its press and the rules we set. The reward of the agent is the difference between the sum of the accumulated waiting time of all vehicles in two consecutive signal cycles. The SUMO is used to simulate two traffic scenarios. We selected two types of evaluation indicators and compared four methods to verify the effectiveness of EP-D3QN. The experimental results show that EP-D3QN has superior performance in light and heavy traffic flow scenarios, which can reduce the waiting time and travel time of vehicles, and improve the traffic efficiency of an intersection.

Keywords: traffic light control; deep reinforcement learning

Citation: Wang, B.; He, Z.; Sheng, J.; Chen, Y. Deep Reinforcement Learning for Traffic Light Timing Optimization. *Processes* **2022**, *10*, 2458. https://doi.org/10.3390/pr10112458

Academic Editors: Jie Zhang and Meihong Wang

Received: 27 October 2022
Accepted: 17 November 2022
Published: 20 November 2022

1. Introduction

Traffic congestion has increasingly become one of the major problems in cities. Traffic light control can effectively alleviate traffic congestion and improve traffic efficiency in urban intersections. The existing traffic light control methods are divided into timing control and adaptive traffic signal control (ATSC) [1]. The timing control is FixedTime [2], and the most representative ATSC is SCOOT [3] and SCATS [4]. Self-organizing traffic lights (SOTL) [5] and max pressure (MP) control [6] aim to maximize the global throughput from observation of traffic states.

These conventional methods are not effective as the complexity of the traffic network increases. Recently, reinforcement learning (RL) has been widely used for traffic light control. Reinforcement learning defines traffic light control as a Markov decision process (MDP) and learns an optimal control strategy through continuous iteration with the environment. Reinforcement learning based on table Q-learning can only deal with discrete intersection states [7]. Deep reinforcement learning (DRL) can deal with discrete or continuous intersection states. Some DRL-based methods have shown better performance than many traditional methods in specific scenarios, which can be used to control traffic lights and improve the traffic efficiency of the intersection [8,9].

Most DRL-based methods focus on learning a strategy to switch the current traffic lights phase in a signal cycle [10]. The duration of traffic lights is a fixed-length interval, which is not flexible enough to cope with changing traffic conditions. Liang et al., tried to control the traffic light duration in a cycle based on deep reinforcement learning and the extracted information from vehicular networks [11]. It shows good performance in specific scenarios, but it will become unstable with the increase of intersection complexity. Hua et al., also attempted to introduce the concept of max pressure from the traffic field as the reward for control model optimization [12–14].

This paper proposes the EP-D3QN algorithm based on 3DQN [11], MP [6] algorithm, and SOTL [5] algorithm. In EP-D3QN, the road entering the intersection is divided into grids, and each grid stores the information of each vehicle. The matrix formed by these grids serves as the state of the agent, whose action is the combination of different traffic light phases in a cycle, and the reward is the difference between the sum of accumulated waiting times of all vehicles in two consecutive cycles. The EP-D3QN can dynamically adjust the duration of traffic lights in a signal cycle and activate the effective phase during phase switching. The experiment uses the Simulation of Urban Mobility (SUMO) [15] to simulate two traffic scenarios to verify the effectiveness of EP-D3QN.

The remainder of this paper is organized as follows. The related work is introduced in Section 2. Section 3 shows the problem statement and the details of the EP-D3QN. Section 4 shows the result of the experiment and analysis. Finally, the paper is concluded in Section 5.

2. Related Work

Traffic light control approaches can be divided into two types: traditional methods and RL-based methods.

2.1. Traditional Methods

Early research works on traffic light control mainly are Fixed-time Traffic Light Control (FT) [2], which fix the duration of traffic lights according to historical traffic information. Subsequently, SCOOT [3], SCATS [4], and other adaptive traffic signal control methods emerged. These methods are still widely deployed in many cities. SOTL [5] and MP [6] followed. Both MP and SOTL switch the traffic lights phase based on current traffic conditions. In the SOTL method, whether the traffic light phases is switched depends on the current observed traffic conditions and the rules defined in advance. Compared with fixed time, it is more flexible. The MP method introduces the concept of max pressure. The pressure is defined as the difference between the number of vehicles on incoming lanes and the number of vehicles on outgoing lanes. When the traffic light phases are switched, the phase with the max pressure is preferentially activated. Hua et al., introduced the concept of maximum pressure as a reward function, so as to learn more optimal strategies [12–14]. Wu et al., also optimized the MP algorithm and proposed the efficient MP [16]. Liang et al., also integrated the advantages of SOTL and MaxPressure and applied them in traffic light control [17]. Despite the high performance of the MP-based control, it lacks flexibility as the duration of traffic lights is a fixed-length interval.

2.2. RL-Based Methods

RL-based traffic light control has attracted wide attention from both academia and industry in the last two decades. Traditional RL-based methods mainly use table Q-learning, which can only handle discrete intersection state representation [7]. Later, deep reinforcement learning (DRL) appeared, which can deal with discrete or continuous intersection states. Compared with traditional traffic light methods, some DRL-based methods show better performance in certain situations [18–20]. However, DRL-based methods will overestimate Q-value, and due to the complexity of traffic conditions, DRL-based methods will become unstable. Some scholars also proposed multiple optimization elements to improve the DRL's performance, such as double Q-learning network [21], dueling Q-learning network [22], and prioritized experience replay [23]. Liang et al., incorporates these optimization technologies and tried to dynamically control the duration of traffic lights in a cycle [11], which showed good performance in light traffic flow scenarios, but became unstable in heavy traffic flow scenarios.

Inspired by the above work, this paper proposes the EP-D3QN algorithm based on MP, SOTL, and 3DQN algorithms. The advantage of double dueling deep Q-network can dynamically calculate the duration of the traffic light in each signal cycle. It is more flexible, and the advantages of MP and SOTL just overcome the instability of deep Q-network, which make EP-D3QN more adaptive.

3. Method

3.1. Problem Statement

Traffic congestion often occurs at key intersections [24], so this paper focuses on an intersection. The agent corresponding to the intersection can receive its observation and obtain the duration of the traffic lights, and control the traffic light phases switching in a cycle. Each traffic light has three colors: red, yellow, and green. Each traffic light phases is a set of permissible traffic movements, and each intersection has four phases, and the combination of all phases is called a signal cycle. The vehicles always travel across an intersection from one incoming road to one outgoing road. The signal phases of a cycle are played in a fixed sequence to control the traffic flow.

Our problem is to learn a strategy based on DRL to dynamically adjust the duration of traffic lights in a cycle, and to switch the phase based on the rules we set, so as to control the traffic flow. The state of the agent is the information matrix formed by all vehicles entering the intersection, and its action is the set of all phases in a cycle. Its reward is the difference between the accumulated waiting time and the sum of all vehicles in two consecutive cycles. The goal of the agent is to maximize the reward. In each time step, the agent will obtain the observation of the intersection, and then select an action based on its own strategy. The action is a set of phases in a cycle, and the effective phase will be activated first in each time step. After the execution of the action, the agent will get the reward, and then the state of the intersection will change. The agent finally learns to get a high reward by reacting to different traffic scenarios.

3.2. Agent Design

In EP-D3QN, each intersection corresponds to an agent. In order to better introduce EP-D3QN into traffic light control. First, we need to design the state, action, and reward of the agent.

3.2.1. Intersection State

For each intersection, we divide the road entering the intersection into grids, and each grid can only accommodate one vehicle, as shown in Figure 1.

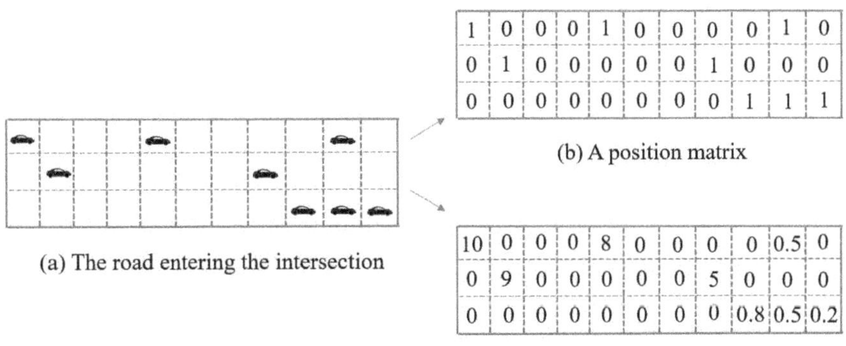

(a) The road entering the intersection

(b) A position matrix

1	0	0	0	1	0	0	0	0	1	0
0	1	0	0	0	0	0	1	0	0	0
0	0	0	0	0	0	0	0	1	1	1

(c) A speed matrix

10	0	0	0	8	0	0	0	0	0.5	0
0	9	0	0	0	0	0	5	0	0	0
0	0	0	0	0	0	0	0	0.8	0.5	0.2

Figure 1. The process to build the state matrix. (**a**) is the snapshot of the road entering the intersection. (**b**) is the position matrix of vehicles at the current moment. (**c**) is the speed matrix of vehicles at the intersection at the current moment.

Among them, (b) is the position information of the vehicle. A grid with 1 means a vehicle, and 0 means no vehicle. In (b), the more grids with 1, the more vehicles stay at the intersection at the current moment. (c) represent the speed information of the vehicle. Floating data is used to represent the speed of the vehicle. Each grid is the actual speed of the vehicle in meters/second. Therefore, all the lanes entering the intersection can be represented as a matrix. This matrix is the state of the intersection.

3.2.2. Agent Actions

The action of the agent is defined as $a_i \in [1, 2, \ldots, 9]$, where $a_i = <NS, NSL, WE, WEL>$, as shown in Figure 2b, NS, NSL, WE, and WEL represent the four traffic lights phases in a cycle, which indicates going straight from north to south, turning left from north to south, going straight from east to west, and turning left from east to west, respectively. We set the longest duration of the traffic light to 60 s and the shortest duration to 0 s.

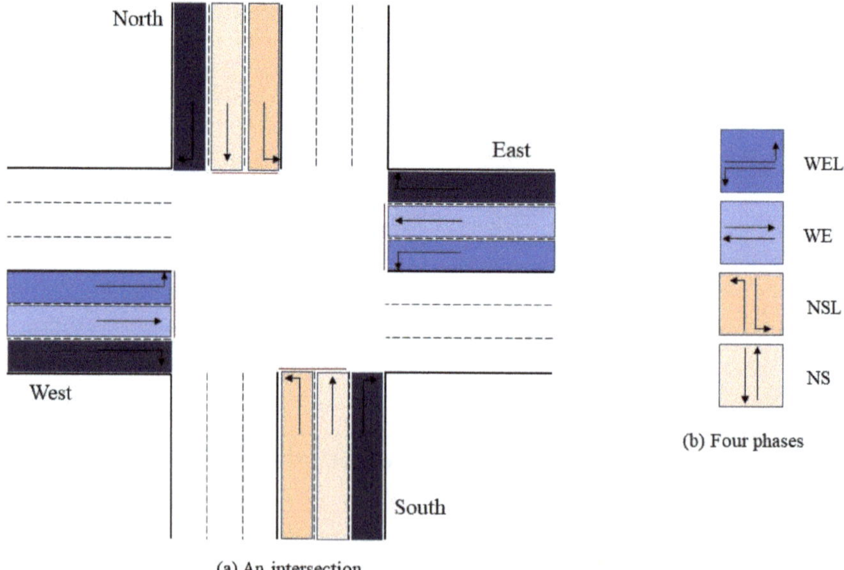

Figure 2. The framework of EP-D3QN.

At each time step, the agent will choose an action from the nine actions to act on the traffic light controller at the intersection. For example, the current action is $a1 = <NS, NSL, WE, WEL>$, and the next legal action is $<NS5, NSL, WE, WEL>$, $<NS, NSL5, WE, WEL>$, $<NS, NSL, WE5, WEL>$ and $<NS, NSL, WE, WEL5>$.

3.2.3. Reward

The reward of the agent is crucial for the deep reinforcement learning model. An appropriate reward can guide the agent to get better training results. Therefore, based on previous research work [7,25,26], the reward of the agent is defined as follows:

$$r_t = \sum_{i=1}^{n} w_{t-1}^i - \sum_{j=1}^{m} w_t^j \tag{1}$$

Among them, w_{t-1}^i represents the waiting time of i-th vehicle in the $(t-1)$-th cycle, w_t^j represents the waiting time of j-th vehicle in the t-th cycle, and n and m represent the number of vehicles entering the intersection in the $(t-1)$-th and t-th cycle, respectively.

3.3. Effective-Pressure with Double Dueing Deep Q-Network for Traffic Light Control

In this paper, we propose the EP-D3QN algorithm based on MP, SOTL, and 3DQN algorithms. In EP-D3QN, the agent corresponding to the intersection can receive its observation and choose an action to execute. After receiving its observation, the agent encodes its observation by the convolution layer and the fully connected layer, and then obtains Q-value of each action. The greedy strategy is used to select the action with the largest Q-value, and the effective traffic lights phase is preferentially activated in a cycle during each time step. The complete process is shown in Figure 3.

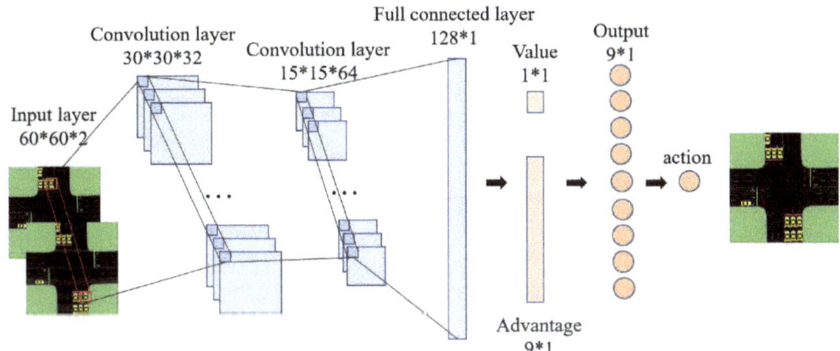

Figure 3. The synthetic intersection during all training episodes.

In the EP-D3QN, the state of the intersection is a vehicle information matrix with a size of $60*60$, and each grid in the matrix stores the location and speed of the vehicle. There are two convolutional layers in total. The first convolutional layer contains 32 filters, the size of each filter is $4*4$, and its moving stride is $2*2$. The second convolutional layer contains 64 filters with size $2*2$, and its moving stride is also $2*2$. The fully connected layer is responsible for integrating the information extracted by the convolutional layer. After the fully-connected layer, the data are split into two parts. The first part is then used to calculate the value and the second part is for the advantage. The advantage of action means how well it can achieve by taking an action over all the other actions. The formula can be expressed in Equation (2).

$$Q(o,a;\theta) = V(o;\theta) + (A(o,a;\theta)\frac{1}{|A|}\sum_{a'} A(o,a';\theta)) \tag{2}$$

Among them, the value of a state $V(o;\theta)$ denotes the overall expected rewards by taking probabilistic actions in future steps. The $A(o,a;\theta)$ is the advantage that corresponds to every action. Each action corresponds to a Q-value. The target Q-value is calculated as follows:

$$Q_{target}(o,a) = r + \gamma Q_{target}(o', argmax_{a'}(Q(o',a';\theta)),\theta') \tag{3}$$

where r is the reward of the agent, γ is the discount factor, θ and θ' are the parameters of the main network and target network, o' and a' are the state of intersection and the action of the agent at the next time step respectively.

After the agent obtains Q-value through its main network, the main network is updated by TD-error as follows:

$$J(\theta) = \frac{1}{m}\sum_{i=1}^{m}[Q_{target}(o_i,a_i;\theta') - Q(o_i,a_i;\theta)]^2 \tag{4}$$

Among them, m represents the sample size extracted by the replay buffer. The parameters of the target network θ' are updated as Equation (5).

$$\theta' = \varepsilon\theta' + (1 - \alpha)\theta \tag{5}$$

Among them, ε is the update rate from the main network to the target network and α is the learning rate of the main network.

The detailed steps of the EP-D3QN algorithm are shown in Algorithm 1. First, to initialize the parameters of the main network and target network. Meanwhile, to initialize discount factor γ, target network update rate ε, replay buffer D and threshold v_{min} and v_{max}; Then, in each time step, the agent receives state o, select action a. During the action a be executed, preferentially activates the effective phase in a cycle, then get reward r, and new state o', and stored the experience (o, a, r, o') in the replay buffer (lines 4 to 13); Next, for the agent, sample a minibatch of step episodes experience trajectories (o, a, r, o') from the replay buffer (line 14). Finally, update the parameters of the main network by TD-error (line 19), and then update the parameters of the target network (line 20). This process is repeated until it converges.

Algorithm 1 EP-D3QN for traffic light control.

Input:
    Intersections' state o
Output:
    Action a
Initialize:
    The parameters of main network θ and target network θ', discount factor γ,target network update rate ε, replay buffer D, threshold v_{min} and v_{max}

1: **for** *episode* $= 1$ to M **do**
2:     Initialize observation o and $t = 1$
3:     **for** $t < T$ **do**
4:         The agent select an action a
5:         Calculate pressure p_{NS} and p_{WE} for the phases NS and WE
6:         Calculate vehicles approaching red phase v_r
7:         Calculate vehicles approaching green phase v_g
8:         **if** $v_g < v_{min}$ and $v_r > v_{max}$ **then**
9:             **if** phase $=$ WE and $p_{NS} > p_{WE}$ or phase $=$ NS and $p_{WE} > p_{NS}$ **then**
10:            switch light
11:           **end if**
12:        **end if**
13:        Then get reward r and new observation o'
14:        Store (o, a, o', r) in D
15:        $o \leftarrow o'$
16:        **if** $T_{update} > minSteps$ **then**
17:           Sample random minibatch of step (o, a, o', r) from D
18:           Calculate $Q_{target}(o, a)$ for the agent with Equation (3)
19:           Update the main network θ with Equation (4)
20:           Update target network θ' with Equation (5)
21:        **end if**
22:     **end for**
23: **end for**

---

4. Experiment and Analysis

4.1. Experimental Setup

In the experiments, we use Simulation of Urban Mobility (SUMO), an open-source, micro, multi-model traffic simulation software [15], to simulate light and heavy traffic flow scenarios respectively. The intersection created by sumo is with a two-way six-lane, as shown in Figure 2a. Each direction has 6 lanes. Each lane is 180 m, and the length of the vehicles traveling is 5 m.

Vehicles entering the intersection can go straight, and turn left or right. The distance between the adjacent vehicles is set to 1 m. The maximum speed of the car is 13.89 m/s. The intersection has four traffic light phases, as shown in Figure 2b. In the light traffic flow scenario, the traffic flow is generated using the Bernoulli distribution with a probability of 0.2 (two vehicles arriving at the intersection every 10 s). In the heavy traffic flow scenario, the traffic flow is also generated using the Bernoulli distribution with a probability of 0.4.

4.2. Evaluation Metrics

Following previous research work [25–27], we select two types of evaluation indicators, the first is the average reward, the second is average waiting time (AWT), average queue length (AQL), and average travel time (ATT). The average reward is used to reflect the performance of EP-D3QN, and the bigger its value, the better its performance. AVWT, AQL and ATT are used to reflect the traffic conditions of the intersection. The smaller the average, the higher the traffic efficiency of the intersection, and vice versa.

4.3. Compared Methods

In order to verify the performance of EP-D3QN, we compare it with the following methods.

- **Fixed-time Traffic Light Control (FT)**. FT [2] is the most widely used traffic light control method. Each intersection sets a fixed sequence of signal phases, and the duration of traffic lights is also fixed
- **Self-organizing traffic lights (SOTL)**. In the SOTL [5], whether the traffic light phases is switched depends on the observed traffic conditions and the rules defined in advance. Compared to FT, SOTL is very flexible
- **MaxPressure (MP)**. In the MP [6], the traffic light controller activates the traffic light phases with max pressure in a cycle. The MP introduces the concept of pressure, which is the difference between the number of vehicles on incoming lanes and the number of vehicles on outgoing lanes. At each time step, the pressure of each phase is calculated and compared. Finally, the phase with the max pressure is activated.
- **Double dueling Deep Q-network (3DQN)**. The 3DQN [11], incorporates multiple optimization elements to improve the performance of traffic light control, such as dueling network, target network, double Q-learning network, and prioritized experience replay.

4.4. Result and Analysis

4.4.1. Light Traffic Flow Scenario

The simulation results are shown in Figure 4. We plotted the average waiting time during all the episodes. The red line shows the EP-D3QN, the blue line is the 3DQN, and the green line is the FixTime method, SOTL and MaxPressure correspond to the brown and orange lines, respectively. As can be seen from the figure, there is little difference in the average waiting time between SOTL and MP, but both are better than the FT method. That's because in the MP algorithm, the traffic light phases with the max pressure will be activated at each time step, while in the SOTL algorithm, whether to activate the phase according to the set threshold. Compared with 3DQN, it can be seen that EP-D3QN has faster convergence speed and stronger stability. That's because EP-D3QN activates the effective traffic phase preferentially when the action is performed at each time step.

Table 1 shows the results of the comprehensive evaluation. As can be seen from the table, compared with other methods, EP-D3QN shows better performance, its AWT, AQL, and ATT are relatively small, and the average reward is the largest. The AQL and ATT of FT are the largest, followed by MP and SOTL, and EP-D3QN is the smallest. The AWT of 3DQN and SOTL is at a medium level, the AWT of FT and MP is the highest, while the AWT of EP-D3QN is the lowest. It shows that EP-D3QN can dynamically control traffic lights more effectively, so as to improve the traffic efficiency of the intersection.

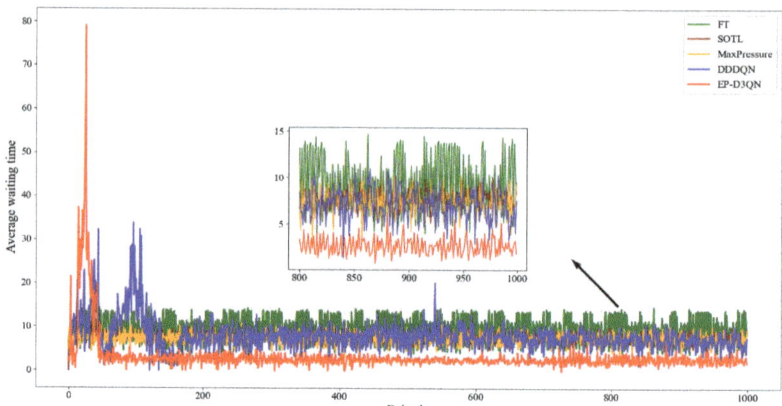

Figure 4. The average waiting time in the light traffic flow scenario during all training episodes.

Table 1. Comprehensive evaluation in the light traffic flow scenario.

Algorithm	Reward	AQL	AWT	ATT
FT	0.005	0.555	7.729	27.538
MaxPressure	0.172	0.504	7.240	15.631
SOTL	4.410	0.386	5.341	9.331
3DQN	4.229	0.351	6.535	6.775
EP-D3QN	**6.332**	**0.322**	**3.911**	**4.982**

4.4.2. Heavy Traffic Flow Scenario

Figure 5 shows the average waiting time during all training episodes in the heavy traffic flow scenario. From the figure, we can see that at the beginning, 3DQN and EP-D3QN were unstable. That's because, during the initial training episodes, the agent randomly selects actions. When the training episodes reach about 200, EP-D3QN starts to converge, and in the later training episodes, the AWT of both EP-D3QN and 3DQN could be maintained within a fixed range, but EP-D3QN showed better performance. The AWT of FT, MP, and SOTL keeps fluctuating in a fixed range, and they are all worse than EP-D3QN, but SOTL performs better than D3QN, which indicates that the traditional method is not flexible for heavy traffic flow scenarios due to the complexity of the traffic conditions.

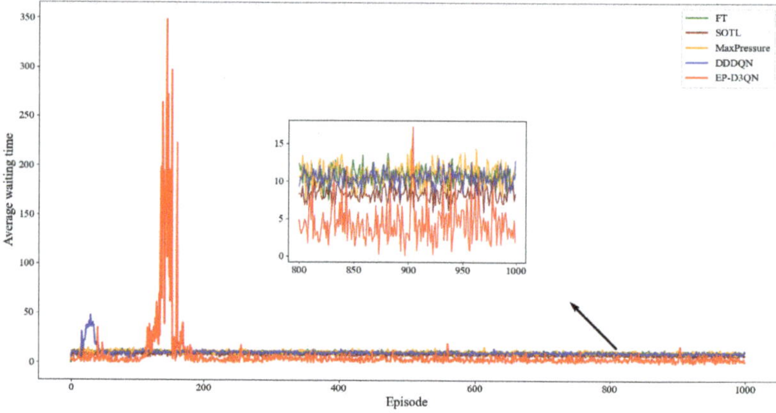

Figure 5. The average waiting time in the heavy traffic flow scenario during all training episodes.

Table 2 is the result of the comprehensive evaluation in the heavy traffic flow scenario. As can be seen from the table, AQL and ATT of FT are the largest, AWT of MP is the largest, while the three indexes of EP-D3QN are the smallest. Moreover, the average reward of EP-D3QN is the largest. That's because EP-D3QN incorporates the advantages of SOTL and MP. Compared with SOTL and MP, AWT is significantly reduced, and ATT is also significantly reduced compared with 3DQN.

Table 2. Comprehensive evaluation in the heavy traffic flow scenario.

Algorithm	Reward	AQL	AWT	ATT
FT	0.009	3.130	6.648	16.030
MaxPressure	0.007	2.740	9.626	6.377
SOTL	0.058	2.243	6.452	10.420
3DQN	9.588	2.703	6.232	7.532
EP-D3QN	**11.658**	**1.385**	**4.110**	**3.519**

In conclusion, the EP-D3QN can perform better performance in both light and heavy traffic flow scenarios, ensuring that vehicles entering the intersection spend less waiting time and pass the intersection quickly, thus effectively improving the traffic efficiency of the intersection.

5. Conclusions

In this paper, we study the problem of how to control the traffic light duration in a cycle based on deep reinforcement learning in an intersection and propose an EP-D3QN algorithm based on 3DQN, MP, and SOTL algorithms. In EP-D3QN, the intersection corresponds an agent. The agent can receive its own observation and choose an action. Its state is the information matrix of the vehicles entering the intersection. The action of the agent is the traffic light phases in a cycle. During the action being executed, the traffic lights phase with the effective pressure will be activated preferentially. The reward of the agent is the sum of the accumulated waiting time of all the vehicles in two consecutive cycles. We use SUMO to simulate the traffic scenarios and verify the effectiveness of EP-D3QN. The experimental results show that EP-D3QN significantly outperforms other methods in both light and heavy traffic flow scenarios, which can improve the traffic efficiency of the intersection and relieve traffic pressure.

Author Contributions: Conceptualization, B.W. and J.S.; methodology, Z.H.; validation, Z.H. and Y.C.; formal analysis, Z.H. and J.S. writing—original draft preparation, Z.H.; writing—review and editing, J.S., Z.H. and Y.C.; visualization, Z.H.; supervision, B.W.; funding acquisition, J.S. and B.W. All authors have read and agreed to the published version of the manuscript.

Funding: This work is supported by the National Key Research and Development Program of China under grant No. 2018YFB1003602.

Institutional Review Board Statement: Not applicable.

Informed Consent Statement: Not applicable.

Data Availability Statement: Not applicable.

Conflicts of Interest: The authors declare no conflict of interest.

References

1. Noaeen, M.; Naik, A.; Goodman, L.; Crebo, J.; Abrar, T.; Abad, Z.S.H.; Bazzan, A.L.; Far, B. Reinforcement learning in urban network traffic signal control: A systematic literature review. *Expert Syst. Appl.* **2022**, *199*, 116830. [CrossRef]
2. Li, L.; Wen, D. Parallel systems for traffic control: A rethinking. *IEEE Trans. Intell. Transp. Syst.* **2016**, *17*, 1179–1182. [CrossRef]
3. Robertson, D.I.; Bretherton, R.D. Optimizing networks of traffic signals in real-time SCOOT method. *IEEE Trans. Veh. Technol.* **1991**, *40*, 11–15. [CrossRef]
4. Sims, A. The Sydney coordinated adaptive traffic (SCAT) system philosophy and benefits. *IEEE Trans. Veh. Technol.* **1980**, *29*, 130–137. [CrossRef]

5. Cools, S.B.; Gershenson, C.; D'Hooghe, B. Self-organizing traffic lights: A realistic simulation. In *Advances in Applied Self-Organizing Systems*; Springer: London, UK, 2013; pp. 45–55.

6. Varaiya, P. Max pressure control of a network of signalized intersections. *Transp. Res. Part C Emerg. Technol.* **2013**, *36*, 177–195. [CrossRef]

7. Haydari, A.; Yilmaz, Y. Deep Reinforcement Learning for Intelligent Transportation Systems: A Survey. *IEEE Trans. Intell. Transp. Syst.* **2022**, *23*, 11–32. [CrossRef]

8. Li, L.; Lv, Y.; Wang, F.Y. Traffic Signal Timing via Deep Reinforcement Learning. *IEEE-CAA J. Autom. Sin.* **2016**, *3*, 247–254.

9. Mousavi, S.; Schukat, M.; Howley, E. Traffic Light Control Using Deep Policy-Gradient and Value-Function Based Reinforcement Learning. *IET Intell. Transp. Syst.* **2017**, *11*, 417–423. [CrossRef]

10. Genders, W.; Razavi, S. Policy Analysis of Adaptive Traffic Signal Control Using Reinforcement Learning. *J. Comput. Civ. Eng.* **2020**, *34*, 19–46. [CrossRef]

11. Liang, X.; Du, X.; Wang, G.; Han, Z. A deep reinforcement learning network for traffic light cycle control. *IEEE Trans. Veh. Technol.* **2019**, *68*, 1243–1253. [CrossRef]

12. Wei, H.; Zheng, G.; Yao, H.; Li, Z. IntelliLight: A Reinforcement Learning Approach for Intelligent Traffic Light Control. In Proceedings of the 24th ACM SIGKDD International Conference on Knowledge Discovery and Data Mining, London, UK, 19–23 August 2018; pp. 2496–2505.

13. Wei, H.; Chen, C.; Zheng, G.; Wu, K.; Li, Z. PressLight: Learning Max Pressure Control to Coordinate Traffic Signals in Arterial Network. In Proceedings of the 25th ACM SIGKDD International Conference on Knowledge Discovery and Data Mining, Anchorage, AK, USA, 4–8 August 2019; pp. 1290–1298.

14. Chen, C.; Wei, H.; Xu, N.; Zheng, G.; Yang, M.; Xiong, Y.; Xu, K.; Li, Z. Toward a Thousand Lights: Decen-tralized Deep Reinforcement Learning for Large-Scale Traffic Signal Control. In Proceedings of the 33rd AAAI Conference on Artificial Intelligence (AAAI'19), Honolulu, HI, USA, 27 January–1 February 2019; pp. 3414–3421.

15. Krajzewicz, D.; Erdmann, J.; Behrisch, M.; Bieker, L. Recent development and applications of sumo simulation of urban mobility. *Int. J. Adv. Syst. Meas.* **2012**, *5*, 128–138.

16. Wu, Q.; Zhang, L.; Shen, J.; Lu, L.; Du, B.; Wu, J. Efficient pressure: Improving efficiency for signalized intersections. *arXiv* **2021**, arXiv:2112.02336.

17. Zhang, L.; Wu, Q.; Jun, S.; Lu, L.; Du, B.; Wu, J. Expression might be enough: Representing pressure and demand for re-inforcement learning based traffic signal control. In Proceedings of the 39th International Conference on Machine Learning, Baltimore, MD, USA, 17–23 July 2022; pp. 26645–26654.

18. Shabestary, S.M.A.; Abdulhai, B. Deep learning vs. discrete reinforcement learning for adaptive traffic signal control. In Proceedings of the 2018 21st International Conference on Intelligent Transportation Systems (ITSC), Maui, HI, USA, 4–7 November 2018; pp. 286–293.

19. Zeng, J.; Hu, J.; Zhang, Y. Adaptive Traffic Signal Control with Deep Recurrent Q-learning. In Proceedings of the 2018 IEEE Intelligent Vehicles Symposium (IV), Changshu, China, 26–30 June 2018; pp. 1215–1220.

20. Chen, P.; Zhu, Z.; Lu, G. An Adaptive Control Method for Arterial Signal Coordination Based on Deep Rein-forcement Learning. In Proceedings of the 2019 IEEE Intelligent Transportation Systems Conference (ITSC), Auckland, New Zealand, 27–30 October 2019; pp. 3553–3558.

21. Van Hasselt, H.; Guez, A.; Silver, D. Deep reinforcement learning with double q-learning. In Proceedings of the 29th AAAI Conference on Artificial Intelligence (AAAI'15), Austin, TX, USA, 25–30 January 2015; pp. 2094–2100.

22. Wang, Z.; Schaul, T.; Hessel, M.; van Hasselt, H.; Lanctot, M.; de Freitas, N. Dueling network architectures for deep reinforcement learning. In Proceedings of the 33rd International Conference on Machine Learning (ICML'16), New York, NY, USA, 19–24 June 2016; pp. 1995–2003.

23. Schaul, T.; Quan, J.; Antonoglou, I.; Silver, D. Prioritized experience replay. In Proceedings of the 4th International Conference on Learning Representations (ICLR'16), San Juan, PR, USA, 2–4 May 2016.

24. Xu M.; Wu J.; Huang L.; Zhou, R.; Wang, T.; Hu, D. Network-wide traffic signal control based on the discovery of critical nodes and deep reinforcement learning. *J. Intell. Transport. Syst.* **2020**, *24*, 1–10. [CrossRef]

25. Shashi, F.I.; Md Sultan, S.; Khatun, A.; Sultana, T.; Alam, T. A Study on Deep Reinforcement Learning Based Traffic Signal Control for Mitigating Traffic Congestion. In Proceedings of the 2021 IEEE 3rd Eurasia Conference on Biomedical Engi-neering, Healthcare and Sustainability (ECBIOS), Tainan, Taiwan, 28–30 May 2021; pp. 288–291.

26. Wei, H.; Zheng, G.; Gayah, V.; Li, Z. Recent Advances in Reinforcement Learning for Traffic Signal Control: A Survey of Models and Evaluation. *ACM SIGKDD Explor. Newsl.* **2022**, *22*, 12–18. [CrossRef]

27. Liu, J.; Qin, S.; Luo, Y.; Wang, Y.; Yang, S. Intelligent Traffic Light Control by Exploring Strategies in an Optimised Space of Deep Q-Learning. *IEEE Trans. Veh. Technol.* **2022**, *71*, 5960–5970. [CrossRef]

Article

Dual Input Fuzzy Logic Controllers for Closed Loop Hemorrhagic Shock Resuscitation

David Berard [1], Saul J. Vega [1], Guy Avital [1,2,3] and Eric J. Snider [1,*]

[1] U.S. Army Institute of Surgical Research, JBSA Fort Sam Houston, San Antonio, TX 78234, USA

[2] Trauma & Combat Medicine Branch, Surgeon General's Headquarters, Israel Defense Forces, Ramat-Gan 52620, Israel

[3] Division of Anesthesia, Intensive Care & Pain Management, Tel-Aviv Sourasky Medical Center, Sackler Faculty of Medicine, Tel-Aviv University, Tel-Aviv 64239, Israel

* Correspondence: eric.j.snider3.civ@health.mil; Tel.: +1-210-539-8721

Abstract: Hemorrhage remains a leading cause of preventable death in emergency situations, including combat casualty care. This is partially due to the high cognitive burden that constantly adjusting fluid resuscitation rates can require, especially in austere or mass casualty situations. Closed-loop control systems have the potential to simplify hemorrhagic shock resuscitation if properly tuned for the application. We have previously compared 4 different controller types using a hardware-in-loop test platform that simulates hemorrhagic shock conditions, and we found that a dual input—(1) error from target and (2) rate of error change—fuzzy logic (DFL) controller performed best. Here, we highlight a range of DFL designs to showcase the tunability the controller can have for different hemorrhage scenarios. Five different controller setups were configured with different membership function logic to create more and less aggressive controller designs. Overall, the results for the different controller designs ranged from reaching the setup rapidly but often overshooting the target to more conservatively approaching the target, resulting in not reaching the target during high active hemorrhage rates. In conclusion, DFL controllers are well-suited for hemorrhagic shock resuscitation and can be tuned to meet the response rates set by clinical practice guidelines for this application.

Keywords: control systems; hemorrhagic shock; fluid resuscitation; fuzzy logic; closed-loop; fluid resuscitation; hardware-in-loop

Citation: Berard, D.; Vega, S.J.; Avital, G.; Snider, E.J. Dual Input Fuzzy Logic Controllers for Closed Loop Hemorrhagic Shock Resuscitation. *Processes* **2022**, *10*, 2301. https://doi.org/10.3390/pr10112301

Academic Editors: Jie Zhang and Meihong Wang

Received: 23 September 2022
Accepted: 31 October 2022
Published: 5 November 2022

1. Introduction

Hemorrhage is the most common cause of preventable death in both civilian [1] and military [2] trauma casualties. The main pillars of care for these patients are expeditious hemorrhage control and volume resuscitation—the restoration of blood volume, preferentially using whole blood or blood components, to restore oxygen delivery to the end organs [3]. In cases where definite control of the hemorrhage is not immediately achievable, most experts recommend the "damage control resuscitation" (DCR) approach, which prompts goal-directed volume resuscitation balancing the need for restoring perfusion on one hand, while avoiding exacerbation of the hemorrhage on the other [3]. However, this can require constant monitoring of the patient's condition and frequent adjusting of the infusion rate.

As this task can be described as controlling a variable (e.g., blood pressure) towards a setpoint (i.e., the resuscitation goal), it is not surprising that several attempts have been made to automize this task in a closed-loop controlled fashion [4]. They vary in the approaches taken, secondary to the intended use case. A variety of approaches, including complex mathematical modeling [5,6] and adaptive controls [7,8] were described for the purpose of hemodynamic control through fluid management. However, DCR in its most basic form, which resembles current clinical (manual) practice, can be described as a single input (e.g., blood pressure)—single output (infusion flow rate). Hence, simpler controllers, such as decision tables, proportional-integral-derivative (PID) and fuzzy logic (FL) controllers [9] should at least be considered for this purpose.

We have previously developed a hardware-in-loop automated test platform for resuscitation controllers (HATRC) for comparing the performance between closed-loop controller designs across a wide range of hemorrhage resuscitation scenarios [10]. With this, we recently compared various controller logic types and determined that a dual-input fuzzy logic controller design performed best [11]. This was determined across various subject variability runs and four hemorrhage scenarios, using aggregate performance metrics tied to the intensity of the resuscitation, stability of the subject, and resource efficiency. In this work, we expand on this previous study to compare a range of dual-input fuzzy logic controller types to highlight the controller capabilities based on tuning for hemorrhagic shock resuscitation.

2. Materials and Methods

2.1. Overview of HATRC Platform

We previously developed the Hardware-in-loop Automated Testbed for Resuscitation Controllers (HATRC) for the purpose of high throughput testing of physiological closed-loop controllers designed to control fluid infusion, particularly for hemorrhagic shock resuscitation [10,12]. Water was circulated in a closed-loop by a peristaltic pump (Masterflex L/S, Masterflex Bioprocessing, Vernon Hills, IL, USA) while pressure was monitored and recorded using LabChart (PowerLab, ADinstruments, Sydney, Australia) via pressure transducer (ICU Medical, San Clemente, CA, USA). A key component of the system was the PhysioVessel (PV) model, a customizable fluidic reservoir that provides a volume-responsive hydrostatic pressure [13]. Analysis of a large animal hemorrhage model revealed a linear pressure–volume response for the administration of whole blood in swine who underwent a spleen injury following a controlled hemorrhage. Though alternative pressure–volume curves were found to characterize other fluids, like crystalloids, only whole blood was used as the simulated infusate during the hemorrhage scenarios in this study. The whole blood-tuned PV (PV_{WB}) was connected to two additional peristaltic pumps. One pump provided outflow comprised of a basal urine rate and a hemorrhage rate determined by the current hemorrhage scenario (see Section 2.2). The other pump provided an infusion whose rate was controlled by the resuscitation controller being evaluated. MATLAB (MathWorks, Natick, MA, USA) was used to run the hemorrhage scenario, determine inflow rates based on resuscitation controller algorithms, and control the corresponding pumps through an RS232 USB-to-serial adapter (CoolGear, Clearwater, FL, USA) configured as indicated by the pumps' manufacturer.

2.2. Hemorrhage Scenarios for Controller Performance

For a previous study, we designed 11 simulated hemorrhage scenarios to evaluate the performance of fluid resuscitation controllers by challenging them to operate against a variety of bleeding rates and initial arterial pressures [12]. Given the similarities found in controllers' performances in several scenarios during that study, here we focus on four distinct whole-blood scenarios to assess the new set of fuzzy logic controllers. Throughout, a target pressure of 65 mmHg mean arterial pressure (MAP) was the goal controllers were seeking during resuscitation.

Scenario 1 was the only scenario to last 62 min, and it simulated a compressible bleed that was already under control by the time resuscitation started. During the first half of this scenario, the fluid controller attempted to resuscitate the simulated subject from an initial MAP of 45 mmHg up to a target MAP of 65 mmHg without an active hemorrhage. At the 30-min mark, however, a high-rate bleed lasting 2 min was triggered, simulating a loosening and re-tightening of a tourniquet. Afterwards, hemorrhage was stopped, and controllers were given an additional 30 min to restabilize at target MAP.

The remaining three scenarios all lasted 30 min and simulated non-compressible hemorrhages. Both Scenarios 2 and 3 allowed natural coagulation to affect the simulation—the only difference was in the initial MAP: Scenario 2 started in a state of simulated compensated shock at 65 mmHg, while Scenario 3 started in a state of decompensation at 45 mmHg. Finally, Scenario 4 mimicked a subject with an initial MAP of 45 mmHg, who starts to experience a gradual degradation of their internal hemostatic mechanisms 5 min into the resuscitation.

2.3. Fuzzy-Logic Controller Design

Fuzzy logic controllers are widely used in industries such as manufacturing [14,15], automobile operation [16–18], and even space exploration [19,20], and their utility in various areas of medical care has been a subject of ongoing research and development [21–24]. Fuzzy logic takes a discrete input value and classifies it into a non-discrete linguistic, or descriptive, term using a set of membership functions. These functions map the input to a value of 0–1 which is its degree of membership for each class within the linguistic set. A set of logical rules then evaluate the fuzzified input(s) to determine the corresponding output. This approach is particularly advantageous when precise classifications are not easily determined, making Boolean-based logic suboptimal. The nonlinear, time-varying nature of the cardiovascular system makes it a prime candidate for fuzzy logic control.

We previously tested multiple types of hemorrhagic shock resuscitation controllers on HATRC that included two different versions of decision table, PID, single-input fuzzy logic (SFL), and dual-input fuzzy logic (DFL) controllers. Based on a comparative analysis using select controller performance metrics and a set of three aggregate metrics, described in further detail in Section 2.4, we determined that the DFL controllers demonstrated the best balance of Intensity, Stability, and Resource Efficiency [12]. We kept the two original DFL controller configurations and included an additional three DFL controllers with a wider range of tuning variations for a total of 5 in a comparative study using HATRC. The MATLAB Fuzzy Logic Designer toolbox was used to develop all the controllers evaluated here, and the infusion flow rate was the single output to the system. The first input to the controllers was the error expressed as a percentage of the measured system pressure divided by the setpoint, with a value of 1 representing the target being reached and was titled *PerformanceError* (Figure 1). The second input was the rate of change in the error over time taken as the slope of a linear regression across the last three samples and was titled *(d/dt)PerformanceError* (Figure 1).

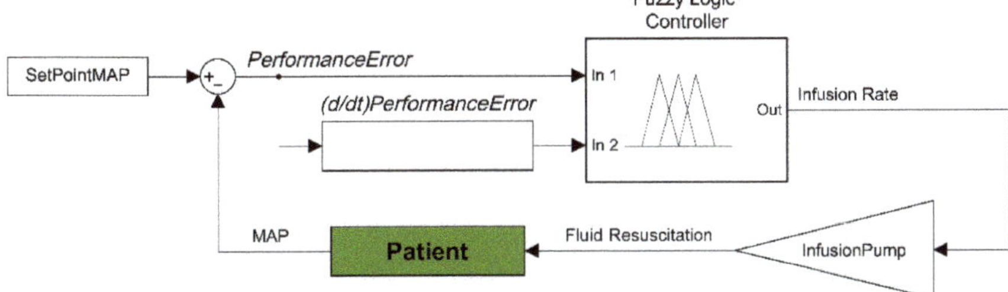

Figure 1. Diagram of the dual input fuzzy logic controller for hemorrhagic shock resuscitation. Two inputs to the fuzzy logic controller are derived from input pressure readings and the distance from set point mean arterial pressure—performance error and rate of performance error change. These two inputs are used to determine an infusion rate output for providing fluid to resuscitate and stabilize pressure. Each controller was set with the same types of membership functions, but with varying constants. DFL 1−4 classified *PerformanceError* into three fuzzy sets: *VeryLow*, *Low*, and *Set*. DFL 5 used these same three with an additional set called *Over*. All controllers used z-shaped membership functions for mapping *PerformanceError* into *VeryLow* and s-shaped membership functions for *Set*. DFL 1-3 used simple Gaussian curves while 4 and 5 used generalized bell-shaped membership functions for mapping *PerformanceError* into *Low*. DFL 5 also used an s-shaped membership function

to map *Over*. Smooth and Gaussian curves were selected as the membership functions for input 1 due to their lower computational cost and guaranteed continuity when compared to trapezoidal functions [25,26]. They have also been shown to be easier to optimize using evolutionary computational algorithms in type-2 fuzzy controllers which will be important for future iterations. Parameters for the functions used here were informed by expert feedback and current DCR guidelines. The membership functions for both inputs of all controllers are shown in Figure 2.

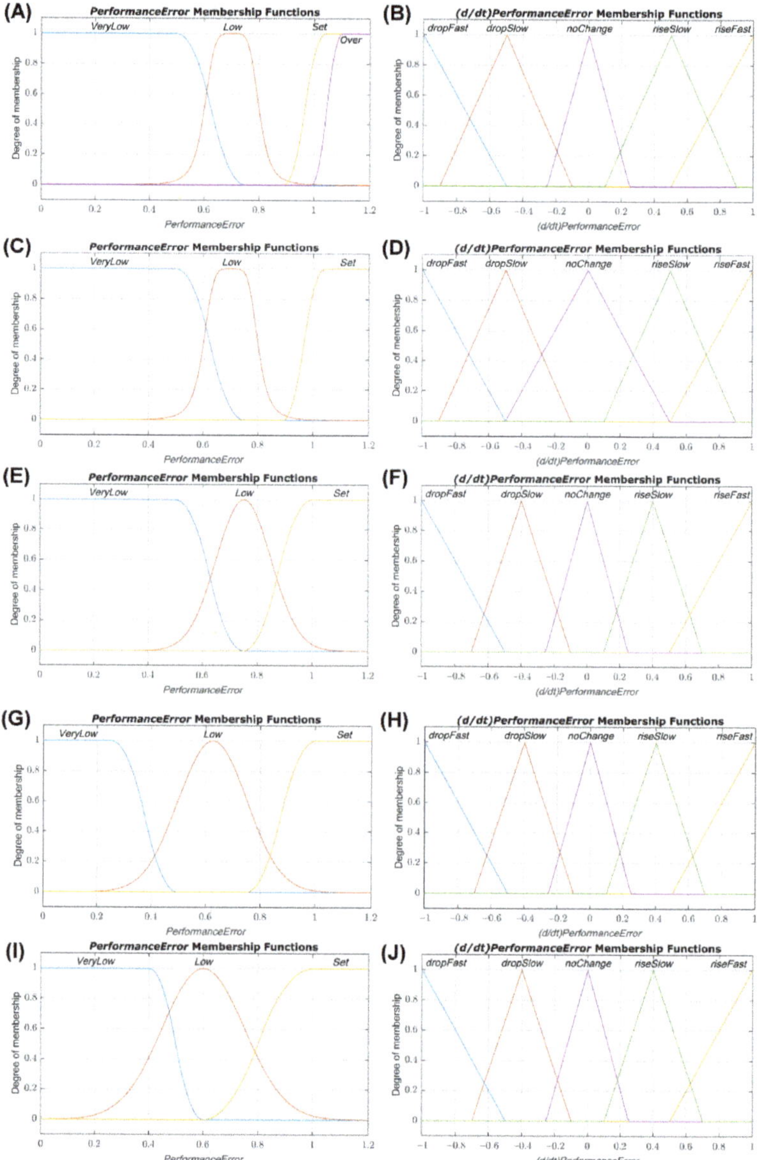

Figure 2. Membership function plots for dual-input fuzzy logic controllers. Plots of the *PerformanceError* and *(d/dt)PerformanceError* input membership functions for dual-input fuzzy logic controllers 1 (**A**,**B**), 2 (**C**,**D**), 3 (**E**,**F**), 4 (**G**,**H**), and 5 (**I**,**J**).

All controllers used the same fuzzy sets and membership function types for *(d/dt)PerformanceError*. Five fuzzy sets were defined: *dropFast*, *dropSlow*, *noChange*, *riseSlow*, and *riseFast*. Linear z-shaped membership functions were used to map *(d/dt)PerformanceError* into *dropFast*, triangular membership functions were used for *dropSlow*, *noChange*, and *riseSlow*, and linear s-shaped membership functions were used to map *riseFast*. Parameters of the membership functions for both inputs were tuned for each controller to produce a range of performance (e.g., prioritizing reaching the set point quickly vs. prioritizing minimum overshoot of the set point). Distinct rules were created for each controller using a similar ethic, and plots of the resulting rule surfaces are presented in Figure 3. The output, titled *InfusionRate*, was broken into the fuzzy sets *Off*, *Med*, and *Max* which utilized linear functions mapping to the output values of 0, 250 mL/min, and 500 mL/min, respectively. All controllers were type-1 Sugeno systems and used the following implication methods: a product AND, probabilistic OR, minimum Implication, maximum Aggregation, and a weighted average defuzzification method.

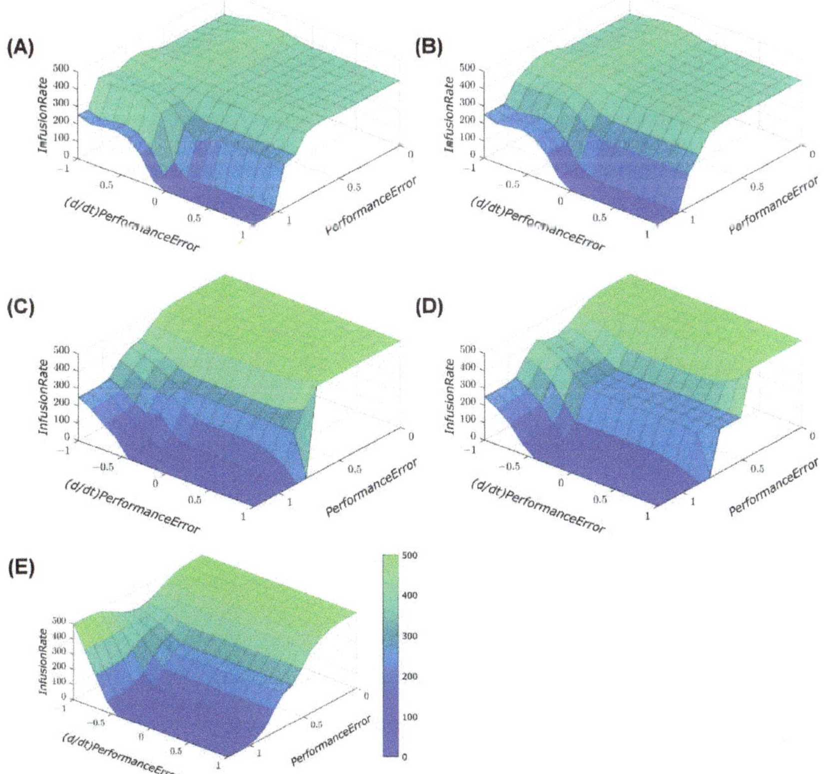

Figure 3. Rule surface plots for dual-input fuzzy logic controllers 1 (**A**), 2 (**B**), 3 (**C**), 4 (**D**), and 5 (**E**).

2.4. Controller Performance Metrics

A total of 12 individual metrics were used to evaluate the performance of each fluid resuscitation controller during the simulated hemorrhage scenarios. Additionally, with the goal of making all these measurements more useful for reaching conclusions about the controllers, 3 aggregate combinations of the individual metrics were also calculated. All of these measurements have been described previously [5,11,27,28].

A number of the individual metrics used are derived from measurements of performance error (PE)—that is, the difference between the measured pressure at a given time and the target pressure, as a percentage of the target. In summary, these metrics were:

- Median performance error (MDPE): median value of all the PEs;
- Median absolute performance error (MDAPE): median of the absolute values of all the PEs;
- MDAPE at steady state (MDAPE_{SS}): MDAPE after system has reached steady state;
- Target overshoot: maximum positive PE value, relative to the target pressure;
- Effectiveness: percent of time that the pressure remained within 5 mmHg of the target value
- Wobble: median of the absolute values of the differences between each PE and MDPE;
- End-state divergence: expressed as a percentage, this is the slope of the linear regression of PE vs. time during the final 10% of the test scenario, multiplied by the total duration of the scenario;
- Percent rise time: amount of time required for the measured MAP to reach 90% of the target, relative to the total duration of the scenario;
- Volume efficiency: ratio of total volume of fluid infused over the output volume;
- Areas above and below target: expressed as a percentage, these are the total areas delimited by the target pressure line and the measured MAP-vs-time curve, both above and below said line, respectively, relative to the target pressure and further normalized by scenario time duration;
- Mean infusion rate: mean rate of infusion as a percentage of the maximum infusion rate allowed by the controller (500 mL/min);
- Infusion rate variability: the averaged standard deviations of the infusion rates as a percentage of the mean infusion rate.

The aggregate metrics derived from the aforementioned individual ones were used to aide in evaluating the controllers' overall performances in three areas, as follows:

- Intensity: the controller's ability to effectively treat hypotension; it is the product of Percent rise time and Area below target, divided by the Effectiveness.
- Stability: the controller's propensity for stable performance and reduced overshooting; it is the product of Wobble, the absolute value of End-state divergence, the squared value of MDAPE_{SS}, and the sum of Area above target and Target overshoot.
- Resource efficiency: the controller's capacity for reduced fluid consumption and hardware wearing; it is the product of Mean infusion rate, Infusion rate variability and Volume efficiency.

It should be noted that whenever any of the measurements listed above are evaluated, except for "Effectiveness", lower values are generally considered better.

2.5. Statistical Analysis

For each controller, three subject variability experiments were conducted for all the test scenarios. Each metric was made unitless as described in Section 2.4. Metrics were averaged across all test scenarios for each subject variability and normalized to the median value for each metric to make the weights for each metric similar. Aggregate metrics for Intensity, Stability, Resource Efficiency, and an average of each were calculated. Results throughout are reported as mean $\pm$ standard deviation. For evaluating statistical significance between aggregate scores, one-way analysis of variance (ANOVA) was used, post hoc Tukey's test, for each metric to evaluate differences between the five controllers. Significance was defined as $p < 0.05$.

3. Results

3.1. Scenario 1: Low Initial MAP with Momentary Severe Hemorrhage Results

The first scenario tested began with a low MAP of 45 mmHg with no active hemorrhage. An intense hemorrhage was then introduced after 30 min, simulating a complication such as an extremity tourniquet failure, and lasted for 2 min. This was followed by an additional 30-min period with no active hemorrhage. This scenario evaluated the controllers' ability to resuscitate a patient without complications and test how quickly the controllers responded to an acute but brief hemorrhage. Plots of the MAP vs. Time and Flow Rate vs. Time for a single run of each controller are presented in Figure 4A with positive flow rate values representing the infusion rate outputs of the controllers and negative flow rate values being a representative plot of the outflow (a combination of basal urine rate and hemorrhage). Percent Area Above Target and Absolute End-State Divergence are shown in Figure 4B,C, respectively, and results for all performance metrics can be found in Table A1. While all controllers achieved less than 5% Area Above Target relative to the target pressure and total scenario time, DFL 4 demonstrated the best overshoot performance in this scenario with a near 0% result. DFL 1 and 2 technically performed the worst, both overshooting around 3%. None of the controllers exceeded the overshoot limit of 5% of the target to cause a re-bleed event. DFL 1 and 2 had the lowest End-State Divergence relative to total scenario time with values of 0.10% and 0.15%, respectively. These both were significantly lower than the other three controllers which all were above 0.5% with DFL 5 having the highest value of 1.04%.

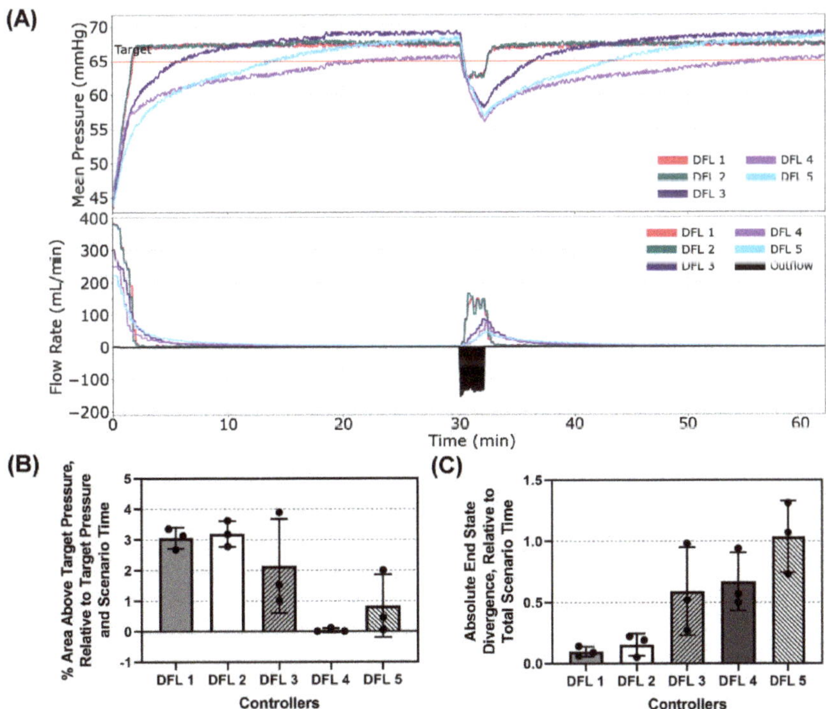

Figure 4. Dual-input fuzzy logic results for Scenario 1. Scenario 1 began with a MAP of 45 mmHg with no active hemorrhage for 30 min, followed by a fast hemorrhage for 2 min and then no hemorrhage for the remaining 30 min. (**A**) Five controller designs' MAP and infusion rate vs. time are shown for one replicate run. A single representative outflow vs. time result is shown. (**B**) Area above target pressure and (**C**) Absolute end state divergence performance metrics for each controller design are shown as mean values from three subject variability runs, with error bars denoting standard deviation.

3.2. Scenario 2: Target Initial MAP with Coagulating Hemorrhage Results

Scenario 2 presented the patient with a MAP starting at the targeted 65 mmHg but with an active hemorrhage that gradually reduced over time simulating an internal re-bleed accompanied by coagulation. This tested the controllers' responsiveness to perturbations to the system after reaching the set point. Plots of the MAP vs. Time and Flow Rate vs. Time for a single run of each controller are presented in Figure 5A. Percent Area Below Target and Percent Infusion Rate Variability are shown in Figure 5B,C, respectively, and results for all performance metrics can be found in Table A2. DFL 1 and 2 were the most responsive to a drop in MAP while near the target pressure with % areas below target of 0.25% and 0.27%, respectively. DFL 3–5 all had significantly higher areas below the target with DFL 4 having the highest value (8.5%). It should be noted that DFL 1 and 2, as well as 5, ended up overshooting the target, though not enough to trigger a re-bleed penalty (Table A2). DFL 5 had the lowest % Infusion Rate Variability (7.24%) while DFL 1 had the highest (27.9%). This can be visually observed when looking at the varying magnitudes of the peaks in the Flow Rate vs. Time plot for DFL 1 and DFL 2 which had the second highest % Infusion Rate Variability (21.9%).

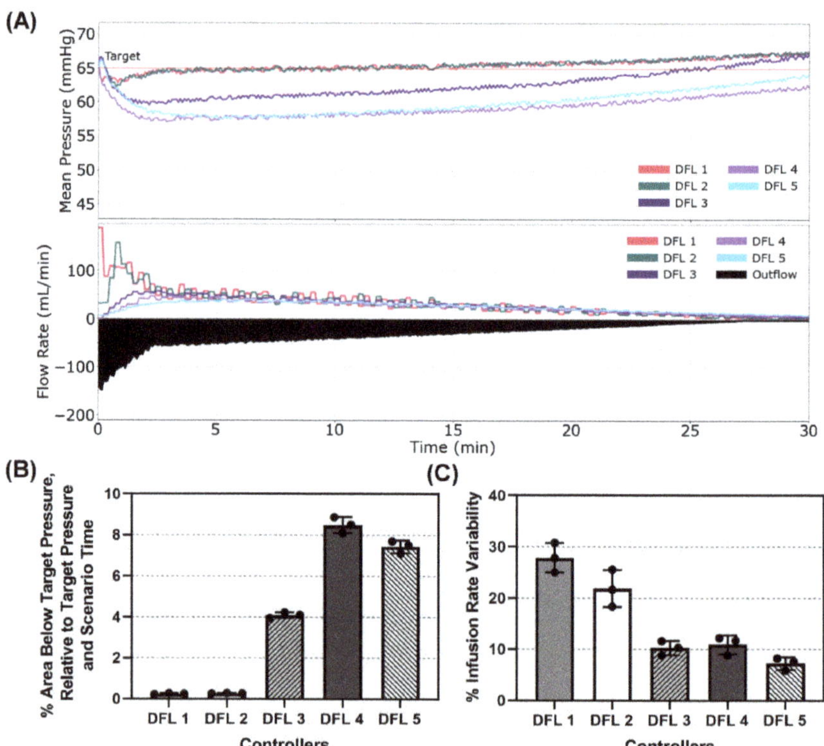

Figure 5. Dual-input fuzzy logic results for Scenario 2. Scenario 2 began with a MAP of 65 mmHg with an active hemorrhage that slows with time to mimic coagulation for 30 min, with a resuscitation target of 65 mmHg. (**A**) Five controller designs' MAP and infusion rate vs. time are shown for one replicate run. A single representative outflow vs. time result is shown. (**B**) Area below target pressure and (**C**) infusion rate variability performance metrics for each controller design are shown as mean values from three subject variability runs, with error bars denoting standard deviation.

3.3. Scenario 3: Low Initial MAP with Coagulating Hemorrhage Results

Scenario 3 began with a low MAP of 45 mmHg like Scenario 1 but included an ongoing hemorrhage with accompanying coagulation effects like Scenario 2. This scenario evaluated how effectively the controllers resuscitated a patient against complications like an internal, non-compressible hemorrhage. Plots of the MAP vs. Time and Flow Rate vs. Time for a single run of each controller are presented in Figure 6A. Percent Rise Time and % Effectiveness are shown in Figure 6B,C, respectively, and results for all performance metrics can be found in Table A3. DFL 4 and 5 had extremely high % Rise times compared to the other three controllers (44.7% and 41.3%, respectively) with DFL 1 and 2 performing almost identically with the lowest rise times (6.30% and 6.20%, respectively). This inversely correlates with the % Effectiveness with all 5 controllers maintaining the same relative rankings with respect to each other (DFL 1 = 93.1%, DFL 2 = 92.9%, DFL 3 = 81.4%, DFL 5 = 24.5%, and DFL 4 = 34.0%).

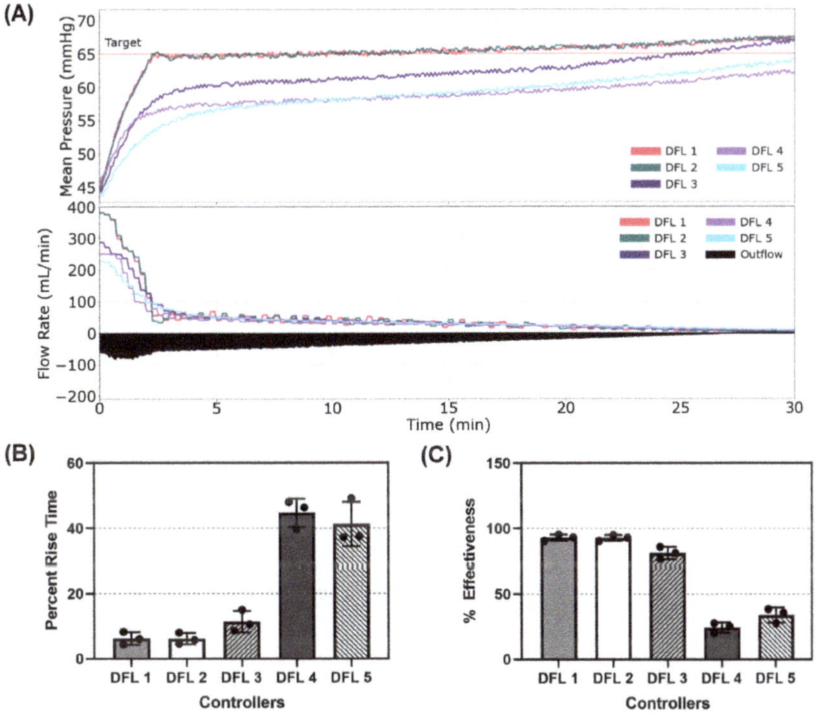

Figure 6. Dual-input fuzzy logic results for Scenario 3. Scenario 3 began with a MAP of 45 mmHg with an active hemorrhage slowing with time to mimic coagulation for 30 min, with a resuscitation target of 65 mmHg. (**A**) Five controller designs' MAP and infusion rate vs. time are shown for one replicate run. A single representative outflow vs. time result is shown. (**B**) Percent rise time and (**C**) effectiveness performance metrics for each controller design are shown as mean values from three subject variability runs, with error bars denoting standard deviation.

3.4. Scenario 4: Low Initial MAP with Coagulopathic Hemorrhage

Lastly, Scenario 4 provided the most complications of the scenarios tested. The patient began with a low MAP of 45 mmHg and presented with an ongoing non-compressible hemorrhage. This hemorrhage gradually slowed over time as the result of coagulation, but after 5 min, simulated coagulopathy was introduced gradually accelerating the hemorrhage until reaching a maximum rate of ~125 mL/min. This scenario was designed to tease out weaknesses of the controllers resulting in equilibrating infusion and outflow at a steady state that significantly deviates from the target. Plots of the MAP vs. Time and Flow Rate vs. Time for a single run of each controller are presented in Figure 7A. Percent Area Below Target and % MDAPE at Steady-State are shown in Figure 7B,C, respectively, and results for all performance metrics can be found in Table A4. These two metrics correlate closely in this scenario with the controller ranking and metric values nearly equal between the two. DFL 2 performed the best (3.56% area below target, 2.69% MDAPE at Steady State) and DFL 1 was nearly identical (3.56% area below target, 2.70% MDAPE at Steady State). DFL 3 (10.9% area below target, 10.8% MDAPE at Steady State) and DFL 4 (14.6% area below target, 14.4% MDAPE at Steady State) were next, and DFL 5 performed the worst in these two metrics (16.6% area below target, 16.8% MDAPE at Steady State).

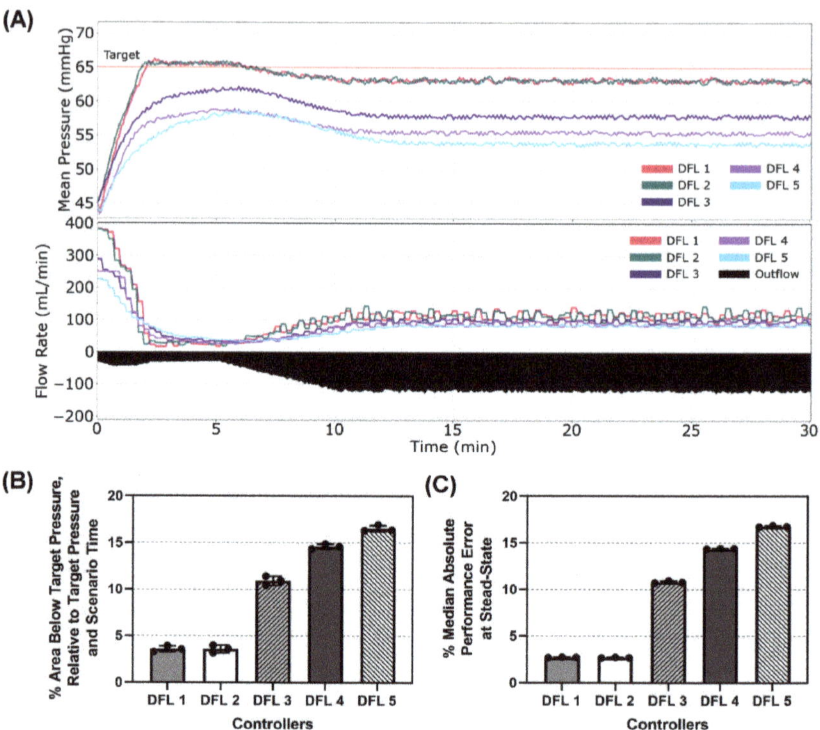

Figure 7. Dual-input fuzzy logic results for Scenario 4. Scenario 4 began with a MAP of 45 mmHg with an active hemorrhage slowing for the first 5 min and then accelerating to a maximum hemorrhage rate to mimic coagulopathy, with a resuscitation goal of 65 mmHg MAP. (**A**) Five controller designs MAP and infusion rate vs. time are shown for one replicate run. A single representative outflow vs. time result is shown. (**B**) The area below target pressure and (**C**) median absolute performance error at steady state performance metrics for each controller design are shown as mean values from three subject variability runs, with error bars denoting standard deviation.

3.5. Controller Performance in Aggregate Performance Metrics

We compiled the average score across all four scenarios of each aggregate performance metric for the 5 DFL controllers and then took the average of the three aggregate performance metrics (Figure 8). DFL 1 and 2 had the lowest two scores for the Intensity (both at 0.121) and Stability (0.314 and 0.299, respectively) metrics while holding the highest two scores in Resource Efficiency (2.02 and 1.88, respectively). DFL 4 had the highest score in the Intensity aggregate metric (10.18) with the second lowest score in Resource Efficiency (0.836). DFL 5 had the second highest score for Intensity (8.84) and held the lowest score for Resource Efficiency (0.604), though it had the highest score in the Stability aggregate (5.08). DFL 2 had the lowest average aggregate score (0.765) followed closely by DFL 1 (0.817) while DFL 4 had the second highest (3.88) and DFL 5 had the highest average aggregate (4.84). DFL 3 did not have the highest nor lowest of any aggregate score and had the median average aggregate score of the 5 controllers (1.49). A summary of one-way ANOVA statistical analyses comparisons for each aggregate metrics are shown in Table A5.

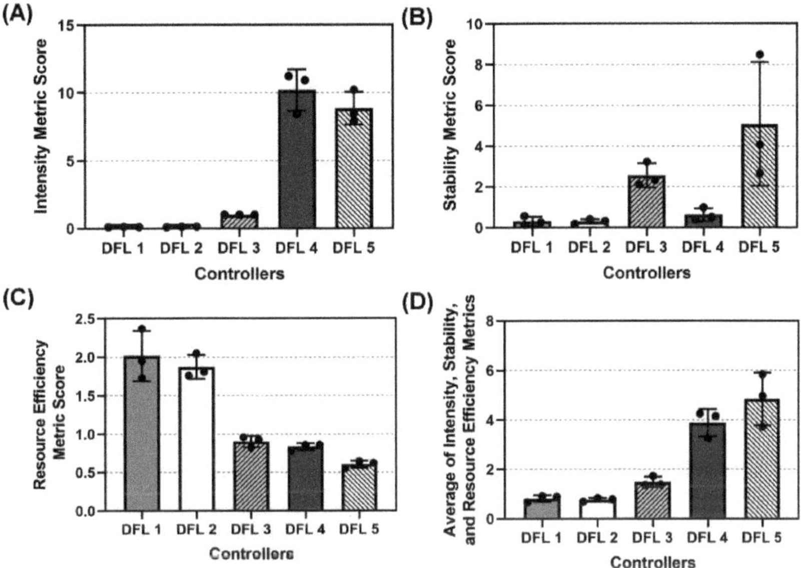

Figure 8. Aggregate metric results for dual-input fuzzy logic controller configurations. Aggregate (**A**) Intensity, (**B**) Stability, (**C**) Resource Efficiency, and (**D**) Average across the three metric results are shown as mean values from three subject variability runs, with error bars denoting standard deviation.

4. Discussion

Hemorrhagic shock resuscitation remains a challenging aspect of emergency medicine and trauma care. This is especially true in the context of resource and care provider-limited environments or mass casualty incidents where the attention needed to properly monitor their status and adjust therapy accordingly can quickly exceed current capabilities. The implementation of automated systems to provide DCR in these environments can lighten this load on care providers and potentially improve patient outcomes. Previously, multiple physiological closed-loop controllers were evaluated, and DFL controllers were shown to have the best performance in an assortment of performance and aggregated metrics [11].

This work compared an expanded set of five DFL controllers to identify which one performed favorably using the HATRC system. Three aggregate metrics were used to evaluate the controllers along the criteria of Intensity, Stability, and Resource Efficiency. Four hemorrhage scenarios were used in the comparison to assess how the controllers performed against an array of patient conditions including low MAP, with and without ongoing hemorrhage, the introduction of sudden, acute hemorrhage, and the introduction of ongoing hemorrhage. We also tested a scenario particularly designed to challenge the controllers' ability to overcome steady state error.

DFL 1 and 2 had the best performance in Intensity and Stability, indicating they were best suited to quickly reach the set point with the lowest degree of long-term oscillations in MAP. Although they tended to overshoot the target, especially when there was no ongoing hemorrhage, they did not exceed the overshoot threshold allowed, and the amount of overshoot was minimal with respect to the other metrics considered when calculating Stability. They performed worst in Resource Efficiency indicating they placed a high demand on equipment and consumed the most amount of fluid. This is understandable when observing the larger peak-to-trough magnitudes in infusion rate and the fact that a contributor to hemorrhage rate in HATRC was the MAP—i.e., sustaining a higher MAP for a larger proportion of the scenario time resulted in a larger accumulative hemorrhage, requiring more infused volume to compensate. DFL 4 and 5 showed the worst performance in Intensity but had the best scores in Resource Efficiency. This illustrates the compromise taking place between fast, aggressive resuscitation aimed at rapid restoration of oxygen delivery to the tissues, and a more gradual approach. It also emphasizes the gap in current clinical knowledge regarding which method is optimal for patients' outcomes. There may be conditions that would be better treated by one versus the other, but there has not been a robust enough study to draw conclusions based on established patient outcomes. All that said, in this study DFL 2 had the best average score in our tests, seemingly offering the most comprehensive balance of the metrics evaluated. This controller will be further investigated for potential optimizations and considered for testing in other models.

The current study does possess certain limitations that should be considered. The HATRC platform was designed based on empirical data and does not contain the degree of complexity and unpredictability of an in vivo model. There are alternative models, such as in silico simulations that may be useful for further evaluating the capabilities of automated controllers [29,30]. The intent here was to provide real-world performance data when physical hardware was used, and we believe the results show promise within the limitations of the empirical data-guided platform. The controllers investigated used MAP as the sole input, but real-time values provided by invasive measurements may not always be available. The reduced feedback frequency of current non-invasive arterial blood pressure measurement techniques may greatly hinder the performance of these controllers. Improvement opportunities exist in using other physiological variables for inputs as an alternative, such as cardiac output [31], a photoplethysmography waveform [32], or tissue oxygen saturation. Although the membership functions and their corresponding parameters were selected based on feedback from subject matter experts in anesthesiology, surgery, and military medicine, the lack of universal agreement within the medical community on the best resuscitation profile and the unpredictable nature of the physiologic response makes this tuning difficult and requires further refinement. This could be offered by more complex models that cover a wider range of physiologic states, such as septic shock, which would also make it possible to expand into type-2 fuzzy logic systems and iteratively optimize using simulation techniques [33–35]. As previously mentioned, the aggregate performance metrics used do not account for the full scope of physiological responses present when systemic trauma is experienced and may require weighting of some metrics over others. There are also unknowns regarding multiple simultaneous injuries, incapacitation of certain physiological systems, and the impact of chemical therapies that may be present. These interactions and their potential effect on arterial blood pressure, such as distributive or cardiogenic shock, were outside the scope of the current study but will be addressed in future in vivo ones.

5. Conclusions

Hemorrhagic shock goal-directed resuscitation can be facilitated in both emergency and military medicine by automating the constant fluid rate changes required to adequately stabilize a patient. Dual input fuzzy logic controllers have a wide range of tunability for managing various hemorrhagic shock resuscitation scenarios. As advancements in hemorrhagic shock resuscitation standard of care develop, DFL controllers have demonstrated the flexibility to be adapted to meet physiological demands that can promote the most desired patient outcomes. Through aggregate performance metric scores, a single DFL controller was identified as performing best which will be further evaluated in large animal hemorrhagic shock studies. This will bring closed loop control for acute hemorrhage resuscitation closer to reality to help improve the patient's recovery and stabilization while lessening the cognitive burden for the medical provider.

Author Contributions: Conceptualization, D.B., S.J.V., E.J.S. and G.A.; methodology, D.B. and E.J.S.; data collection, D.B. and S.J.V.; formal analysis D.B., E.J.S. and S.J.V.; writing (original draft preparation), D.B., S.J.V., E.J.S. and G.A.; writing (review and editing), D.B., S.J.V., E.J.S. and G.A.; supervision and project administration, E.J.S. All authors have read and agreed to the published version of the manuscript.

Funding: This work was funded by the U.S. Army Medical Research and Development Command (Proposal Number IS220008).

Institutional Review Board Statement: Not applicable.

Informed Consent Statement: Not applicable.

Data Availability Statement: The datasets generated during and/or analyzed during the current study are available from the corresponding author upon reasonable request.

Conflicts of Interest: The authors declare no conflict of interest.

DoD Disclaimer: The views expressed in this article are those of the authors and do not reflect the official policy or position of the U.S. Army Medical Department, Department of the Army, DoD, or the U.S. Government.

Appendix A

Table A1. Summary of performance metrics for Scenario 1. Performance metrics for each of five DFL controllers is shown as mean values for three subject variability runs.

	DFL 1	DFL 2	DFL 3	DFL 4	DFL 5
MDPE (%)	3.25%	3.41%	2.18%	−4.04%	−1.15%
MDAPE (%)	3.32%	3.46%	3.16%	4.04%	3.12%
MDAPE_SS (%)	3.29%	3.43%	3.03%	3.17%	2.42%
Target Overshoot (%)	4.25%	4.59%	5.00%	0.46%	3.79%
Effectiveness (%)	97.14%	97.23%	94.01%	83.29%	86.06%
Wobble (%)	0.43%	0.42%	0.94%	1.20%	1.18%
End-State Divergence (%)	0.10%	0.15%	0.59%	0.67%	1.04%
Percent Rise Time (%)	2.64%	2.64%	4.12%	6.02%	8.47%
Volume Efficiency	310.93%	311.83%	321.43%	297.47%	312.70%
Area Above Target Pressure (%)	3.06%	3.19%	2.14%	0.04%	0.84%
Area Below Target Pressure (%)	0.73%	0.72%	1.73%	5.21%	3.80%
Mean Infusion (%)	3.72%	3.75%	3.77%	3.26%	3.58%
Variable Infusion (%)	40.34%	41.03%	22.27%	24.58%	16.72%

Table A2. Summary of performance metrics for Scenario 2. Performance metrics for each of five DFL controllers is shown as mean values for three subject variability runs.

	DFL 1	DFL 2	DFL 3	DFL 4	DFL 5
MDPE (%)	0.55%	0.56%	−4.61%	−9.03%	−8.07%
MDAPE (%)	0.94%	0.93%	4.61%	9.03%	8.07%
MDAPE_SS (%)	0.94%	0.93%	3.60%	8.27%	6.78%
Target Overshoot (%)	3.54%	3.52%	1.70%	0.51%	1.30%
Effectiveness (%)	100.28%	100.28%	98.61%	32.04%	46.67%
Wobble (%)	0.90%	0.92%	1.39%	1.14%	1.20%
End-State Divergence (%)	0.54%	0.51%	1.06%	1.15%	1.72%
Percent Rise Time (%)	NA	NA	NA	NA	NA
Volume Efficiency	105.00%	102.70%	97.67%	82.87%	87.50%
Area Above Target Pressure (%)	0.88%	0.92%	0.09%	0.00%	0.01%
Area Below Target Pressure (%)	0.25%	0.27%	4.10%	8.50%	7.44%
Mean Infusion (%)	6.69%	6.55%	5.72%	4.52%	4.86%
Variable Infusion (%)	27.87%	21.88%	10.25%	10.91%	7.24%

Table A3. Summary of performance metrics for Scenario 3. Performance metrics for each of five DFL controllers is shown as mean values for three subject variability runs.

	DFL 1	DFL 2	DFL 3	DFL 4	DFL 5
MDPE (%)	0.45%	0.55%	−4.88%	−9.96%	−9.50%
MDAPE (%)	1.06%	1.07%	4.88%	9.96%	9.50%
MDAPE_SS (%)	0.92%	0.92%	1.89%	7.44%	5.36%
Target Overshoot (%)	3.49%	3.54%	1.48%	0.00%	0.00%
Effectiveness (%)	93.06%	92.87%	81.39%	24.54%	33.98%
Wobble (%)	0.92%	0.89%	1.25%	1.11%	1.16%
End-State Divergence (%)	0.49%	0.54%	1.16%	1.41%	1.36%
Percent Rise Time (%)	6.30%	6.20%	11.39%	44.72%	41.30%
Volume Efficiency	186.47%	188.07%	191.40%	179.70%	192.83%
Area Above Target Pressure (%)	0.88%	0.92%	0.09%	0.00%	0.00%
Area Below Target Pressure (%)	1.67%	1.65%	5.91%	10.39%	10.47%
Mean Infusion (%)	10.83%	10.96%	9.88%	8.63%	8.96%
Variable Infusion (%)	23.39%	21.76%	13.36%	15.42%	11.34%

Table A4. Summary of performance metrics for Scenario 4. Performance metrics for each of five DFL controllers is shown as mean values for three subject variability runs.

	DFL 1	DFL 2	DFL 3	DFL 4	DFL 5
MDPE (%)	−2.76%	−2.75%	−10.86%	−14.43%	−16.85%
MDAPE (%)	2.76%	2.75%	10.86%	14.43%	16.85%
MDAPE_SS (%)	2.70%	2.69%	10.81%	14.38%	16.79%
Target Overshoot (%)	1.18%	0.94%	0.00%	0.00%	0.00%
Effectiveness (%)	93.54%	93.63%	15.51%	0.00%	0.00%
Wobble (%)	0.47%	0.48%	0.58%	0.55%	0.68%
End-State Divergence (%)	0.18%	0.09%	0.16%	0.11%	0.20%
Percent Rise Time (%)	5.83%	5.93%	9.54%	15.42%	19.44%
Volume Efficiency	120.67%	121.13%	117.40%	117.43%	113.60%
Area Above Target Pressure (%)	0.06%	0.05%	0.00%	0.00%	0.00%
Area Below Target Pressure (%)	3.56%	3.56%	10.94%	14.61%	16.56%
Mean Infusion (%)	22.87%	22.92%	18.92%	17.35%	15.91%
Variable Infusion (%)	14.32%	14.99%	9.32%	11.51%	8.05%

Table A5. Summary of statistical analysis for aggregate metrics averaged across all tested scenarios. $p < 0.05$ indicated statistical significance. Values are italicized when this threshold was reached for the particular comparison pairing.

	Statical Analysis for Intensity Aggregate Scores				
	DFL 1	**DFL 2**	**DFL 3**	**DFL 4**	**DFL 5**
DFL 1					
DFL 2	> 0.99				
DFL 3	0.7297	0.7323			
DFL 4	*<0.0001*	*<0.0001*	*<0.0001*		
DFL 5	*<0.0001*	*<0.0001*	*<0.0001*	0.3829	

	Statical Analysis for Stability Aggregate Scores				
	DFL 1	**DFL 2**	**DFL 3**	**DFL 4**	**DFL 5**
DFL 1					
DFL 2	>0.9999				
DFL 3	0.3469	0.3405			
DFL 4	0.9986	0.9983	0.4788		
DFL 5	*0.0128*	*0.0125*	0.2509	*0.0194*	

	Statical Analysis for Resource Efficiency Aggregate Scores				
	DFL 1	**DFL 2**	**DFL 3**	**DFL 4**	**DFL 5**
DFL 1					
DFL 2	0.8266				
DFL 3	*<0.0001*	*0.0002*			
DFL 4	*<0.0001*	*0.0001*	0.9841		
DFL 5	*<0.0001*	*<0.0001*	0.2548	0.4824	

	Statical Analysis for Average Aggregate Scores				
	DFL 1	**DFL 2**	**DFL 3**	**DFL 4**	**DFL 5**
DFL 1					
DFL 2	0.6678				
DFL 3	0.2177	0.1343			
DFL 4	*0.043*	*0.0384*	*0.0426*		
DFL 5	0.0726	0.0705	0.0941	0.2818	

References

1. Goolsby, C.; Rouse, E.; Rojas, L.; Goralnick, E.; Levy, M.J.; Kirsch, T.; Eastman, A.L.; Kellermann, A.; Strauss-Riggs, K.; Hurst, N. Post-Mortem Evaluation of Potentially Survivable Hemorrhagic Death in a Civilian Population. *J. Am. Coll. Surg.* **2018**, *227*, 502–506. [CrossRef]
2. Eastridge, B.J.; Mabry, R.L.; Seguin, P.; Cantrell, J.; Tops, T.; Uribe, P.; Mallett, O.; Zubko, T.; Oetjen-Gerdes, L.; Rasmussen, T.E.; et al. Death on the Battlefield (2001–2011): Implications for the Future of Combat Casualty Care. *J. Trauma Acute Care Surg.* **2012**, *73*, S431–S437. [CrossRef] [PubMed]
3. Jenkins, D.H.; Rappold, J.F.; Badloe, J.F.; Berséus, O.; Blackbourne, L.; Brohi, K.H.; Butler, F.K.; Cap, A.P.; Cohen, M.J.; Davenport, R.; et al. THOR Position Paper on Remote Damage Control Resuscitation: Definitions, Current Practice and Knowledge Gaps. *Shock. Augusta Ga* **2014**, *41*, 3–12. [CrossRef] [PubMed]
4. Avital, G.; Snider, E.J.; Berard, D.; Vega, S.J.; Hernandez Torres, S.I.; Convertino, V.A.; Salinas, J.; Boice, E.N. Closed-Loop Controlled Fluid Administration Systems: A Comprehensive Scoping Review. *J. Pers. Med.* **2022**, *12*, 1168. [CrossRef] [PubMed]
5. Mirinejad, H.; Parvinian, B.; Ricks, M.; Zhang, Y.; Weininger, S.; Hahn, J.-O.; Scully, C.G. Evaluation of Fluid Resuscitation Control Algorithms via a Hardware-in-the-Loop Test Bed. *IEEE Trans. Biomed. Eng.* **2020**, *67*, 471–481. [CrossRef]
6. Bighamian, R.; Kim, C.-S.; Reisner, A.T.; Hahn, J.-O. Closed-Loop Fluid Resuscitation Control Via Blood Volume Estimation. *J. Dyn. Syst. Meas. Control* **2016**, *138*, 111005. [CrossRef]
7. Rinehart, J.; Alexander, B.; Manach, Y.L.; Hofer, C.K.; Tavernier, B.; Kain, Z.N.; Cannesson, M. Evaluation of a Novel Closed-Loop Fluid-Administration System Based on Dynamic Predictors of Fluid Responsiveness: An in Silico Simulation Study. *Crit. Care* **2011**, *15*, R278. [CrossRef]
8. Gholami, B.; Haddad, W.M.; Bailey, J.M.; Geist, B.; Ueyama, Y.; Muir, W.W. A Pilot Study Evaluating Adaptive Closed-Loop Fluid Resuscitation during States of Absolute and Relative Hypovolemia in Dogs. *J. Vet. Emerg. Crit. Care* **2018**, *28*, 436–446. [CrossRef]
9. Kramer, G.C.; Kinsky, M.P.; Prough, D.S.; Salinas, J.; Sondeen, J.L.; Hazel-Scerbo, M.L.; Mitchell, C.E. Closed-Loop Control of Fluid Therapy for Treatment of Hypovolemia. *J. Trauma* **2008**, *64*, S333–S341. [CrossRef]
10. Snider, E.J.; Berard, D.; Vega, S.J.; Hernandez Torres, S.I.; Avital, G.; Boice, E.N. An Automated Hardware-in-Loop Testbed for Evaluating Hemorrhagic Shock Resuscitation Controllers. *Bioengineering* **2022**, *9*, 373. [CrossRef]

11. Snider, E.J.; Berard, D.; Vega, S.J.; Ross, E.; Knowlton, Z.J.; Avital, G.; Boice, E.N. Hardware-in-Loop Comparison of Physiological Closed-Loop Controllers for the Autonomous Management of Hypotension. *Bioengineering* **2022**, *9*, 420. [CrossRef]

12. Snider, E.J.; Berard, D.; Vega, S.J.; Avital, G.; Boice, E.N. Evaluation of a Proportional–Integral–Derivative Controller for Hemorrhage Resuscitation Using a Hardware-in-Loop Test Platform. *J. Pers. Med.* **2022**, *12*, 979. [CrossRef] [PubMed]

13. Berard, D.; Vega, S.J.; Torres, S.I.H.; Polykratis, I.A.; Salinas, J.; Ross, E.; Avital, G.; Boice, E.N.; Snider, E.J. Development of the PhysioVessel: A Customizable Platform for Simulating Physiological Fluid Resuscitation. *Biomed. Phys. Eng. Express* **2022**, *8*, 035017. [CrossRef] [PubMed]

14. Haber, R.E.; Alique, A.; Ros, S.; Haber, R.H. Application of Knowledge-Based Systems for Supervision and Control of Machining Processes. In *Handbook of Software Engineering and Knowledge Engineering*; World Scientific Publishing Company: Singapore, 2002; pp. 673–709, ISBN 978-981-02-4974-8.

15. Chen, J.-H.; Ho, S.-Y. A Novel Approach to Production Planning of Flexible Manufacturing Systems Using an Efficient Multi-Objective Genetic Algorithm. *Int. J. Mach. Tools Manuf.* **2005**, *45*, 949–957. [CrossRef]

16. Panse, P.; Singh, A.; Satsangi, C. Adaptive Cruise Control Using Fuzzy Logic. *Int. J. Digit. Appl. Contemp. Res.* **2015**, *3*, 7.

17. Mauer, G.F. A Fuzzy Logic Controller for an ABS Braking System. *IEEE Trans. Fuzzy Syst.* **1995**, *3*, 381–388. [CrossRef]

18. Rastelli, J.P.; Peñas, M.S. Fuzzy Logic Steering Control of Autonomous Vehicles inside Roundabouts. *Appl. Soft Comput.* **2015**, *35*, 662–669. [CrossRef]

19. Ismail, Z.; Varatharajoo, R.; Chak, Y.-C. A Fractional-Order Sliding Mode Control for Nominal and Underactuated Satellite Attitude Controls. *Adv. Space Res.* **2020**, *66*, 321–334. [CrossRef]

20. Lea, R.N.; Jani, Y. Fuzzy Logic in Autonomous Orbital Operations. *Int. J. Approx. Reason.* **1992**, *6*, 151–184. [CrossRef]

21. Alkholy, E.M.N.; Aboutabl, A.E.; Haggag, M.H. A Proposed Fuzzy Model for Diseases Diagnosis. *Int. J. Eng. Adv. Technol.* **2020**, *9*. [CrossRef]

22. Bates, J.H.T.; Young, M.P. Applying Fuzzy Logic to Medical Decision Making in the Intensive Care Unit. *Am. J. Respir. Crit. Care Med.* **2003**, *167*, 948–952. [CrossRef] [PubMed]

23. Yardimci, A.; Hadimioglu, N. An Intraoperative Fluid Therapy Fuzzy Logic Control System for Renal Transplantation. *Eur. J. Control* **2005**, *11*, 572–585. [CrossRef]

24. Suha, S.A.; Akhtaruzzaman, M.; Sanam, T.F. A Fuzzy Model for Predicting Burn Patients' Intravenous Fluid Resuscitation Rate. *Healthc. Anal.* **2022**, *2*, 100070. [CrossRef]

25. Wu, D. Twelve Considerations in Choosing between Gaussian and Trapezoidal Membership Functions in Interval Type-2 Fuzzy Logic Controllers. In Proceedings of the 2012 IEEE International Conference on Fuzzy Systems, Brisbane, QLD, Australia, 10–15 June 2012; pp. 1–8.

26. Sadollah, A. Introductory Chapter: Which Membership Function Is Appropriate in Fuzzy System? In *Fuzzy Logic Based in Optimization Methods and Control Systems and Its Applications*; Sadollah, A., Ed.; IntechOpen: Rijeka, Croatia, 2018.

27. Varvel, J.R.; Donoho, D.L.; Shafer, S.L. Measuring the Predictive Performance of Computer-Controlled Infusion Pumps. *J. Pharmacokinet. Biopharm.* **1992**, *20*, 63–94. [CrossRef]

28. 14:00-17:00 IEC 60601-1-10:2007. Available online: https://www.iso.org/cms/render/live/en/sites/isoorg/contents/data/standard/04/41/44104.html (accessed on 25 March 2022).

29. Jin, X.; Bighamian, R.; Hahn, J.-O. Development and In Silico Evaluation of a Model-Based Closed-Loop Fluid Resuscitation Control Algorithm. *IEEE Trans. Biomed. Eng.* **2018**. [CrossRef] [PubMed]

30. Alsalti, M.; Tivay, A.; Jin, X.; Kramer, G.C.; Hahn, J.-O. Design and In Silico Evaluation of a Closed-Loop Hemorrhage Resuscitation Algorithm with Blood Pressure as Controlled Variable. *J. Dyn. Syst. Meas. Control* **2021**, *144*, 021001. [CrossRef]

31. Nguyen, L.S.; Squara, P. Non-Invasive Monitoring of Cardiac Output in Critical Care Medicine. *Front. Med.* **2017**, *4*, 200. [CrossRef] [PubMed]

32. Kurylyak, Y.; Lamonaca, F.; Grimaldi, D. A Neural Network-Based Method for Continuous Blood Pressure Estimation from a PPG Signal. In Proceedings of the 2013 IEEE International Instrumentation and Measurement Technology Conference (I2MTC), Minneapolis, MN, USA, 6–9 May 2013; pp. 280–283.

33. Arslan, A.; Kaya, M. Determination of Fuzzy Logic Membership Functions Using Genetic Algorithms. *Fuzzy Sets Syst.* **2001**, *118*, 297–306. [CrossRef]

34. Esmin, A.A.A.; Aoki, A.R.; Lambert-Torres, G. Particle Swarm Optimization for Fuzzy Membership Functions Optimization. In Proceedings of the IEEE International Conference on Systems, Man and Cybernetics, Yasmine Hammamet, Tunisia, 17–20 October 2002; Volume 3, p. 6.

35. Nayagam, V.L.G.; Jeevaraj, S.; Sivaraman, G. Total Ordering Defined on the Set of All Intuitionistic Fuzzy Numbers. *J. Intell. Fuzzy Syst.* **2016**, *30*, 2015–2028. [CrossRef]

processes

Article

SOC Estimation of E-Cell Combining BP Neural Network and EKF Algorithm

Yun Gao *, Wujun Ji and Xin Zhao

College of Automotive Engineering, Henan Polytechnic, Zhengzhou 450046, China
* Correspondence: gyauto@126.com

Abstract: Power lithium battery is an important core component of electric vehicles (EV), which provides the main power and energy for EV. In order to improve the estimation accuracy of the state of charge (SOC) of the electric vehicle battery (E-cell), the extended Kalman filter (EKF) algorithm, and backpropagation neural network (BPNN) are used to build the SOC estimation model of the E-cell, and the self-learning characteristic of BP neural network is used to correct the error and track the SOC of the E-cell. The results show that the average error of SOC estimation of BP-EKF model is 0.347%, 0.0231%, and 0.0749%, respectively, under the three working conditions of constant current discharge, pulse discharge, and urban dynamometer driving schedule (UDDS). Under the influence of different initial value errors, the average estimation errors of BP-EKF model are 0.2218%, 0.0976%, and 0.5226%. After the noise interference is introduced, the average estimation errors of BP-EKF model under the three working conditions are 1.2143%, 0.2259%, and 0.5104%, respectively, which proves that the model has strong robustness and stability. Using the BP-EKF model to estimate and track the SOC of E-cell can provide data reference for vehicle battery management and is of great significance to improve the battery performance and energy utilization of EV.

Keywords: back propagation neural network; extended Kalman filter; electric vehicle; state of charge

Citation: Gao, Y.; Ji, W.; Zhao, X. SOC Estimation of E-Cell Combining BP Neural Network and EKF Algorithm. *Processes* **2022**, *10*, 1721. https://doi.org/10.3390/pr10091721

Academic Editors: Jie Zhang and Meihong Wang

Received: 5 August 2022
Accepted: 25 August 2022
Published: 29 August 2022

Publisher's Note: MDPI stays neutral with regard to jurisdictional claims in published maps and institutional affiliations.

1. Introduction

Since the 21st century international energy consumption is increasing, the earth's oil resources are scarce, and there are serious environmental pollution problems, the global oil energy can not meet the needs of industry and automobiles and other aspects [1,2]. In recent years, due to the increasing sales of cars in China, the country's energy consumption has also shown a rapid growth trend. China is gradually becoming a net importer of non-renewable energy sources such as crude oil, coal, and natural gas in the world, and the external demand for energy is rising year by year, which has seriously affected China's energy security. In order to reduce environmental pollution and achieve the global goal of energy saving and emission reduction of greenhouse gases, the world and China's energy structure is also undergoing a series of changes, with the traditional fossil energy-based energy system gradually changing into a new supply system with renewable energy as the main source [3,4]. Traditional cars use petroleum as the main source of energy supply and emit a large amount of toxic exhaust gas during operation, further aggravating environmental pollution and degradation, so new energy vehicles powered by clean energy have become an important development direction for the automotive industry in recent years.

The remaining battery energy of an electric vehicle is an important parameter for battery management. The battery management system of an electric vehicle will use the remaining battery energy data as the basis to equalise the individual batteries, so as to improve the performance of battery use while ensuring stable battery operation. Accurate estimation of the remaining battery energy of EVs helps to reasonably allocate and plan the battery energy, develop scientific energy allocation strategies, ensure the driving range of EVs, and is of great value in extending the service life of EV batteries [5,6]. On the other hand, the estimation of the remaining energy of EV can also play a protective role for the battery, avoiding the phenomenon of overcharging and over discharging, which accelerates the aging and elimination of the battery. Therefore, in order to further improve the service performance of electric vehicle batteries and strengthen the battery management, the extended Kalman filter algorithm and BP neural network are used to estimate the battery SOC and realize the accurate tracking of the battery SOC. It is expected to provide data support for the battery management system.

2. Related Works

The E-cell SOC of an electric vehicle directly affects the battery usage and daily driving of the vehicle, but the E-cell SOC is non-linear and cannot be measured directly, so many researchers have conducted research on the E-cell SOC estimation problem. A deep neural network with different number of hide tiers was trained to predict the E-cell SOC under different driving cycles. It was found that the deep neural network with four hide tiers could accurately predict the SOC under various operating conditions such as Dynamic Stress Test (DST) and Federal Urban Driving Schedule (FUDS) [7]. Xiong R and his team proposed a fractional-order discrimination model for E-cell based on least squares and nonlinear optimization by combining the Butler–Volmer equation and fractional-order calculus and used the model as the basis for estimating the E-cell SOC using the fractional-order traceless Kalman filter algorithm, which was further processed by singular value decomposition [8]. Sarrafan K et al. proposed a combined estimation model based on battery energy, adaptive forgetting factor recursive least squares for the non-linear E-cell SOC estimation problem, which solves the battery parameter rate variation problem while reducing the computational difficulty and fully considers the traffic environment and other conditions; after laboratory tests, the results showed that the combined SOC estimation model has high estimation accuracy [9]. He Z's group proposed to use the adaptive extended Kalman filter algorithm and adaptive recursive least squares for parameter identification and constructed a first-order RC equivalent circuit to identify the battery parameters using the estimation model with forgetting factors and dynamically adjust the system noise to improve the estimation accuracy of the model [10]. Wadi A et al. combined the extended Kalman filter algorithm and the smoothed variable structure filter algorithm to appraise the SOC state of lithium batteries, combining the robustness and noise sequence approximation advantages of the two algorithms to improve the accuracy of SOC estimation; experimental tests on different data sets showed that the model has high estimation accuracy and effectively reduces the computational complexity of the algorithm [11]. Many researchers use different methods to estimate the SOC state of electric vehicle batteries, but the estimation accuracy of the existing research needs to be improved. The research is expected to explore different estimation methods to further accurately track the battery SOC.

Kalman filter (KF) is a commonly used algorithm in the field of control, which has advantages in solving a variety of problems. whl A et al. proposed a wind speed estimation model based on a Gaussian process regression model and an extended KF for the rotor effective wind speed estimation of wind turbines, taking into account the effects of other dynamics and atmospheric changes on the wind speed estimation; the results show that the model can effectively reduce the estimation errors arising from wind speed and other factors, and has high estimation accuracy [12]. Han F and his team proposed to use KF and random matrix theory for smart grid data-driven event detection in order to improv e the power system's grid situational awareness and use dynamic KF to process the phasor measurement unit data; the research results show that the model has strong robustness [13]. Zhang Y et al. proposed a denoising algorithm based on composite Kalman and least squares curve fitting for the noise problem of marine sensors, using least squares to eliminate the nonlinear factors of the system and combining wavelet transform for real-time tracking of noise, using a combined model based on composite KF for sensor denoising to improve the temperature measurement accuracy of sensors [14]. Zhou T et al. proposed a hybrid pairwise KF algorithm for predictive analysis of short-time traffic flow, modelled on the basis of the difference propagation between the traditional Kalman algorithm and the random wandering algorithm, and compensated for the prediction by means of error calibration, and the results of the study showed that the model has advantages on parametric and non-parametric models [15]. Li Y and his group members used magneto-optical imaging. The finite element model of the weld is combined with image characterisation of the weld to analyse the distribution of the leakage field under different conditions; in order to reduce the impact of weld coupling noise on the accuracy of laser weld analysis, the KF algorithm is used to identify the weld centre [16].

In summary, there are many studies on the SOC state estimation problem of E-cell, but the estimation accuracy needs to be further improved. Therefore, the study introduces BPNN and extended KF in the E-cell SOC estimation, expecting to further improve the SOC estimation accuracy of E-cell and provide help to improve the energy utilization efficiency of batteries.

3. Study on SOC Estimation of E-Cell Combining BP and EKF

3.1. E-Cell Modelling

The battery system of the electric vehicle is the basis of the driving and operation of the electric vehicle. The battery management system of the vehicle can monitor and analyze the state of the vehicle battery, monitor and collect the voltage, temperature, and other parameters of the vehicle battery, and monitor the state of charge (SOC) of the battery, so as to facilitate the management of the battery energy and ensure the safe and stable operation of the power battery pack of the electric vehicle [17–19]. SOC is the remaining capacity data of the automobile battery. Accurate battery SOC data can provide a safe range for the charging and discharging of the automobile, promote the extension of the service life of the automobile battery, and directly affect the accuracy of the automobile mileage data [20,21]. The research studies the SOC estimation strategy of electric vehicles. Firstly, the lithium battery model of electric vehicles is established to provide a model basis for the subsequent SOC estimation of electric vehicles.

The research simulates the external dynamic response of the battery by establishing the equivalent circuit model of the automobile battery. In order to fully reflect the voltage current characteristics of the electric vehicle battery, the research uses the eneral nonlinear (GNL) equivalent circuit model to build the battery model of the electric vehicle. The study considers that although the high-order number model can improve the accuracy to a certain extent, it greatly increases the computational workload and complexity, so the study simplifies the high-order GNL model and uses the second-order RC ring model to simulate the battery circuit [22]. The second-order RC ring model of the battery is shown in Figure 1. Compared with the traditional GNL circuit, the second-order RC circuit has one less RC circuit, which can effectively maintain the non-linear characteristics of the battery and reduce the computational difficulty. The equivalent resistance of the battery is R_0, the open circuit voltage is E_o, the terminal voltage is v_0, C_1 and C_2 represent the battery electrochemical polarisation capacitance and the concentration difference polarisation capacitance, respectively, R_1 and R_2 represent the battery electrochemical polarisation resistance and the concentration difference polarisation resistance, respectively, and V represents the battery open circuit voltage.

The terminal voltage and current functions for the second-order RC model of the battery are shown in Equation (1).

$$\begin{cases} v_0 = E_o + v_1 + v_2 + iR_0 \\ i = \frac{v}{R} + C\frac{dv}{dt} \end{cases} \tag{1}$$

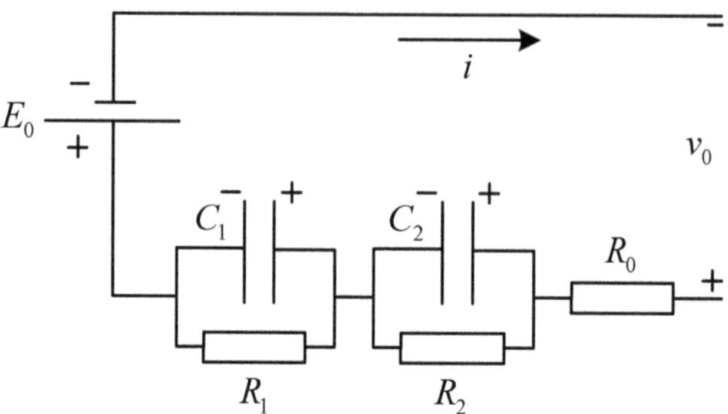

Figure 1. Second order RC ring model of battery.

In Equation (1), v_1 and v_2 represent the voltages at the ends of C_1 and C_2, respectively. Using the E-cell SOC and capacitor terminal voltages as the state vectors of the system, the battery state space equations are shown in Equation (2).

$$\dot{x} = \begin{bmatrix} -\frac{1}{R_1 C_1} & 0 & 0 \\ 0 & -\frac{1}{R_2 C_2} & 0 \\ 0 & 0 & 0 \end{bmatrix} x + \begin{bmatrix} \frac{1}{C_1} \\ \frac{1}{C_2} \\ -\frac{1}{Q} \end{bmatrix} i \tag{2}$$

In Equation (2), $x = \begin{bmatrix} v_1 & v_2 & SOC \end{bmatrix}$, and Q represents the battery capacity. The Eulerian method is used to discretize the continuous state equation, and the discretized state equation and system output equation are shown in Equation (3).

$$\begin{cases} x_k = \begin{bmatrix} -\frac{1}{R_1 C_1}\Delta t + 1 & 0 & 0 \\ 0 & -\frac{1}{R_2 C_2}\Delta t + 1 & 0 \\ 0 & 0 & 1 \end{bmatrix} x_{k-1} + \begin{bmatrix} \frac{\Delta t}{C_1} \\ \frac{\Delta t}{C_2} \\ -\frac{\Delta t}{Q} \end{bmatrix} i_{k-1} \\ v_{0,k} = E_{o,k} + v_{1,k} + v_{2,k} + i_k R_0 \end{cases} \tag{3}$$

3.2. EKF-Based E-Cell SOC Estimation

The Kalman filter (KF) algorithm is widely used in control applications and has a significant accuracy advantage in system estimation when the system noise satisfies the Gaussian distribution [23,24]. However, the classical Kalman filter is less suitable for non-linear systems and more suitable for linear systems, while EV have non-linear characteristics. The operation principle of the EKF model is shown in Figure 2. The EKF model is shown in Figure 2. The initialisation process is carried out first, then the battery state and error covariance are appraised, and the state appraises and error covariance values are updated by the extended Kalman filter gain calculation. Due to the strong robustness and adaptiveness of the EKF model, although the EKF model does not require a high degree of accuracy in terms of input battery parameters and initial SOC values, the EKF model can quickly achieve automatic convergence of the E-cell SOC during the computational update process, making it converge to the initial value, so the EKF model has a good and wide applicability for EV E-cell SOC estimation.

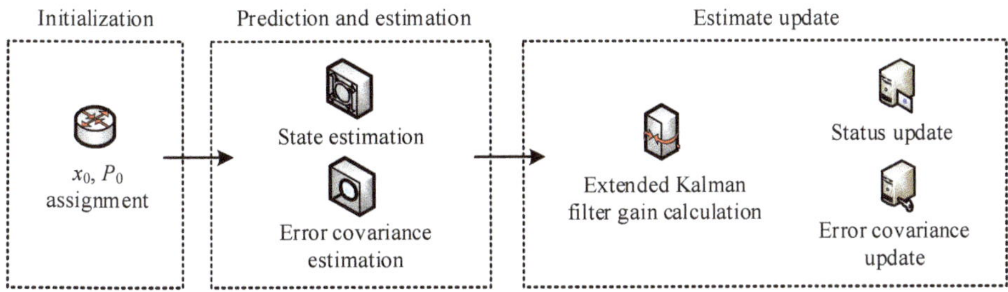

Figure 2. Operation principle of EKF model.

Let the inputs and outputs of the discrete system be u_k and y_k, respectively, and the system state quantities be x_k. The state transfer matrix of the system is A_k. The relationship between the inputs and states of the system is represented by B_k, and the observation and feedforward matrices of the system are C_k and D_k, respectively. Let the measurement noise and process noise of the system be v_k and w_k, respectively, and the noise is assumed to be Gaussian white noise with a mean value of 0. The discrete system function is shown in Equation (4).

$$\begin{cases} x_{k+1} = A_k x_k + B_k u_k + w_k \\ y_k = C_k x_k + D_k u_k + v_k \end{cases} \tag{4}$$

The EKF model is used for EV E-cell SOC estimation, where the state and observation equations of the system are Taylor expanded at the optimal estimation point, the higher order terms are discarded, the non-linear system is linearised, and the traditional KF algorithm is used for estimation. The state and observation equations of the system are shown in Equation (5).

$$\begin{cases} x_{k+1} = f(x_k, u_k) + w_k \\ y_k = h(x_k, u_k) + v_k \end{cases} \tag{5}$$

In Equation (5), $f(x_k, u_k)$ and $h(x_k, u_k)$ are the non-linear state transfer function and observation function, respectively. The state vector of the model is $x = \begin{bmatrix} v_1 & v_2 & SOC \end{bmatrix}^T$ and the input vector is $u_k = i_k$. The state transfer matrix and observation matrix of the system are shown in Equation (6).

$$\begin{cases} A_k = \dfrac{\partial f(x_k, u_k)}{\partial x_k} = \begin{bmatrix} -\dfrac{1}{R_1 C_1}\Delta t + 1 & 0 & 0 \\ 0 & -\dfrac{1}{R_2 C_2}\Delta t + 1 & 0 \\ 0 & 0 & 1 \end{bmatrix} \\ C_k = \dfrac{\partial h(x_k, u_k)}{\partial x_k} = \begin{bmatrix} 1 & 1 & \dfrac{\partial E_0}{\partial SOC} \end{bmatrix} \end{cases} \tag{6}$$

Further estimation is performed using the KF algorithm by first initialising, at this point $k = 0$, assigning values to the system state initial value x_0 and the error writing covariance P_0. The a priori values of the system state quantities are then appraised using the model inputs and the best appraise of the system state at the previous moment, and the estimation function for the system state quantities at the current moment is shown in Equation (7).

$$x_k^- = A_{k-1} x_{k-1} + B_{k-1} u_{k-1} \tag{7}$$

The initial value error will have an impact on the accuracy of the a priori appraise of the system state. To ensure that the a priori appraise accurately tracks the true value, the degree of dispersion between the appraise and the true value is analysed through the calculation of the error covariance. The current moment value is appraised using the error covariance of the previous moment, and the error covariance calculation function is shown in Equation (8).

$$P_k^- = A_{k-1} P_{k-1} A_{k-1}^T + Q_{k-1} \tag{8}$$

In Equation (8), the covariance matrix of the process noise is represented. The correction factor for the appraised value is obtained from the error covariance, i.e., the Kalman gain matrix. The extended Kalman gain calculation function is shown in Equation (9).

$$L_k = P_k^- C_k^T \left[P_k^- C_k^T + R_w \right]^{-1}$$

(9)

In Equation (9), Q_k is the covariance matrix of the measurement noise. The a priori appraises are updated in combination with the gain to obtain the best a priori appraise, and the system state appraise update function is shown in Equation (10).

$$x_k = x_k^- + L_k \left(y_k - C_k x_k^- - D_k u_k \right)$$

(10)

In Equation (10), y_k is the measured value of the system state quantity at the current moment. The error covariance of the system is updated, and the update function is shown in Equation (11).

$$P_k = (I - L_k C_k) P_k^-$$

(11)

In Equation (11), I is the unit matrix.

3.3. Optimization of Estimation Models Based on BP Neural Networks

In order to improve the estimation accuracy of the EKF model, the BPNN was used to compensate for the error, and the self-learning characteristics of the BPNN and its ability to approximate nonlinear functions were used to further optimize the EKF model. The EKF model is shown in Figure 3. The SOC state quantities at k, v_1, and v_2 and the gain values calculated by the EKF model are used as the input values of the BPNN, and the BPNN is used to output the error compensation values to correct the appraised values.

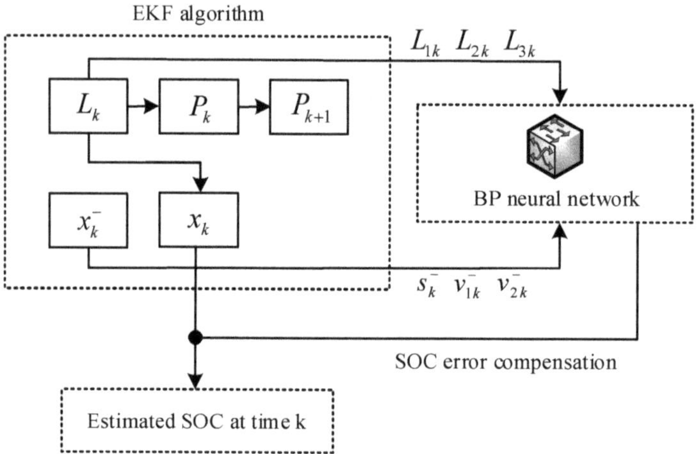

Figure 3. Structure of BP-EKF model.

Due to the variation in the range of E-cell SOC, voltage and gain values, and the different magnitudes and orders of magnitude of the individual input and output data, the input and output data are first normalised so that the data are between [1] to ensure that the sample data are weighted differently by the influence in training and to avoid the problem of neuron saturation, which affects the training effect of the network model. The normalisation function for the input and output data is shown in Equation (12).

$$\begin{cases} \overline{x}_i = \frac{x_i - x_{min}}{x_{max} - x_{min}} \\ \overline{x}_{ii} = 2 \times \frac{x_{ii} - x_{min}}{x_{max} - x_{ii}} - 1 \end{cases}$$

(12)

Using the battery state quantity and Kalman filter coefficients as the input values of the BPNN, the input quantity of the BPNN is $x_i = \begin{bmatrix} v_1 & v_2 & SOC & k_1 & k_2 & k_3 \end{bmatrix}$. The input quantity of the output tier of the BPNN x_l is the weighted sum of the output of the hide tier, while the excitation function of the output quantity x'_l is the tansig function with the function value range of $(-1, 1)$. The input quantity and output quantity of the output tier are shown in Equation (13).

$$\begin{cases} x_l = \sum w_{jl} x'_j \\ x'_l = \frac{2}{1+e^{-2x_l}} - 1 \end{cases} \tag{13}$$

The algorithm for learning the connection weights between the implicit and import tiers of the BPNN, and the connection weights to the output tier, is shown in Equation (14).

$$\begin{cases} \Delta w_{ij} = \eta \cdot e \cdot w_{jl} \cdot f'\left(x'_j\right) \cdot x_i \\ \Delta w_{jl} = \eta \cdot e \cdot x'_j \end{cases} \tag{14}$$

In Equation (14), η denotes the learning rate, $\eta \in [0,1]$, and e denotes the error of the output value from the ideal value. In order to solve the problem of model oscillation or slow convergence of BPNN in weight learning, the momentum factor α is introduced for adjustment, $\alpha \in [0,1]$, fully considering the influence of the previous moment weights on the present weights, then the weights w_{ij} and w_{jl} are shown in Equation (15).

$$\begin{cases} w_{ij}(k+1) = w_{ij}(k) + \Delta w_{ij} + \alpha\left(w_{ij}(k) - w_{ij}(k-1)\right) \\ w_{jl}(k+1) = w_{jl}(k) + \Delta w_{jl} + \alpha\left(w_{jl}(k) - w_{jl}(k-1)\right) \end{cases} \tag{15}$$

The role of the implicit layer of the BPNN is to extract and store the laws in the sample data. The implicit layer directly affects the mapping ability of the network; too many nodes in the implicit layer may lead to overfitting problems and affect the network generalization ability, while too few nodes will affect the extraction of the sample laws. The estimation function of the number of nodes in the hide tier is shown in Equation (16).

$$\begin{cases} m = \sqrt{n+1} + \alpha \\ m = \log_2 n \\ m = \sqrt{nl} \end{cases} \tag{16}$$

In Equation (16), n and l denote the number of nodes in the input and output tiers, respectively, α is a constant, and $\alpha \in [1,10]$.

4. Analysis of the Application Effect of the BP-EKF Based E-Cell SOC Estimation Model

4.1. Validation of Battery Modelling Effects

In order to verify the response effect of the electric vehicle battery model constructed in the research on the dynamic characteristics of the battery, the equivalent circuit model of the battery was built in the Simulink environment. The input and output of the battery model were circuit current and terminal voltage, respectively. The rated capacity of the battery was 20 Ah, the maximum discharge current was 100 A, the maximum charging voltage was 4.15 V, and the cut-off voltage was 3 V. The electrochemical polarization resistance and concentration difference polarization resistance of the battery are 0.0042 Ω and 0.0020 Ω, respectively, the equivalent resistance of the battery is 0.0013 Ω, and the electrochemical polarization capacitance and concentration difference polarization capacitance of the battery are 17,111 F and 440.56 F, respectively. The output data of the battery model under DST working condition is compared with the actual data of the battery, and the reflecting ability of the dynamic characteristics of the battery model is analyzed. The comparison between the output data of the battery model and the actual data is shown in Figure 4.

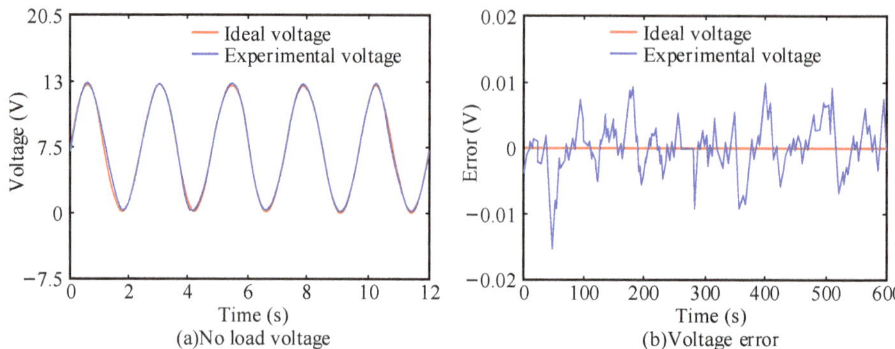

Figure 4. No load discharge test of battery.

Figure 4a shows the results of the battery no-load voltage test, and it can be seen that the ideal no-load voltage required for the study is 13 V, and the voltage value exhibited by the battery during the experiment is also 13 V, i.e., the state of the experimental voltage under no-load is consistent with the ideal value. Figure 4b shows the error variation between the experimental voltage and the ideal voltage in the battery discharge test. It can be seen that there is significant error variation between the experimental voltage and the ideal voltage during the 600 s test time, but it is known from the error magnitude of the experimental voltage that the maximum error voltage is only 0.015 V, and the error stays within 0.005 V for most of the time. The above results show that the proposed BP-EKF battery has a high agreement with the ideal voltage under no-load condition and during the discharge process, indicating that the BP-EKF-based battery design is reasonable.

4.2. Analysis of the Estimation Accuracy of the BP-EKF Model

In order to verify the accuracy of the BP-EKF model constructed in the study for E-cell SOC estimation and to investigate the validity and feasibility of the model, the study was carried out in the MATLAB/SIMULINK working environment to simulate the battery operating mode for experiments. The battery is assumed to be fully charged at the initial moment, the initial value of SOC is set to 1, the battery current is 0, and the terminal voltage at both ends of the RC circuit is 0. Therefore, the initial values of the states are $\begin{bmatrix} 0 & 0 & 1 \end{bmatrix}$ and $R = 1000$. The operating effects of the model under constant current discharge, pulse discharge, and UDDS conditions are analysed, respectively. The SOC estimation and error of the model under constant current discharge conditions are shown in Figure 5.

As can be seen from Figure 5, the model's SOC appraises are very close to the true values under constant current discharge conditions, and the model is able to converge accurately and quickly to approximate the true values, with a maximum error of 0.22% and an average error of 0.347% between the model's SOC appraises and the true values' accuracy. The SOC estimation results and errors of the model under pulse discharge conditions are shown in Figure 6.

As can be seen in Figure 6, the difference between the appraised and true values of the EV E-cell SOC using the model under pulse discharge conditions is small, and the model is able to converge to the true value quickly, with a maximum error of 1.0367% and an average error of 0.0231% in the model SOC estimation. The results and errors of the model E-cell SOC estimation under UDDS conditions are shown in Figure 7.

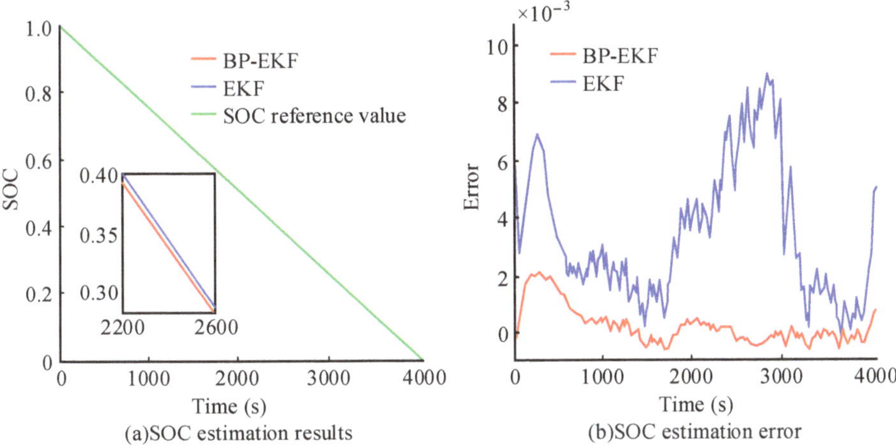

Figure 5. SOC estimation and error of the model under constant current discharge condition.

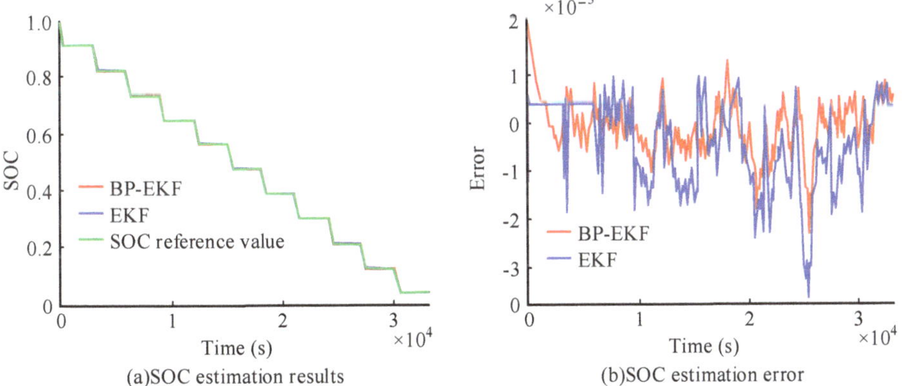

Figure 6. SOC estimation results and errors of the model under pulse discharge conditions.

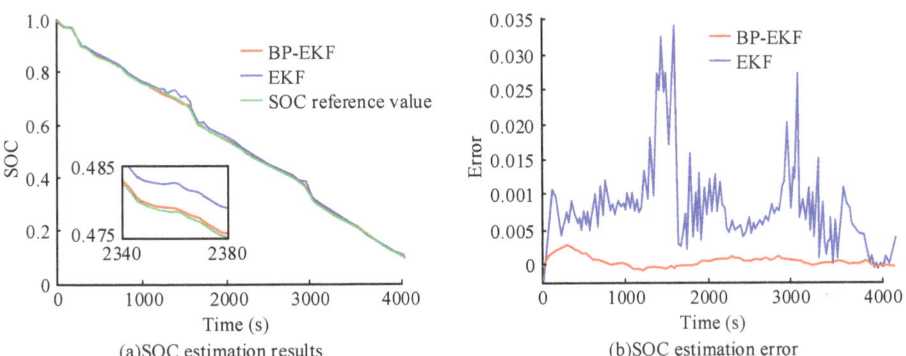

Figure 7. SOC estimation results and errors of model battery under UDDS conditions.

As can be seen from Figure 7, the model converges quickly to the true value in the UDDS condition, and the difference between the appraised SOC and the true value is small. Compared with the constant current discharge condition and the pulse discharge condition, the estimation error of the model under UDDS condition increases slightly, but still maintains a good estimation accuracy. The BP-EKF model was compared with the unscented Kalman filter (UKF) model and the EKF model, and the absolute value of the estimation error of the E-cell SOC for the DST condition is shown in Figure 8.

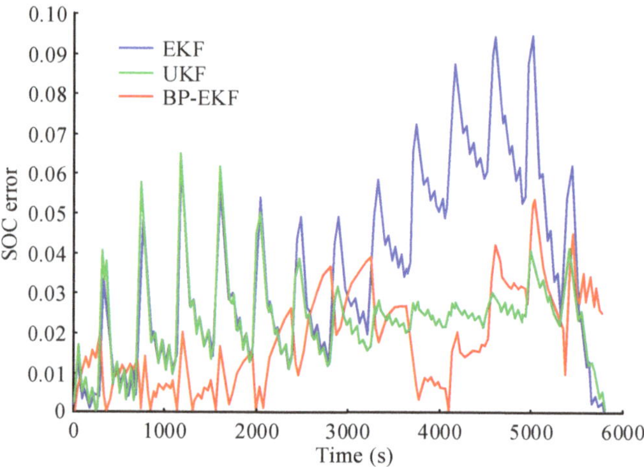

Figure 8. Absolute value curve of E-cell SOC estimation error of three models under DST operating mode.

As can be seen in Figure 8, during the time period 0–3000 s, when the battery is in a stable state of low current charging and discharging, all three models have a small error in estimating the E-cell SOC and can better appraise the E-cell SOC situation accurately. At 3000 s, when the battery has a sudden change in current, the EKF and UKF models are unable to track the state quantity quickly and accurately, resulting in a sudden increase in the SOC estimation error, and the SOC estimation error curves of the EKF and UKF models show large fluctuations and rising trends. The estimation error of the EKF model gradually becomes larger and the estimation performance is less stable. In contrast, the BP-EKF model, with the effect of BPNN and extended Kalman filter gain, always maintains a low level of E-cell SOC estimation error, and after a sudden current change at 3000 s, the estimation error of the model falls back quickly, and the error curve fluctuates less.

4.3. Robustness Analysis of the BP-EKF Model

During the initial start-up of an electric vehicle, it is difficult for the battery management system to obtain the initial value of the E-cell SOC. The E-cell SOC estimation model is required to provide fast and accurate tracking of the true value of the SOC when the initial value of the SOC is clear, and also needs to converge quickly to the true value when the initial value is unknown to ensure the accuracy and efficiency of the SOC value estimation. In order to investigate the adaptability of the BP-EKF model proposed in the study to the case of unknown initial SOC values and to analyse the convergence capability of the model, the study analyses the estimation performance of the model under different initial SOC values. Ensuring that other parameters are the same, the range of SOC initial values is set to [0.5, 1], the division interval is 0.1, and the model estimation results for six SOC initial value cases are shown in Figure 9.

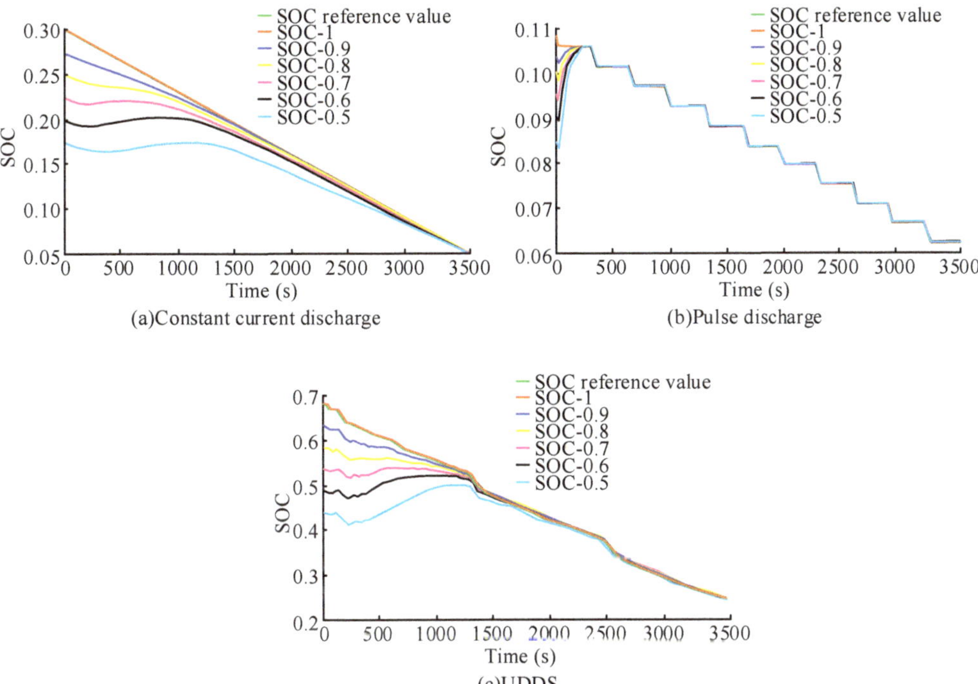

Figure 9. Model estimation effect under six initial SOC values.

It can be seen from Figure 9 that under the constant current discharge condition, the model proposed by the research is used to estimate the SOC of the battery. When the initial value of SOC is different from the actual value to different degrees, the average SOC estimation error of the model is 0.2218% under the six initial values of SOC, while the average SOC estimation error under the pulse discharge condition is 0.0976%. Under UDDS working condition, the average error of SOC estimation of the model battery is 0.5226%, which can meet the actual application requirements of battery management of electric vehicles. It proves that the BP-EKF model can effectively track the SOC of the battery quickly and accurately.

In order to verify the adaptability of the proposed model to current noise, the SOC estimation results of the BP-EKF model and the EKF model under three operating conditions were compared and analysed in a simulation environment with a white noise with a mean square deviation of $10A^2$ to investigate the robustness and optimisation of the model.

From Figure 10, it can be seen that after adding noise to the current signal, the average estimation error values of the BP-EKF model are 1.2143%, 0.2259%, and 0.5104%, respectively, for the three operating conditions, while the average error values of the EKF model are 2.2416%, 0.9968%, and 2.1864%, respectively, which proves that the BP-EKF model can effectively cope with the noise interference, and the SOC estimation error of the BP-EKF model is smaller with the error compensation of the BPNN.

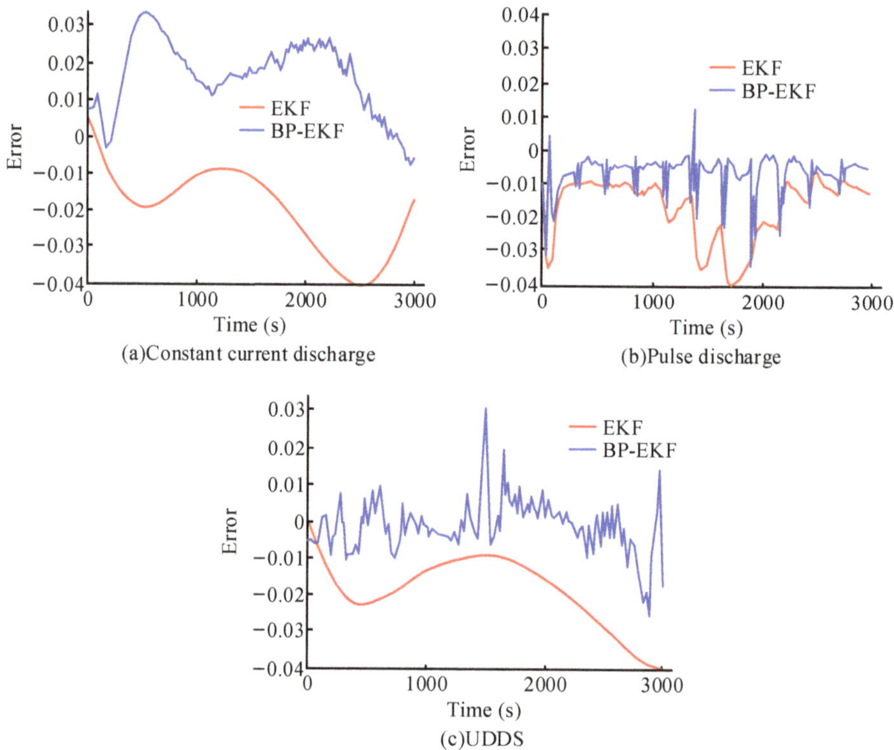

Figure 10. Model estimation effect under six initial SOC values (EKF and BP-EKF model).

5. Conclusions

In order to make accurate estimation of E-cell SOC, the study uses the extended Kalman filter algorithm to construct an E-cell SOC estimation model and introduces BPNN to compensate the error of the SOC estimation model to achieve accurate and fast tracking of E-cell SOC. The study conducts simulation experiments in the MATLAB/SIMULINK working environment. The experimental results show that the average errors between the BP-EKF model SOC estimation and the real value are 0.347%, 0.0231%, and 0.0749% under the three operating conditions of constant current discharge, pulse discharge, and UDDS. The average SOC estimation errors of the BP-EKF model for different initial values were 0.2218%, 0.0976%, and 0.5226%, proving that the model is more applicable to different initial value errors. The average estimation errors of the BP-EKF model under noise disturbance were 1.2143%, 0.2259%, and 0.5104%, proving that the model has strong robustness. The study analyses the model performance under three operating conditions, and in the future, the influence of temperature and other factors can be further considered to optimise the E-cell SOC estimation under a variety of operating conditions.

Author Contributions: Conceptualization, Y.G.; data curation, X.Z.; formal analysis, W.J.; investigation, W.J.; methodology, Y.G.; project administration, Y.G.; software, W.J.; validation, X.Z.; writing—original draft, Y.G.; writing—review & editing, Y.G. All authors have read and agreed to the published version of the manuscript.

Funding: This research received no external funding.

Institutional Review Board Statement: Not applicable.

Informed Consent Statement: Not applicable.

Data Availability Statement: The data used to support the findings of this study are available from the corresponding author upon request.

Conflicts of Interest: The authors declare no conflict of interest.

References

1. Bi, J.; Wang, Y.; Sai, Q.; Ding, C. Estimating remaining driving range of battery EV based on real-world data: A case study of Beijing, China. *Energy* **2019**, *169*, 833–843. [CrossRef]
2. Chen, X.; Lei, H.; Xiong, R.; Shen, W.; Yang, R. A novel approach to reconstruct open circuit voltage for state of charge estimation of lithium ion batteries in EV. *Appl. Energy* **2019**, *255*, 113758. [CrossRef]
3. How, D.N.; Hannan, M.A.; Lipu, M.H.; Ker, P.J. State of charge estimation for lithium-ion batteries using model-based and data-driven methods: A review. *IEEE Access* **2019**, *7*, 136116–136136. [CrossRef]
4. Kim, W.; Lee, P.Y.; Kim, J.; Kim, K.S. A robust state of charge estimation approach based on nonlinear battery cell model for lithium-ion batteries in EV. *IEEE Trans. Veh. Technol.* **2021**, *70*, 5638–5647. [CrossRef]
5. Zhang, Y.; Chu, L.; Ding, Y.A.N.; Xu, N.A.N.; Guo, C.; Fu, Z.; Xu, L.; Tang, X.; Liu, Y. A hierarchical energy management strategy based on model predictive control for plug-in hybrid EV. *IEEE Access* **2019**, *16*, 2303–2308.
6. Wang, C.; Yang, R.; Yu, Q. Wavelet transform based energy management strategies for plug-in hybrid EV considering temperature uncertainty. *Appl. Energy* **2019**, *256*, 113928. [CrossRef]
7. How, D.N.; Hannan, M.A.; Lipu, M.S.H.; Sahari, K.S.; Ker, P.J.; Muttaqi, K.M. State-of-charge estimation of li-ion battery in EV: A deep neural network approach. *IEEE Trans. Ind. Appl.* **2020**, *56*, 5565–5574. [CrossRef]
8. Xiong, R.; Tian, J.; Shen, W.; Sun, F. A novel fractional order model for state of charge estimation in lithium ion batteries. *IEEE Trans. Veh. Technol.* **2019**, *68*, 4130–4139. [CrossRef]
9. Sarrafan, K.; Muttaqi, K.M.; Sutanto, D. Real-time estimation of model parameters and state-of-charge of li-ion batteries in EV using a new mixed estimation model. *IEEE Trans. Ind. Appl.* **2020**, *56*, 5417–5428. [CrossRef]
10. He, Z.; Yang, Z.; Cui, X.; Li, E. A method of state-of-charge estimation for EV power lithium-ion battery using a novel adaptive extended Kalman filter. *IEEE Trans. Veh. Technol.* **2020**, *69*, 14618–14630. [CrossRef]
11. Wadi, A.; Abdel-Hafez, M.F.; Hussein, A.A.; Alkhawaja, F. Alleviating dynamic model uncertainty effects for improved e-cell soc estimation of EVs in highly dynamic environments. *IEEE Trans. Veh. Technol.* **2021**, *70*, 6554–6566. [CrossRef]
12. Whl, A.; Al, A.; Fm, A. Real-time rotor effective wind speed estimation using Gaussian processregression and Kalman filtering. *Renew. Energy* **2021**, *169*, 670–686.
13. Han, F.; Ashton, P.M.; Li, M.; Pisica, I.; Taylor, G.; Rawn, B.; Ding, Y. A data driven approach to robust event detection in smart grids based on random matrix theory and kalman filtering. *Energies* **2021**, *14*, 2166. [CrossRef]
14. Zhang, Y.; Wang, R.; Li, S.; Qi, S. Temperature sensor denoising algorithm based on curve fitting and compound kalman filtering. *Sensors* **2020**, *20*, 1959. [CrossRef]
15. Zhou, T.; Jiang, D.; Lin, Z.; Han, G.; Xu, X.; Qin, J. Hybrid dual Kalman filtering model for short-term traffic flow forecasting. *IET Intell. Transp. Syst.* **2019**, *13*, 1023–1032. [CrossRef]
16. Mo, Y.; Song, Z.; Li, H.; Jiang, Z. A hierarchical safety control strategy for exoskeleton robot based on maximum correntropy kalman filter and bounding box. *Robotica* **2019**, *37*, 2165–2175. [CrossRef]
17. Li, Y.; Gao, X.; Chen, Y.; Zhou, X.; Zhang, Y.; You, D. Modeling for tracking micro gap weld based on magneto-optical sensing and kalman filtering. *IEEE Sens. J.* **2020**, *21*, 11598–11614. [CrossRef]
18. Esfandyari, M.J.; Esfahanian, V.; Yazdi, M.H.; Nehzati, H.; Shekoofa, O. A new approach to consider the influence of aging state on Lithium-ion battery state of power estimation for hybrid EV. *Energy* **2019**, *176*, 505–520. [CrossRef]
19. Yu, D.; Ma, Z.; Wang, R. Efficient smart grid load balancing via fog and cloud computing. *Math. Probl. Eng.* **2022**, *2022*, 3151249. [CrossRef]
20. Wang, S.; Lu, L.; Han, X.; Ouyang, M.; Feng, X. Virtual-battery based droop control and energy storage system size optimization of a DC microgrid for electric vehicle fast charging station. *Appl. Energy* **2020**, *259*, 114146. [CrossRef]
21. Singh, B.; Kushwaha, R. Power factor pre-regulation in interleaved luo converter fed e-cell charger. *IEEE Trans. Ind. Appl.* **2021**, *57*, 2870–2882. [CrossRef]
22. Yu, D.; Wu, J.; Wang, W.; Gu, B. Optimal performance of hybrid energy system in the presence of electrical and heat storage systems under uncertainties using stochastic p-robust optimization technique. *Sustain. Cities Soc.* **2022**, *83*, 103935. [CrossRef]
23. Li, X.; Wang, Z.; Zhang, L. Co-estimation of capacity and state-of-charge for lithium-ion batteries in EV. *Energy* **2019**, *174*, 33–44. [CrossRef]
24. Hannan, M.A.; How, D.N.; Lipu, M.H.; Ker, P.J.; Dong, Z.Y.; Mansur, M.; Blaabjerg, F. SOC estimation of li-ion batteries with learning rate-optimized deep fully convolutional network. *IEEE Trans. Power Electron.* **2020**, *36*, 7349–7353. [CrossRef]

processes

MDPI

Article

Multi-Objective Optimization of a Crude Oil Hydrotreating Process with a Crude Distillation Unit Based on Bootstrap Aggregated Neural Network Models

Wissam Muhsin and Jie Zhang *

School of Engineering, Merz Court, Newcastle University, Newcastle upon Tyne NE1 7RU, UK; wissam2000muhsin@gmail.com
* Correspondence: jie.zhang@newcastle.ac.uk; Tel.: +44-191-2087240

Abstract: This paper presents the multi-objective optimization of a crude oil hydrotreating (HDT) process with a crude atmospheric distillation unit using data-driven models based on bootstrap aggregated neural networks. Hydrotreating of the whole crude oil has economic benefit compared to the conventional hydrotreating of individual oil products. In order to overcome the difficulty in developing accurate mechanistic models and the computational burden of utilizing such models in optimization, bootstrap aggregated neural networks are utilized to develop reliable data-driven models for this process. Reliable optimal process operating conditions are derived by solving a multi-objective optimization problem incorporating minimization of the widths of model prediction confidence bounds as additional objectives. The multi-objective optimization problem is solved using the goal-attainment method. The proposed method is demonstrated on the HDT of crude oil with crude distillation unit simulated using Aspen HYSYS. Validation of the optimization results using Aspen HYSYS simulation demonstrates that the proposed technique is effective.

Keywords: crude oil refining; crude oil hydrotreating; bootstrap aggregated neural networks; multi-objective optimization

Citation: Muhsin, W.; Zhang, J. Multi-Objective Optimization of a Crude Oil Hydrotreating Process with a Crude Distillation Unit Based on Bootstrap Aggregated Neural Network Models. *Processes* **2022**, *10*, 1438. https://doi.org/10.3390/pr10081438

Academic Editor: Enrique Rosales-Asensio

Received: 21 June 2022
Accepted: 20 July 2022
Published: 22 July 2022

Publisher's Note: MDPI stays neutral with regard to jurisdictional claims in published maps and institutional affiliations.

1. Introduction

Oil and gas are among the most widely utilized natural resources in modern society. Crude oil is a mixture of different hydrocarbons and small quantities of sulphur, nitrogen, and some metal elements [1]. Refined oil products provide fuels for modern transportation, such as automobiles, airplanes, and ships. In addition to being used as fuels, oil products also provide raw materials for the chemical industry in the production of a wide range of products with the most well known as various types of plastics. Crude oil needs to be split into more valuable products by distillation processes in the oil refinery. The purpose of the oil-refining process is to separate crude oil (raw material) into different types of oil products, such as light gas oil, jet fuels, light naphtha, kerosene, etc. Figure 1 shows a flowsheet of a typical crude oil refinery [2,3], which is a complex process containing many processing units, such as crude distillation units (CDU), reformer units, hydrotreating units (HDT), fluid catalytic cracking (FCC), vacuum distillation units (VDU), hydrocracking units (HCU), and others. In addition, there are some offsite facilities, such as tank farms, pipe systems, traps and depots. Refineries are large and complex processes and consume large amounts of energy and water.

Crude oil from different oil fields around the world varies in composition as well as containing undesirable impurities, such as sulphur and nitrogen. The undesirable impurities in oil products need to be removed or reduced due to strict environmental regulations for limiting sulphur and nitrogen contents in oil products. As a matter of fact, oil products with high sulphur content can lead to corrosion, pollution, and poisoning to catalyst. Thus, it is important to reduce the sulphur content in oil products. In crude oil refineries, the process for reducing sulphur compounds is known as hydrodesulphurization (HDS) while that for reducing nitrogen compounds is known as hydrodenitrogenation (HDN) [4]. The hydrodemetallization (HDM) process is responsible for removing nickel and vanadium contaminants from the heavy feed. The hydrotreating (HDT) process is employed to remove sulphur, nitrogen, and aromatic saturation compounds [5].

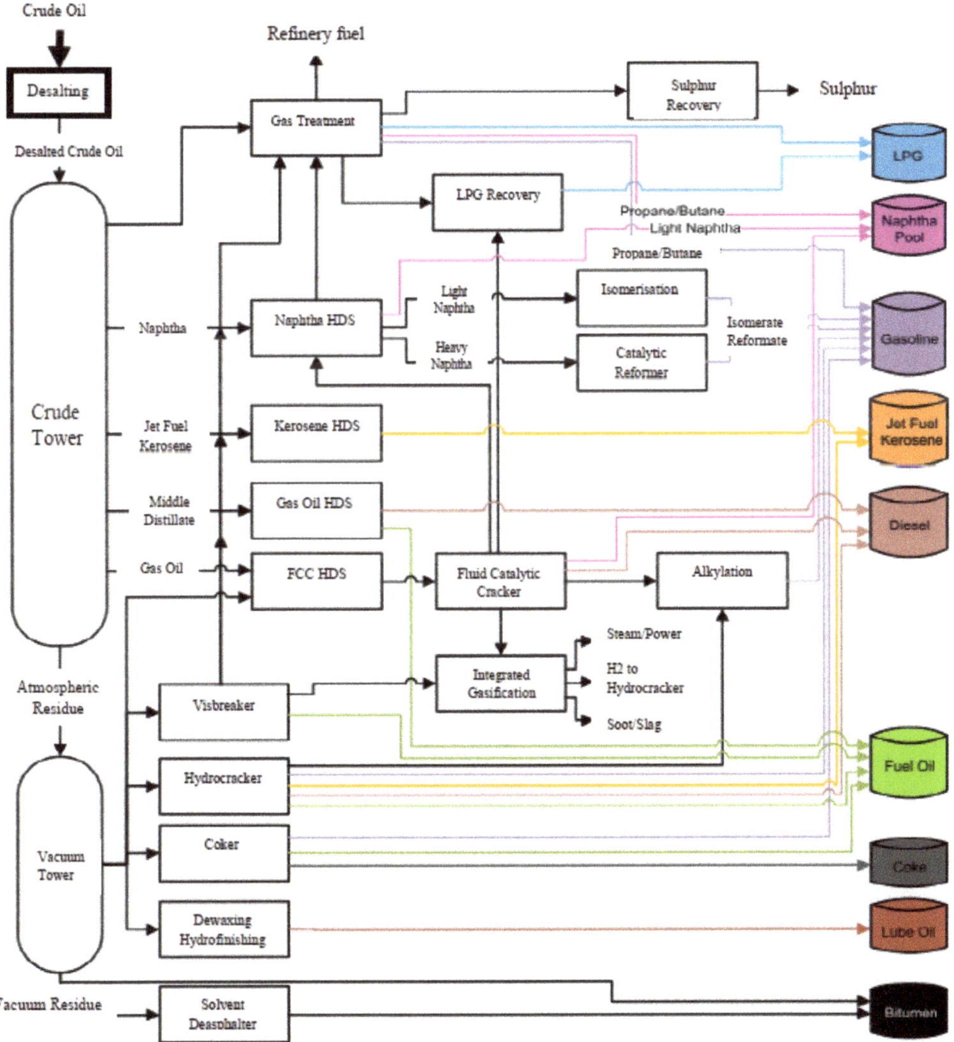

Figure 1. Flow sheet diagram of a crude oil refinery.

The conventional approach is to apply HDT on some individual oil products. In order to improve refinery efficiency, a few researchers have investigated applying HDT to the crude oil instead of individual oil products recently [6,7]. HDT of crude oil is a new hydrotreating process that has not been extensively reported and it has economic benefit compared to the conventional HDT process [8]. Jarullah et al. [6] shows that this new hydrotreating process can significantly improve middle distillate yields. Moreover, the desulphurization of whole crude oil has the potential to more effectively meet the need of environmental legislation for decreasing sulphur content and producing clean fuels in the refining processes and is expected to become more common in the future [7].

Optimal operating conditions for crude oil HDT process need to be determined in order to obtain the best process operation performance. A number of process variables can be considered as the decision variables in the optimization of crude oil HDT process. Table 1 shows the operating conditions of different HDT technologies applied to different oil products [9]. The process variables shown in Table 1, pressure (P), temperature (T), liquid hourly space velocity (LHSV), and hydrogen-to-hydrocarbon (H_2/HC) ratio can be considered as the decision variables in optimizing crude oil HDT process.

Table 1. Operating conditions of various HDT processes [9].

Hydrotreating Process	T ($^\circ$C)	P_{H2} (MPa)	LHSV (h^{-1})	H_2/Oil (Nm^3/m^3)
Naphtha	320	1–2	3–8	60
Kerosene	330	2–3	2–5	80
Gasoil	340	2.5–4	1.5–4	140
VGO	360	5–9	1–2	210
Atmospheric Residue	370–410	8–13	0.2–0.5	>525
Hydrocracking VGO	380–430	9–20	0.5–1.5	1000–2000
Vacuum Residue	400–440	12–21	0.1–0.5	1000–2000

Optimizing crude oil HDT processes will require an accurate process model. It is generally very difficult to obtain accurate mechanistic models for crude oil HDT processes due to the complexity of the material and process involved. To overcome this difficulty, data-driven models obtained from process operation data should be utilized. In recent years, there has been an increasing interest in computational intelligence, particularly in the area of machine learning, which has contributed significantly to data-driven modelling. Among the machine-learning tools, neural networks are a very powerful tool for data-driven modelling. They attempt to mimic the way in which the human brain functions. A neural network consists of a number of information-processing units named neurons which are arranged into layers. Neurons in adjacent layers are connected through weighted connections. During neural network training, the neural network weights are adjusted so that the neural network can learn the relationship between the input and output data. Advanced versions of neural networks, such as stacked neural networks, extreme learning machines, and deep learning, have also been applied to nonlinear process modelling [10–14].

In this work, bootstrap aggregated neural networks are employed to model a crude oil HDT with the CDU process and then used to optimize the process operation. Simulated process operation data are obtained from Aspen HYSYS simulation and used for developing bootstrap aggregated neural network models. Note that simulated process operation data are used here to represent real plant operation data. When real plant operation data are available, they can be directly used to build neural network models. The developed bootstrap aggregated neural network models are then utilized in a multi-objective optimization framework. In order to enhance the reliability of optimization results, minimizing the widths of model prediction confidence bounds is considered as a further optimization objective. The multi-objective optimization problem is solved by using the goal-attainment method. In some cases, even the Aspen HYSYS models are available, and their neural network surrogate models are often used in the optimization due to the short computation time of neural network models. This type of surrogate modelling approach has been getting popular in recent years [15]. A recent study on reducing the computational burden of mechanistic models using analytical simplifications could enable real-time optimization with mechanistic models in some cases [16].

The novelty of this study lies in the following two areas. First, reliable data-driven models are developed from a limited amount of simulated process operation data through utilizing bootstrap aggregated neural networks. Secondly, a reliable multi-objective optimization framework for a crude oil HDT with the CDU process is developed through utilizing the model prediction confidence bounds.

This paper is structured as follows: Section 2 provides process description for the crude oil HDT process with CDU. Also, the crude oil feedstock and products specifications are given in this section. Modelling of a crude oil HDT with the CDU process using bootstrap aggregated neural networks is detailed in Section 3. Section 4 presents multi-objective optimization of the crude oil HDT with the CDU process using the goal-attainment method. The last section gives some concluding remarks.

2. A Crude Oil HDT Process with CDU

2.1. Process Description

The process flow diagram of crude oil HDT process with CDU is illustrated in Figure 2. At the beginning, crude oil is taken from storage tanks and is joined by a stream of hydrogen gas. The mixture is fed to a train of heat-exchangers where they are pre-heated. After that, the mixture enters the convection and radiation sections of a furnace where they are heated to the required reaction temperature and then are fed to a reactor containing catalysts (HBED in Figure 2). The reactor effluent is employed to preheat the charge (crude oil) by the heat-exchanging system. Then the effluent is cooled in a cooler. Following that, the mixture of liquid and gases is fed into a high-pressure separator where gases, such as hydrogen sulphide and unreacted hydrogen, are removed from the liquid. The gases are then compressed and sent to a vessel whereas the liquid passes to a low-pressure separator to further remove gases not being removed in the high-pressure separator. After that, the hydrotreated crude oil is fed to a CDU process. Aspen HYSYS with the Peng–Robinson fluid package is employed to simulate the process. The CDU is designed to be able to cope with the maximum crude feed rate considered to prevent column flooding so that the optimization study will not be affected by the hydraulic limitation. The CDU considered here has 29 trays with 0.5 m tray spacing.

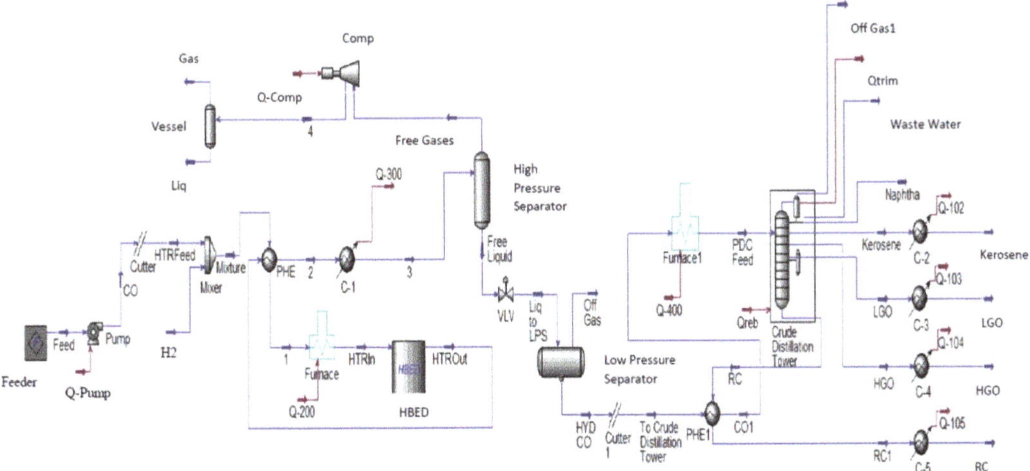

Figure 2. HDT for crude oil with CDU.

In general, the function of the crude distillation unit is to distil and split the feedstock into different types of oil products, such as off gas, naphtha (N), kerosene (K), light gas oil (LGO), heavy gas oil (HGO), and reduced crude (RC). The hydrotreated crude oil is preheated in the train of heat-exchangers and then fed to the furnace where it is finally heated to the required temperature and vaporized. The mixed liquid and vapor charge flows to the flash zone of the crude distillation tower. Liquid from the flash zone flows across many stripping trays in the bottom section of the tower. Additionally, stripping steam is injected to increase vaporization and reduce volatile content and in this way to remove lighter compounds. Vapors leaving the flash zone pass through the wash section of the tower and they are further condensed and fractionated on the trays of a fraction section with two pump-around sections to yield side-drawn products. The total naphtha leaves the column via the column top and accumulates in the overhead drum after condensation. From the overhead drum, part of the naphtha is recycled back to the tower top and the rest is pumped to the naphtha stabilizer as a product. The side products of the distillation tower (K, LGO, and HGO) flow to the stripper tower sections where they are individually steam-stripped to remove dissolved lighter components which are returned to the tower. Each side product (K, LGO, and HGO) is cooled and sent by pump to the storage tanks. Finally, RC from the tower bottom is used to preheat the hydrotreated crude oil (charge) and further cooled by the cooler and then sent to the storage tanks.

2.2. Feed and Products Specifications

Generally, there are specific chemical and physical properties for each type of crude oil. These characteristics include refinery-related specifications, such as PONA analysis, specific gravity, pour point, kinematic viscosity, and sulphur and nitrogen contents. The data involved in a petroleum assay includes yields produced from the physical distillate and residue properties [17]. The feed and product specifications are given in Table 2 [18] and Table 3, respectively. In this work, the products' specifications were taken from the Midland Refineries Company (Daura Refinery). Crude oil products and the ranges of hydrocarbons in each fraction are illustrated in Table 4 [19].

Table 2. Petroleum Essay.

No.	Property	Bulk Value
1	Sulphur By (Wt.%)	2.63
2	Std Liquid Density (kg/m^3)	867.5162
3	Watson K	11.4279
4	Pour Point (°C)	21.8696
5	Total Acid Number (mg KOH/g)	0.171
6	Kinematic Viscosity (cSt)@ 20 (°C)	13.0798
7	Kinematic Viscosity (cSt)@ 37.78 (°C)	7.7831
8	Kinematic Viscosity (cSt)@ 37.78 (°C)	7.7831
9	Kinematic Viscosity (cSt)@ 50 (°C)	5.697
10	Kinematic Viscosity (cSt)@ 60 (°C)	4.5238
11	Kinematic Viscosity (cSt)@ 80 (°C)	2.9883
12	Kinematic Viscosity (cSt)@ 100 (°C)	2.0967
13	NaCl By (Wt.%)	0.002
14	Mercaptan Sulphur By (Wt.%)	0.0217
15	Conradson Carbon By (Wt.%)	6.0699
16	Asphaltene By (Wt.%)	2.3412
17	Nickel By (Wt.%)	0.0008
18	Vanadium By (Wt.%)	0.0037
19	Iron By (Wt.%)	0.0001
20	Gross Heating Value (kJ/kg)	44,157.58
21	Net Heating Value (kJ/kg)	41,482.25
22	Cut Yield By (Wt.%)	100
23	Cut Yield By (Vol.%)	100
24	Nitrogen By (Wt.%)	0.1113

Table 2. *Cont.*

No.	Property	Bulk Value
25	Paraffins By (Vol.%)	30.5540
26	Naphthenes By (Vol.%)	40.8213
27	Arom By (Vol.%)	28.6245
28	N + 2A (%)	98.0705
29	Smoke Pt (m)	0.0156
30	Freeze Point (°C)	79.3312
31	Basic Nitrogen By (Wt.%)	0.0378
32	Cloud Point (°C)	38.6010
33	CtoH Ratio By Wt	6.6651

Table 3. Crude distillation products.

Cut Oils	Yield (Wt.%)	Specific Gravity at 15 °C	Flash Point (°C)	Color	TBP (°C)
Fuel gases	0.01	–	–	–	–
LPG	0.12	–	–	–	–
LN	8.98	0.665–0.680	–	–	35–120
HN	12.40	0.735–0.750	–	–	90–178
Ker	10.80	0.785–0.800	40 min.	30 min.	135–250
LGO	17.70	0.825–0.840	70 min.	0.5 max.	200–350
HGO	3.68	0.880–0.890	90 min.	2.5 max.	335–355
RC	46.31	0.965–0.980	120 min.	–	355+

Table 4. Crude oil hydrocarbon ranges.

Petroleum Products	Carbon Range
Fuel gases	C_1–C_2
LPG	C_3–C_4
LN and HN	C_5–C_{12}
Ker	C_{12}–C_{16}
LGO and HGO	C_{12}–C_{20}
Lubricating oil	C_{20}–C_{50}
RC	$>C_{50}$

3. Modelling of the Crude Oil HDT Process with CDU Using Bootstrap Aggregated Neural Networks

HDT process optimization should be carried out in order to enhance process efficiency. Accurate process models are essential for process optimization. Process models can be broadly divided into mechanistic models and data-driven models. The development of detailed mechanistic models is typically very time-consuming. Furthermore, optimization using mechanistic models in the form of differential and algebraic equations is also computationally demanding. In some cases, where even a mechanistic model is available, data-driven surrogate models are used in process optimization [15,20]. In building such data-driven surrogate models, detailed mechanistic models are used to simulate process operation under various operating conditions and the simulated process operation data are used in the development of data-driven surrogate models which are computationally efficient in solving process optimization problems.

The goal of HDT of crude oil with the CDU scheme is to minimize undesirable impurities, for instance, sulphur and some other compounds in the treated kerosene produced from the main atmospheric column. To overcome the difficulties in developing detailed mechanistic models, as well as using them in process optimization, neural network models are developed from process operation data. An Aspen HYSYS-based process simulator was used to produce various process operation data under different operating conditions of the HDT of crude oil with the CDU process. When sufficient plant operation data are available, then data reconciliation can be applied so that the Aspen HYSYS can match with real plant operation data [21]. Then, the data were utilized to construct neural network models. It should be noted that when real plant operation data are available, then they can be directly used in building neural network models.

As the main purpose of the developed neural network models is for process optimization, the neural network inputs and outputs should be selected so that they can be used in process optimization, i.e., they should be related to the optimization objective function and decision variables. In this work, the neural network inputs are selected as the flow rates of crude oil and hydrogen, and the pressure and temperature of the reactor. These are important process operation variables and can be measured and adjusted during process operation. The neural network outputs are selected as the contents of sulphur and nitrogen in the kerosene produced from the CDU, which will be minimized in the optimization problem. In this work, two neural network models are developed and they are represented by the following equations:

$$S = f_1(x_1, x_2, x_3, x_4) \tag{1}$$

$$N = f_2(x_1, x_2, x_3, x_4) \tag{2}$$

where S and N are the contents of sulphur and nitrogen, respectively, in the kerosene produced from the CDU, and x_1 to x_4 are, respectively, flow rates of crude oil and hydrogen, reactor pressure, and temperature.

The development of neural network models for predicting sulphur and nitrogen contents in the treated kerosene comprise four essential steps. The first step is the collection of data for model building. The second step is data normalization and data partition into training data, testing data, and unseen validation data. The third step is to select the structure of neural networks, such as the number of hidden neurons, layers, and the type of transfer functions. The fourth step is the training and validation of the neural networks.

In this study, 197 data samples are generated from the Aspen HYSYS simulation of a crude oil HDT process with CDU to develop neural network models. The data samples are generated by varying the crude oil flow rate, hydrogen flow rate, and reactor pressure and temperature within their constraints and they cover the range of inputs over which the optimization is carried out. The lower and upper bounds of these variables are given in Table 5. To represent the practical situations where process operation data are limited, a relatively small amount of simulated plant operation data are produced through simulation. To address the issue of different magnitudes in the model input and output variables, all input and output data are scaled to zero mean and unit variance before they are used in network training. In order to represent practical situations, simulated measurement noises are added to the simulated plant operation data. The simulated measurement noises follow normal distribution with zero means. In this study, the standard deviations for the measurement noises on feed flow rate, H_2 molar flow rate, reactor pressure and temperature, and sulphur and nitrogen contents are, respectively, 0.3 m^3/h, 1.5 kgmole/h, 0.5 bar, 0.3 °C, 0.003 Wt.%, and 1.5 ppmwt. Note that these measurement noises are only added to the outputs from the Aspen HYSYS simulation, not to any inputs to the Aspen HYSYS simulation. The data are split into three groups: training data (56%), testing data (23%), and unseen validation data (21%). The networks are trained on the training data. The testing data are utilized for the determination of network structure and early termination of the training process to avoid over-fitting. The final developed model is evaluated on the unseen validation data. As mentioned earlier, the simulated process operation data represent the practical situations where the available process operation data may not be abundant. Therefore, a relatively large portion of the data are used as training data. On the other hand, if the neural network models are used as surrogate models for the mechanistic model (Aspen HYSYS model), then plenty of data can be generated and a large portion of the data should be used as testing and validation data.

Table 5. The lower and upper bounds of the process operation variables.

Variables	Units	Lower Bounds	Upper Bounds
crude oil flow rate	m^3/h	40	70
hydrogen flow rate	kgmole/h	700	1000
reactor pressure	bar	70	130
reactor temperature	°C	330	380

3.1. Single Neural Network Models

In this work, single neural network models are developed first for the purpose of comparison. The networks have a single hidden layer as a single hidden layer network can approximate any continuous nonlinear function [22]. The activation function in the hidden neurons is the sigmoid whereas that in the output layer is the linear activation function. The networks are trained using the Levenberg–Marquardt training algorithm with regularization and early stopping to avoid over-fitting. During the process of network training, network errors on the testing data are continuously monitored and training is terminated when testing errors stop decreasing. The initial network weights are taken as random values uniformly distributed in the range (−0.1, 0.1), and the regularization parameter is selected as 0.1. The number of hidden neurons is determined by trying a number of neural networks with a range of hidden neurons (from 2 to 30) and examining their sum of squared errors (SSE) on the testing data. The network with the least SSE on the testing data is considered as having the appropriate number of hidden neurons. Figure 3 depicts the neural network model performance on the training, testing, and unseen validation data for modelling the contents of sulphur and nitrogen in treated kerosene from the HDT of crude oil with CDU. It can be seen that there are a few noticeable errors on the unseen validation data although model errors on training and testing data seem to be small. This indicates that single neural network models are not very accurate.

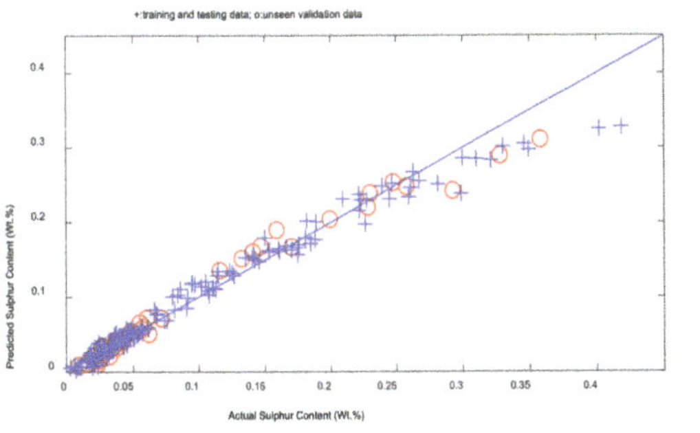

(a)

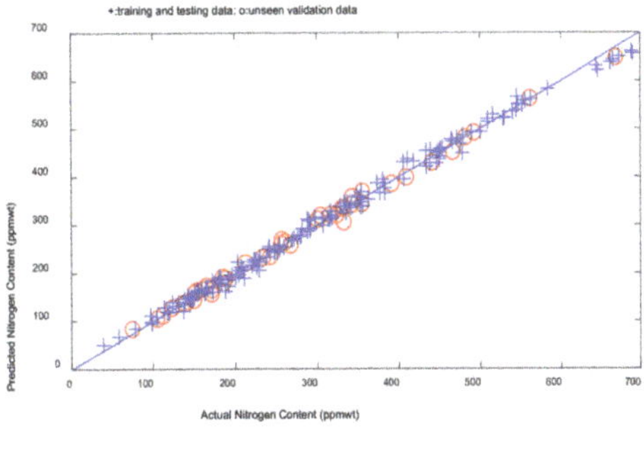

(b)

Figure 3. Network model performance on the training, testing, and validation data for sulphur content (**a**) and nitrogen content (**b**).

3.2. Bootstrap Aggregated Neural Networks

The developed neural network models to be used in the optimization of crude oil hydrotreating with CDU are required to be accurate and reliable. The drawback of a single neural network is the lack of generalization when applied to unseen validation data. In other words, a single neural network giving good performance on the training data, however, can give poor performance on the unseen validation data which is not utilized in network training [23]. Different methods have been used to improve neural network generalization, such as network training with regularization [24], Bayesian learning [25], and aggregating multiple neural networks [26–28]. It was noticed that the approach of aggregating multiple neural networks usually provides better performance than other techniques [29,30].

A simple diagram of bootstrap aggregated neural networks is illustrated in Figure 4, where several neural networks are developed to model the same relationship between model inputs and outputs and are then aggregated together [31,32].

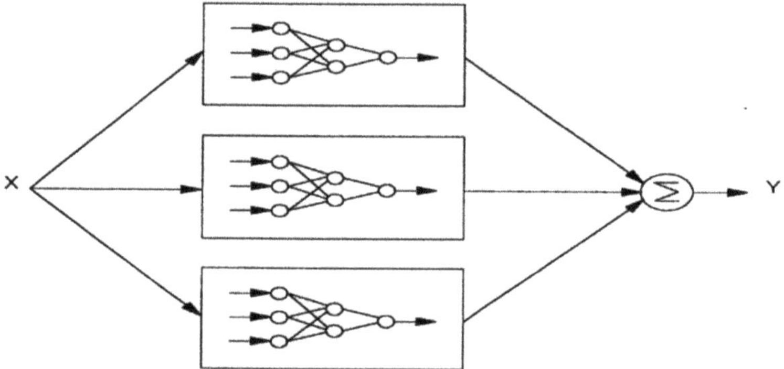

Figure 4. A bootstrap aggregated neural network.

The bootstrap aggregated neural network can be represented as:

$$f(x) = \sum_{i=1}^{n} w_i f_i(x) \tag{3}$$

where $f(x)$ is the bootstrap aggregated network model, $f_i(x)$ is the ith neural network model, w_i is the aggregating weight for the ith neural network, n is the number of networks included in the aggregated networks, and x is a vector of model inputs.

In this study, each of the developed bootstrap aggregated neural network contains 35 single hidden layer networks. The original training and testing data are put together and re-sampled via bootstrap re-sampling with replacement to generate 35 replications of the original data. Each resampled data set is then randomly partitioned into training data (70%) and testing data (30%). A single hidden layer neural network is developed on each set of resampled data. These networks are trained by utilizing the Levenberg–Marquardt training method with regularization and early stopping. The initial network weights are taken as random values uniformly distributed in the range (−0.1, 0.1), and the regularization parameter is selected as 0.1. Cross-validation is used to determine the number of hidden neurons in each individual network. In this study, 29 networks with the number of hidden neurons ranging from 2 to 30 are trained on the training data and then tested on the testing data. The network giving the smallest SSE on the testing data is considered as having the appropriate number of hidden neurons. The number of hidden neurons determined for the 35 single neural networks for sulphur and nitrogen contents are shown in Figure 5. Then, these 35 neural networks are aggregated, instead of choosing the "best" individual neural network. Finally, the combined prediction from the 35 neural networks is taken as the final model prediction [28].

Figure 6 shows the mean squared errors (MSE) of the individual networks for predicting sulphur content on the training, testing, and unseen validation data. It can be seen that the single networks provide inconsistent performance on the training, testing, and unseen validation data. For example, the MSE on the validation data by the 12th network (0.0342) is the second largest. On the other hand, the same network gives better performance (0.0200) than many of the single networks on the training and testing data. This demonstrates that the single neural network models are not reliable.

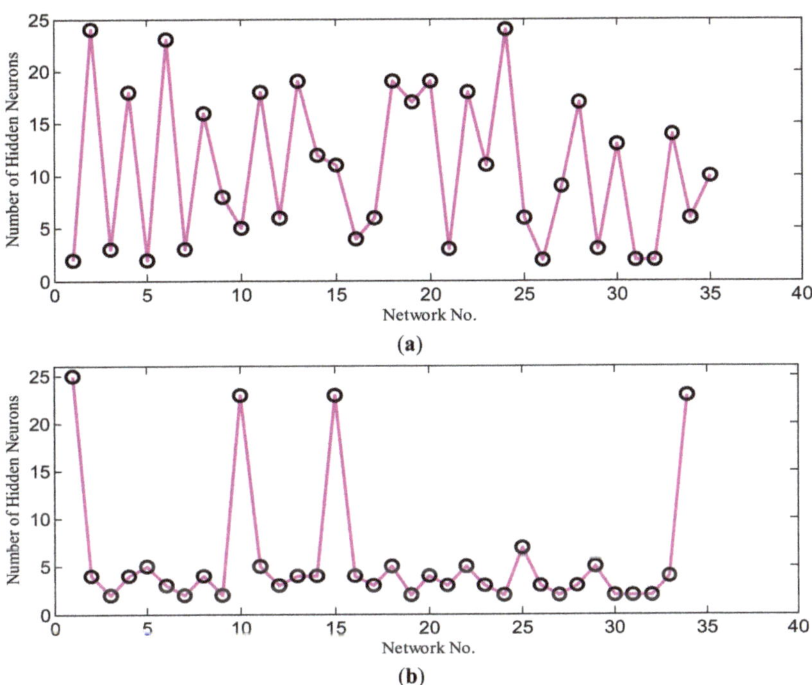

Figure 5. Number of hidden neurons of single neural networks for sulphur content (**a**) and nitrogen content (**b**).

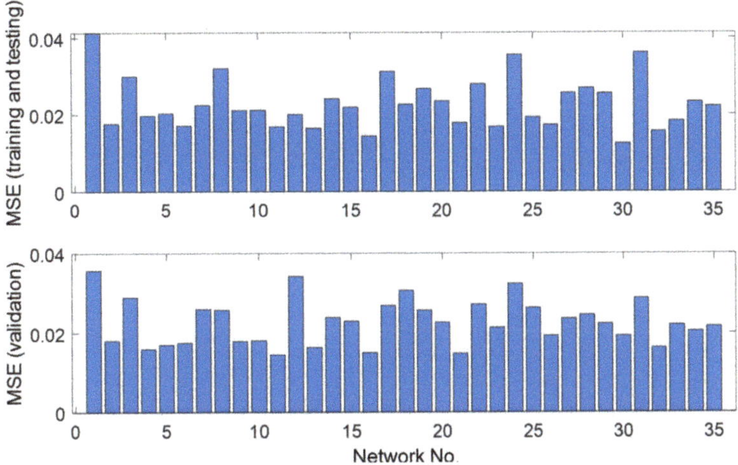

Figure 6. Errors of signal neural networks for estimating sulphur content.

Figure 7 shows the MSE of bootstrap aggregated neural networks when stacking different numbers of networks in predicting sulphur content on the training, testing, and unseen validation data. The first bar in both plots in Figure 7 is the first single neural network shown in Figure 6, the second bar represents aggregating the first two single neural networks in Figure 6, and the last bar in Figure 7 represents aggregating all the single neural network models in Figure 6. It can be seen from Figure 7 that the MSE values on the training, testing, and unseen validation data decrease with the number of networks being aggregated and then remain stable. This clearly indicates that bootstrap aggregated neural networks are more reliable and give more accurate model prediction performance than individual neural networks.

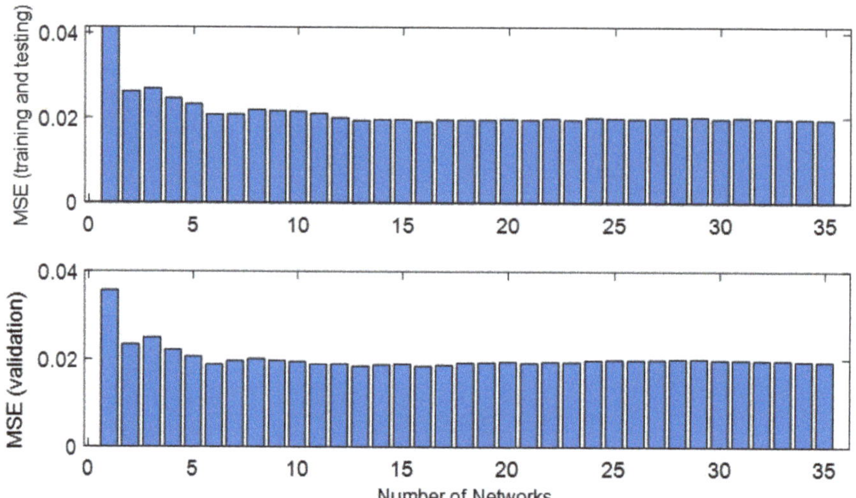

Figure 7. Errors of stacked neural networks for estimating sulphur content.

Figure 8 shows the bootstrap aggregated neural network model predictions and actual values for sulphur and nitrogen contents on the training and testing data, and the unseen validation data. The training and testing data are represented by '+', and the unseen validation data are represented by 'o'. It can be seen from both plots in Figure 8 that the model predictions correlate well with the true values for most of the samples. The bootstrap aggregated neural network model prediction performance is better than that of single neural networks shown in Figure 3. For the sulphur prediction model, there are a few samples with large errors when the sulphur content is high. This is probably due to fewer training samples in this region. However, as the optimization objective is to minimize sulphur content, the large error at the high sulphur content region has no impact on the optimization.

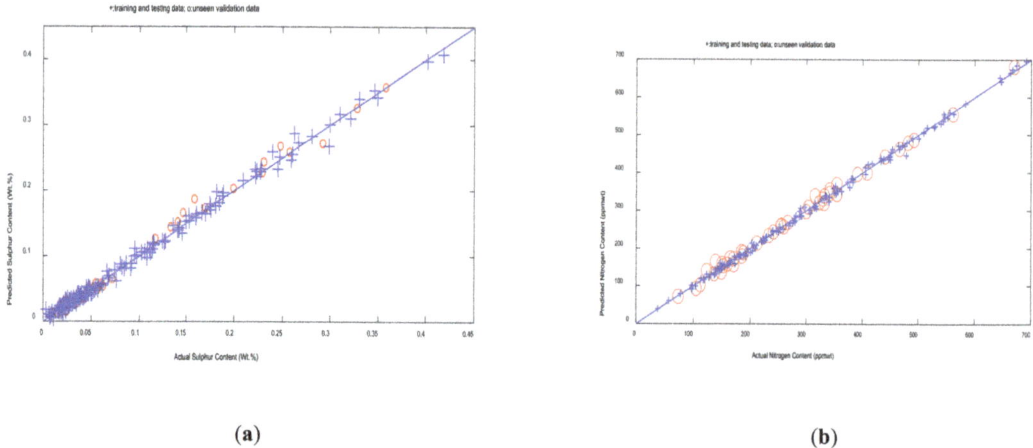

(**a**) (**b**)

Figure 8. Stacked networks prediction and real values for sulphur content (**a**) and nitrogen content (**b**).

Figure 9 shows the MSE values of single neural networks for the prediction of nitrogen content on the training, testing, and unseen validation data. It can be seen from this figure that the 10th network gives much worse performance than any other networks on all the data sets. Thus, this network is removed. Note that the deletion of this network is purely based on its very poor performance on the training and testing data. Figure 10 shows the performance of the remaining networks. The performance of individual networks on the training and testing data is not in agreement with that on the unseen validation data. It can be seen from Figure 10 that the second network gives better performance than the third network on the training and testing data. However, its performance on the unseen validation data is worse than that of the third network. This clearly shows that the single networks are not reliable.

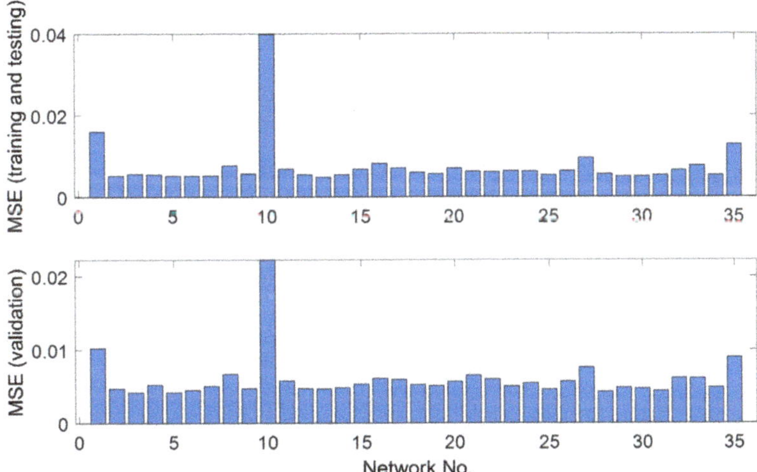

Figure 9. Errors of signal neural networks for estimating nitrogen content (35 networks).

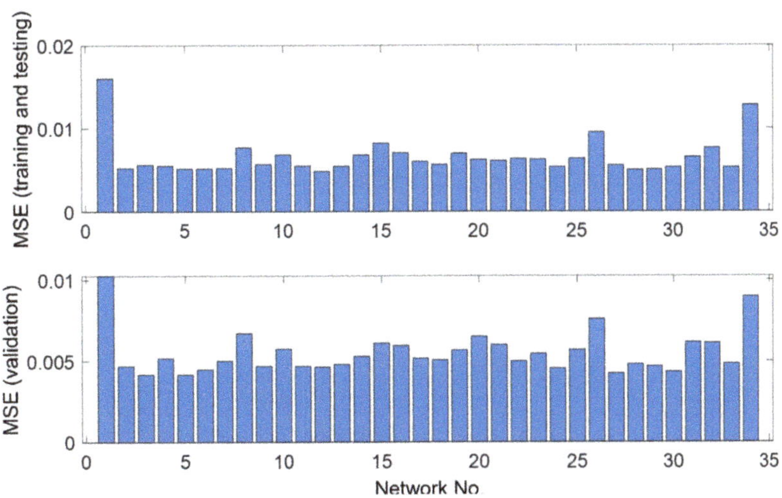

Figure 10. Errors of signal neural networks for estimating nitrogen content (34 networks).

Figure 11 shows the MSE values of predicting nitrogen content on the training, testing, and unseen validation data by aggregating several numbers of single neural networks. In Figure 11, the first bar in both plots is the first single network in Figure 10, the second bar is aggregating the first two single networks in Figure 10, and the last bar represents aggregating all the single neural network models in Figure 10. It can be seen from Figure 11 that aggregated networks produce consistent performance on the training and testing data and on the unseen validation data. The MSE values of bootstrap aggregated neural networks decrease and remain stable on the training, testing, and unseen validation data as more networks are aggregated. Furthermore, it can be observed that bootstrap aggregated neural networks are more accurate and reliable than single neural networks. Figures 7 and 11 indicate that model errors level off after combining about 10 networks. Although combining more networks does not further improve model accuracy, the estimation of model prediction confidence bounds (to be discussed in Section 3.3) would not be accurate if too few networks are used.

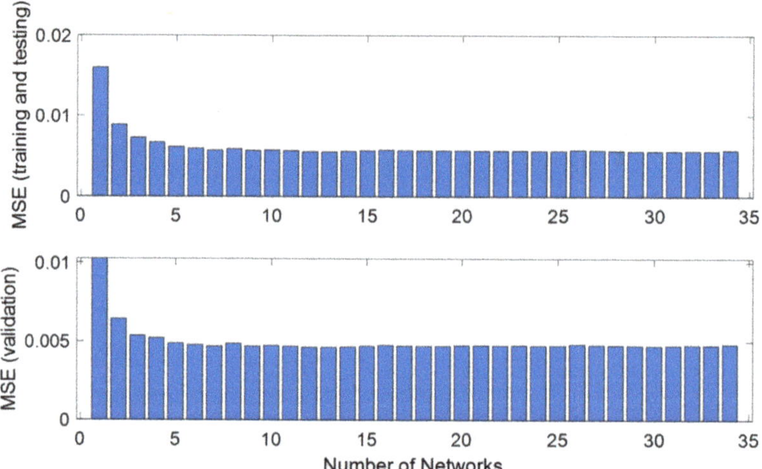

Figure 11. Errors of stacked neural networks for estimating nitrogen content.

3.3. Neural Network Model Prediction Confidence Bounds

One major advantage with the use of bootstrap aggregated neural networks is that model prediction confidence bounds can be easily estimated from the predictions of individual neural networks [31]. Confidence bounds reveal how confident the associated prediction is.

The standard error of the individual network predictions can be estimated as:

$$\sigma_e = \left\{ \frac{1}{n-1} \sum_{i=1}^{n} [f_i(x) - y(x)]^2 \right\}^{\frac{1}{2}} \tag{4}$$

where n is the number of neural networks in the aggregated neural network and $y(x) = \sum_{i=1}^{n} f_i(x)/n$. The 95% confidence bounds for the prediction corresponding to an input x is estimated as $y(x) \pm 1.96\sigma_e$. A lower σ_e, i.e., a narrower confidence bound, means that the model prediction is more reliable.

Figure 12a,b show the 95% model prediction confidence bounds for predicting sulphur and nitrogen contents on the unseen validation data by aggregated neural network models, respectively. The actual values are represented by "o", the predicted values from the aggregated network models are represented by "+", and 95% confidence intervals are represented by the green dashed lines. When the confidence bounds are tight, the reliability of the model predictions will be high. It can be seen that model predictions using bootstrap aggregated models are quite close to the real values for most of the samples. Furthermore, the confidence bounds are quite narrow for most of the samples indicating reliable model predictions. It can be concluded that the bootstrap aggregated neural network models for sulphur and nitrogen contents give very good performance.

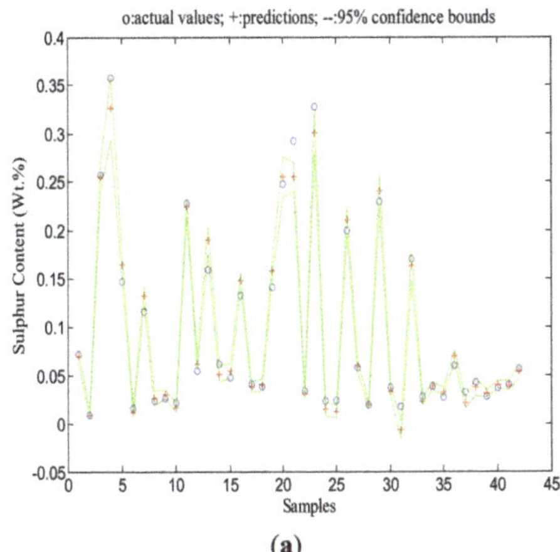
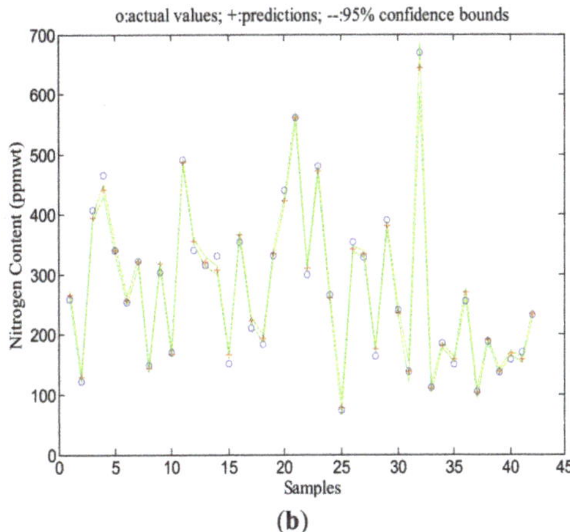

Figure 12. Stacked neural network predictions of sulphur content (**a**) and nitrogen content (**b**) on the unseen validation data.

4. Multi-Objective Optimization of the Process Using the Goal-Attainment Technique

Multi-objective optimization is a field of multiple criteria decision-making, which is concerned with mathematical optimization problems with conflicting objectives [33]. A single objective function in many cases with various constraints cannot adequately represent the multi-criteria decision-making problem, such as balancing results between profit and energy costs [34]. When the number of objectives rises, trade-offs become complicated. Multi-objective optimization includes minimizing or maximizing various objectives which are subject to a number of constraints. It is concerned with the creation of non-inferior solutions which are also named as efficient or Pareto optimum solutions [35]. According to a formal definition provided by [36], "a non-inferior solution is one in which no decrease can be obtained in any of the objectives without causing a simultaneous increase in at least one of the other objectives". A non-inferior solution is also known as Pareto front or Pareto optimal.

Some common methods for multi-objective optimization include: goal-attainment, minimax, and multi-objective genetic algorithm. In this study, the multi-objective optimization problem for the crude oil hydrotreating process with the crude distillation unit is solved using the goal-attainment method.

4.1. Goal-Attainment Method

The goal-attainment method is a powerful tool which can be used to find the best solution in a multi-objective optimization problem. In this method, the decision-maker specifies a goal for each of the objectives. This method includes a set of goals $F(x) = [F_1(x), F_2(x), F_3(x), \ldots, F_n(x)]$ which are associated with a set of objectives $Y(x) = [Y_1(x), Y_2(x), Y_3(x), \ldots, Y_n(x)]$. Also, a set of weighting factors $W(x) = [W_1(x), W_2(x), W_3(x), \ldots, W_n(x)]$ is used to control the degree of goal achievement [37]. Figure 13 shows the goal-attainment method with two objectives, Y_1 and Y_2.

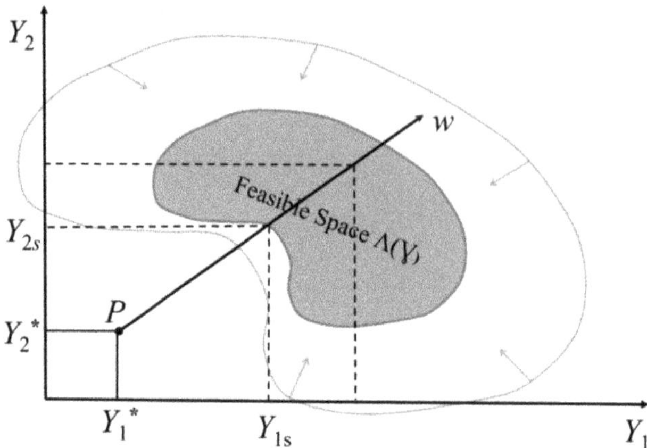

Figure 13. The goal-attainment method for a two-dimensional problem.

It can be seen from Figure 13 that the goal point P is defined by goals (Y_1^* and Y_2^*) corresponding to the two objectives Y_1 and Y_2, respectively, while the weighting factors W determines the direction of search from the goal point P to the feasible space $\Lambda(y)$. The set of nonlinear solutions can be obtained by changing W over Λ during the optimization.

In this work, the multi-objective optimization problem deals with three process operation objectives, namely, minimization of sulphur content, minimization of nitrogen content, and maximization of production rate. Four decision variables are selected and they are feedstock flow rate, H_2 molar flow rate, reactor temperature, and pressure. These four decision variables are also the neural network model inputs.

The multi-objective optimization problem considered in this paper can be represented as follows:

$$Y = \begin{bmatrix} S \\ N \\ -f \end{bmatrix} \tag{5}$$

$$\min_{x, \gamma} \gamma$$
$$s.t.$$
$$Y_i(x) - W_i\, \gamma \leq F_i \quad i = 1,\ 2,\ 3 \tag{6}$$
$$LB_i \leq x_i \leq UB_i \quad i = 1,\ 2,\ 3,\ 4$$
$$\text{Equations (1) and (2)}$$

In the above equation, Y is a vector of the objectives, S and N are, respectively, the predicted contents of sulphur and nitrogen in the kerosene produced from the CDU, f is the crude oil feed rate, $x = [x_1,\ x_2,\ x_3,\ x_4]$ is a vector of decision variables which are the neural network model inputs, LB_i and Ub_i are the lower and upper bounds for x_i respectively and are given in Table 5, W_i is the weighting factor for the ith objective, γ is a slack variable, and F_i is the desired goal for the ith objective. The three objectives in Equation (5) are minimizing the contents of sulphur and nitrogen in the kerosene product, and maximizing the refinery throughput.

Table 6 shows two cases of the multi-objective optimization results for two sets of goals. In Case 1, the goals for sulphur content, nitrogen content, and feed flow rate were selected as 0.04 Wt.%, 140.0 ppmwt, and 70 m^3/h, respectively. The weighing factors (W) were selected as 0.5, 5.0, and 0.1 for sulphur content, nitrogen content, and feed flow rate, respectively. A smaller weighting means the associated goal is more important. As can be seen from Table 6 (Case 1), all the three goals have been met according to neural network model predictions. The neural network predicted sulphur and nitrogen contents are 0.0329 Wt.% and 140.0 ppmwt, respectively. However, when the optimal process operating conditions are implemented on HYSYS simulation, the actual sulphur content decreases to 0.0300 Wt.% and the actual nitrogen content increases to 143.0 ppmwt. In Case 2, the goals for sulphur content, nitrogen content, and feed flow rate were selected as 0.03 Wt.%, 130.0 ppmwt, and 70 m^3/h, respectively. The weighing factors (W) for sulphur content, nitrogen content, and feed flow rate are kept the same as those in Case 1. It can be seen from Table 6 (Case 2) that all the three goals have been met according to neural network model predictions. The neural network predicted sulphur and nitrogen contents are 0.0292 Wt.% and 130.0 ppmwt, respectively. On the other hand, when the optimal process operating conditions are implemented on HYSYS simulation, the real sulphur and nitrogen contents increase to 0.0300 Wt.% and 134.5 ppmwt, respectively.

Table 6. Multi-objective optimization results without confidence bounds.

Case	Goals	$C_b(S)$	$C_b(N)$	W	x	Stacked Network	HYSYS	Absolute Error
1	$\begin{pmatrix} 0.04 \\ 140 \\ -70 \end{pmatrix}$	0.0177	0.0149	$\begin{pmatrix} 0.5 \\ 5 \\ 0.1 \end{pmatrix}$	$\begin{pmatrix} 70 \\ 802.66 \\ 123.89 \\ 377.68 \end{pmatrix}$	S: 0.0329 N: 140.0000	S: 0.0300 N: 143.0000	0.0029 3.0000
2	$\begin{pmatrix} 0.03 \\ 130 \\ -70 \end{pmatrix}$	0.0168	0.0171	$\begin{pmatrix} 0.5 \\ 5 \\ 0.1 \end{pmatrix}$	$\begin{pmatrix} 70 \\ 802.83 \\ 125.99 \\ 377.60 \end{pmatrix}$	S: 0.0292 N: 130.0000	S: 0.0300 N: 134.5000	0.0008 4.5000

The actual nitrogen content exceeds its goal value in both cases. This performance degradation is due to the model plant mismatch. The absolute errors shown in Table 6 are calculated as the difference between bootstrap aggregated neural network predictions and HYSYS simulation.

4.2. Reliable Multi-Objective Optimization through Incorporating Model Prediction Confidence Bounds

The reliability of optimization results is affected by the reliability of model predictions. If the model predictions are not reliable, then the optimization results based on these predictions are not likely to be reliable. Incorporating model prediction reliability in the optimization objectives could improve the reliability of optimization results. In order to improve the reliability of multi-objective optimization, minimization of the widths of model prediction confidence bounds is incorporated as additional optimization objectives. The reliable multi-objective optimization problem is given as follows:

$$Y = \begin{bmatrix} S \\ N \\ -f \\ C_b(S) \\ C_b(N) \end{bmatrix} \tag{7}$$

$$\min_{x,\gamma} \gamma$$
$$s.t.$$
$$Y_i(x) - W_i\, \gamma \leq F_i \quad i = 1,\, 2,\, 3,\, 4,\, 5 \tag{8}$$
$$LB_i \leq x_i \leq UB_i \quad i = 1,\, 2,\, 3,\, 4$$
$$\text{Equations (1) and (2)}$$

where $C_b(S)$ and $C_b(N)$ are the widths of model prediction confidence bounds for sulphur and nitrogen contents, respectively. The purpose of minimizing the width of model prediction confidence bounds is to make the model prediction more reliable leading to reliable optimization results. The goals and weights specify the relative importance of various process objectives and model prediction reliability.

Tables 7 and 8 show the optimization results and HYSYS simulation by incorporating model prediction confidence bounds in the optimization objectives for Case 1 and Case 2, respectively. As can be seen from these tables, the goals for sulphur content, nitrogen content, and feed flow rate are kept the same as those in the corresponding cases in Table 6. Two additional goals on the widths of model prediction confidence bounds for sulphur and nitrogen contents are added here.

Table 7. Multi-objective optimization results with confidence bounds (Case 1).

Run	Goals	W	x	Stacked Network	HYSYS	Absolute Error
1	$\begin{pmatrix} 0.04 \\ 140 \\ -70 \\ 0.01 \\ 0.01 \end{pmatrix}$	$\begin{pmatrix} 0.5 \\ 5.0 \\ 0.1 \\ 1.0 \\ 1.0 \end{pmatrix}$	$\begin{pmatrix} 69.9993 \\ 802.86 \\ 126.37 \\ 377.18 \end{pmatrix}$	S: 0.0294 N: 132.6498 $C_b(S)$: 0.0165 $C_b(N)$: 0.0165	S: 0.0300 N: 137.7000	0.0006 5.0502
2	$\begin{pmatrix} 0.04 \\ 140 \\ -70 \\ 0.01 \\ 0.01 \end{pmatrix}$	$\begin{pmatrix} 0.5 \\ 5.0 \\ 0.1 \\ 0.5 \\ 0.5 \end{pmatrix}$	$\begin{pmatrix} 69.9987 \\ 802.86 \\ 126.37 \\ 377.18 \end{pmatrix}$	S: 0.0294 N: 132.6510 $C_b(S)$: 0.0165 $C_b(N)$: 0.0165	S: 0.0300 N: 137.7000	0.0006 5.0490
3	$\begin{pmatrix} 0.04 \\ 140 \\ -70 \\ 0.01 \\ 0.01 \end{pmatrix}$	$\begin{pmatrix} 0.5 \\ 5.0 \\ 0.1 \\ 0.05 \\ 0.05 \end{pmatrix}$	$\begin{pmatrix} 69.9922 \\ 861.56 \\ 120.88 \\ 376.90 \end{pmatrix}$	S: 0.0322 N: 139.5789 $C_b(S)$: 0.0139 $C_b(N)$: 0.0139	S: 0.0300 N: 132.6000	0.0022 6.9789

Table 8. Multi-objective optimization results with confidence bounds (Case 2).

Run	Goals	W	x	Stacked Network	HYSYS	Absolute Error
1	$\begin{pmatrix} 0.03 \\ 130 \\ -70 \\ 0.01 \\ 0.01 \end{pmatrix}$	$\begin{pmatrix} 0.5 \\ 5.0 \\ 0.1 \\ 1.0 \\ 1.0 \end{pmatrix}$	$\begin{pmatrix} 69.9993 \\ 802.81 \\ 125.70 \\ 377.74 \end{pmatrix}$	S: 0.0294 N: 132.0352 $C_b(S)$: 0.0170 $C_b(N)$: 0.0170	S: 0.0300 N: 134.1000	0.0006 4.0648
2	$\begin{pmatrix} 0.03 \\ 130 \\ -70 \\ 0.01 \\ 0.01 \end{pmatrix}$	$\begin{pmatrix} 0.5 \\ 5.0 \\ 0.1 \\ 0.5 \\ 0.5 \end{pmatrix}$	$\begin{pmatrix} 69.9986 \\ 802.81 \\ 125.70 \\ 377.74 \end{pmatrix}$	S: 0.0294 N: 130.0704 $C_b(S)$: 0.0170 $C_b(N)$: 0.0170	S: 0.0300 N: 137.7000	0.0006 4.0296
3	$\begin{pmatrix} 0.03 \\ 130 \\ -70 \\ 0.01 \\ 0.01 \end{pmatrix}$	$\begin{pmatrix} 0.5 \\ 5.0 \\ 0.1 \\ 0.05 \\ 0.05 \end{pmatrix}$	$\begin{pmatrix} 69.9878 \\ 836.62 \\ 122.27 \\ 378.00 \end{pmatrix}$	S: 0.0304 N: 130.6100 $C_b(S)$: 0.0160 $C_b(N)$: 0.0160	S: 0.0300 N: 127.0000	0.0004 3.6100

When solving the multi-objective optimization problem, it is expected that different optimal operating policies will be obtained from different goals and weightings. It can be seen from Table 7 (Case 1) that, as the weightings on model prediction confidence bounds are further reduced (i.e., making the model reliability more important) in run 3, model prediction reliability is improved leading to much less actual nitrogen content. The nitrogen content has been reduced from 143.0 ppmwt in all runs in Table 7 (Case 1) in the real process (HYSYS simulation). Table 8 (Case 2) shows that the weightings on model prediction confidence bounds are also further reduced in run 3 and model prediction reliability is improved leading to much fewer absolute errors between the bootstrap aggregated neural network model and HYSYS model. The nitrogen content has been reduced from 134.5 ppmwt to 127.0 ppmwt in run 3 in Table 8 (Case 2) in the real process (HYSYS simulation). This reveals the improved reliability of the proposed reliable multi-objective optimization method. As can be seen from run 3 in Table 8 (Case 2), the bootstrap aggregated neural network model predicted values for sulphur and nitrogen contents are closer to the true values compared to those in Table 6 (Case 2). It can be concluded that run 3 in Table 8 (Case 2) can be selected as the best optimum case with confidence bounds.

Table 9 compares the base case and the optimum cases of the operating conditions for HDT of crude oil with CDU. The optimal case 1 is run 3 in Table 7, while the optimal case 2 is run 3 in Table 8. It can be seen that the crude oil charge is increased significantly from 55 m^3/h in the base case to about 70 m^3/h in the optimum cases 1 and 2. Optimal case 2 has slightly higher sulphur and nitrogen removal than optimal case 1.

Table 9. Comparison of the base case and the optimum cases.

Cases	Feed (m^3/h)	H$_2$ Molar Flow (kgmole/h)	Pressure (bar)	Temperature (°C)	S Removal (Wt.%)	N Removal (Wt.%)
Base	55.00	800.00	90.00	375.00	85.32	88.08
Optimum 1	69.65	865.01	120.78	376.42	88.63	88.18
Optimum 2	69.99	836.62	122.27	378.00	88.64	88.63

5. Conclusions

Modelling and multi-objective optimization of a crude oil hydrotreating process with a crude distillation unit using bootstrap aggregated neural networks is studied in this paper. Hydrotreating of whole crude oil is a new process with economic advantages compared to hydrotreating of individual oil products. Bootstrap aggregated neural networks are employed in this work to improve the reliability and accuracy of data-driven non-linear models. Bootstrap aggregated neural networks can also provide model prediction confidence bounds based on the individual neural network predictions. Minimization of the widths of model prediction confidence bounds is incorporated as additional optimization objectives. It is shown that reliable optimization results are obtained by incorporating model prediction confidence in the optimization objectives. The modelling and optimization results are validated using Aspen HYSYS simulation.

Author Contributions: Conceptualization, W.M. and J.Z.; methodology, W.M. and J.Z.; software, W.M. and J.Z.; validation, W.M.; formal analysis, W.M.; investigation, W.M.; resources, J.Z.; data curation, W.M. and J.Z.; writing—original draft preparation, W.M.; writing—review and editing, J.Z.; visualization, W.M.; supervision, J.Z.; and project administration, J.Z. All authors have read and agreed to the published version of the manuscript.

Funding: This research received no external funding.

Institutional Review Board Statement: Not applicable.

Informed Consent Statement: Not applicable.

Data Availability Statement: Not applicable.

Conflicts of Interest: The authors declare no conflict of interest.

References

1. Speight, J.G. *The Chemistry and Technology of Petroleum*, 5th ed.; CRC Press, Taylor & Francis Group: Boca Raton, FL, USA, 2014.
2. Gary, J.H.; Kaiser, M.J. *Petroleum Refining: Technology and Economics*, 5th ed.; Taylor & Francis: Boca Raton, FL, USA, 2007.
3. Orszulik, S.T. *Environmental Technology in the Oil Industry*, 3rd ed.; Springer: Cham, Switzerland, 2016.
4. Muhsin, W.A.S.; Zhang, J.; Lee, J. Modelling and optimisation of a crude oil hydrotreating process using neural networks. *Chem. Eng. Trans.* **2016**, *52*, 211–216.

5. Rodriguez, M.A.; Ancheyta, J. Modeling of hydrodesulfurization (HDS), hydrodenitrogenation (HDN), and the hydrogenation of aromatics (HDA) in a vacuum gas oil hydrotreater. *Energy Fuels* **2004**, *18*, 789–794. [CrossRef]
6. Jarullah, A.T.; Mujtaba, I.M.; Wood, A.S. Kinetic parameter estimation and simulation of trickle-bed reactor for hydrodesulfurization of crude oil. *Chem. Eng. Sci.* **2011**, *66*, 859–871. [CrossRef]
7. Jarullah, A.T.; Mujtaba, I.M.; Wood, A.S. Whole crude oil hydrotreating from small-scale laboratory pilot plant to large-scale trickle-bed reactor: Analysis of operational issues through modeling. *Energy Fuels* **2012**, *26*, 629–641. [CrossRef]
8. Muhsin, W.; Zhang, J. Modelling and optimal operation of a crude oil hydrotreating process with atmospheric distillation unit utilising stacked neural networks. In *Computer Aided Chemical Engineering*; Antonio Espuña, M.G., Luis, P., Eds.; Elsevier: Amsterdam, The Netherlands, 2017; pp. 2479–2484.
9. Ancheyta, J. *Modeling of Processes and Reactors for Upgrading of Heavy Petroleum*; CRC Press, Taylor & Francis Group: Boca Raton, FL, USA, 2013.
10. Shang, C.; Yang, F.; Huang, D.; Lyu, W. Data-driven soft sensor development based on deep learning technique. *J. Process Control* **2014**, *24*, 223–233. [CrossRef]
11. Chang, H.; Su, Z.; Lu, S.; Zhang, G. Application of deep learning network in bumper warpage quality improvement. *Processes* **2022**, *10*, 1006. [CrossRef]
12. Li, F.; Zhang, J.; Shang, C.; Huang, D.; Oko, E.; Wang, M. Modelling of a post-combustion CO_2 capture process using deep belief network. *Appl. Therm. Eng.* **2018**, *130*, 997–1003. [CrossRef]
13. Zhu, C.; Zhang, J. Developing soft sensors for polymer melt index in an industrial polymerization process using deep belief networks. *Int. J. Autom. Comput.* **2020**, *17*, 44–54. [CrossRef]
14. Chen, B.; Huang, P.; Zhou, J.; Li, M. An enhanced stacking ensemble method for granule moisture prediction in fluidized bed granulation. *Processes* **2022**, *10*, 725. [CrossRef]
15. Ibrahim, D.; Jobson, M.; Li, J.; Guillén-Gosálbez, G. Optimization-based design of crude oil distillation units using surrogate column models and a support vector machine. *Chem. Eng. Res. Des.* **2018**, *134*, 212–225. [CrossRef]
16. Brambilla, A.; Vaccari, M.; Pannocchia, G. Analytical RTO for a critical distillation process based on offline rigorous simulation. In Proceedings of the 13th IFAC Symposium on Dynamics and Control of Process Systems, Including Biosystems (DYCOPS), Busan, Korea, 14–17 June 2022.
17. Aspen Technology. *Assay Management in Aspen HYSYS®Petroleum Refining*; Aspen Technology: Houston, TX, USA, 2015.
18. Aspen Technology. *Kirkuk Crude Oil Assay*; Assay Library: Houston, TX, USA, 2011.
19. Fahim, M.A.; Alsahhaf, T.A.; Elkilani, A. Refinery feedstocks and products. In *Fundamentals of Petroleum Refining*; Elsevier: Amsterdam, The Netherlands, 2010; pp. 11–31.
20. Thrampoulidis, E.; Mavromatidis, G.; Lucchi, A.; Orehounig, K. A machine learning-based surrogate model to approximate optimal building retrofit solutions. *Appl. Energy* **2021**, *281*, 116024. [CrossRef]
21. Vaccari, M.; Pannocchia, G.; Tognotti, L.; Paci, M.; Bonciani, R. A rigorous simulation model of geothermal power plants for emission control. *Appl. Energy* **2020**, *263*, 114563. [CrossRef]
22. Cybenko, G. Approximation by superpositions of a sigmoidal function. *Math. Control Signals Syst.* **1989**, *2*, 303–314. [CrossRef]
23. Herrera, F.; Zhang, J. Optimal control of batch processes using particle swam optimisation with stacked neural network models. *Comput. Chem. Eng.* **2009**, *33*, 1593–1601. [CrossRef]
24. Bishop, C. Improving the generalization properties of radial basis function neural networks. *Neural Comput.* **1991**, *3*, 579–588. [CrossRef] [PubMed]
25. MacKay, D.J.C. Bayesian interpolation. *Neural Comput.* **1992**, *4*, 415–447. [CrossRef]
26. Wolpert, D.H. Stacked generalization. *Neural Netw.* **1992**, *5*, 241–259. [CrossRef]
27. Sridhar, D.V.; Seagrave, R.C.; Bartlett, E.B. Process modeling using stacked neural networks. *AIChE J.* **1996**, *42*, 2529–2539. [CrossRef]
28. Zhang, J.; Martin, E.B.; Morris, A.J.; Kiparissides, C. Inferential estimation of polymer quality using stacked neural networks. *Comput. Chem. Eng.* **1997**, *21*, S1025–S1030. [CrossRef]
29. Zhang, J. Batch-to-batch optimal control of a batch polymerisation process based on stacked neural network models. *Chem. Eng. Sci.* **2008**, *63*, 1273–1281. [CrossRef]
30. Khaouane, L.; Ammi, Y.; Hanini, S. Modeling the retention of organic compounds by nanofiltration and reverse osmosis membranes using bootstrap aggregated neural networks. *Arab. J. Sci. Eng.* **2017**, *42*, 1443–1453. [CrossRef]
31. Zhang, J. Developing robust non-linear models through bootstrap aggregated neural networks. *Neurocomputing* **1999**, *25*, 93–113. [CrossRef]
32. Li, F.; Zhang, J.; Oko, E.; Wang, M. Modelling of a post-combustion CO_2 capture process using extreme learning machine. *Int. J. Coal Sci. Technol.* **2017**, *4*, 33–40. [CrossRef]
33. Koziel, S.; Bekasiewicz, A. *Multi-Objective Design of Antennas Using Surrogate Models*; World Scientific: Singapore, 2016.
34. Vaccari, M.; Capaci, R.B.; Brunazzi, E.; Tognotti, L.; Pierno, P.; Vagheggi, R.; Pannocchia, G. Optimally Managing Chemical Plant Operations: An Example Oriented by Industry 4.0 Paradigms. *Ind. Eng. Chem. Res.* **2021**, *60*, 7853–7867. [CrossRef]
35. Miettinen, K. *Nonlinear Multiobjective Optimization*; Kluwer Academic Publishers: Boston, MA, USA, 1998.

36. Haimes, Y.Y.; Hall, W.A.; Freedman, H.T. *Multiobjective Optimization in Water Resources Systems: The Surrogate Worth Trade-off Method*; Elsevier: Amsterdam, The Netherlands, 2011.
37. Osuolale, F.N.; Zhang, J. Multi-objective optimisation of atmospheric crude distillation system operations based on bootstrap aggregated neural network models. *Comput. Aided Chem. Eng.* **2015**, *37*, 671–676.

Article

Event-Triggered Filtering for Delayed Markov Jump Nonlinear Systems with Unknown Probabilities

Huiying Chen [1,*], Renwei Liu [1], Weifeng Xia [1] and Zuxin Li [2]

[1] School of Engineering, Huzhou University, Huzhou 313000, China; liurenwei1128@163.com (R.L.); xwf212@163.com (W.X.)

[2] School of Science and Engineering, Huzhou College, Huzhou 313000, China; lzx@zjhu.edu.cn

* Correspondence: hychen@zjhu.edu.cn

Abstract: This paper focuses on the problem of event-triggered H_∞ asynchronous filtering for Markov jump nonlinear systems with varying delay and unknown probabilities. An event-triggered scheduling scheme is adopted to decrease the transmission rate of measured outputs. The devised filter is mode dependent and asynchronous with the original system, which is represented by a hidden Markov model (HMM). Both the probability information involved in the original system and the filter are assumed to be only partly available. Under this framework, via employing the Lyapunov–Krasovskii functional and matrix inequality transformation techniques, a sufficient condition is given and the filter is further devised to ensure that the resulting filtering error dynamic system is stochastically stable with a desired H_∞ disturbance attenuation performance. Lastly, the validity of the presented filter design scheme is verified through a numerical example.

Keywords: event-triggered scheduling; Markov jump nonlinear systems(MJNSs); error threshold; partly unknown probabilities; asynchronous filtering

Citation: Chen, H.; Liu, R.; Xia, W.; Li, Z. Event-Triggered Filtering for Delayed Markov Jump Nonlinear Systems with Unknown Probabilities. *Processes* 2022, *10*, 769. https://doi.org/10.3390/pr10040769

Academic Editors: Jie Zhang and Meihong Wang

Received: 1 March 2022
Accepted: 11 April 2022
Published: 14 April 2022

Publisher's Note: MDPI stays neutral with regard to jurisdictional claims in published maps and institutional affiliations.

1. Introduction

Markov jump systems (MJSs), as a kind of significant hybrid stochastic systems, have attracted immense attention in recent decades owing to their wide range of applications in aerospace, electric power systems, communication, economic, traffic and other areas [1–4]. Scholars have put a lot of effort into research on MJSs since they were first proposed by Krasovskii and Lidskii [5] in the 1960s, and many results for MJSs have been released in the literature (see [6–13] and the references therein). Additionally, it is a fact that the nonlinearity in MJSs, which makes the system more complex, is ubiquitous in many real-world applications. Therefore, the research on Markov jump nonlinear systems (MJNSs) has great theoretical significance and practical application value and has been widely examined [14–18]. Among this research, neural network (NN) [16–18] is one of the most popular approaches to deal with nonlinearity. For instance, the exponential stability problem was discussed for multiple-delayed Markov jump NNs (MJNNs) in [16].

Moreover, the filtering or estimation is a very essential issue in the field of cybernetics and has received strong interest from scholars [19], mainly for the reason that it is often a difficult job to obtain the accurate values of system states in engineering practice, and thus, a high-quality filter is essential for state estimation. The problem of filtering or estimation for MJSs has been investigated in [12–14,20–25]. To mention a few such studies, the H_∞ filtering and the dissipative asynchronous filtering for periodic MJSs were investigated in [12,13], respectively. The state estimation problem for a class of MJNSs was explored in [14], which put forward a moving horizon estimation algorithm, and the optimal estimate was obtained by minimizing a quadratic estimation cost function.

On the other hand, due to the increasing complexity of networks, communication constraint is also a serious problem for networked control systems (NCSs), which have been extensively used in real systems in recent years [26]. To the best of our knowledge, the event-triggered (ET) scheduling scheme is a useful and emerging approach to deal with this trouble and has become one of the current research hotspots [27–30]. Event-triggered scheduling means that the information transmission of nodes in the system determine whether to execute or not according to the preset event-triggered conditions. Based on this scheme, the measured outputs are transmitted only when the ET condition holds. Compared with the traditional periodic transmission scheme, it has the advantages of reducing redundant communication and saving energy, and so on. In recent years, some useful results for MJSs on this topic have been reported [20–22,31–33]. Specially, the filtering or estimation problem was addressed in [20–22,33]. For example, the event-based state estimation problem was explored for MJSs considering quantization and stochastic nonlinearity simultaneously in [20], in which both the ET and quantization schemes were introduced into the model of MJSs, then an estimator was devised to ensure that the filtering error system was randomly bounded and satisfied a desired H_∞ performance. The event-based dissipative filtering issue was studied for delayed MJSs [22], by means of the Lyapunov and Wirtinger inequality techniques, the stochastic stability with strict dissipativity of the error system and the co-design design scheme of the ET matrices and filter parameters were presented.

In NCSs, the plant, filter or controller are always geographically scattered and connected through communication network, which will inevitably cause some issues, e.g., network-induced delay and data dropout, and lead to incomplete data transmission among different nodes, thus causing asynchronous problems in MJSs [25]. However, in many of the existing works, this problem is ignored by assuming that the filter/controller is mode independent [34,35] or synchronous [36,37] with the original system. Mode independence implies that there is no use of available mode information, which will bring about more conservatism, and the assumption of synchronization means that the modes of the filter/controller are completely consistent with those of the plant, which is too rigorous. Due to realizing the irrationality of these assumptions, recently, scholars have paid increasing attention to the investigation of asynchronous techniques [23–25,38]. In [23,24], the asynchronous phenomenon was described as a piecewise homogeneous Markov process. In [13,25,38], a hidden Markov model (HMM) was proposed to address the asynchronous issue, which related the filter/controller to the plant with a conditional probability matrix (CPM). Based on this, the asynchronous filtering problem for MJNNs in [25] and the asynchronous control problem for MJSs in [38] were investigated, respectively.

Based on the foregoing discussions, we know that some important results have been released about ET scheduling schemes or asynchronous techniques in MJSs. These works have important theoretical and practical significance. Nevertheless, there are few results concerning the ET asynchronous filtering/control for MJSs or MJNNs, which is one of the motivations for our work. In addition, it should be noted that the above asynchronous results are restricted due to the assumption that both the probability information of the original system and the filter are considered to be fully accessible. However, it is difficult or costly to fulfill in many engineering applications. However, some results on the partly unknown transition probabilities (TPs) case [39,40] have been reported recently, in which it is assumed that the modes of the plant and the filter/controller are synchronous, and the unknown entries only exist in the transition probability matrix (TPM), so they are not suitable for asynchronous cases. For instance, in the HMM-based asynchronous case, there may be unknown entries in both TPM and CPM, which is more complex and challenging. This is another motivation for our work.

This paper will concentrate on the issue of ET asynchronous filter design for discrete-time MJNSs based on a NN model with varying delay and unknown probabilities. An ET scheduling scheme is introduced to decrease the transmission rate of measured outputs. The modes of the devised filter are dependent on and asynchronous with those of the original system, represented by an HMM. It is assumed that both the probability information involved in the original system and the filter are only partly available. By utilizing the Lyapunov–Krasovskii functional (LKF) and matrix inequality transformation techniques, an asynchronous filter is devised to ensure the stochastic stability and a desired H_∞ performance of the error system. The slack matrix technique and Projection lemma are introduced to facilitate the filter design. Lastly, a numerical example is offered to demonstrate the validity of the obtained results. The major contributions of this work are stated as follows:

(1) A more practical scenario is considered, which includes not only the varying delay, partly unknown probabilities and nonlinearity of the original system, but also the network-induced communication constraint and asynchronous problem.

(2) The ET asynchronous filtering problem based on HMM is first explored for discrete-time delayed MJNNs, in which both the TPM of the original system and the CPM of the filter are assumed to be only partly accessible.

(3) The filtering scheme proposed in this paper has strong versatility since the asynchronous strategy based on HMM contains two special cases: mode independence and synchronization, and the case with partly unknown probabilities considered in this paper covers both fully known and fully unknown cases.

2. Preliminaries

In this work, the physical plant, which is a discrete-time MJNN with varying delay, is addressed as below:

$$\mathbb{S}_0 : \begin{cases} x(k+1) = A(\alpha_k)x(k) + A_d(\alpha_k)x(k-d(k)) \\ \quad + E(\alpha_k)g(x(k)) + E_d(\alpha_k)g(x(k-d(k))) \\ \quad + B(\alpha_k)w(k) \\ y(k) = C_1(\alpha_k)x(k) + C_d(\alpha_k)x(k-d(k)) \\ \quad + D_1(\alpha_k)w(k) \\ z(k) = C_2(\alpha_k)x(k) + D_2(\alpha_k)w(k) \\ x(k_0) = \chi(k_0), k_0 = -\tau_2, -\tau_2+1, \cdots, -1, 0 \end{cases} \tag{1}$$

where $x(k) \in R^n$ is the system state with the initial value $\chi(k_0)$, $y(k) \in R^p$ is the output signal, $z(k) \in R^q$ is the target value to be estimated, and $w(k) \in R^r$ is referring to the disturbance with $w(k) \in l_2[0, \infty)$. $g(x(k)) \in R^n$ denotes a nonlinear function. $d(k) \in N^+$ means the system delay with lower bound τ_1 and upper bound τ_2. $A(\alpha_k)$, $A_d(\alpha_k)$, $E(\alpha_k)$, $E_d(\alpha_k)$, $B(\alpha_k)$, $C_1(\alpha_k)$, $C_d(\alpha_k)$, $D_1(\alpha_k)$, $C_2(\alpha_k)$ and $D_2(\alpha_k)$ are known constant matrices with proper dimensions. α_k refers to a Markov chain which regulates the jumps of system($\mathbb{S}_0$) in a set of modes $\mathscr{S}_1 = \{1, 2, \cdots, s_1\}$ with a TPM $\Phi = \{\phi_{ij}\}$, and its TP ϕ_{ij} is defined as

$$\Pr\{\alpha_{k+1} = j | \alpha_k = i\} = \phi_{ij} \tag{2}$$

in which $\phi_{ij} \geq 0$ and $\sum_{j=1}^{s_1} \phi_{ij} = 1$ for $\forall i, j \in \mathscr{S}_1$.

Next, a filter will be devised for estimating $z(k)$ according to measured outputs. Nevertheless, due to the introduction of an ET scheduler, the output signal will be transmitted only when the ET condition holds(see Figure 1). While the deviation between the current measured output and the last transmission signal is bigger than its relative error, the output signal will be transmitted (i.e., $\rho(k) = 1$), otherwise it will not be transmitted (i.e., $\rho(k) = 0$).

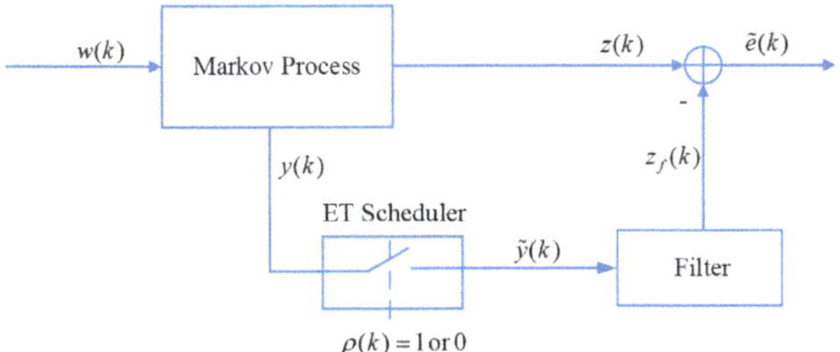

Figure 1. Block diagram of ET asynchronous filtering

Therefore, at the sampling instant k, if the ET condition holds, the filter will receive the latest measured output, otherwise it will keep the last transmission value by zero order holder (ZOH). Based on this scheme, the input of the filter during the period k is addressed as:

$$\tilde{y}_i(k) = \begin{cases} y_i(k) & |\tilde{y}_i(k-1) - y_i(k)| > \delta_i|y_i(k)| \\ \tilde{y}_i(k-1) & |\tilde{y}_i(k-1) - y_i(k)| \leq \delta_i|y_i(k)| \end{cases} \tag{3}$$

where $i = 1, 2, \cdots, p$; $\delta_i \in [0, 1]$ is the error threshold.

Setting $H(k) - diag\{\nabla_1(k), \nabla_2(k), \cdots, \nabla_p(k)\}$, $\nabla_i(k) \in [-\delta_i, \delta_i]$, $i = 1, 2, \cdots, p$, then in accordance with (3) , we can obtain

$$\tilde{y}(k) = (I + H(k))y(k) \tag{4}$$

Remark 1. *Thanks to the introduction of the ET scheduler into MJNSs, the measured outputs need not be transmitted in each sampling period, thus achieving the aim of reducing the data transmission rate. In the following, we introduce a communication performance index of $MTR = \bar{n}_{sent}/n_{total}$, which denotes the mean transmission rate ($\bar{n}_{sent}$ and n_{total} denote the average number of measured output $y(k)$ transmitted with and without the ET scheduler in the simulation time, respectively.). The smaller MTR means better communication performance.*

Based on the ET outputs (4), we will adopt a mode-dependent filter to estimate $z(k)$:

$$\mathbb{S}_f : \begin{cases} x_f(k+1) = A_f(\beta_k)x(k) + B_f(\beta_k)\tilde{y}(k) \\ z_f(k) = C_f(\beta_k)x(k) + D_f(\beta_k)\tilde{y}(k) \end{cases} \tag{5}$$

where $x_f(k) \in R^n$ refers to the filter state, $z_f(k) \in R^q$ denotes the estimated value of $z(k)$. $A_f(\beta_k)$, $B_f(\beta_k)$, $C_f(\beta_k)$ and $D_f(\beta_k)$ are parameters of the filter to be obtained, which are dependent on the filter mode β_k, $\beta_k \in \mathscr{S}_2 = \{1, 2, \cdots, s_2\}$.

In this paper, filter($\mathbb{S}_f$) is mode dependent, and its mode β_k is influenced by the mode α_k of system($\mathbb{S}_0$) via a CPM $\Omega = \{\sigma_{im}\}$, where the conditional probability(CP) σ_{im} is given by

$$\Pr\{\beta_k = m | \alpha_k = i\} = \sigma_{im} \tag{6}$$

which denotes the probability that filter($\mathbb{S}_f$) is in the m-th mode while the plant works in the i-th mode. Obviously, $\sigma_{im} \geq 0$ and $\sum_{m=1}^{s_2} \sigma_{im} = 1$ for $\forall i \in \mathscr{S}_1, m \in \mathscr{S}_2$.

Remark 2. *Notice that the devised filter acts asynchronously with the original system as their jumping processes are controlled by different Markov parameters, β_k and α_k, respectively. However, the parameter β_k is affected by α_k through the CP (6). Thus, the set $(\alpha_k, \beta_k, \Phi, \Omega)$ is addressed as an HMM, linking filter($\mathbb{S}_f$) and system($\mathbb{S}_0$) tightly with a CPM which can reflect the asynchronous degree between them. We should mention that the devised asynchronous filter under this scheme is more general because it includes the synchronous and mode-independent cases [38].*

Considering the complexity of practical systems, in this paper, we assume that the entries of TPM Φ and CPM Ω are partly inaccessible; namely, Φ and Ω may take the forms as follows:

$$\Phi = \begin{bmatrix} \phi_{11} & ? & ? \\ ? & ? & \phi_{23} \\ ? & \phi_{32} & ? \end{bmatrix}, \Omega = \begin{bmatrix} \sigma_{11} & ? & ? \\ ? & ? & \sigma_{23} \\ ? & \sigma_{32} & ? \end{bmatrix} \tag{7}$$

in which "?" refers to the unknown elements. For $\forall i \in \mathscr{S}_1$, define $\mathscr{S}_1 = \mathscr{S}_{1K}^i + \mathscr{S}_{1U}^i$ and $\mathscr{S}_2 = \mathscr{S}_{2K}^i + \mathscr{S}_{2U}^i$, where

$$\begin{cases} \mathscr{S}_{1K}^i = \{j : \phi_{ij} \text{ is known}\} \\ \mathscr{S}_{1U}^i = \{j : \phi_{ij} \text{ is unknown}\} \\ \mathscr{S}_{2K}^i = \{m : \sigma_{im} \text{ is known}\} \\ \mathscr{S}_{2U}^i = \{m : \sigma_{im} \text{ is unknown}\} \end{cases} \tag{8}$$

Remark 3. *In recent years, there have been some research results on the HMM-based asynchronous filtering/control of MJSs, e.g., [25,38], in which all TPs in TPM and CPs in CPM are assumed to be known. Nevertheless, it is very arduous or costly to obtain all the information about TPM or CPM. Hence, a more complex and challenging case where both TPM and CPM are only partly accessible will be explored in this paper. It is worth pointing out that our result under this framework is more general because it contains two special cases: (1) the fully known case, i.e., $\mathscr{S}_{1U}^i = \varnothing$ or $\mathscr{S}_{2U}^i = \varnothing$, which is the most studied case at present; (2) the fully unknown case, i.e., $\mathscr{S}_{1K}^i = \varnothing$ or $\mathscr{S}_{2K}^i = \varnothing$.*

For brevity of notation, in the following, parameters α_k, α_{k+1} and β_k are simplified to i, j and m of the subscript, for example, $A(\alpha_k) \triangleq A_i$, $A_f(\beta_k) \triangleq A_{fm}$.

Selecting the augmented vector $\tilde{x}(k) = [\ x^T(k)\ \ x_f^T(k)\ ]^T$ and the estimated error $\tilde{e}(k) = z(k) - z_f(k)$, and synthesizing (1), (4) and (5), we derive the filtering error dynamic system as follows:

$$\mathbb{S}_e : \begin{cases} \tilde{x}(k+1) = \tilde{A}_{im}\tilde{x}(k) + \tilde{A}_{dim}x(k-d(k)) \\ \qquad\qquad + \tilde{B}_{im}w(k) + \bar{I}[E_i g(x(k)) \\ \qquad\qquad + E_{di}g(x(k-d(k)))] \\ \tilde{e}(k) = \tilde{C}_{im}\tilde{x}(k) + \tilde{C}_{dim}x(k-d(k)) \\ \qquad\qquad + \tilde{D}_{im}w(k) \end{cases} \tag{9}$$

where $\tilde{A}_{im} = \begin{bmatrix} A_i & 0 \\ B_{fm}(I + H(k))C_{1i} & A_{fm} \end{bmatrix}$, $\tilde{A}_{dim} = \begin{bmatrix} A_{di} \\ B_{fm}(I + H(k))C_{di} \end{bmatrix}$,

$\tilde{B}_{im} = \begin{bmatrix} B_i \\ B_{fm}(I + H(k))D_{1i} \end{bmatrix}$, $\bar{I} = \begin{bmatrix} I \\ 0 \end{bmatrix}$, $\tilde{C}_{im} = \begin{bmatrix} C_{2i} - D_{fm}(I + H(k))C_{1i} & -C_{fm} \end{bmatrix}$,

$\tilde{C}_{dm} = \begin{bmatrix} -D_{fm}(I + H(k))C_{di} \end{bmatrix}$, $\tilde{D}_{im} = \begin{bmatrix} D_{2i} - D_{fm}(I + H(k))D_{1i} \end{bmatrix}$.

Next, we will provide some important definitions, assumptions and lemmas that promote the work of this paper.

Definition 1 ([41]). *The filtering error system($\mathbb{S}_e$) with $w(k) = 0$ is said to be stochastically stable if the following condition is satisfied for the arbitrary initial condition $(\tilde{x}(0), \alpha_0)$*

$$\mathbf{E}\left\{ \sum_{k=0}^{\infty} \|\tilde{x}(k)\|^2 |\tilde{x}(0), \alpha_0 \right\} < \infty \tag{10}$$

Definition 2 ([41]). *The filtering error system($\mathbb{S}_e$) with $w(k) \in l_2[0, \infty)$ is said to have an H_∞ disturbance attenuation performance γ, if under the zero initial condition, the error $\tilde{e}(k)$ fulfills the condition as follows:*

$$\sum_{k=0}^{\infty} \mathbf{E}\left\{ \|\tilde{e}(k)\|^2 \right\} < \gamma^2 \sum_{k=0}^{\infty} \|w(k)\|^2 \tag{11}$$

where γ is a positive scalar.

Assumption 1 ([42]). *The continuous nonlinear function $g_i(\bullet)$ in system($\mathbb{S}_0$) is supposed to be bounded, and satisfies the following condition*

$$l_i \le \frac{g_i(x)}{x} \le h_i \; x \ne 0, x \in R$$

where l_i and h_i are constants, $i = 1, 2, \cdots, n$.

Lemma 1 ([42]). *Based on Assumption 1, there is a symmetric matrix $N > 0$, satisfying*

$$\begin{bmatrix} x(k) \\ g(x(k)) \end{bmatrix}^T \begin{bmatrix} Y_1 N & -Y_2 N \\ * & N \end{bmatrix} \begin{bmatrix} x(k) \\ g(x(k)) \end{bmatrix} < 0$$

where $Y_1 = diag\{l_1 h_1, l_2 h_2, \cdots, l_n h_n\}$, $Y_2 = diag\{(l_1+h_1)/2, (l_2+h_2)/2, \cdots, (l_n+h_n)/2\}$.

Lemma 2 (Projection lemma [43]). *For given matrices X, U and V, there exists a matrix Y such that*

$$X + U^T Y V + V^T Y^T U < 0$$

is satisfied, if and only if the inequalities listed below are true

$$U_\perp^T X U_\perp < 0, \; V_\perp^T X V_\perp < 0$$

where $U_\perp$ and U, $V_\perp$ and V are orthogonal complements, respectively.

Based on the above, the objective of this paper is to develop a feasible ET asynchronous filter($\mathbb{S}_f$) for discrete-time delayed MJNSs ($\mathbb{S}_0$) with unknown probabilities, such that the error system ($\mathbb{S}_e$) is stochastically stable and has a desired H_∞ performance γ.

3. Main Results

We will first provide a sufficient condition about the stochastic stability with an H_∞ performance γ of the error system ($\mathbb{S}_e$) in this section, then present a design scheme of a solvable filter.

For brevity, we first introduce the following notations:

$$\bar{A}_{im} = \begin{bmatrix} A_i & 0 \\ B_{fm}C_{1i} & A_{fm} \end{bmatrix}, \bar{A}_{dim} = \begin{bmatrix} A_{di} \\ B_{fm}C_{di} \end{bmatrix}, \bar{B}_{im} = \begin{bmatrix} B_i \\ B_{fm}D_{1i} \end{bmatrix}, \bar{C}_{im} = \begin{bmatrix} C_{2i} - D_{fm}C_{1i} & -C_{fm} \end{bmatrix},$$

$$\bar{C}_{dim} = \begin{bmatrix} -D_{fm}C_{di} \end{bmatrix}, \bar{D}_{im} = \begin{bmatrix} D_{2i} - D_{fm}D_{1i} \end{bmatrix}, \eta_1(k) = \begin{bmatrix} \tilde{x}^T(k) & g^T(x(k)) & x^T(k-d(k)) & g^T(x(k-d(k))) \end{bmatrix}^T,$$

$$\eta(k) = \begin{bmatrix} \eta_1^T(k) & w^T(k) \end{bmatrix}^T, \Lambda = diag\{\delta_1, \delta_2, \cdots, \delta_p\}, Q = \begin{bmatrix} Q_{11} & Q_{12} \\ * & Q_{22} \end{bmatrix}, \tau = \tau_2 - \tau_1 + 1.$$

By use of LKF and H_∞ theory, we can obtain the following conclusions.

Theorem 1. *For a prescribed $\gamma > 0$, the filtering error dynamic system($\mathbb{S}_e$) based on Assumption 1 is stochastically stable with the H_∞ performance γ, if there are matrices A_{fm}, B_{fm}, C_{fm}, D_{fm}, $P_i > 0$, $F_{im} > 0$, and $Q > 0$, and diagonal matrices $N_1 > 0$, $N_2 > 0$, and $W_{im} > 0$, such that the following two conditions are fulfilled for $\forall i \in \mathscr{S}_1, m \in \mathscr{S}_{2U}^i$*

$$\mathcal{F}_i^K + (1 - \sigma_i^K)F_{im} < P_i \tag{12}$$

and for $\forall i \in \mathscr{S}_1, j \in \mathscr{S}_{1U}^i, m \in \mathscr{S}_2$

$$\Pi_{im} = \begin{bmatrix} \Pi_{im}^1 & U_m & Z_i^T \Lambda W_{im} \\ * & -W_{im} & 0 \\ * & * & -W_{im} \end{bmatrix} < 0 \tag{13}$$

where $\mathcal{F}_i^K = \sum\limits_{m \in \mathscr{S}_{2K}} \sigma_{im} F_{im}$, $\sigma_i^K = \sum\limits_{m \in \mathscr{S}_{2K}} \sigma_{im}$, $\bar{P}_i = \mathcal{P}_i^K + (1 - \phi_i^K) P_{ij}$, $\mathcal{P}_i^K = \sum\limits_{j \in \mathscr{S}_{1K}^i} \phi_{ij} P_j$, $\phi_i^K = $

$\sum\limits_{j \in \mathscr{S}_{1K}^i} \phi_{ij}$, $\Pi_{im}^1 = \begin{bmatrix} \Pi_i^{11} & \Pi_{im}^{12} & \Pi_{im}^{13} & \Pi_{im}^{14} \\ * & \tau \bar{Q} - \Pi^{22} - \bar{F}_{im} & 0 & 0 \\ * & * & -Q - \Pi^{33} & 0 \\ * & * & * & -\gamma^2 I \end{bmatrix}$, $\Pi_i^{11} = \begin{bmatrix} -\bar{P}_i^{-1} & 0 \\ 0 & -I \end{bmatrix}$,

$\Pi_{im}^{12} = \begin{bmatrix} \bar{A}_{im} & \bar{I} E_i \\ \bar{C}_{im} & 0 \end{bmatrix}$, $\Pi_{im}^{13} = \begin{bmatrix} \bar{A}_{dim} & \bar{I} E_{di} \\ \bar{C}_{dim} & 0 \end{bmatrix}$, $\Pi_{im}^{14} = \begin{bmatrix} \bar{B}_{im} \\ \bar{D}_{im} \end{bmatrix}$, $\bar{Q} = \begin{bmatrix} \bar{I} Q_{11} \bar{I}^T & \bar{I} Q_{12} \\ * & Q_{22} \end{bmatrix}$,

$\bar{F}_{im} = \begin{bmatrix} F_{im} & 0 \\ 0 & 0 \end{bmatrix}$, $\Pi^{22} = \begin{bmatrix} \bar{I} Y_1 N_1 \bar{I}^T & -\bar{I} Y_2 N_1 \\ * & N_1 \end{bmatrix}$, $\Pi^{33} = \begin{bmatrix} Y_1 N_2 & -Y_2 N_2 \\ * & N_2 \end{bmatrix}$, $U_m = $

$\begin{bmatrix} \begin{bmatrix} 0 & B_{fm}^T \end{bmatrix} & -D_{fm}^T & 0 & 0 & 0 & 0 & 0 \end{bmatrix}^T$, $Z_i = \begin{bmatrix} 0 & 0 & \begin{bmatrix} C_{1i} & 0 \end{bmatrix} & 0 & C_{di} & 0 & D_{1i} \end{bmatrix}$.

Proof. First, we will derive some useful results according to (12) and (13). Equation (12) ensures that

$$\sum_{m=1}^{s_2} \sigma_{im} F_{im} - P_i < 0 \tag{14}$$

holds, because when $\sigma_i^K < 1$,

$$\begin{aligned} \sum_{m=1}^{s_2} \sigma_{im} F_{im} - P_i &= \mathcal{F}_i^K + (1 - \sigma_i^K) \sum_{m \in \mathscr{S}_{2U}^i} \frac{\sigma_{im}}{1 - \sigma_i^K} F_{im} - P_i \\ &= \sum_{m \in \mathscr{S}_{2U}^i} \frac{\sigma_{im}}{1 - \sigma_i^K} \{ \mathcal{F}_i^K + (1 - \sigma_i^K) F_{im} - P_i \} \end{aligned} \tag{15}$$

and when $\sigma_i^K = 1$, obviously, (12) is equivalent to (14).

In terms of the Schur complement, (13) is equivalent to

$$\Pi_{im}^1 + U_m W_{im}^{-1} U_m^T + Z_i^T \Lambda W_{im} \Lambda Z_i < 0 \tag{16}$$

which obtains

$$\Pi_{im}^2 \stackrel{\Delta}{=} \Pi_{im}^1 + U_m H(k) Z_i + Z_i^T H^T(k) U_m^T < 0 \tag{17}$$

and it is easy to derive that

$$\Pi_{im}^2 = \begin{bmatrix} \Pi_i^{11} & \bar{\Pi}_{im}^{12} & \bar{\Pi}_{im}^{13} & \bar{\Pi}_{im}^{14} \\ * & \tau \bar{Q} - \Pi^{22} - \bar{F}_{im} & 0 & 0 \\ * & * & -Q - \Pi^{33} & 0 \\ * & * & * & -\gamma^2 I \end{bmatrix} < 0 \tag{18}$$

where $\bar{\Pi}_{im}^{12} = \begin{bmatrix} \tilde{A}_{im} & \bar{I} E_i \\ \tilde{C}_{im} & 0 \end{bmatrix}$, $\bar{\Pi}_{im}^{13} = \begin{bmatrix} \tilde{A}_{dim} & \bar{I} E_{di} \\ \tilde{C}_{dim} & 0 \end{bmatrix}$, $\bar{\Pi}_{im}^{14} = \begin{bmatrix} \tilde{B}_{im} \\ \tilde{D}_{im} \end{bmatrix}$.

Then, based on the Schur complement and the analysis similar to (14) and (15), we can derive from (18) that

$$\begin{cases} \Pi_{im}^3 \stackrel{\Delta}{=} \upsilon + \mu_{im}^T \tilde{P}_i \mu_{im} < \tilde{F}_{im} \\ \Pi_{im}^4 \stackrel{\Delta}{=} \xi - \varsigma_{im}^T \tilde{\Pi}_i^{11} \varsigma_{im} < \hat{F}_{im} \end{cases} \tag{19}$$

where $\mu_{im} = \begin{bmatrix} \tilde{A}_{im} & \bar{I}E_i & \tilde{A}_{dim} & \bar{I}E_{di} \end{bmatrix}$, $v = \begin{bmatrix} \tau\bar{Q} - \Pi^{22} & 0 \\ * & -Q - \Pi^{33} \end{bmatrix}$,

$\tilde{\Pi}_i^{11} = \begin{bmatrix} -\tilde{P}_i & 0 \\ 0 & -I \end{bmatrix}$, $\xi = diag\{v, -\gamma^2 I\}$, $\varsigma_{im} = \begin{bmatrix} \tilde{\Pi}_{im}^{12} & \tilde{\Pi}_{im}^{13} & \tilde{\Pi}_{im}^{14} \end{bmatrix}$, $\tilde{P}_i = \sum_{j=1}^{s_1} \phi_{ij} P_j$,

$\tilde{F}_{im} = diag\{\bar{F}_{im}, 0\}$, $\hat{F}_{im} = diag\{\bar{F}_{im}, 0, 0\}$.

Next, a mode-dependent LKF is introduced as follows:

$$V(k) = \sum_{l=1}^{2} V_l(k) \tag{20}$$

where $V_1(k) = \tilde{x}^T(k)P_{\alpha_k}\tilde{x}(k)$, $V_2(k) = \sum_{b=-\tau_2+1}^{-\tau_1+1} \sum_{a=k-1+b}^{k-1} \begin{bmatrix} x(a) \\ g(x(a)) \end{bmatrix}^T Q \begin{bmatrix} x(a) \\ g(x(a)) \end{bmatrix}$.

Then, we calculate $\nabla V(k)$ along the locus of the error system $(\mathbb{S}_e)$ and take the expectation. It is easy to find that $\mathbf{E}\{\nabla V(k)\} = \mathbf{E}\{\nabla V_1(k)\} + \mathbf{E}\{\nabla V_2(k)\}$.

$$\begin{aligned}
\mathbf{E}\{\nabla V_1(k)\} &= \mathbf{E}\{V_1(k+1) - V_1(k)|\tilde{x}(k), \alpha_k = i\} \\
&= \mathbf{E}\{\tilde{x}^T(k+1)P_j\tilde{x}(k+1) - \tilde{x}^T(k)P_i\tilde{x}(k)\} \\
&= \mathbf{E}\left\{ \sum_{m=1}^{s_2}\sum_{j=1}^{s_1} \sigma_{im}\phi_{ij}\tilde{x}^T(k+1)P_j\tilde{x}(k+1) - \tilde{x}^T(k)P_i\tilde{x}(k) \right\} \\
&= \mathbf{E}\left\{ \sum_{m=1}^{s_2} \sigma_{im}\tilde{x}^T(k+1)\tilde{P}_i\tilde{x}(k+1) - \tilde{x}^T(k)P_i\tilde{x}(k) \right\} \\
&= \mathbf{E}\left\{ \sum_{m=1}^{\ell_2} \sigma_{im}\eta^T(k)\begin{bmatrix} \mu_{im}^T \\ \tilde{B}_{im}^T \end{bmatrix}\tilde{P}_i\begin{bmatrix} \mu_{im} & B_{im} \end{bmatrix}\eta(k) - \tilde{x}^T(k)P_i x(k) \right\}
\end{aligned} \tag{21}$$

$$\begin{aligned}
\mathbf{E}\{\nabla V_2(k)\} &= \mathbf{E}\{V_2(k+1) - V_2(k)\} \\
&= \mathbf{E}\left\{ \tau\begin{bmatrix} x(k) \\ g(x(k)) \end{bmatrix}^T Q \begin{bmatrix} x(k) \\ g(x(k)) \end{bmatrix} - \right. \\
&\qquad \left. \sum_{a=k-\tau_2}^{k-\tau_1} \begin{bmatrix} x(a) \\ g(x(a)) \end{bmatrix}^T Q \begin{bmatrix} x(a) \\ g(x(a)) \end{bmatrix} \right\} \\
&\leq \mathbf{E}\left\{ \begin{bmatrix} \tilde{x}(k) \\ g(x(k)) \end{bmatrix}^T \tau\bar{Q} \begin{bmatrix} \tilde{x}(k) \\ g(x(k)) \end{bmatrix} - \right. \\
&\qquad \left. \begin{bmatrix} x(k-d(k)) \\ g(x(k-d(k))) \end{bmatrix}^T Q \begin{bmatrix} x(k-d(k)) \\ g(x(k-d(k))) \end{bmatrix} \right\}
\end{aligned} \tag{22}$$

According to Lemma 1, there are diagonal matrices $N_1 > 0$, $N_2 > 0$ such that (23) and (24) are satisfied

$$\begin{bmatrix} \tilde{x}(k) \\ g(x(k)) \end{bmatrix}^T \Pi^{22} \begin{bmatrix} \tilde{x}(k) \\ g(x(k)) \end{bmatrix} \leq 0 \tag{23}$$

$$\begin{bmatrix} x(k-d(k)) \\ g(x(k-d(k))) \end{bmatrix}^T \Pi^{33} \begin{bmatrix} x(k-d(k)) \\ g(x(k-d(k))) \end{bmatrix} \leq 0 \tag{24}$$

Synthesizing (22)–(24), we get that

$$\mathbf{E}\{\nabla V_2(k)\} \leq \mathbf{E}\left\{ \eta_1^T(k)v\eta_1(k) \right\} \tag{25}$$

Next, we will verify that $(\mathbb{S}_e)$ with $w(k) = 0$ is stochastically stable, and that

$$
\begin{aligned}
\mathbf{E}\{\nabla V(k)\} &= \mathbf{E}\{\nabla V_1(k)\} + \mathbf{E}\{\nabla V_2(k)\} \\
&\leq \mathbf{E}\left\{ \sum_{m=1}^{s_2} \sigma_{im}\eta_1^T(k)\Pi_{im}^3\eta_1(k) - \tilde{x}^T(k)P_i\tilde{x}(k) \right\} \\
&< \mathbf{E}\left\{ \eta_1^T(k)\big(\sum_{m=1}^{s_2} \sigma_{im}\tilde{F}_{im} \big)\eta_1(k) - \tilde{x}^T(k)P_i\tilde{x}(k) \right\} \\
&= \mathbf{E}\left\{ \tilde{x}^T(k)\Big(\sum_{m=1}^{s_2} \sigma_{im}F_{im} - P_i \Big)\tilde{x}(k) \right\} \\
&\leq \varepsilon\mathbf{E}\{\tilde{x}^T(k)\tilde{x}(k)\}
\end{aligned}
\tag{26}
$$

where "$<$" is based on (19), $\varepsilon = \lambda_{\max}(\sum_{m=1}^{s_2} \sigma_{im}F_{im} - P_i)$.
$\quad\;\; {}_{i\in\mathscr{S}_1}$

Notice that $\varepsilon < 0$ due to (12) and (14); then,

$$
\mathbf{E}\{\sum_0^\infty \nabla V(k)\} = \mathbf{E}\{V(\infty) - V(0)\} \leq \varepsilon\mathbf{E}\left\{ \sum_0^\infty \tilde{x}^T(k)\tilde{x}(k) \right\}
\tag{27}
$$

therefore,

$$
\mathbf{E}\left\{ \sum_0^\infty \tilde{x}^T(k)\tilde{x}(k) \right\} < \infty
\tag{28}
$$

which conforms to Definition 1, so we have verified the stochastic stability for $(\mathbb{S}_e)$ with $w(k) = 0$.

Next, we will verify that $(\mathbb{S}_e)$ with $w(k) \in l_2[0, \infty)$ has an H_∞ performance γ. Define the performance index as

$$
\begin{aligned}
J &= \sum_{k=0}^\infty \mathbf{E}\{\tilde{e}^T(k)\tilde{e}(k) - \gamma^2 w^T(k)w(k)\} \\
&= \sum_{k=0}^\infty \mathbf{E}\{\tilde{e}^T(k)\tilde{e}(k) - \gamma^2 w^T(k)w(k) + \nabla V(k)\} \\
&\quad + \mathbf{E}\{V(0)\} - \mathbf{E}\{V(\infty)\}
\end{aligned}
\tag{29}
$$

Owing to the zero initial value, we obtain that $V(0) = 0$, whereas $V(\infty) \geq 0$, thus

$$
\begin{aligned}
J &\leq \sum_{k=0}^\infty \mathbf{E}\{\tilde{e}^T(k)\tilde{e}(k) - \gamma^2 w^T(k)w(k) + \nabla V(k)\} \\
&= \sum_{k=0}^\infty \mathbf{E}\left\{ \sum_{m=1}^{s_2} \sigma_{im}\eta^T(k)\Pi_{im}^4\eta(k) - \tilde{x}^T(k)P_i\tilde{x}(k) \right\} \\
&< \sum_{k=0}^\infty \mathbf{E}\left\{ \sum_{m=1}^{s_2} \sigma_{im}\eta^T(k)\hat{F}_{im}\eta(k) - \tilde{x}^T(k)P_i\tilde{x}(k) \right\} \\
&= \sum_{k=0}^\infty \mathbf{E}\left\{ \tilde{x}^T(k)\Big(\sum_{m=1}^{s_2} \sigma_{im}F_{im} - P_i \Big)\tilde{x}(k) \right\} \\
&< 0
\end{aligned}
\tag{30}
$$

in which the two "$<$" are obtained on the basis of (19) and (14), respectively. Then, from (11) and (30), we can readily conclude that the error system $(\mathbb{S}_e)$ has an H_∞ performance γ. Thus, the proof is accomplished. $\quad\square$

Remark 4. *The purpose of introducing the extra matrix F_{im} in Theorem 1 is to simplify matrix inequalities. However, in order to solve the parameters of the filter, the nonlinearity in (13) needs to be further processed so as to transform the matrix inequalities into linear matrix inequalities (LMIs).*

Next, we will devise the filter with the techniques of slack matrix and Projection lemma and obtain Theorem 2.

Theorem 2. *The filtering error dynamic system* $(\mathbb{S}_e)$ *based on Assumption 1 is stochastically stable with an H_∞ performance γ, if there are matrices $\tilde{A}_{fm}, \tilde{B}_{fm}, \tilde{C}_{fm}, \tilde{D}_{fm}$, and G_m, a scalar $\tilde{\gamma} > 0$, diagonal matrices $N_1 > 0$, $N_2 > 0$, and $W_{im} > 0$, and the following matrices*

$$P_i = \begin{bmatrix} P_i^1 & P_i^2 \\ * & P_i^3 \end{bmatrix} > 0,\, F_{im} = \begin{bmatrix} F_{im}^1 & F_{im}^2 \\ * & F_{im}^3 \end{bmatrix} > 0,\, Q = \begin{bmatrix} Q_{11} & Q_{12} \\ * & Q_{22} \end{bmatrix} > 0,$$

such that the following two conditions are fulfilled for $\forall i \in \mathscr{S}_1, m \in \mathscr{S}_{2U}^i$

$$\mathcal{F}_i^K + (1 - \sigma_i^K)F_{im} < P_i \tag{31}$$

and for $\forall i \in \mathscr{S}_1, j \in \mathscr{S}_{1U}^i, m \in \mathscr{S}_2$

$$\Xi_i^T \widehat{\Pi}_{im} \Xi_i < 0,\, \breve{\Pi}_{im} < 0 \tag{32}$$

where

$$\widehat{\Pi}_{im} = \begin{bmatrix} \bar{P}_i - \widehat{G}_m & \widehat{\Pi}_{im}^{12} \\ * & \breve{\Pi}_{im} \end{bmatrix},\, \widehat{G}_m = \begin{bmatrix} 0 & G_m \\ * & G_m + G_m^T \end{bmatrix},\, \breve{\Pi}_{im} = \begin{bmatrix} \breve{\Pi}_{im}^1 & \breve{U}_m & (\breve{Z}_i)^T \Lambda W_{im} \\ * & -W_{im} & 0 \\ * & * & -W_{im} \end{bmatrix},$$

$$\widehat{\Pi}_{im}^{12} = \begin{bmatrix} 0 & \tilde{B}_{fm}C_{1i} & \tilde{A}_{fm} & 0 & \tilde{B}_{fm}C_{di} & 0 & \tilde{B}_{fm}D_{1i} & \tilde{B}_{fm} & 0 \\ 0 & \tilde{B}_{fm}C_{1i} & \tilde{A}_{fm} & 0 & \tilde{B}_{fm}C_{di} & 0 & \tilde{B}_{fm}D_{1i} & \tilde{B}_{fm} & 0 \end{bmatrix},$$

$$\breve{\Pi}_{im}^1 = \begin{bmatrix} -I & \breve{\Pi}_{im}^{12} & 0 & D_{2i} - \tilde{D}_{fm}D_{1t} \\ * & \breve{\Pi}_{im}^{22} & 0 & 0 \\ * & * & -Q - \Pi^{33} & 0 \\ * & * & * & -\tilde{\gamma}I \end{bmatrix},\, \breve{\Pi}_{im}^{12} = \begin{bmatrix} C_{2i} - \tilde{D}_{fm}C_{1i} & -\tilde{C}_{fm} & 0 \end{bmatrix},$$

$$\breve{\Pi}_{im}^{22} = \begin{bmatrix} \tau Q_{11} - Y_1 N_1 - F_{im}^1 & -F_{im}^2 & \tau Q_{12} + Y_2 N_1 \\ * & -F_{im}^3 & 0 \\ * & * & \tau Q_{22} - N_1 \end{bmatrix},\, \breve{U}_m = \begin{bmatrix} -\tilde{D}_{fm}^T & 0 & 0 & 0 & 0 & 0 & 0 \end{bmatrix}^T,$$

$$\breve{Z}_i = \begin{bmatrix} 0 & C_{1i} & 0 & 0 & C_{di} & 0 & D_{1i} \end{bmatrix},\, \bar{\Xi}_i = \begin{bmatrix} 0 & 0 & A_i & 0 & E_i & A_{di} & E_{di} & B_i & 0 & 0 \end{bmatrix},$$

$$\Xi_i = \begin{bmatrix} \bar{\Xi}_i^T & I_{(6n+2p+q+r)} \end{bmatrix}^T,\, \tilde{\gamma} = \gamma^2.$$

In addition, if (31) and (32) are solvable, the filter matrices of (5) can be gained by

$$\begin{cases} A_{fm} = (G_m)^{-1}\tilde{A}_{fm}, B_{fm} = (G_m)^{-1}\tilde{B}_{fm} \\ C_{fm} = \tilde{C}_{fm}, D_{fm} = \tilde{D}_{fm} \end{cases} \tag{33}$$

Proof. In order to verify Theorem 2, (13) is rewritten as

$$\Pi_{im} = \begin{bmatrix} -(\bar{P}_i)^{-1} & \Psi_{im}^1 \\ * & \Psi_{im}^2 \end{bmatrix} < 0 \tag{34}$$

By comparing (13) and (34), it is easy to obtain Ψ_{im}^1 and Ψ_{im}^2, which is omitted here to save space. To handle the nonlinearity $(\bar{P}_i)^{-1}$ in (34), an invertible slack matrix G_{im} is introduced as follows:

$$G_{im} \triangleq \begin{bmatrix} G_{im}^1 & G_m \\ G_{im}^2 & G_m \end{bmatrix} \tag{35}$$

where G_{im}^1, G_{im}^2, G_m are n-dimensional square matrices. Then, (34) is pre-multiplied and post-multiplied by $diag\{G_{im}, I\}$ and its transpose; hence, one has

$$\begin{bmatrix} -G_{im}(\bar{P}_i)^{-1}G_{im}^T & G_{im}\Psi_{im}^1 \\ * & \Psi_{im}^2 \end{bmatrix} < 0 \tag{36}$$

On the other hand, according to the fact that $(\bar{P}_i - G_{im})(\bar{P}_i)^{-1}(\bar{P}_i - G_{im})^T \geq 0$, we can readily obtain that

$$\bar{P}_i - G_{im} - G_{im}^T \geq -G_{im}(\bar{P}_i)^{-1}G_{im}^T \tag{37}$$

Combining (13), (36) and (37), we know that the following condition

$$\tilde{\Pi}_{im} \triangleq \begin{bmatrix} \bar{P}_i - G_{im} - G_{im}^T & G_{im}\Psi_{im}^1 \\ * & \Psi_{im}^2 \end{bmatrix} < 0 \tag{38}$$

is sufficient for (13). Moreover, we define

$$\begin{cases} \tilde{A}_{fm} \triangleq G_m A_{fm}, \tilde{B}_{fm} \triangleq G_m B_{fm} \\ \tilde{C}_{fm} \triangleq C_{fm}, \tilde{D}_{fm} \triangleq D_{fm} \end{cases} \tag{39}$$

and substitute them into (38).

Then, we define

$$\Theta_i = \begin{bmatrix} -I_n & \bar{\Xi}_i \end{bmatrix}, \Gamma = \begin{bmatrix} I_{2n} & 0_{(2n)\times(5n+2p+q+r)} \end{bmatrix}, \Gamma_\perp = \begin{bmatrix} 0_{(2n)\times(5n+2p+q+r)} \\ I_{(5n+2p+q+r)} \end{bmatrix},$$

$$\bar{G}_{im} = \begin{bmatrix} G_{im}^1 \\ G_{im}^2 \end{bmatrix}$$

We can readily derive that Ξ_i and Θ_i, $\Gamma_\perp$ and Γ are orthogonal complements, respectively. Then, $\tilde{\Pi}_{im}$ is decomposed into the following form

$$\tilde{\Pi}_{im} = \widehat{\Pi}_{im} + \Gamma^T \bar{G}_{im}\Theta_i + \Theta_i^T \bar{G}_{im}^T \Gamma \tag{40}$$

In accordance with lemma 2 (i.e., Projection lemma), $\tilde{\Pi}_{im} < 0$ is equivalent to

$$\Xi_i^T \widehat{\Pi}_{im}\Xi_i < 0, \; \Gamma_\perp^T \widehat{\Pi}_{im}\Gamma_\perp < 0 \tag{41}$$

which is obviously equivalent to (32). Furthermore, it can be inferred from (38) that G_{im} and G_m are both nonsingular, so we can deduce (33) from (39). Thus, we have accomplished the proof. $\square$

Remark 5. *In Theorem 2, a filter design scheme is provided such that the error system $(\mathbb{S}_e)$ is stochastically stable with an H_∞ performance γ. γ means the H_∞ performance level, a smaller γ indicates a better performance. The optimal performance $\gamma^* = \sqrt{\tilde{\gamma}_{\min}}$ can be yielded by solving the problem of convex optimization as follows:*

$$\begin{cases} \min & \tilde{\gamma} \\ s.t. & (31),(32) \end{cases} \tag{42}$$

Remark 6. *The number of LMIs $\mathcal{N}_l$ in Theorem 2 is*

$$\mathcal{N}_l = \sum_{i=1}^{s_1} \max\left\{1, \left|\mathscr{S}_{2U}^i\right|\right\} + s_2 \cdot \sum_{i=1}^{s_1} \max\left\{1, \left|\mathscr{S}_{1U}^i\right|\right\} + s_1 \cdot s_2 \tag{43}$$

where $\left|\mathscr{S}_{1U}^i\right|$ and $\left|\mathscr{S}_{2U}^i\right|$ represent the number of elements in the set $\mathscr{S}_{1U}^i$ and $\mathscr{S}_{2U}^i$, respectively. From (43), we clearly find that as the number of unknown entries for TPM Φ and CPM Ω increases, so does the number of LMIs required, thus aggravating the computational burden.

4. Numerical Example

This section will introduce a numerical example to verify the validity of the presented method. A three-mode MJNN($\mathbb{S}_0$) is considered with the parameters as follows, which are partly borrowed from [25]:

Mode 1 :

$$A_1 = \begin{bmatrix} 0.2 & 0 \\ 0 & 0.2 \end{bmatrix}, \quad A_{d1} = \begin{bmatrix} 0.05 & 0 \\ 0 & 0.05 \end{bmatrix}, \quad E_1 = \begin{bmatrix} 0.3 & -0.2 \\ 0.1 & 0.3 \end{bmatrix},$$

$$E_{d1} = \begin{bmatrix} 0.1 & -0.2 \\ 0.1 & 0.15 \end{bmatrix}, B_1 = \begin{bmatrix} 0.1 \\ 0.2 \end{bmatrix}, C_{11} = \begin{bmatrix} 0.17 & 0.18 \end{bmatrix}, C_{d1} = \begin{bmatrix} 0.1 & 0.1 \end{bmatrix}, D_{11} = 0.1,$$

$$C_{21} = \begin{bmatrix} 0.2 & 0.35 \end{bmatrix}, D_{21} = 0.1.$$

Mode 2:

$$A_2 = \begin{bmatrix} 0.1 & 0 \\ 0 & 0.3 \end{bmatrix}, \quad A_{d2} = \begin{bmatrix} 0.1 & 0 \\ 0 & -0.1 \end{bmatrix}, \quad E_2 = \begin{bmatrix} 0.3 & 0.1 \\ 0 & 0.2 \end{bmatrix}, \quad E_{d2} = \begin{bmatrix} 0.1 & -0.2 \\ 0 & 0.1 \end{bmatrix},$$

$$B_2 = \begin{bmatrix} 0.6 \\ 0.3 \end{bmatrix}, \quad C_{12} = \begin{bmatrix} 0.42 & 0.90 \end{bmatrix}, \quad C_{d2} = \begin{bmatrix} -0.1 & -0.1 \end{bmatrix}, \quad D_{12} = 0.5,$$

$$C_{22} = \begin{bmatrix} 0.1 & 0.15 \end{bmatrix}, D_{22} = 0.15.$$

Mode 3:

$$A_3 = \begin{bmatrix} 0.2 & 0 \\ 0 & 0.4 \end{bmatrix}, \quad A_{d3} = \begin{bmatrix} 0.05 & 0 \\ 0 & -0.15 \end{bmatrix}, \quad E_3 = \begin{bmatrix} 0.2 & -0.1 \\ 0 & 0.1 \end{bmatrix}, \quad E_{d3} = \begin{bmatrix} 0.1 & -0.1 \\ 0.1 & 0.1 \end{bmatrix},$$

$$B_3 = \begin{bmatrix} 0.4 \\ 0.2 \end{bmatrix}, \quad C_{13} = \begin{bmatrix} 0.12 & 0.5 \end{bmatrix}, \quad C_{d3} = \begin{bmatrix} -0.05 & -0.1 \end{bmatrix}, \quad D_{13} = 0.3,$$

$$C_{23} = \begin{bmatrix} 0.2 & 0.2 \end{bmatrix}, D_{23} = 0.2.$$

The nonlinear function is chosen as $g(x) = \tanh(x)$ with the bounds of $l_1 = l_2 = 0$ and $h_1 = h_2 = 1$; the delay $d(k) \in \{1,2,3\}$ is time-varying and random with $\tau = 3$, and the error threshold of the ET scheduler is $\delta = 0.3$.

In the sequel, four different TPM Φ^i and CPM Ω^i ($i \in \{1,2,3,4\}$) will be considered.

$$\Phi^1 = \begin{bmatrix} 0.85 & 0.05 & 0.1 \\ 0.2 & 0.5 & 0.3 \\ 0.5 & 0.1 & 0.4 \end{bmatrix}, \Phi^2 = \begin{bmatrix} 0.85 & ? & ? \\ 0.2 & 0.5 & 0.3 \\ 0.5 & 0.1 & 0.4 \end{bmatrix}, \Phi^3 = \begin{bmatrix} 0.85 & ? & ? \\ ? & ? & 0.3 \\ 0.5 & 0.1 & 0.4 \end{bmatrix}, \Phi^4 = \begin{bmatrix} ? & ? & ? \\ ? & ? & ? \\ ? & ? & ? \end{bmatrix}.$$

$$\Omega^1 = \begin{bmatrix} 0.9 & 0.05 & 0.05 \\ 0.1 & 0.9 & 0 \\ 0.1 & 0.1 & 0.8 \end{bmatrix}, \Omega^2 = \begin{bmatrix} 0.9 & ? & ? \\ 0.1 & 0.9 & 0 \\ 0.1 & 0.1 & 0.8 \end{bmatrix}, \Omega^3 = \begin{bmatrix} 0.9 & ? & ? \\ ? & ? & 0 \\ 0.1 & 0.1 & 0.8 \end{bmatrix}, \Omega^4 = \begin{bmatrix} ? & ? & ? \\ ? & ? & ? \\ ? & ? & ? \end{bmatrix}.$$

Notice that Φ^1 and Ω^1 are fully known; Φ^4 and Ω^4 are fully unknown; Φ^3 and Ω^3 have more unknown elements than Φ^2 and Ω^2, respectively.

Firstly, in accordance with Theorem 2 and Remark 5, we can achieve the optimal H_∞ performance for different combinations (Φ^i, Ω^j), $(i, j \in \{1,2,3,4\})$, listed in Table 1.

From Table 1, we can clearly observe that, for a given Φ (or Ω), the optimal γ^* increases gradually when varying Ω from Ω^1 to Ω^4 (or Φ from Φ^1 to Φ^4). In addition, for $(\Phi, \Omega) = (\Phi^1, \Omega^1)$, which denotes the fully known case, γ^* is the smallest, which means that the H_∞ performance is the best. On the contrary, for $(\Phi, \Omega) = (\Phi^4, \Omega^4)$, which represents the fully unknown case, γ^* is the largest, i.e., the H_∞ performance is the worst. Therefore, we can conclude that the less probability information of TPM Φ or CPM Ω is available, the worse the H_∞ performance is. What is more interesting is that for each case of $\Omega = \Omega^4$, we find that the designed filter parameters are the same, e.g., when $(\Phi, \Omega) = (\Phi^2, \Omega^4)$, the solved filter parameters are as follows:

$$A_{fm} = \begin{bmatrix} 0.2266 & -0.9508 \\ 0.0551 & 0.7867 \end{bmatrix}, B_{fm} = \begin{bmatrix} -1.0299 \\ -0.2585 \end{bmatrix}, C_{fm} = \begin{bmatrix} -0.0912 & -0.1673 \end{bmatrix}, D_{fm} = 0.1963.$$

for $m = 1,2,3$, which indicates that the filter is mode independent when Ω is fully unknown.

Table 1. Optimal H_∞ performance for different Φ and Ω with unknown elements.

γ^*		CPM Ω			
		Ω^1	Ω^2	Ω^3	Ω^4
TPM Φ	Φ^1	0.5437	0.5524	0.6300	0.6386
	Φ^2	0.5498	0.5588	0.6393	0.6489
	Φ^3	0.5806	0.5884	0.6733	0.6784
	Φ^4	0.6319	0.6379	0.7440	0.7441

Furthermore, when $(\Phi, \Omega) = (\Phi^3, \Omega^3)$, the designed filter parameters can be obtained as follows:

Filter 1 :

$$A_{f1} = \begin{bmatrix} -0.1645 & -1.3380 \\ 0.1701 & 0.9285 \end{bmatrix}, \quad B_{f1} = \begin{bmatrix} -1.5449 \\ -0.0437 \end{bmatrix}, \quad C_{f1} = \begin{bmatrix} -0.0896 & -0.1532 \end{bmatrix},$$
$$D_{f1} = 0.1879.$$

Filter 2:

$$A_{f2} = \begin{bmatrix} 0.0037 & -0.8974 \\ 0.0714 & 0.6686 \end{bmatrix}, \quad B_{f2} = \begin{bmatrix} -1.2739 \\ -0.2034 \end{bmatrix}, \quad C_{f2} = \begin{bmatrix} -0.0901 & -0.1549 \end{bmatrix},$$
$$D_{f2} = 0.1867.$$

Filter 3:

$$A_{f3} = \begin{bmatrix} 0.3964 & -0.0578 \\ -0.0550 & 0.2278 \end{bmatrix}, \quad B_{f3} = \begin{bmatrix} 0.7787 \\ -1.8741 \end{bmatrix}, \quad C_{f3} = \begin{bmatrix} -0.0792 & -0.1296 \end{bmatrix},$$
$$D_{f3} = 0.2777.$$

We further assume that the initial values of filter ($\mathbb{S}_f$) and system ($\mathbb{S}_0$) are $x_f(0) = \begin{bmatrix} 0 & 0 \end{bmatrix}^T$ and $x(k_0) = \begin{bmatrix} 0.2 & -0.2 \end{bmatrix}^T$, $k_0 = -3, -2, -1, 0$, $\alpha_0 = 1$, and the external disturbance is $w(k) = 0.9^k \sin(k)$. Based on the above parameters, a simulation is made with the presented ET asynchronous filtering scheme. The mode jumps of the original plant and the filter are plotted in Figure 2 to show the asynchronization between them.

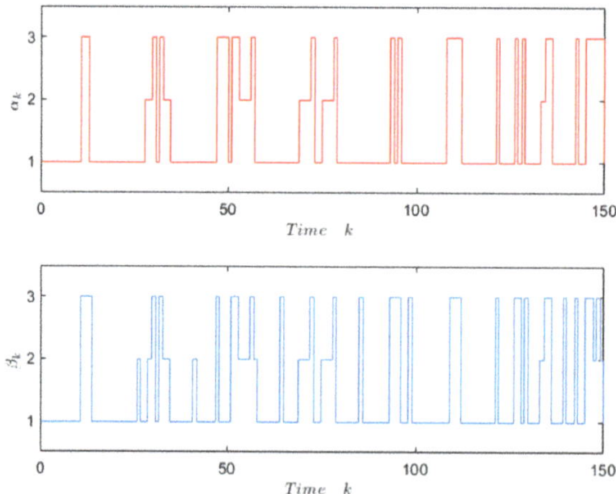

Figure 2. Mode jumps of the original plant and the filter.

The response curves of $z(k)$ and $z_f(k)$, and $\tilde{e}(k) = z(k) - z_f(k)$ are shown in Figures 3 and 4, from which we observe that the filtering error system is stochastically stable. In addition, we obtain $MTR = 0.84$ via calculation with the threshold $\delta = 0.3$, which implies that the ET scheduler can effectively decrease the data-transmission rate of measured outputs. Therefore, it can be observed that the effect of the devised ET filter in Theorem 2 is fine.

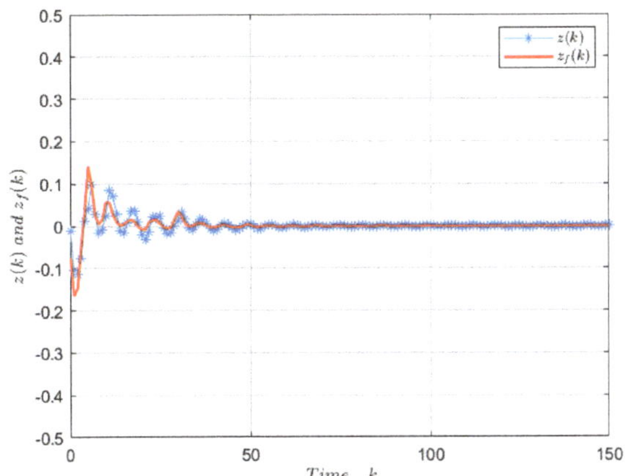

Figure 3. The response curves of $z(k)$ and $z_f(k)$.

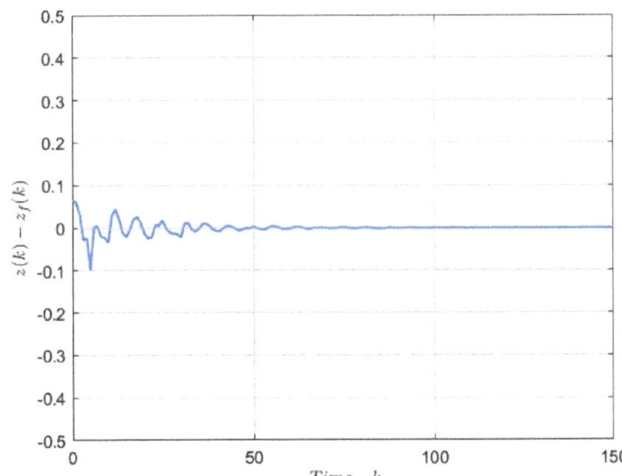

Figure 4. The curve of estimation error $\tilde{e}(k)$.

In our research, the asynchronous issue is characterized as a HMM, the core of which is the CPM, reflecting the asynchronous degree between filter ($\mathbb{S}_f$) and system ($\mathbb{S}_0$). Next, four different CPM Ω^i ($i \in \{a, b, c, d\}$) are chosen to exhibit the influence of asynchronous features on the H_∞ performance of the error system ($\mathbb{S}_e$):

$$\Omega^a = \begin{bmatrix} 1 & 0 & 0 \\ 0 & 1 & 0 \\ 0 & 0 & 1 \end{bmatrix}, \Omega^b = \begin{bmatrix} 1 & 0 & 0 \\ 0 & 1 & 0 \\ 0.1 & 0.1 & 0.8 \end{bmatrix}, \Omega^c = \begin{bmatrix} 1 & 0 & 0 \\ 0.1 & 0.9 & 0 \\ 0.1 & 0.1 & 0.8 \end{bmatrix}, \Omega^d = \Omega^1.$$

which represent four different cases: synchronization, weak asynchronization, strong asynchronization and full asynchronization. In addition, in order to compare the results of the fully known TPs case and the partly unknown TPs case, we choose TPM Φ as Φ^1 and Φ^3, respectively. By solving the convex optimization in (42) with the LMI toolbox of Matlab, we use the ***mincx*** function to calculate the corresponding optimal γ^*, as shown in Table 2. We can easily see from Table 2 that, for a given Φ, with the increase in asynchronous degree between filter ($\mathbb{S}_f$) and system ($\mathbb{S}_0$), γ^* becomes larger, which implies the decline of the H_∞ performance.

Finally, we will investigate the influence of the ET feature on the H_∞ performance and communication performance with the varying threshold of δ in the ET scheduler. We keep the other parameters fixed, and only vary the threshold parameter δ. The evolution curves of the corresponding H_∞ performance γ^* and communication performance MTR for the cases of (Φ^1, Ω^1) and (Φ^3, Ω^3) are shown in Figure 5. We can easily find that as the parameter δ increases, γ^* becomes larger, which implies that the H_∞ performance decreases, whereas the MTR value shows a trend of getting smaller, which means that the communication performance of measured outputs is becoming better. Considering the trade-off between the H_∞ performance and communication performance, thus we can choose a compromise error threshold of the ET scheduler to achieve a more satisfactory comprehensive performance in practical applications.

Table 2. Optimal H_∞ performance for Ω with different asynchronous features.

γ^*		CPM Ω			
		Ω^a	Ω^b	Ω^c	Ω^d
TPM Φ	Φ^1	0.4791	0.4912	0.5249	0.5437
	Φ^3	0.5104	0.5219	0.5644	0.5806

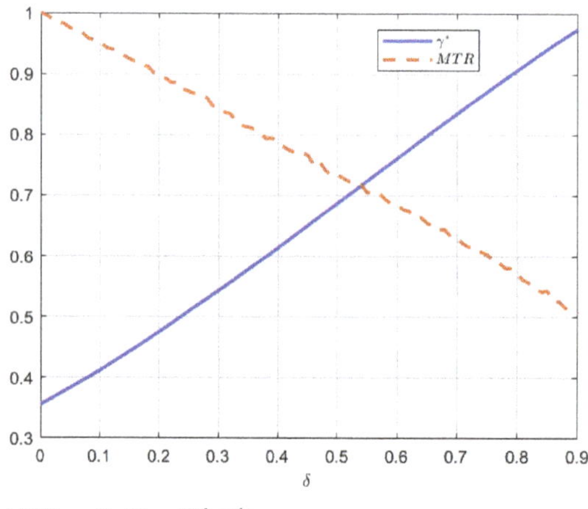

(a) When $(\Phi, \Omega) = (\Phi^1, \Omega^1)$

Figure 5. *Cont.*

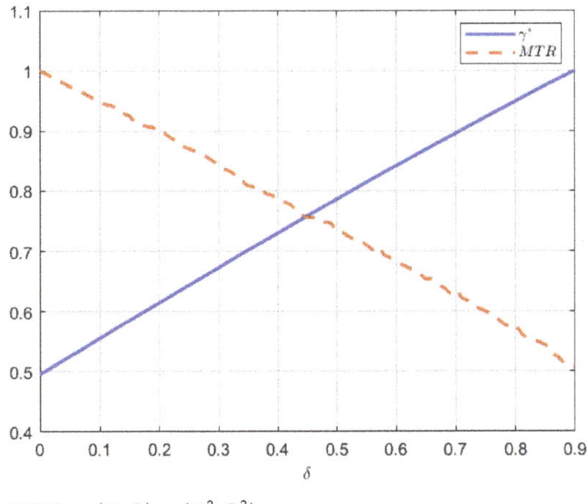

(**b**) When $(\Phi, \Omega) = (\Phi^3, \Omega^3)$

Figure 5. The H_∞ performance and communication performance with varying δ.

5. Conclusions

In this paper, the study of the ET H_∞ asynchronous filtering issue was explored for MJNSs with varying delay and unknown probabilities. An ET scheduling strategy was adopted to decrease the transmission rate of measured outputs, and the filter was mode dependent and asynchronous with the original MJNS, represented by an HMM. Both the TPM of the original system and the CPM of the filter were assumed to be only partly accessible. Under this framework, based on Lyapunov stability and H_∞ theory, a sufficient condition was derived, in which the nonlinearity of the matrix inequalities was further dealt with and a feasible filter was achieved with the techniques of slack matrix and Projection lemma. Lastly, the relationship between the H_∞ performance and the unknown elements of TPM and CPM, the relationship between the H_∞ performance and the asynchronous feature of CPM, and the relationship among the H_∞ performance, communication performance and the ET threshold were discussed and exhibited through a numerical example. The simulation results sufficiently validated the availability of our developed filtering scheme, which will contribute to the further research involving this subject, e.g., control and fault detection. This can also be extended to other dynamic systems, such as singular MJNSs and 2-D MJNSs.

Author Contributions: Conceptualization, investigation and methodology: H.C., R.L. and W.X.; simulation and writing: H.C. and R.L.; supervision: Z.L. All authors have read and agreed to the published version of the manuscript.

Funding: This work was partly supported by the National Natural Science Foundation of China under Grant 61603133, and the Zhejiang Provincial Public Welfare Technology Application Research Project of China under Grant LGG21E020001 and LGG22F030023.

Institutional Review Board Statement: Not applicable.

Informed Consent Statement: Not applicable.

Data Availability Statement: Not applicable.

Conflicts of Interest: The authors declare no conflicts of interest.

Abbreviations

The following abbreviations are used in this manuscript:

HMM	hidden Markov model
MJNSs	Markov jump nonlinear systems
MJSs	Markov jump systems
NN	neural network
MJNNs	Markov jump neural networks
NCSs	networked control systems
ET	event-triggered
CPM	conditional probability matrix
TPs	transition probabilities
TPM	transition probability matrix
LKF	Lyapunov–Krasovskii functional
ZOH	zero-order holder
CP	conditional probability
LMIs	linear matrix inequalities

References

1. He, H.F.; Qi, W.H.; Kao, Y.G. HMM-based adaptive attack-resilient control for Markov jump system and application to an aircraft model. *Appl. Math. Comput.* **2021**, *392* 1–12. [CrossRef]
2. Zhang, L.X.; Yang, T.; Shi, P.; Zhu, Y.Z. *Analysis and Design of Markov Jump Systems with Complex Transition Probabilitie*; Springer: Berlin/Heidelberg, Germany, 2016; ISBN: 978-3-319-28846-8.
3. Costa, O.L.V.; Araujo, M.V. A generalized multi-period mean-variance portfolio optimization with Markov switching parameters. *Automatica* **2008**, *44*, 2487–2497. [CrossRef]
4. Zhang, L.; Prieur, C. Stochastic stability of Markov jump hyperbolic systems with application to traffic flow control. *Automatica* **2017** , *86*, 29–37. [CrossRef]
5. Krasovskii, N.M.; Lidskii, E.A. Analytical design of controllers in systems with random attributes. *Autom. Remote Control* **1961**, *22*, 1021–2025.
6. Bolzern, P.; Colaneri, P.; De Nicolao, G. Stochastic stability of positive Markov jump linear systems. *Automatica* **2014**, *50*, 1181–1187. [CrossRef]
7. Hou, T.; Ma, H. Exponential stability for discrete-time infinite Markov jump systems. *IEEE Trans. Autom. Control* **2016**, *61*, 4241–4246. [CrossRef]
8. Huang, H.; Long, F.; Li, C. Stabilization for a class of Markovian jump linear systems with linear fractional uncertainties. *Int. J. Innov.Comput. Inf. Control* **2015**, *11*, 295–307.
9. Wang, B.; Zhu, Q.X. Stability analysis of discrete-time semi-Markov jump linear systems with partly unknown semi-Markov kernel. *Syst. Control Lett.* **2020**, *140*, 104688. [CrossRef]
10. Shen, M.; Park, J.H.; Ye, D. A separated approach to control of Markov jump nonlinear systems with general transition probabilities. *IEEE Trans. Cybern.* **2016**, *46*, 2010–2018. [CrossRef]
11. Oliveira, A.M.D.; Costa, O.L.V. Mixed $H_2/H\infty$ control of hidden Markov jump systems. *Int. J. Robust Nonlinear Control* **2018**, *28*, 1261–1280. [CrossRef]
12. Aberkane, S.; Dragan, V. H∞ filtering of periodic Markovian jump systems: Application to filtering with communication constraints. *Automatica* **2012**, *48*, 3151–3156. [CrossRef]
13. Shen, Y.; Wu, Z.G.; Shi, P.; Su, H.Y.; Lu, R. Dissipativity-based asynchronous filtering for periodic Markov jump systems. *Inf. Sci.* **2017**, *420*, 505–516. [CrossRef]
14. Sun, Q.; Lim, C.; Shi, P.; Liu, F. Moving horizon estimation for Markov jump systems. *Inf. Sci.* **2016**, *367*, 143–158. [CrossRef]
15. Tao, J.; Lu, R.; Shi, P.; Su, H.; Wu, Z.G. Dissipativity-based reliable control or fuzzy Markov jump systems with actuator faults. *IEEE Trans. Cybern.* **2017**, *47*, 2377–2388. [CrossRef]
16. Zhu, Q.; Chao, J. Exponential stability of stochastic neural networks with both Markovian jump parameters and mixed time delays. *IIEEE Trans. Syst. Man Cybern. Part B* **2011**, *41*, 341–353. [CrossRef]
17. Cheng, J.; Park, J.H.; Karimi, H.R.; Shen, H. A flexible terminal approach to sampled-data exponentially synchronization of Markovian neural networks with time-varying delayed signals. *IEEE Trans. Cybern.* **2017**, *48*, 2232–2234. [CrossRef]
18. Dai, M.C.; Xia, J.W.; Xia, H.; Shen, H. Event-triggered passive synchronization for Markov jump neural networks subject to randomly occurring gain variations. *Neurocomputing* **2018**, *331*, 403–411. [CrossRef]
19. Li, Z.X.; Su, H.Y.; Gu, Y.; Wu, Z.G. H∞ filtering for discrete-time singular networked systems with communication delays and data missing. *Int. J. Syst. Sci.* **2013**, *44*, 604–614. [CrossRef]
20. Zha, L.; Fang, J.; Liu, J.; Tian, E. Event-based finite-time state estimation for Markovian jump systems with quantizations and randomly occurring nonlinear perturbations. *ISA Trans.* **2017**, *66*, 77–85. [CrossRef]

21. Chen, H.Y.; Li, Z.X.; Xia, W.F. Event-triggered dissipative filter design for semi-Markovian jump systems with time-varying delays. *Math. Probl. Eng.* **2020**, *2020*, 8983403. [CrossRef]
22. Xia, W.F.; Xu, S.Y.; Lu, J.W.; Li, Y.M.; Chu, Y.M.; Zhang, Z.Q. Event-triggered filtering for discrete-time Markovian jump systems with additive time-varying delays. *Appl. Math. Comput.* **2021**, *391*, 125630. [CrossRef]
23. Zhang, L.; Zhu, Y.; Shi, P.; Zhao, Y. Resilient asynchronous H∞ filtering for Markov jump neural networks with unideal measurements and multiplicative noises. *IEEE Trans. Cybern.* **2015**, *45*, 2840–2852. [CrossRef]
24. Xu, Y.; Lu, R.; Peng, H.; Xie, K.; Xue, A. Asynchronous dissipative state estimation for stochastic complex networks with quantized jumping coupling and uncertain measurements. *IEEE Trans. Neural Netw. Learn. Syst.* **2017**, *28* , 268–277. [CrossRef]
25. Shen, Y.; Wu, Z.G.; Shi, P.; Huang, T.W. Asynchronous filtering for Markov jump neural networks with quantized outputs. *IEEE Trans. Syst. Man Cybern. Syst.* **2019**, *49*, 433–443. [CrossRef]
26. Mahmoud, M.S.; Hamdan, M.M. Fundamental issues in networked control systems. *IEEE/CAA J. Autom. Sin.* **2018**, *5*, 902–922. [CrossRef]
27. Lunze, J.; Lehmann, D. A state-feedback approach to event-based control. *Automatica* **2010**, *46*, 211–215. doi: 10.1016/j.automatica.2009.10.035. [CrossRef]
28. Guan, Y.; Han, Q.; Ge, X. On asynchronous event-triggered control of decentralized networked systems. *Inf. Sci.* **2018**, *425*, 127139. [CrossRef]
29. Zhang, X.; Han, Q. Event-based H∞ filtering for sampled data systems. *Automatica* **2015**, *51*, 55–69. doi: 10.1016/j.automatica.2014.10.092. [CrossRef]
30. Ding, D.;Wang, Z.; Han, Q. A set-membership approach to event-triggered filtering for general nonlinear systems over sensor networks. *IEEE Trans. Autom. Control* **2020**, *65*, 1792–1799. [CrossRef]
31. Li, H.; Zuo, Z.; Wang, Y. Event triggered control for Markovian jump systems with partially unknown transition probabilities and actuator saturation. *J. Frankl. Inst.* **2016**, *358*, 1848–1861. [CrossRef]
32. Ma, R. J.; Shao, X. G.; Liu, J. X.; Wu, L. Event-triggered sliding mode control of Markovian jump systems against input saturation. *Syst. Control. Lett.* **2019**, *134*, 104525. [CrossRef]
33. Wang, H. J.; Shi, P.; Agarwal, R. K. Network-based event-triggered filtering for Markovian jump systems. *Int. J. Control* **2016**, *89*, 1096–1110. [CrossRef]
34. Wu, H. N.; Cai, K. Y. Mode-independent robust stabilization for uncertain Markovian jump nonlinear systems via fuzzy control. *IEEE Trans. Cybern.* **2005**, *36*, 509–519. [CrossRef]
35. Souza, C. E. De.; Trofino, A.; Barbosa, K. A. Mode-independent H∞ filters for Markovian jump linear systems. *IEEE Trans. Autom. Control* **2006**, *51*, 1837–1841. [CrossRef]
36. Shi, P.; Yin, Y.; Liu, F.; Zhang, J. Robust control on saturated markov jump systems with missing information. *Inf. Sci.* **2014**, *265*, 123–138. [CrossRef]
37. Shen, H.; Li, F.; Wu, Z.; Park, J. Finite-time L_2-L∞ tracking control for Markov jump repeated scalar nonlinear systems with partly usable model information. *Inf. Sci.* **2016**, *332*, 153–166. [CrossRef]
38. Wu, Z.G.; Shi, P.; Shu, Z.; Su, H.; Lu, R. Passivity-based asynchronous control for Markov jump systems. *IEEE Trans. Autom. Control* **2017**, *62*, 2020–2025. [CrossRef]
39. Yao, D.Y.; Lu, R.Q.; Xu, Y.; Wang, L.J. Robust H∞ filtering for Markov jump systems with mode-dependent quantized output and partly unknown transition probabilities. *Signal Processing* **2017**, *137*, 328–338. [CrossRef]
40. Wang, Y.Q.; An, Q.C.; Wang, R.H.; Zhang, S.Y. Reliable control for event-triggered singular Markov jump systems with partly unknown transition probabilities and actuator faults. *J. Frankl. Inst.* **2019**, *356*, 1828–1855. [CrossRef]
41. Zhang, L.X.; Boukas, E.K. Mode-dependent H∞ filtering for filtering for discrete-time Markovian jump linear systems with partly unknown transition probabilities. *Automatica* **2009**, *45*, 1462–1467. [CrossRef]
42. Liu, Y.; Wang, Z.; Liang, J.; Liu, X. Stability and synchronization of discrete-time Markovian jumping neural networks with mixed mode-dependent time delays. *IEEE Trans. Neural Netw.* **2009**, *20*, 1102–1116. [CrossRef] [PubMed]
43. Boyd, S.; Ghaoui, L.; Feron, E.; Balakrishnan, V. Linear Matrix Inequalities in System and Control Theory. *Proc. IEEE* **1994**, *86*, 2473–2474. [CrossRef]

Article

Identification of Control Parameters for Converters of Doubly Fed Wind Turbines Based on Hybrid Genetic Algorithm

Linlin Wu [1], Hui Liu [1], Jiaan Zhang [2,*], Chenyu Liu [3,*], Yamin Sun [1], Zhijun Li [2] and Jingwei Li [3]

[1] State Grid Jibei Electric Power Co., Ltd., Research Institute, Beijing 100045, China; wulin226@163.com (L.W.); liuhtj@163.com (H.L.); sunyamin8888@163.com (Y.S.)
[2] State Key Laboratory of Reliability and Intelligence of Electrical Equipment, Hebei University of Technology, Tianjin 300130, China; zhijun_li@263.net
[3] College of Artificial Intelligence and Data Science, Hebei University of Technology, Tianjin 300401, China; tpcnljwyx@163.com
* Correspondence: 2011086@hebut.edu.cn (J.Z.); liuchenyux@163.com (C.L.)

Abstract: The accuracy of doubly fed induction generator (DFIG) models and parameters plays an important role in power system operation. This paper proposes a parameter identification method based on the hybrid genetic algorithm for the control system of DFIG converters. In the improved genetic algorithm, the generation gap value and immune strategy are adopted, and a strategy of "individual identification, elite retention, and overall identification" is proposed. The DFIG operation data information used for parameter identification considers the loss of rotor current, stator current, grid-side voltage, stator voltage, and rotor voltage. The operating data of a wind farm in Zhangjiakou, North China, were used as a test case to verify the effectiveness of the proposed parameter identification method for the Maximum Power Point Tracking (MPPT), constant speed, and constant power operation conditions of the wind turbine.

Keywords: wind power; doubly fed induction generator; parameter identification; immune algorithm; genetic algorithm

Citation: Wu, L.; Liu, H.; Zhang, J.; Liu, C.; Sun, Y.; Li, Z.; Li, J. Identification of Control Parameters for Converters of Doubly Fed Wind Turbines Based on Hybrid Genetic Algorithm. *Processes* **2022**, *10*, 567. https://doi.org/10.3390/pr10030567

Academic Editors: Jie Zhang and Meihong Wang

Received: 27 January 2022
Accepted: 12 March 2022
Published: 14 March 2022

Publisher's Note: MDPI stays neutral with regard to jurisdictional claims in published maps and institutional affiliations.

1. Introduction

Considering the depletion of fossil fuels and the threat that greenhouse gas emissions pose to the global climate, the proportion of renewable energy will continue to expand [1], and wind power is poised to be a major contributor to this expansion. Owing to the different structures, types, and capacities of wind turbines, the control system strategy and parameters will also be different, resulting in different power generation characteristics [2]. As large-scale wind turbine integration will greatly affect the stability of the power system, the accuracy of the power system model has become an important technical issue in the operation, which needs to be consistent with the physical system, and the accuracy of the parameters is the key to ensuring model correctness. Parameter identification is a feasible method for model acquisition during the test and operation of wind turbines.

At present, the doubly fed induction generator (DFIG) is one of the main wind turbine types used on the market. The research on parameter identification of the DFIG has mainly focused on electrical parameters [3] and parameters of the converter control system. Classified from the perspective of algorithms, it can be divided into two types: traditional statistical algorithms and intelligent algorithms. In [4], an Extended Kalman Filter (EKF) was proposed for parameter estimation of DFIG in wind turbine systems. Belmokhtar et al. [5] explored the recursive least-squares (RLS) online parameter identification of a DFIG operating in a wind energy conversion system. Based on the RLS, a two-stage identification method was applied in [6], and the correctness of the method was verified by simulation. Wang et al. [7] applied the damped least-squares algorithm to identify the parameters of a variable speed DFIG-based wind turbine generator for wind power dynamic analysis. Takahashi et al. [8] proposed a recursive least-squares sensorless identification method for online identification of permanent magnet synchronous generator parameters, and it effectively detects characteristic changes during aging and degradation. A new decoupled weighted recursive least-squares (DWRLS) method, proposed in [9], improves the modeling accuracy by separately estimating the parameters of the fast and slow dynamics. Xia et al. [10] improved the model's parameter discrimination accuracy based on forgetting the factor recursive least squares for the state-of-charge (SOC) estimation of a battery management system.

Most traditional identification methods require that the input signal is known and varies significantly. For some situations or systems, it may not be possible to obtain all necessary input signals accurately, making the method less adaptable. Especially for nonlinear systems, this often leads to low identification accuracy or poor global search ability. Therefore, some bionic optimization algorithms, which have been developed and applied to research on parameter identification, have gradually formed the current intelligent parameter identification method [11]. These methods include the artificial neural network, particle swarm optimization algorithm, and genetic algorithm.

Based on the artificial neural network method, Rong et al. [12] proposed a step-by-step identification strategy to get the electrical parameters of a generator. In [13], a performance evaluation model was constructed with long short-term memory (LSTM) neural units and auto-encoder (AE) networks to evaluate the degree of abnormal performance of wind turbines, and an adaptive threshold estimation method was established to identify key condition-monitoring parameters. Based on the particle swarm algorithm, Li et al. [14] realized the dynamic equivalence of multiple units and simplified the equivalent values of the electrical parameters for multi-wind turbines. In [15], a new method for estimating the parameters of a wind turbine DFIG and drivetrain system was proposed, and the global optimal estimation result was obtained based on the local estimation and the coordinated estimation method under different types of disturbances. In [16], a symbolic regression method was introduced to identify models of a horizontal-axis wind turbine with evolutionary multi-objective optimization. In [17], wind turbine structural parameters such as inertial parameters, the damping coefficient, axial strength, and gearbox damping ratio were identified based on the genetic algorithm. In [18], different control modes of wind turbines under different lower voltage levels were identified based on the genetic algorithm. In [19], the objective function was to minimize the active output power error between the equivalent model and the actual wind farm, and an improved genetic algorithm was used to identify the key parameters of a permanent magnet synchronous generator. In [20], a set of fan parameters was identified with fault record data and the genetic algorithm to identify, and the evaluation function was to calculate the total deviation between the original signal and the simulation result.

The local search ability of the traditional genetic algorithm is insufficient and prone to premature convergence. Consequently, the relevant parameters of an excitation system were decomposed into multiple sets in [21], and a niche genetic algorithm with a fitness-sharing mechanism was proposed to overcome the local convergence for parameters identification. In view of the particularity of the DFIG structure and the complexity of its inverter control system, in this paper, the generation gap value and immune strategy are applied, and a strategy to improve the parameter identification based on the genetic algorithm is introduced, namely "individual identification, elite retention and overall identification", so as to establish a hybrid genetic algorithm (HGA) suitable for parameter identification of the DFIG converter control system. It is also considered that data variables used for DFIG parameter identification, such as rotor current, stator current, grid voltage, stator voltage, and rotor voltage, may be missing during the operation of the wind farm.

2. DFIG Converter Control Model and Parameters to Be Identified

2.1. Converter Control Model and its Parameters to Be Identified

DFIG has been widely used in wind power systems and is mainly composed of a wind turbine, transmission chain system, wound induction generator, and control system [22]. Figure 1 shows the structure of the DFIG-based wind power generation system, where the stator windings are directly connected to the grid and the rotor windings are directly connected to the external power grid through back-to-back converters [23]. The back-to-back converter provides three-phase rotor excitation power with adjustable amplitude, frequency, and phase, and ensures that the slip power can flow in both directions.

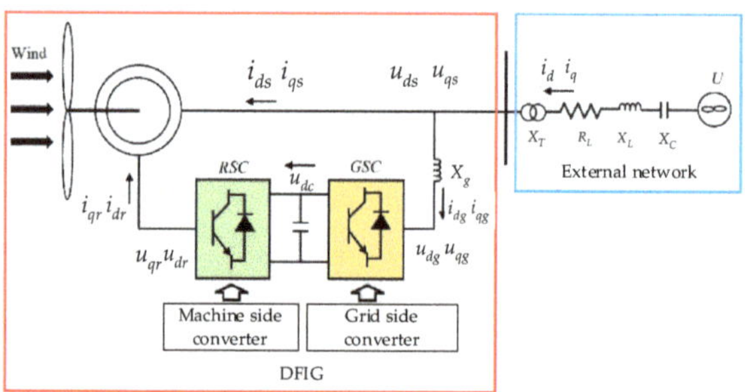

Figure 1. Structure of doubly fed induction generator (DFIG)-based wind power generation system.

The converter of the doubly fed wind turbine consists of a rotor-side converter (RSC) and a grid-side converter [23]. The RSC realizes the variable speed and constant frequency operation of the DFIG by controlling the rotor excitation current, while the grid side converter (GSC) maintains the DC bus voltage constant by controlling the power output. Because the control objectives of the RSC and the GSC are different, the RSC adopts the stator flux linkage-oriented control method, while the GSC adopts the vector control method based on the grid voltage orientation. Figure 2a shows a block diagram of a typical control system of the RSC of a doubly fed wind turbine, which is a double closed-loop structure composed of an outer power loop and an inner current loop to realize active and reactive power decoupling control. Figure 2b shows the typical control system block diagram of the GSC of the DFIG, which is a double closed-loop structure of the DC voltage outer loop and the grid-connected current inner loop. The DC voltage outer loop is used to realize the DC voltage. The stable control of the innercurrent loop is used to achieve fast tracking of active current and reactive current.

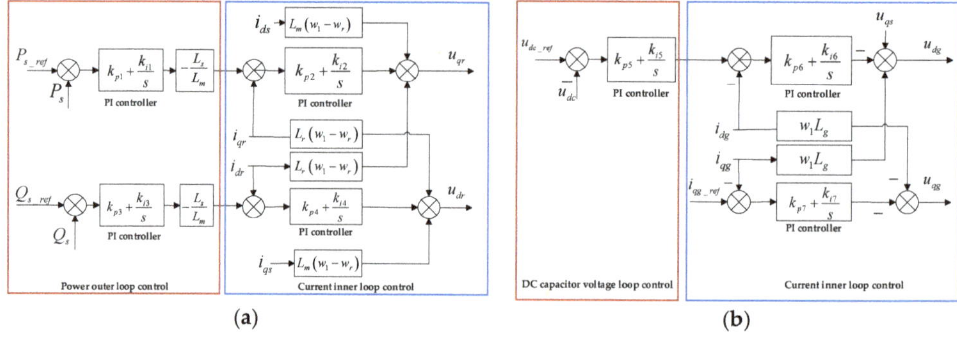

Figure 2. Control strategy diagram of the converter. (**a**) Typical control system block diagram of the rotor-side converter (RSC); (**b**) Typical control system block diagram of the grid side converter (GSC).

It can be known from the current inner loop control flow in Figure 2a that the output expression of the double closed-loop PI controller of the rotor-side converter is as follows:

$$\begin{cases} u_{dr} = \frac{dx_4}{dt}(k_{p4} + \frac{k_{i4}}{s}) - L_r(\omega_1 - \omega_r)i_{qr} - L_m(\omega_1 - \omega_r)i_{qs} \\ u_{qr} = \frac{dx_2}{dt}(k_{p2} + \frac{k_{i2}}{s}) + L_r(\omega_1 - \omega_r)i_{dr} + L_m(\omega_1 - \omega_r)i_{ds} \end{cases} \tag{1}$$

where k_{p2} and k_{p4} are the proportional coefficients, while k_{i2} and k_{i4} are the integral coefficients, which are in the two PI controllers of RSC. L_r and L_m are the inductance of the rotor winding and the mutual inductance between the stator winding and the rotor winding, respectively. w_1 and w_r are the synchronous speed and the rotor speed of the wind turbine, respectively. i_{dr} and i_{qr} are respectively the d-axis and q-axis components of the rotor current in the dp coordinate system. x_4 and x_2 are state variables.

As the current inner loop control flow in Figure 2b shows, the output expression of the double closed-loop PI controller of the grid-side converter is as follows:

$$\begin{cases} u_{dg} = u_{qs} - \frac{dx_6}{dt}(k_{p6} + \frac{k_{i6}}{s}) + \omega_1 L_g i_{qg} \\ u_{qg} = -\frac{dx_7}{dt}(k_{p7} + \frac{k_{i7}}{s}) - \omega_1 L_g i_{dg} \end{cases} \tag{2}$$

where k_{p6}, k_{p7} and k_{i6}, k_{i7} are the proportional and integral parameters of the two PI controllers of the grid-side converter, respectively. L_g is the filter inductor on the grid side. i_{dg} and i_{qg} are the d-axis and q-axis components of the grid-side current in the dp coordinate system, respectively. x_6 and x_7 are state variables.

In Equations (1) and (2), the expressions of the state variables, x_2, x_4, x_6, and x_7 are shown in Equations (3) and (4), representing the rotor side and grid side, respectively.

$$\begin{cases} \frac{dx_1}{dt} = P_{s\_ref} - P_s \\ \frac{dx_2}{dt} = \frac{dx_1}{dt}(k_{p1} + \frac{k_{i1}}{s})(-\frac{L_s}{L_m}) - i_{qr} \\ \frac{dx_3}{dt} = Q_{s\_ref} - Q_s \\ \frac{dx_4}{dt} = \frac{dx_3}{dt}(k_{p3} + \frac{k_{i3}}{s})(-\frac{L_s}{L_m}) + \frac{U_s}{\omega_1 L_m} - i_{dr} \end{cases} \tag{3}$$

$$\begin{cases} \frac{dx_5}{dt} = u_{dc\_ref} - u_{dc} \\ \frac{dx_6}{dt} = \frac{dx_5}{dt}(k_{p5} + \frac{k_{i5}}{s}) - i_{dg} \\ \frac{dx_7}{dt} = i_{qg\_ref} - i_{qg} \end{cases} \tag{4}$$

In Equation (3), k_{p1} and k_{i1} are the proportional and integral parameters of the active power outer loop PI controller, respectively. k_{p3} and k_{i3} are the proportional and integral parameters of the reactive power outer loop PI controller, respectively. P_{s_ref}, P_s, Q_{s_ref}, and Q_s are the systemic active power reference value, active power output value, reactive power reference value, and reactive power output value, respectively. U_s represents the stator winding voltage, $u_{ds}^2 + u_{qs}^2 = U_s^2$. x_1, x_2, x_3, and x_4 are intermediate state variables.

In Equation (4), k_{p5} and k_{i5} are the proportional and integral parameters of the DC bus voltage outer loop PI control, respectively. u_{dc_ref} and u_{dc} are the DC bus voltage reference value and the DC bus voltage output value, respectively. i_{dg}, i_{qg_ref}, and i_{qg} are the d-axis components of the grid-side current, and the reference value of the q-axis components and the q-axis components are in the dp coordinate system. x_5, x_6, and x_7 are intermediate state variables.

In the converter parameter identification process, the known parameters include P_{s_ref}, P_s, Q_{s_ref}, Q_s, u_{dc_ref}, u_{dc}, i_{dr}, i_{qr}, i_{ds}, i_{qs}, u_{ds}, u_{qs}, u_{dr}, u_{qr}, u_{dg}, and u_{qg}. The parameters to be identified include electrical parameters (L_s, L_r, L_m, and L_g), rotor winding resistance (R_r), stator winding resistance (R_s), and parameters of the converter PI controller.

It is shown that PI controller parameters are important for sensitivity analysis [24,25]. Therefore, the parameters to be identified for the DFIG converter control system are k_{p1},

$k_{p2}\ldots k_{p7}$ and $k_{i1}, k_{i2}\ldots k_{i7}$ in Equations (3) and (4). In the DFIG control system, parameters are generally set as follows: $k_{p2} = k_{p4}$, $k_{i2} = k_{i4}$, $k_{p6} = k_{p7}$, $k_{i6} = k_{i7}$ and $k_{i6} = k_{i7}$, so this setting is also used in this paper. The parameters to be identified for the back-to-back converter control system are shown in Tables 1 and 2.

Table 1. Parameters to be identified for the rotor-side converter (RSC).

Parameters to be Identified	Parameter Description
k_{p1}	Proportional parameters of the active power outer loop PI controller
ki_1	Integral parameters of the active power outer loop PI controller
k_{p2}	Proportional parameters of the rotor q-axis current inner loop PI controller
ki_2	Integral parameters of the rotor q-axis current inner loop PI controller
k_{p3}	Proportional parameters of the reactive power outer loop PI controller
ki_3	Integral parameters of the reactive power outer loop PI controller
k_{p4}	Proportional parameters of the rotor d-axis current inner loop PI controller
ki_4	Integral parameters of the rotor d-axis current inner loop PI controller

Table 2. Parameters to be identified for grid side converter (GSC).

Parameters to be Identified	Parameter Description
k_{p5}	Proportional parameters of the DC bus voltage outer loop PI controller
ki_5	Integral parameters of the DC bus voltage outer loop PI controller
k_{p6}	Proportional parameters of the grid-side current d-axis PI controller
ki_6	Integral parameters of the grid-side current d-axis PI controller
k_{p7}	Proportional parameters of the grid-side current q-axis PI controller
ki_7	Integral parameters of the rotor q-axis current inner loop PI controller

There are various operating modes of the DFIG, such as maximum power point tracking (MPPT) mode, constant speed mode, and constant power mode. In these modes, the control structure of the RSC and GSC are familiar, but the parameters may be different, which should be considered in the parameter identification method. Especially in the constant power mode, the blade angle of the DFIG will change. Because the blade controller only operates under specific wind speed conditions, even if the blade controller model is introduced here, the complexity of parameter identification method will not increase. Therefore, this paper only discusses the identification method of relevant parameters for the converter control system. Moreover, for generator parameter identification, it is mentioned in literature [17] that the greater the degree of disturbance, the higher are the accuracy of the identification results. Therefore, a certain disturbance should be applied to the DFIG at the beginning of parameter identification, and the data obtained in this way will make the identification results more accurate.

2.2. Identification of Converter Control Parameters in the Absence of Certain Variables

Data for certain variables corresponding to some key working conditions for parameter identification may be missing or contain errors occasionally. The parameter identification method in this case has to be considered. In a group of variables, there is usually a situation where one type of variable is not available, such as rotor current, stator current, stator voltage, or rotor-side or grid-side voltage.

In actual working conditions, because the excitation voltage on the rotor side of the doubly fed fan is not collected, the rotor side voltage is missing. Figure 2 shows that the loss of rotor-side voltage will only affect the solution of the RSC equations. During the parameter identification of DFIG, in the description equation of the RSC, the flux linkage equation, the voltage equation, and the state equation are substituted into the output equation so as to eliminate the variable of the rotor-side voltage. Expressions of the rotor-side simultaneous equations are shown in Equation (5).

$$\begin{cases} \frac{dx_1}{dt} = P_{s\_ref} - P_s \\ \frac{dx_2}{dt} = \frac{dx_1}{dt}(k_{p1} + \frac{k_{i1}}{s})(-\frac{L_s}{L_m}) - i_{qr} \\ \frac{dx_3}{dt} = Q_{s\_ref} - Q_s \\ \frac{dx_4}{dt} = \frac{dx_3}{dt}(k_{p3} + \frac{k_{i3}}{s})(-\frac{L_s}{L_m}) + \frac{U_s}{\omega_1 L_m} - i_{dr} \\ \frac{dx_4}{dt}(k_{p4} + \frac{k_{i4}}{s}) = \frac{d\psi_{dr}}{dt} + R_r i_{dr} \\ \frac{dx_2}{dt}(k_{p2} + \frac{k_{i2}}{s}) = \frac{d\psi_{qr}}{dt} + R_r i_{qr} \\ \psi_{dr} = L_r i_{dr} + L_m i_{ds} \\ \psi_{qr} = L_r i_{qr} + L_m i_{qs} \end{cases} \tag{5}$$

According to Equation (5), the parameters to be identified are electrical parameters (L_m, L_r, L_s, R_r) and controller parameters $(k_{p1}, k_{i1}, k_{p2}, k_{i2}, k_{p3}, k_{i3}, k_{p4}, k_{i4})$. Other types of variables missing can also be handled in a similar way.

3. Hybrid Genetic Algorithm for Converter Control System Identification

At present, the swarm intelligence algorithms mainly used in wind power parameter identification research problems include the genetic algorithm, particle swarm algorithm, ant colony algorithm, differential evolution algorithm. The use of these algorithms alone in parameter identification will inevitably lead to some shortcomings and limitations. For example, the genetic algorithm easily falls into the local optimum when solving the objective function in parameter identification. The particle swarm algorithm highlights strong robustness in the process of use; however, the initial parameter setting relies on experience and experimentation. The differential evolution algorithm highlights the characteristics of strong robustness during use and exhibits a high degree of parallelism, but it easily falls into premature convergence. Moreover, the early convergence of the ant colony algorithm is relatively low. It is fast and has high precision, but it lacks effective mutation measures, the convergence is slow in the later stage, and the algorithm easily falls into the local optimum. In addition, the parameter identification model becomes complicated in the absence of key information such as rotor current, rotor voltage, stator current, stator voltage, grid-side current, and grid-side voltage, which makes the identification process easier for a single group of intelligent algorithms stuck in a local optimum. Based on this, an identification method for the control parameters of the DFIG converter based on the hybrid genetic algorithm is proposed in this paper.

3.1. Hybrid Genetic Algorithm

Genetic algorithm (GA) is a global optimization search algorithm based on evolutionary mechanisms such as good and bad selection and genetic variation in the process of biological survival [26]. GA mainly realizes the problem-solving process through four basic operations: reproduction, mutation, competition, and selection. Compared with the least-squares method, GA does not need to consider the influence of the initial value of the function for the identification results, and only needs to get the form of all objective functions to obtain the optimal solution [27].

The immune algorithm (IM) is proposed based on the diversity of the immune system and the learning and memory mechanism, which can be simulated by the recognition and binding between antibodies and antigens in the immune system and the production of antibodies [28].

The hybrid genetic algorithm proposed in this paper introduces an immune strategy and generation gap value, and retains the elite individuals in the population in the memory bank. The elite individuals are no longer genetically manipulated to ensure that each chromosome has immune memory function. Then, the excellent genes in the individual are retained in the iterative process so as to improve the convergence speed and avoid falling into local solutions. The specific steps of the hybrid genetic algorithm include population initialization, calculating the population fitness value, calculating affinity, calculating the concentration of antibodies, and calculating the expected reproduction rate of antibodies.

3.1.1. Calculating Affinity

Affinity represents the binding degree between the antigen and antibody, which correspond to the objective function and feasible solution of the objective function, respectively [25]. Obtaining the affinity between the antibody and antigen in the genetic algorithm is done to calculate the individual fitness value (F_v). The greater the value of F_v, the greater the affinity. The affinity between the antibody and antibody in the hybrid genetic algorithm is the approximation between two chromosomes.

Whether the absolute error between the fitness values of the two individuals is less than a certain threshold (ε) is used to judge the similarity between two antibodies. As shown in Equation (6), if it is less than ε, the two individuals are considered to be approximately the "same", so $S_{v,s} = 1$; otherwise, the two individuals are different, so $S_{v,s} = 0$.

$$S_{i,j} = \begin{cases} 1, & \left|F_{vi} - F_{vj}\right| \leq \varepsilon \\ 0, & \left|F_{vi} - F_{vj}\right| > \varepsilon \end{cases}. \tag{6}$$

Here, F_{vi} and F_{vj} represent the fitness values of two individuals.

3.1.2. Calculation of the Concentration of Antibody

The antibody concentration C_v represents the ratio of individuals and their similar individuals in the current population to the total number of individuals [17], as shown in Equation (7).

$$C_v = \frac{\sum\limits_{j \in N} S_{v,s}}{N}. \tag{7}$$

Here, N is the total number of antibodies, and j is the number of similar individuals of an individual. $S_{v,s}$ represents the degree of similarity between individual v and s. If the value of Equation (7) is 1, it means that the two individuals are similar; otherwise, if it is 0, they are different.

3.1.3. Calculation of the Expected Reproduction Rate of the Antibody

The expected reproduction rate of antibodies is used to promote the inheritance and variation of superior antibodies and ensure the diversity of antibodies. The expected reproduction probability of the antibody is determined by the fitness value (F_v) and the concentration (C_v) of the antibody; that is,

$$\begin{cases} P = \alpha \frac{\Delta y}{\sum \Delta y} + (1 - \alpha) \frac{C_v}{\sum C_v} \\ \Delta y = \frac{1}{F_v} \end{cases}, \tag{8}$$

where α is a constant value, and the deviation value (Δy) is the inverse value of the fitness value (F_v). According to Equation (8), the individual fitness value, deviation value, and individual concentration will affect the expected reproduction probability.

3.2. Application of Hybrid Genetic Algorithm for DFIG Control System Identification

The parameter identification of the DFIG control system needs to fit the actual output curve and model output curve. The smaller the difference between the two sets of curves, the closer the identified parameters are to the actual values. Therefore, in the parameter identification of the DFIG converter control system, the fitness calculation function used is as follows:

$$\begin{cases} F_v = \frac{1}{\Delta y} \\ \Delta y = \sqrt{\dfrac{\sum\limits_{i=1}^{n}\left[y_d(i)-y_{d_{data}}(i)\right]^2}{n}} + \sqrt{\dfrac{\sum\limits_{i=1}^{n}\left[y_q(i)-y_{q_{data}}(i)\right]^2}{n}} \end{cases}, \tag{9}$$

where y_d and y_q are respectively the d-axis and q-axis response curves of the actual controller output; $y_{d_{data}}$ and $y_{q_{data}}$ are respectively the *d*-axis and *q*-axis output simulation curves of the identified model; and n represents the number of curve data.

As shown in Figure 2, both the RSCs and GSCs include current inner loop control blocks containing parameters $k_{p4} = k_{p2}$, $k_{i4} = k_{i2}$, $k_{p7} = k_{p6}$ and $k_{i7} = k_{i6}$. Therefore, the strategy of "individual identification, elite retention, and overall identification" will be adopted for parameter identification.

When identifying the parameters of the DFIG converter control system, the real number coding method is chosen to initialize the population. Taking the RSC as an example, the flow chart of the hybrid genetic algorithm in parameter identification of RSC of the DFIG is shown in Figure 3, and the corresponding steps are as follows:

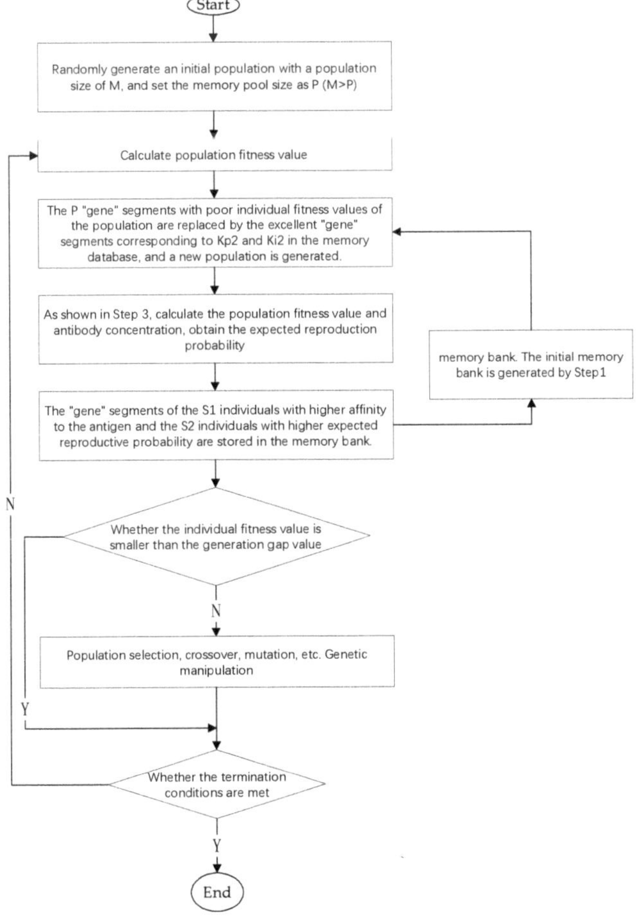

Figure 3. Flow chart for parameter identification of rotor-side converter.

Step 1: A separate fitting strategy for the d-axis and q-axis is adopted for the RSC. The evolutionary generations in the genetic algorithm are set as m_1 ($m_1 > 5$). The elite retention strategy is adopted, and the top five individuals with the highest fitness in the population are selected as elite individuals after one operation. The "gene" fragments corresponding to k_{p2} and k_{i2} in these five individuals are extracted and averaged to obtain a group of excellent "gene" fragments. Thus, n_1 groups of excellent "gene" fragments will be obtained from the d-axis and q-axis. These "gene" fragments will be used as the initial memory bank of the algorithm. The memory bank is set to be composed of $P = 2 * n_1$ individuals.

Step 2: The overall identification strategy of the d-axis and q-axis is adopted. The evolutionary algebra in the genetic algorithm [29] is set to M_1, and the initial population has M ($M > P$) randomly generated individuals. The fitness value of these M individuals is calculated by Equation (9). The "gene" fragments corresponding to k_{p2} and k_{i2} in the P individuals with the lowest fitness are replaced with excellent "gene" fragments in the memory bank.

Step 3: The expected reproductive probability of the antibody in the parent population is calculated. First, the concentration of the antibody is calculated by Equations (6) and (7). Then, the expected reproduction probability of the antibody is obtained with Equation (8).

Step 4: Progeny populations are produced. When updating the memory bank, the "gene" fragments corresponding to k_{p2} and k_{i2} in s_1 individuals with higher fitness values are first stored in the memory bank. According to the expected reproduction probability of the remaining individuals in the population, the "gene" segments corresponding to k_{p2} and k_{i2} in the $s_2 = P\text{-}s_1$ individuals with a high expected reproduction probability are stored in the memory bank.

Step 5: The fitness value and deviation value (Δ_y) of all individuals in the current population are calculated. Comparing Δ_y with the generation gap (σ), those with $\Delta y < \sigma$ are set as elite individuals and directly retained, while those with $\Delta y > \sigma$ go to the next genetic operation.

Step 6: The non-elite individuals in step 5 are put in order. Non-elite individuals evolve through selection, crossover, and mutation [29]. In the selection operation, the details are shown in Equation (8).

Step 7: If the termination conditions are met, the current individual value is output; otherwise, step 2 is followed.

Step 8: The above processes from step 2 to step 6 are run for n_1 times with the initial memory unchanged, and the running results to reduce the error are averaged.

The structure of the GSC is basically similar to that of the RSC, and so are the identification steps.

4. Test Case

The following test was used to evaluate the effectiveness of the proposed hybrid genetic algorithm. The operating conditions of MPPT, constant power, and constant speed were tested based on the simulation model (Section 4.1). The missing variables of rotor current, rotor voltage, stator current, stator current, stator voltage, and grid-side voltage were tested based on the simulation model (Section 4.2). Recorded data from a wind farm in Zhangjiakou, China, under the rotor voltage variable missing condition, were examined under power oscillation conditions (Section 4.3).

4.1. Simulation Test of Converter Parameter Identification under Three Operating Conditions

DFIG data with a wind speed of 8 m/s, 10.5 m/s, and 12.5 m/s, and fluctuation value of 1 m/s, 0.4 m/s, and 1 m/s were used for the test of MPPT, constant speed, and constant power operation conditions, respectively. The data of rotor current, rotor voltage, stator current, stator voltage, grid-side current, and grid-side voltage were used.

Taking the PI controller of the RSC as an example, the identification strategy of "individual identification, elite retention, and overall identification" was adopted. The basic parameters of the hybrid genetic algorithm were set as follows: evolutionary algebra

$m_1 = 20$, number of runs $n_1 = 5$, number of memory bank individuals $P = 10$, $M = 30$, threshold $\varepsilon = 1$; $s_1 = 7$, $s_2 = 3$, and generation gap $\sigma = 0.05$.

The basic parameter settings for parameter identification of GSCs were basically the same as those of RSCs. In the running state of MPPT, the comparison of the identification results between genetic algorithm and hybrid genetic algorithm is shown in Figure 4. The identification results of the converter in the constant speed region and the constant power region also converged when the evolutionary algebra was 20. As shown in Figure 4, in the subsequent iterations, because the fitness of the new offspring was not as good as the fitness value of the parent, the algorithm would retain the parent population and discard the offspring population, so the final identification curve was a straight line, representing the optimal value searched by the algorithm at this time. Because the genetic algorithm had problems such as local optimality, after a certain number of iterations, the genetic algorithm could not jump out of the local optimality and could not find the global optimal solution, which produced a large error in the parameter identification result.

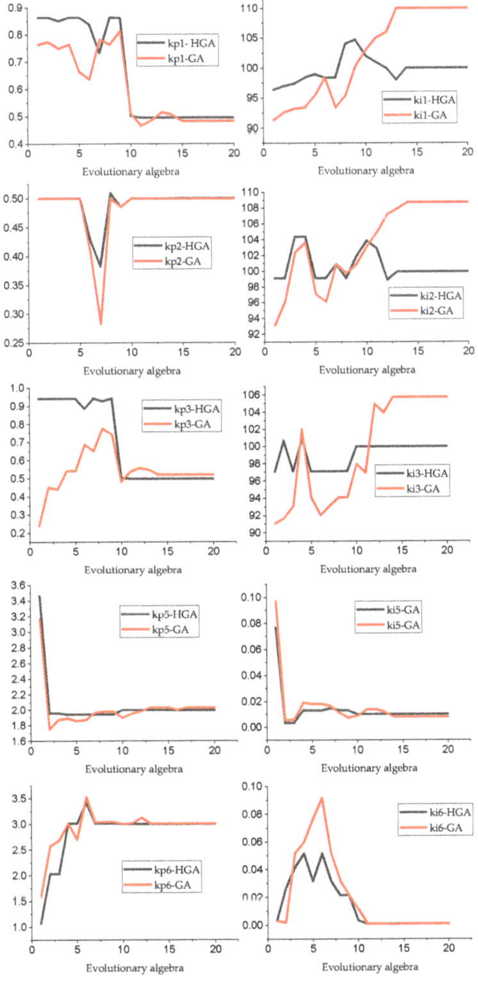

Figure 4. Comparison of identification results of the converter at the lower side of Maximum Power Point Tracking (MPPT).

Parameter identification results of the DFIG, RSC, and GSC under different operation conditions are shown in Tables 3 and 4, respectively.

Table 3. Identification results of RSC parameters in different control modes.

Operating Area	Algorithm	k_{p1}	k_{i1}	k_{p2}	k_{i2}	k_{p3}	k_{i3}
	actual value	0.5	100	0.5	100	0.5	100
MPPT	GA	0.4831	110.1348	0.4916	108.7231	0.5195	105.7642
	HGA	0.5001	99.9948	0.5000	99.9555	0.4995	100.0036
constant speed	GA	0.5063	103.0831	0.5011	99.5462	0.4969	100.1028
	HGA	0.5003	100.0700	0.5000	99.9641	0.4997	100.0049
constant power	GA	0.4824	102.7250	0.5113	108.2004	0.4893	98.0537
	HGA	0.4997	100.0030	0.5000	100.0073	0.5006	99.9974

Table 4. Identification results of GSC parameters in different control modes.

Operating Area	Algorithm	k_{p5}	k_{i5}	k_{p6}	k_{i6}
	actual value	2	0.01	3	0.001
MPPT	GA	2.004	0.0080	3.010	0.00087651
	HGA	2.000	0.01	3.000	0.00099918
constant speed	GA	2.001	0.0105	3.004	0.0009228
	HGA	2.000	0.01	3.000	0.0009458
constant power	GA	2.010	0.0103	3.006	0.0009022
	HGA	2.000	0.01	3.000	0.0009913

Among them, the parameters $k_{p4} = k_{p2}$, $k_{i4} = k_{i2}$; $k_{p7} = k_{p6}$, $k_{i7} = k_{i6}$.

It can be seen from Tables 3 and 4 that the hybrid genetic algorithm could identify the key parameters of the DFIG converter control systems effectively, as well as whether DFIG was in MPPT operation condition, constant speed operation condition or constant power operation condition. In addition, for the same set of control system parameters, the identification results are not consistent under different operation conditions. As the identification result of MPPT operation condition is closest to the actual value, the operation data in MPPT condition can be preferentially selected for practical engineering application.

4.2. Simulation Test of Converter Parameter Identification in the Absence of Variables

From the analysis in Section 4.1, it is known that the parameter identification results obtained by applying the data of DFIG in MPPT operation conditions are more accurate. Therefore, MPPT is used as a typical operating condition to test the effect of converter parameter identification under the variable missing condition. Based on the simulation model, the data of rotor current, rotor voltage, stator current, stator voltage, grid-side current, and grid-side voltage corresponding to the wind turbine in the MPPT operation condition were generated. Then, by simulating the missing data of rotor current, rotor voltage, stator current, stator voltage, grid-side current, or grid-side voltage data, the hybrid genetic algorithm proposed in this paper was used to identify the parameters of the DFIG converter.

4.2.1. Missing Rotor Current

When the rotor currents i_{dr} and i_{qr} were missing, according to Figure 2, the absence of rotor current did not affect the parameter identification of the GSC, but did affect the parameter identification process of the machine-side converter. Taking the unknown electrical parameters L_s and L_r and the L_m converter parameters k_{p1} k_{i1}, k_{p2}/k_{p4}, k_{i2}/k_{i4}, k_{p3}, and k_{i3} as unknown important parameters, the hybrid genetic algorithm was used to identify these unknown parameters. The identification results of electrical parameters and controller parameters are shown in Tables 5 and 6, respectively.

Table 5. Identification results of electrical parameters.

Electrical Parameter	L_s	L_r	L_m
actual value (pu)	0.0049	0.0049	0.0045
identification value (pu)	0.0051	0.0047	0.0044

Table 6. Identification of rotor-side control parameters.

Control Parameter	k_{p1}	k_{i1}	k_{p2}	k_{i2}	k_{p3}	k_{i3}
actual value	0.5	100	0.5	100	0.5	100
identification value	0.4962	102.5231	0.4899	99.8759	0.5043	103.4254

Among them, the default $k_{p2} = k_{p4}$, $k_{i2} = k_{i4}$.

4.2.2. Missing Stator Current

When the stator current i_{ds} and i_{qs} were missing, the situation was basically the same as the rotor current missing. When the stator current was missing, it is shown in Figure 2 that it did not affect the identification process of GSC parameters, only the process of machine-side converter parameter identification. The electrical unknown parameters were L_s, L_r, and L_m, and the remaining unknown parameters were the controller parameters k_{p1}, k_{i1}, k_{p2}/k_{p4}, k_{i2}/k_{i4}, k_{p3}, and k_{i3}. Therefore, a strategy of identifying the two sets of parameters as unknown parameters was adopted. The specific identification results are shown in Tables 7 and 8.

Table 7. Identification results of electrical parameters.

Electrical Parameters	L_s	L_r	L_m
actual value (pu)	0.0049	0.0049	0.0045
identification value (pu)	0.0043	0.0043	0.0048

Table 8. Identification of rotor-side control parameters.

Control Parameter	k_{p1}	k_{i1}	k_{p2}	k_{i2}	k_{p3}	k_{i3}
actual value	0.5	100	0.5	100	0.5	100
identification value	0.4912	99.6518	0.5145	99.1584	0.4899	99.5214

Among them, the default $k_{p2} = k_{p4}$, $k_{i2} = k_{i4}$.

4.2.3. Missing Grid-Side Converter Voltage

When the grid-side converter voltages u_{dg} and u_{qg} were missing, it can be seen from Figure 2 that the situation at this time was different from the situation in which the stator and rotor currents were missing. The GSC voltage loss would only affect the GSC. The identification of the parameters would not affect the identification of the parameters of the machine-side converter. When fitting the unknown parameters, only the grid-side filter inductance L_g was unknown for the electrical parameters, and the other unknown parameters were the controller parameters k_{p5}, k_{i5}, k_{p6}/k_{p7}, k_{i6}/k_{i7}, so the grid-side filter inductance L_g was the same as the grid-side controller. The parameters were identified together as unknown parameters. The results of parameter identification are shown in Tables 9 and 10.

Table 9. Identification results of electrical parameters.

Electrical Parameters	L_g
actual value (pu)	0.0040
identification value (pu)	0.0038

Table 10. Identification of grid-side control parameters.

Control Parameter	k_{p5}	k_{i5}	k_{p6}	k_{i6}
actual value	2	0.01	3	0.001
identification value	1.999	0.0098	3.001	0.00099

Among them, the default $k_{p7} = k_{p6}$, $k_{i7} = k_{i6}$.

4.2.4. Missing Stator Voltage

When the stator voltages u_{ds} and u_{qs} were missing, according to the structure of the machine-side converter and the structure of the grid-side converter in Figure 2, the absence of the stator voltage would affect the parameter identification of the machine-side converter and the GSC. Therefore, it was necessary to replace the stator voltage u_{ds} and u_{qs} in the output items of the machine-side converter and the GSC so as to complete the parameter identification research of the converter control system of the DFIG based on the absence of the stator voltage. As the structure of the GSC is simpler than that of the machine-side converter, when the parameters of the grid-side converter were first identified, at this time, the unknown parameters were the electrical parameters (L_m, L_r, L_s, and L_g) and controller parameters (k_{p5}, k_{i5}, k_{p6}/k_{p7}, k_{i6}/k_{i7}). The identification results are shown in Tables 11 and 12.

Table 11. Identification results of electrical parameters.

Electrical Parameters	L_s	L_g	L_m	L_r
actual value (pu)	0.0049	0.0040	0.0045	0.0049
identification value (pu)	0.0043	0.0048	0.0044	0.0049

Table 12. Identification of grid-side control parameters.

Control Parameter	k_{p5}	k_{i5}	k_{p6}	k_{i6}
actual value	2	0.01	3	0.001
identification value	1.898	0.0091	3.105	0.0015

Then, the machine-side converter was identified, and the parameters to be identified were the controller parameters k_{p1}, k_{i1}, k_{p2}/k_{p4}, k_{i2}/k_{i4}, k_{p3}, and k_{i3}. The identification results are shown in Table 13.

Table 13. Identification of rotor-side control parameters.

Control Parameter	k_{p1}	k_{i1}	k_{p2}	k_{i2}	k_{p3}	k_{i3}
actual value	0.5	100	0.5	100	0.5	100
identification value	0.4756	97.3581	0.5205	105.1554	0.5199	104.9518

Among them, the default $k_{p2} = k_{p4}$, $k_{i2} = k_{i4}$.

4.2.5. Missing Rotor-Side Voltage

When the rotor-side voltages u_{dr} and u_{qr} were missing, according to Figure 2, the absence of rotor-side voltages u_{dr} and u_{qr} did not affect the parameter identification of the GSC, but only affected the parameters of the machine-side converter. At this time, new unknown parameters would appear as stator resistance (Rs) and stator winding inductance (L_s). The electrical parameters L_m, R_s, L_s, identification results, and process are shown in Table 14.

Table 14. Identification results of electrical parameters.

Electrical Parameters	L_m	L_s	R_s
actual value (pu)	0.0045	0.0049	0.002
identification value (pu)	0.0046	0.0050	0.00199

Then, the strategy of identifying R_r and L_r together with the controller parameters k_{p1}, k_{i1}, k_{p2}/k_{p4}, k_{i2}/k_{i4}, k_{p3} and k_{i3} was adopted. The identification results are shown in Table 15.

Table 15. Identification results of electrical parameters.

Electrical Parameters	R_r	L_r
actual value (pu)	0.001	0.0049
identification value (pu)	0.0012	0.0049

Then, the machine-side converter was identified, and the parameters to be identified were the controller parameters k_{p1}, k_{i1}, k_{p2}/k_{p4}, k_{i2}/k_{i4}, k_{p3}, and k_{i3}. The identification results are shown in Table 16.

Table 16. Identification of rotor-side control parameters.

Control Parameter	k_{p1}	k_{i1}	k_{p2}	k_{i2}	k_{p3}	k_{i3}
actual value	0.5	100	0.5	100	0.5	100
identification value	0.5011	99.9878	0.5002	99.6645	0.4992	100.0125

Among them, the default $k_{p2} = k_{p4}$, $k_{i2} = k_{i4}$.

4.3. Engineering Application

A DFIG wind farm in Zhangjiakou of North China was selected as the test case. Figure 5 shows recorded data of the DFIG under the condition of power oscillation, where the time length was 21 s, and the sampling period was 0.01 s, wherein the rotor-side voltages u_{dr} and u_{qr} were missing.

In the absence of the rotor-side voltages u_{dr} and u_{qr}, the identification results are shown in Tables 17–19, in which the assumption of $k_{p4} = k_{p2}$, $k_{i4} = k_{i2}$, $k_{p7} = k_{p6}$ and $k_{i7} = k_{i6}$ was still adopted.

Table 17. Identification of DFIG generator parameters.

Unknown Parameter	R_s	L_s	L_m	R_r	L_r
identification value (pu)	0.022	4.857	4.68	0.026	4.796

Table 18. Identification of rotor-side control parameters.

Unknown Parameter	k_{p1}	k_{i1}	k_{p2}	k_{i2}	k_{p3}	k_{i3}
identification value (pu)	0.001	5.425	0.0242	2.001	0.001	4.996

Table 19. Identification of grid-side control parameters.

Unknown Parameter	k_{p5}	k_{i5}	k_{p6}	k_{i6}
identification value (pu)	2.158	20.751	5.056	2.000

Applying the identified parameters, a set of curves was obtained from the DFIG simulation model, and the comparison made with the original recorded data is shown in

Figure 5. It can be seen from Figure 6 that the amplitude and phase of the original data and the simulated power curve were extremely close, indicating that the identification results are relatively close to the actual values.

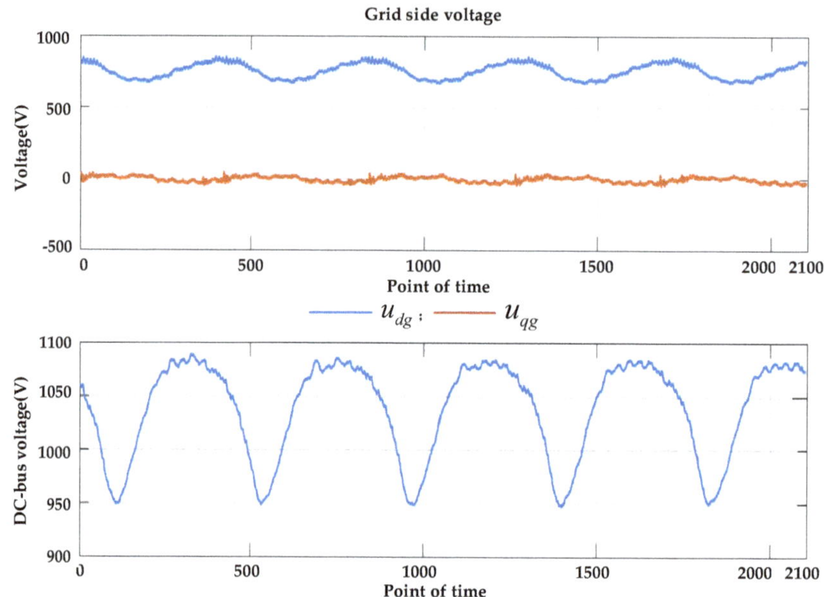

(**a**) voltage variables of U_{dg}, U_{qg}, and DC-bus voltage U_{dc}

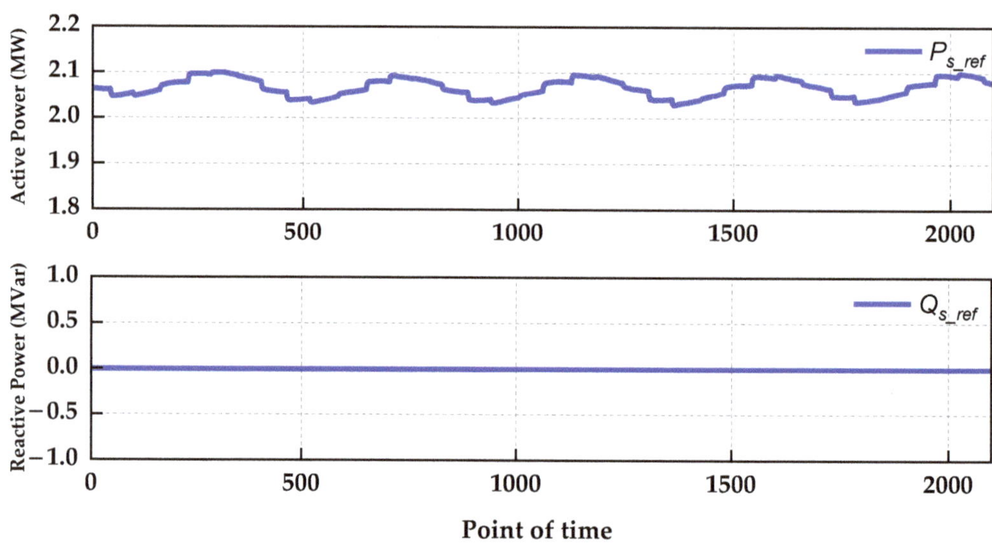

(**b**) reference variables of P and Q

Figure 5. *Cont.*

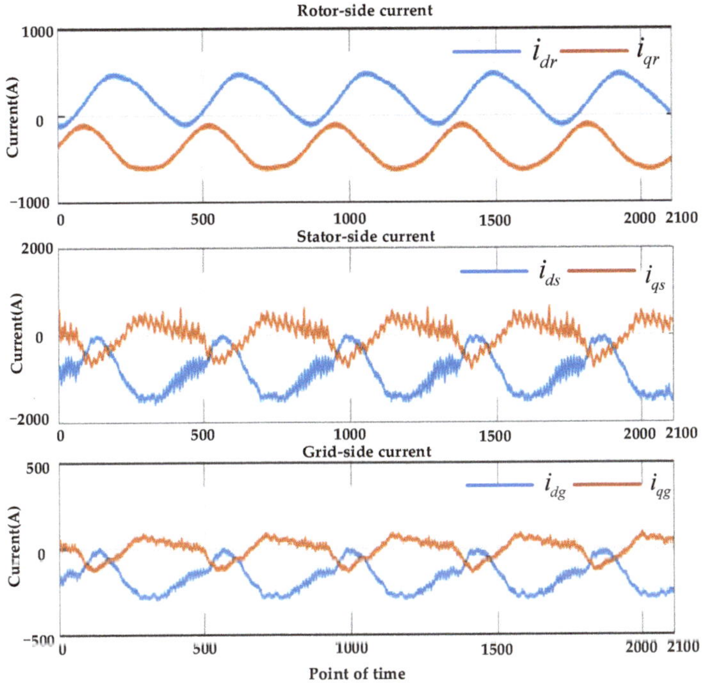

(**c**) current variables of I_{dg}, I_{qg}, I_{ds}, I_{qs}, I_{dr}, and I_{qr}

Figure 5. DFIG variables under the condition of power oscillation.

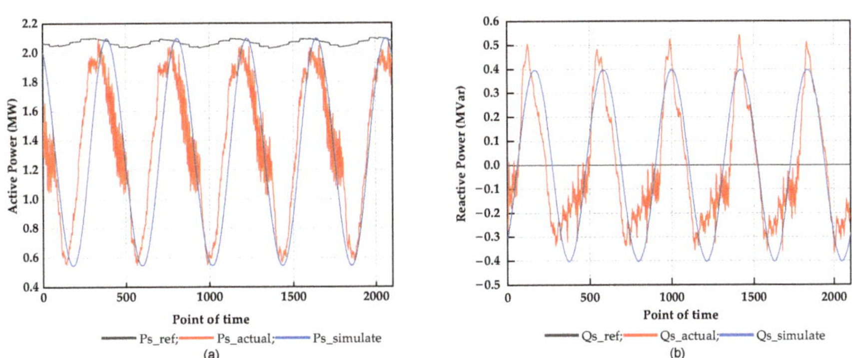

Figure 6. Comparison chart of the actual power curve and simulated power curve. (**a**) Comparison chart of the actual active power curve and simulated active power curve; (**b**) Comparison chart of the actual reactive power curve and simulated reactive power curve.

5. Conclusions

A hybrid genetic algorithm for parameter identification of the DFIG control system was proposed. The algorithm introduces the generation gap value and immune strategy, and adopts the identification strategy of "individual identification, elite retention, and overall identification." The test case showed that the identification results are different under different DFIG operating conditions, and the identification results under the MPPT operating condition are preferable.

According to the engineering requirements of parameter identification, we discussed the lack of DFIG variables, such as rotor voltage, rotor current, stator voltage, stator current, and grid-side voltage. If one of the variables is missing, the DFIG converter control system parameters can still be identified by the proposed method.

Author Contributions: Conceptualization, methodology, validation, writing—review and editing, L.W.; methodology, data curation, validation, writing—review and editing, H.L.; investigation, validation, supervision, writing—review and editing, J.Z.; software, visualization, writing—review and editing, C.L.; data curation, investigation, writing—review and editing, Y.S.; supervision, writing—review and editing, Z.L.; investigation, software, validation, J.L. All authors have read and agreed to the published version of the manuscript.

Funding: This work was supported by the science and technology project "Electromagnetic Transient Simulation Initialization Program and Simulation Data Visualization Software Development" of North China Electric Power Research Institute Co., Ltd, Beijing, China [grant number SGTYHT/21-JS-225].

Institutional Review Board Statement: Not applicable.

Informed Consent Statement: Not applicable.

Data Availability Statement: Not applicable.

Conflicts of Interest: The authors declare no conflict of interest.

References

1. Jia, K.; Gu, C.; Li, L.; Xuan, Z.; Bi, T.; Thomas, D. Sparse voltage amplitude measurement based fault location in large-scale photovoltaic power plants. *Appl. Energy* **2018**, *211*, 568–581. [CrossRef]
2. Kim, D.; El-Sharkawi, M.A. Dynamic equivalent model of wind power plant using parameter identification. *IEEE Trans. Energy Convers.* **2016**, *31*, 37–45. [CrossRef]
3. Jin, Y.; Lu, C.; Ju, P.; Rehtanz, C.; Wu, F.; Pan, X. Probabilistic Preassessment Method of Parameter Identification Accuracy with an Application to Identify the Drive Train Parameters of DFIG. *IEEE Trans. Power Syst.* **2020**, *35*, 1769–1782. [CrossRef]
4. Abdelrahem, M.; Hackl, C.; Kennel, R. Application of extended Kalman filter to parameter estimation of doubly-fed induction generators in variable-speed wind turbine systems. In Proceedings of the International Conference on Clean Electrical Power (ICCEP), Taormina, Italy, 16–18 June 2015.
5. Belmokhtar, K.; Ibrahim, H.; Merabet, A. Online parameter identification for a DFIG driven wind turbine generator based on recursive least squares algorithm. In Proceedings of the IEEE 28th Canadian Conference on Electrical and Computer Engineering, Halifax, NS, Canada, 3–6 May 2015.
6. Kong, M.; Sun, D.; He, J.; Nian, H. Control Parameter Identification in Grid-side Converter of Directly Driven Wind Turbine Systems. In Proceedings of the 12th IEEE PES Asia-Pacific Power and Energy Engineering Conference (APPEEC), Nanjing, China, 20–23 September 2020.
7. Wang, X.; Xiong, J.; Geng, L.; Zheng, J.; Zhu, S. Parameter identification of doubly-fed induction generator by the Levenberg-Marquardt-Fletcher method. In Proceedings of the IEEE Power & Energy Society General Meeting, Vancouver, BC, Canada, 21–25 July 2013.
8. Takahashi, K.; Matayoshi, H.; Senjyu, T.; Takahashi, H.; Howlader, A.M. Online Parameter identification of PMSG Wind turbine for Output Power control. In Proceedings of the TENCON 2019–2019 IEEE Region 10 Conference (TENCON), Kochi, India, 17–20 October 2019.
9. Zhang, C.; Allafi, W.; Dinh, Q.; Ascencio, P.; Marco, J. Online estimation of battery equivalent circuit model parameters and state of charge using decoupled least squares technique. *Energy* **2018**, *142*, 678–688. [CrossRef]
10. Xia, B.; Lao, Z.; Zhang, R.; Tian, Y.; Chen, G.; Sun, Z.; Wang, W.; Sun, W.; Lai, Y.; Wang, M.; et al. Online parameter identification and state of charge estimation of Lithium-Ion batteries based on forgetting factor recursive least squares and nonlinear kalman filter. *Energies* **2018**, *11*, 3. [CrossRef]
11. Zhao, Q.; Wu, L.; Wang, X.; Sheng, S.; Feng, Z. Overview of research on modelling and parameter identification of wind power generator. In Proceedings of the 16th IET International Conference on AC and DC Power Transmission (ACDC 2020), Online Conference, 2–3 July 2020.
12. Rong, Y.; Wang, H.; Yang, W.; Qi, H. Artificial neural network in the application of the doubly-fed type wind power generator parameter identification. In Proceedings of the IEEE Conference and Expo Transportation Electrification Asia-Pacific (ITEC Asia-Pacific), Beijing, China, 31 August–3 September 2014.
13. Chen, H.; Liu, H.; Chu, X.; Liu, Q.; Xue, D. Anomaly detection and critical SCADA parameters identification for wind turbines based on LSTM-AE neural network. *Renew. Energy* **2021**, *172*, 829–840. [CrossRef]
14. Li, Y.; Yang, J.; Yi, B.; Fang, R.; Zhang, D. Dynamic equivalence of doubly-fed wind turbines based on parameter identification and optimization. In Proceedings of the 4th International Conference on Mechatronics and Computer Technology Engineering (MCTE 2021), Xi'an, China, 15–17 October 2021.

15. Pan, X.; Ju, P.; Wu, F.; Jin, Y. Hierarchical parameter estimation of DFIG and drive train system in a wind turbine generator. *Front. Mech. Eng.* **2017**, *12*, 367–376.
16. La Cava, W.; Danai, K.; Spector, L.; Fleming, P.; Wright, A.; Lackner, M. Automatic identification of wind turbine models using evolutionary multiobjective optimization. *Renew. Energy* **2016**, *87*, 892–902. [CrossRef]
17. Liu, J.Z.; Guo, J.L.; Hu, Y.; Wang, J.; Liu, H. Dynamic modeling of wind turbine generation system based on grey-box identification with genetic algorithm. In Proceedings of the 36th Chinese Control Conference (CCC), Dalian, China, 26–28 July 2017.
18. Junxian, H.; Xiangyu, T.; Shi, Z.; Hong, S.; Haiyan, T.; Tao, L.; Peng, Z. The dynamic simulation model and parameter identification method of DFIG type wind generator for power system elec-tro-mechanic simulation. In Proceedings of the IEEE PES Asia-Pacific Power and Energy Engineering Conference (APPEEC), Hong Kong, China, 7–10 December 2014.
19. Gu, R.; Dai, J.; Zhang, J.; Miao, F.L.; Tang, Y. Research on Equivalent Modeling of PMSG-based Wind Farms using Parameter Identification method. In Proceedings of the 12th IEEE PES Asia-Pacific Power and Energy Engineering Conference (APPEEC), Nanjing, China, 2–3 July 2020.
20. Zhang, J.A.; Liu, H.F.; Liu, H.; Wu, L.; Wang, Y.H. Estimation of wind turbine parameters with piecewise trends identification. In Proceedings of the International Conference on Mechatronic Science, Electric Engineering and Computer (MEC), Shenyang, China, 20–22 December 2013.
21. Liu, X.; Yan, L.; Liu, Y.; Zhao, L.; Jie, J. Improved niche genetic algorithm based parameter identification of excitation system considering parameter identifiability. In Proceedings of the 14th IET International Conference on AC and DC PowerTransmission (ACDC 2018), Chengdu, China, 28–29 June 2018.
22. Li, S.; Haskew, T.A.; Jackson, J. Integrated power characteristic study of DFIG and its frequency converter in wind power generation. *Renew. Energy* **2010**, *35*, 42–51. [CrossRef]
23. Taveiros, F.E.V.; Barros, L.S.; Costa, F.B. Back-to-back converter state-feedback control of DFIG (doubly-fed induction generator)-based wind turbines. *Energy* **2015**, *89*, 896–906. [CrossRef]
24. Shen, W.; Li, H. A Sensitivity-Based Group-Wise parameter identification algorithm for the electric model of Li-Ion battery. *IEEE Access* **2017**, *5*, 4377–4387. [CrossRef]
25. Moshksar, E.; Ghanbari, T. Adaptive estimation approach for parameter identification of photovoltaic modules. *IEEE J. Photovolt.* **2017**, *7*, 614–623. [CrossRef]
26. Wasilewski, J.; Wiechowski, W.; Bak, C.L. Harmonic domain modeling of a distribution system using the DIgSILENT PowerFactory software. In Proceedings of the International Conference on Future Power Systems, Amsterdam, The Netherlands, 18 November 2005.
27. Li, W.; Chai, Z.; Tang, Z. A decomposition-based multi-objective immune algorithm for feature selection in learning to rank. *Knowl.-Based Syst.* **2021**, *234*, 107577.
28. Su, Y.; Luo, N.; Lin, Q.; Li, X. Many-objective optimization by using an immune algorithm. *Swarm Evol. Comput.* **2022**, *69*, 101026. [CrossRef]
29. Ganthia, B.P.; Barik, S.K.; Nayak, B. Genetic Algorithm Optimized and Type-I fuzzy logic controlled power smoothing of mathematical modeled Type-III DFIG based wind turbine system. *Mater. Today Proc.* 2021, *in press*.

Article

A Novel Fault Detection Scheme Based on Mutual k-Nearest Neighbor Method: Application on the Industrial Processes with Outliers

Jian Wang [1], Zhe Zhou [1,*], Zuxin Li [2] and Shuxin Du [1]

[1] School of Engineering, Huzhou University, Huzhou 313000, China; Wangjian1191020@163.com (J.W.); shxdu@zjhu.edu.cn (S.D.)
[2] School of Science and Engineering, Huzhou College, Huzhou 313000, China; lzx@zjhu.edu.cn
* Correspondence: zzhou@zjhu.edu.cn; Tel.: +86-572-232-0682

Abstract: The k-nearest neighbor (kNN) method only uses samples' paired distance to perform fault detection. It can overcome the nonlinearity, multimodality, and non-Gaussianity of process data. However, the nearest neighbors found by kNN on a data set containing a lot of outliers or noises may not be actual or trustworthy neighbors but a kind of pseudo neighbor, which will degrade process monitoring performance. This paper presents a new fault detection scheme using the mutual k-nearest neighbor (MkNN) method to solve this problem. The primary characteristic of our approach is that the calculation of the distance statistics for process monitoring uses MkNN rule instead of kNN. The advantage of the proposed approach is that the influence of outliers in the training data is eliminated, and the fault samples without MkNNs can be directly detected, which improves the performance of fault detection. In addition, the mutual protection phenomenon of outliers is explored. The numerical examples and Tenessee Eastman process illustrate the effectiveness of the proposed method.

Keywords: k-nearest neighbor; outliers; pseudo-neighbors; mutual nearest neighbor; fault detection; process monitoring

Citation: Wang, J.; Zhou, Z.; Li, Z.; Du, S. A Novel Fault Detection Scheme Based on Mutual k-Nearest Neighbor Method: Application on the Industrial Processes with Outliers. *Processes* **2022**, *10*, 497. http://doi.org/10.3390/pr10030497

Academic Editors: Massimiliano Barolo

Received: 25 January 2022
Accepted: 27 Februray 2022
Published: 1 March 2022

Publisher's Note: MDPI stays neutral with regard to jurisdictional claims in published maps and institutional affiliations.

1. Introduction

Data are being generated all the time in industrial processes. Since industry became a separate category from social production, data collection and use in industrial production has gradually increased. In this context, data-driven multivariate statistical process monitoring (MSPM) methods have developed leaps and bounds [1,2], where principal component analysis (PCA) methods are the most widely used [3–6]. However, there are cases where PCA-based fault detection methods do not perform well. For example, the detection threshold of Hotelling-T^2 and squared prediction error (SPE) are calculated based on the premise that process variables satisfy a normal or Gaussian distribution. Due to the nonlinearity, non-Gaussianity, and multimodality in industrial processes, it is not easy to meet this assumption in practice [7–11]. Therefore, the traditional PCA-based process monitoring method has poor monitoring performance when facing the above problems [12–16].

He and Wang [11] proposed a non-parametric lazy fault detection method based on the k-nearest neighbor rule (FD-kNN) to deal with the above problems. The main idea is to measure the difference between samples by distance; that is the online normal samples and training samples are similar, but fault samples and training samples are significantly different. It only uses samples' paired distance to perform fault detection and has no strict requirements for data distribution. Hence, this method provide an alternative way to overcome the nonlinearity, non-Gaussianity, multimodality in industrial processes.

However, the data collected in the actual industrial process usually contain a certain amount of noise and even outliers, and the quality of the data cannot be guaranteed [17,18].

Outliers are generally those samples that are far from the normal training samples and tend to behave statistically inconsistent with the other normal samples [19,20]. In the actual industrial processes, outliers are usually introduced when measurement or recording errors are made. In addition, the considerable process noise is also one of the main reasons for the generation of outliers [20].

The neighbors of the samples found by kNN from a data set containing noises or outliers may not be actual neighbors but a pseudo-nearest neighbor (PNN). For example, in Figure 1, the samples x_2, x_3, and x_4 are the 3-NNs of x_1, but sample x_1 is not one of the 3-NNs of x_2, x_3, and x_4. In other words, x_2, x_3, and x_4 are the PNNs of x_1. This interesting phenomenon can be explained with an example from human interaction: I regard you as one of my best friends, but I am not among your best friends. As can be seen from Figure 1, the sample x_1 is far away from its pseudo-neighbors so that the detection threshold calculated by the pseudo-neighbors in the training phase will have a significant deviation, which will seriously degrade the detection performance of the FD-kNN.

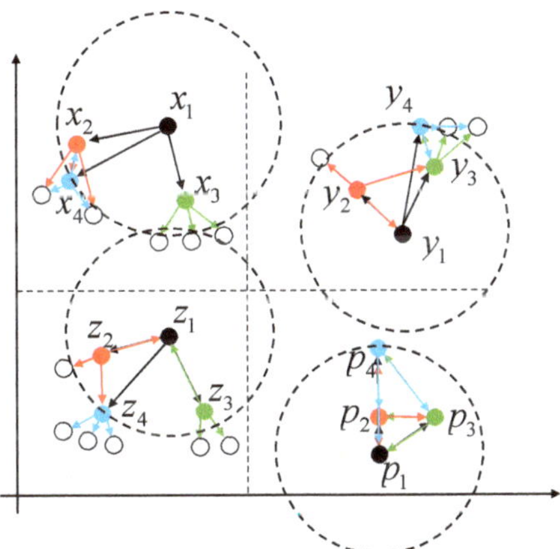

Figure 1. Samples x_1, y_1, z_1, p_1 and their 3-nearest neighbors.

While there are many techniques for removing outliers, these data preprocessing methods make the model building extraordinarily time-consuming and labor-intensive [21,22].

In this paper, a novel fault detection method using the mutual kNN rule (FD-MkNN) is proposed. Finding nearest neighbors using the mutual k-nearest neighbor rule will exclude the influence of PNNs (see Section 2.2.1 for the definition of mutual k-nearest neighbor (MkNN)). Before the model is established, the outliers in the training set are eliminated by the MkNN method, and the data quality for monitoring is improved. In the stage of fault detection, if the test sample does not have mutual neighbors, it is judged to be faulty. For test samples with mutual neighbors, the corresponding distance statistics are calculated to perform process monitoring. Compared with the FD-kNN, MkNN uses more valuable and truthful information (i.e., neighbors of the sample's neighbors), which improves the performance of process monitoring. The main contributions of this paper are as follows:

- To our best knowledge, the MkNN method is proposed to perform fault detection of industrial processes with outliers for the first time;
- The proposed method simultaneously realizes the elimination of outliers and the fault detection;
- The mutual protection problem of outliers is solved.

This paper will proceed as follows. In Section 2, the FD-kNN method is first briefly reviewed and then the proposed FD-MkNN approach is presented in detail. In Section 3, the experiments on numerical examples and Tenessee Eastman process (TEP) illustrate the superiority of the proposed monitoring method. Sections 4 and 5 are Discussion and Conclusions, respectively.

2. Methods

2.1. Process Monitoring Based on kNN Rule

The kNN method is widely used in pattern classification due to its simplicity. In December 2006, the top ten classic algorithms in data mining included kNN. FD-kNN was first proposed by He and Wang [11]. The main principle is to measure the difference between samples by distance; that is, normal samples and training samples are similar, but fault samples and training samples are significantly different.

- Training phase (determine the detection control limit):

  (1) Use Euclidean distance to get the kNNs of each training sample.

  $$d_{p,q} = \|x_p - x_q\|, p = 1,\ldots,n, q \neq p \tag{1}$$

  (2) Calculate the distance statistic D_p^2.

  $$D_p^2 = \frac{1}{k} \sum_{q=1}^{k} d_{p,q}^2 \tag{2}$$

  where D_p^2 represents the average squared distance between the pth sample and its k neighbors, $d_{p,q}^2$ denotes the squared Euclidean distance between the pth sample and its qth nearest neighbor.

  (3) Establish the control limit D_α^2 for fault detection. There are many ways to estimate D_α^2, such estimation using a noncentral chi-square distribution [11], kernel density estimation (KDE). The method proposed in this paper uses the $(1 - \alpha)$-empirical [23] quartile of D_p^2 as the threshold.

  $$D_\alpha^2 = D_{(\lfloor n(1-\alpha)\rfloor)}^2 \tag{3}$$

- Detection phase:

  (1) For a sample x to be tested, find its kNNs from the training set.
  (2) Calculate D_x^2 between x and its k neighbors using Equation (2).
  (3) Compare D_x^2 with the threshold D_α^2. If $D_x^2 > D_\alpha^2$, x is considered abnormal. Otherwise, it is normal.

2.2. Fault Detection Based on Mutual kNN Method

Since the nearest neighbors found by the kNN rule in the training set containing outliers may be pseudo-nearest neighbors, the fault detection threshold seriously deviates from the average level, resulting in the degradation or even failure of the monitoring performance of FD-kNN. To overcome the above problems, the concept of the mutual k-nearest neighbor (MkNN) is introduced. This section first defines MkNN and then provides the detailed steps of the proposed fault detection method.

2.2.1. MkNN

The MkNN of sample x can be defined by Equation (4). Given a sample x, if x has x_i in its kNNs, x_i should also have x in its kNNs [18]. According to the above definition, in Figure 1, $M_3(x_1) = \Phi$, $M_3(y_1) = \{y_2\}$, $M_3(z_1) = \{z_2, z_3\}$ and $M_3(p_1) = \{p_2, p_3, p_4\}$.

$$M_k(x) = \{x_i \in D | x_i \in N_k(x) \wedge x \in N_k(x_i)\} \tag{4}$$

where $N_k(x)$ denotes the kNNs of x, $N_k(x_i)$ represent the kNNs of x_i. If $M_k(x) = \Phi$, that is, x does not exist mutual kNNs. In other words, the kNNs of x are all pseudo-neighbors, and x is an outlier.

2.2.2. Proposed Fault Detection Scheme Based on Mutual kNN Method (FD-MkNN)

Before the model was established, outliers in the training set were eliminated by the MkNN method. This improves the data quality for modeling. In the fault detection stage, the relationship between samples was determined by looking for mutual nearest neighbors. If a test sample did not have mutual neighbors, this test sample was judged to be faulty. For test samples with mutual neighbors, the corresponding distance statistics were calculated to perform process monitoring. Compared with the kNN method, the proposed method uses more valuable and truthful information, improving fault detection performance. The flow chart of the proposed fault detection method is shown in Figure 2.

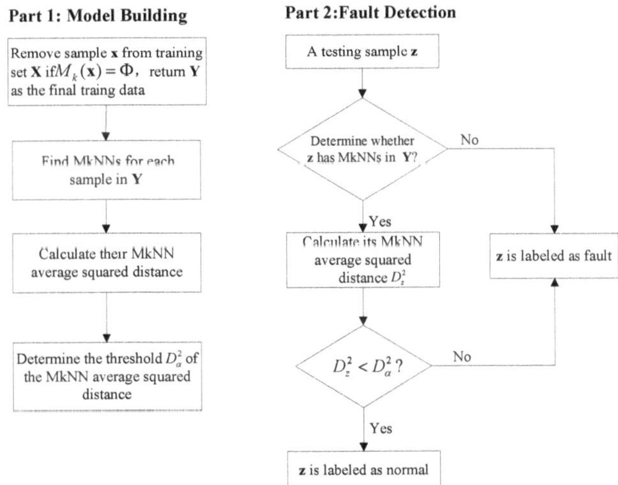

Figure 2. Flow chart of proposed fault detection method.

- Model building:
  (1) Finding MkNNs for each sample in the training data set **X**. Eliminate the training samples that do not have any MkNNs in **X** using Equation (4). For example, if $M_k(\mathbf{x}) = \Phi$, remove sample x from **X** and return **Y** as the final training data.
  (2) Calculate the MkNN average squared distance statistics of each sample in **Y** using Equation (2).
  (3) Determine the threshold D_α^2 for fault detection using Equation (3).
- Fault detection:
  (1) For a sample **z** to be tested, determine whether **z** has MkNNs in **Y** using Equation (4).
  (2) If **z** has no MkNNs, **z** is judged as a fault sample; otherwise, go to the next step.
  (3) Calculate $D_\mathbf{z}^2$ between **z** and its MkNNs using Equation (2).
  (4) Compare $D_\mathbf{z}^2$ with the threshold D_α^2. If $D_\mathbf{z}^2 > D_\alpha^2$, **z** is considered faulty. Otherwise, **z** is detected as a normal sample.

2.3. Remarks

- If x is in the q_1th nearest neighbor of y, y is in the q_2th nearest neighbor of x and $k = \max(q_1, q_2)$, x is the kth MNN of y and y is the kth MNN of x [24].
- The number of kNNs of sample x is k, and the number of MkNNs of x is an integer between $[0, k]$. Therefore, the average cumulative distance is used to calculate the

distance statistics. The values of k in the outlier elimination process and fault detection stages are different, denoted as k_1 and k_2, respectively. The k_1 and k_2 are chosen according to the best cross-validation [11]. Since the value of k is more significant, the probability that the sample has MkNNs is higher. Therefore, MkNN can more easily identify outliers when the value of k_1 is generally smaller than k_2.

3. Results

In this section, numerical examples and TEP are used to explore the effectiveness of the proposed method in fault detection. In addition, the mutual protection phenomenon of outliers is explored and solved using the elbow method to improve the detection performance of FD-MkNN.

3.1. Numerical Simulation

The number of generated training samples is 300. The outliers follow the Gaussian with mean 2 and variance 2 [25], the proportion of outliers compared to the training samples is set to 0%, 1%, 2%, 3%, 4%, and 5%, respectively. In addition, there are 100 testing samples, of which the first 50 samples are normal, and the rest are faulty.

$$x = t_1 + e_1,$$
$$y = t_2 + e_2 \tag{5}$$

where $t_i, i = 1, 2$ is a latent variable with zero mean and unit variance, and $e_i, i = 1, 2$ is a zero-mean noise with variance 10^{-4}.

FD-kNN is first applied to detect the faults in the data set. The number of nearest neighbors is 3. At the confidence level of 99%, the detection result is shown in Figure 3. It can be seen that, as the proportion of outliers increases, the detection performance of the FD-kNN method degrades seriously. As shown in Table 1, when the ratio of outliers is 5%, the fault detection rates (FDR) of the FD-kNN approach is only 20.00%. Due to outliers in the training samples, part of the neighbors of the samples found using kNN rule in the training phase are pseudo-neighbors. These pseudo-neighbors seriously affect the determination of the control threshold (that is, the control limit will be much greater than the average level) and result in poor fault detection performance.

For FD-MkNN, the parameters k_1 and k_2 are set to 3 and 5, respectively. At the same confidence level (that is, 99%), the detection result is shown in Figure 4. As shown in Table 1, when the proportion of outliers increases from 0 to 2%, the detection performance of the FD-MkNN method is not significantly affected, and the FDR always remains above 90%. When the proportion of outliers increases from 2% to 5%, the FDR of the FD-MkNN method is significantly reduced but the FDR is always better than that of FD-kNN.

The false alarm rates (FAR) of the two methods are shown in Table 2 (Note that the FAR is obtained based on the normal training samples). Due to outliers, the control limit or threshold of the FD-kNN method seriously deviates from the average level. Therefore, the FAR of the FD-kNN method is all zero.

The reason why the fault detection superiority of the FD-MkNN is better than that of the FD-kNN is as follows:

- Before the training phase, part of the outliers in the training samples are removed so that the outliers will not affect the determination of the control limit in the training phase;
- In the fault detection phase, MkNN carries more valuable and reliable information than kNN. Furthermore, the effect of PNN is eliminated.

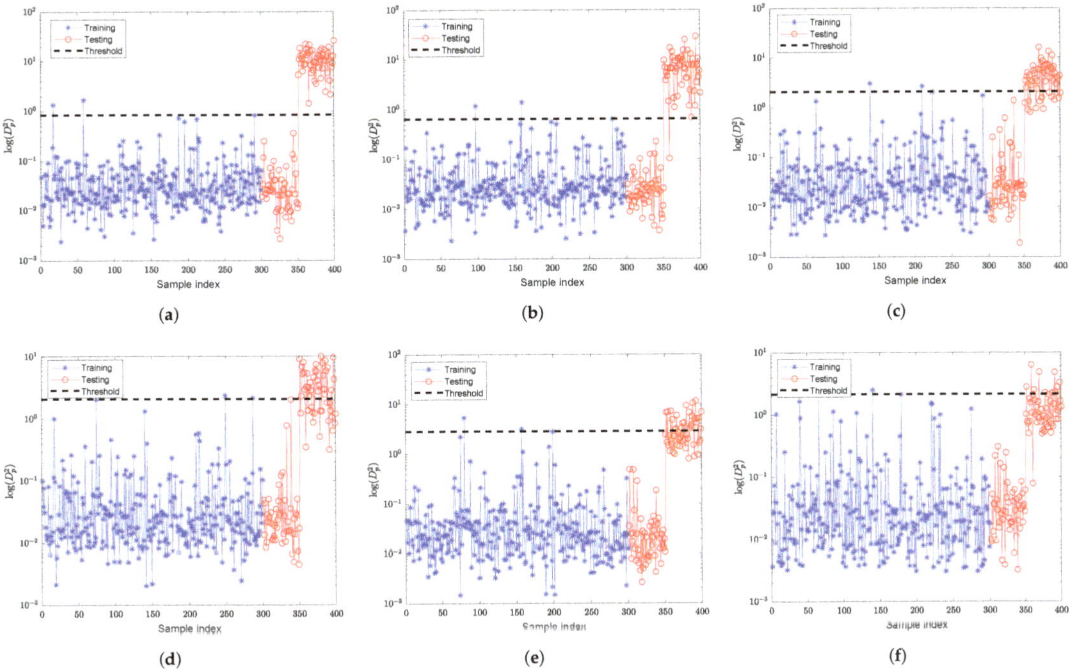

Figure 3. Fault detection performance of FD-kNN for the numerical example with different proportions of outliers. (**a**) no outlier; (**b**) 1%; (**c**) 2%; (**d**) 3%; (**e**) 4%; (**f**) 5%.

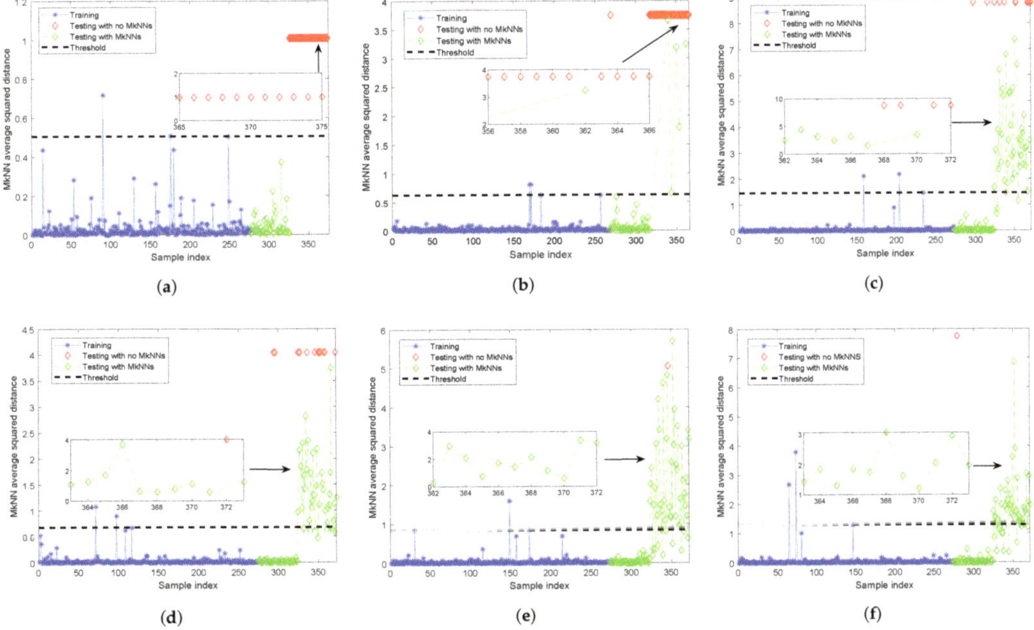

Figure 4. Fault detection performance of FD-MkNN for the numerical example with different proportions of outliers. (**a**) no outlier; (**b**) 1%; (**c**) 2%; (**d**) 3%; (**e**) 4%; (**f**) 5%.

Table 1. Fault detection rates (FDR) (%) of FD-kNN and FD-MkNN for the numerical example.

Method	No Outlier	1% Outliers	2% Outliers	3% Outliers	4% Outliers	5% Outliers
FD-kNN	100.00	98.00	74.00	62.00	46.00	20.00
FD-MkNN	100.00	100.00	94.00	84.00	80.00	76.00

Table 2. False alarm rates (FAR) (%) of FD-kNN and FD-MkNN for the numerical example.

Method	No Outlier	1% Outliers	2% Outliers	3% Outliers	4% Outliers	5% Outliers
FD-kNN	0.00	0.00	0.00	0.00	0.00	0.00
FD-MkNN	0.00	2.00	4.00	4.00	0.00	2.00

3.1.1. Experimental Results of FD-MkNN with Different Values of k

The values of k in the outlier elimination and fault detection stages are different and can be denoted as k_1 and k_2, respectively. The larger the value of k, the higher the probability that the query sample finds its mutual neighbors. Therefore, MkNN can more easily identify outliers when the value of k_1 is generally smaller than k_2. However, the value of k_1 cannot be too small because the MkNN method will misidentify the normal training samples as outliers and eliminate them. For example, as shown in Table 3, when the value of k_1 is set to 1, the MkNN method will eliminate all 300 training samples (the actual proportion of outliers introduced is 5%), resulting in the failure of the MkNN fault detection stages. As the value of k_1 increases, the number of outliers removed decreases, which makes the monitoring threshold deviate from the normal level, and the FDR decreases seriously.

Table 3. Fault detection results of MkNN with different values of k for the numerical example.

k_1	k_2	The Number of Outliers Removed	FDR	FAR
1	3	300	-	-
3	5	33	98.00	4.00
5	7	5	86.00	0.00
7	9	2	64.00	0.00
9	11	1	52.00	0.00

3.1.2. Mutual Protection Phenomenon of Outliers

As shown in Figure 5, when two outliers are relatively close, an interesting phenomenon will appear: they will become each other's mutual nearest neighbors. Therefore, the MkNN rule cannot identify them as outliers. For example, in Figure 5, b_1, b_2, and b_3 are protected by 1, 2, and 3 outliers, respectively. When the outliers far from the normal training samples are kept in the training set due to mutual protection, it will cause the threshold or control limit calculated in the training phase to deviate seriously from the average level. We call this phenomenon the "Mutual Protection of Outliers (MPO)", which is also the main reason why the detection performance of the FD-MkNN method decreases when the proportion of outliers increases from 2% to 5%.

It can be observed from Figure 5 that, for outliers with mutual protection, the corresponding MkNN distance statistic is significantly larger than that of the normal training sample. Therefore, the elbow method [26] is used to eliminate outliers with mutual protection: first, arrange the MkNN distance statistics of the training samples in descending order, then determine all samples before the elbow position as outliers with mutual protection, and finally eliminate these outliers from the training set.

As shown in Figure 6, the outliers with mutual protection can be identified according to the elbow method, that is, all samples before the elbow point. After determining the outliers with mutual protection, these outliers need to be removed from the training set. Finally, the process monitoring method was repeated. The detection results are shown in Figure 7. After eliminating outliers with mutual protection, the recalculated threshold (that is, the red dotted line in Figure 7) is more reasonable, and the FDR has reached 100.00%, as shown in Table 4.

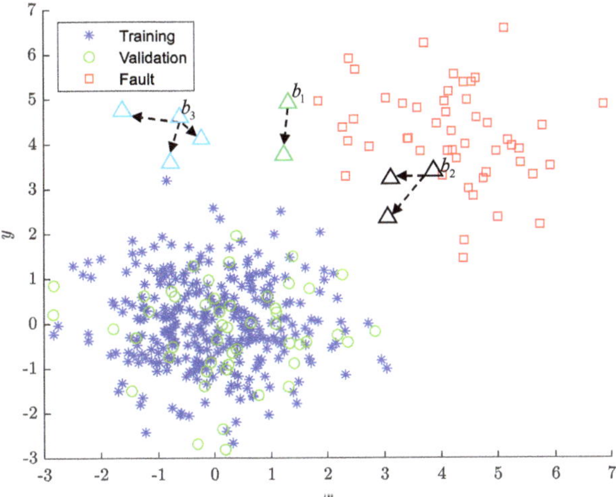

Figure 5. Mutual protection of outliers (MPO) (triangles represent outliers).

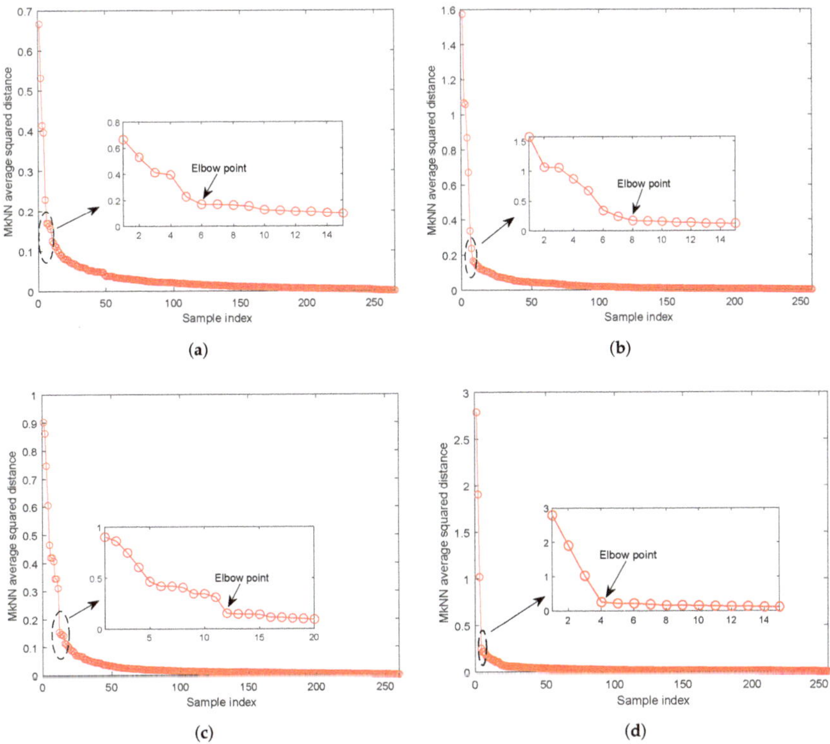

Figure 6. The descent curve of distance statistic for the numerical example with different proportions of outliers. (**a**) 2%; (**b**) 3%; (**c**) 4%; (**d**) 5%.

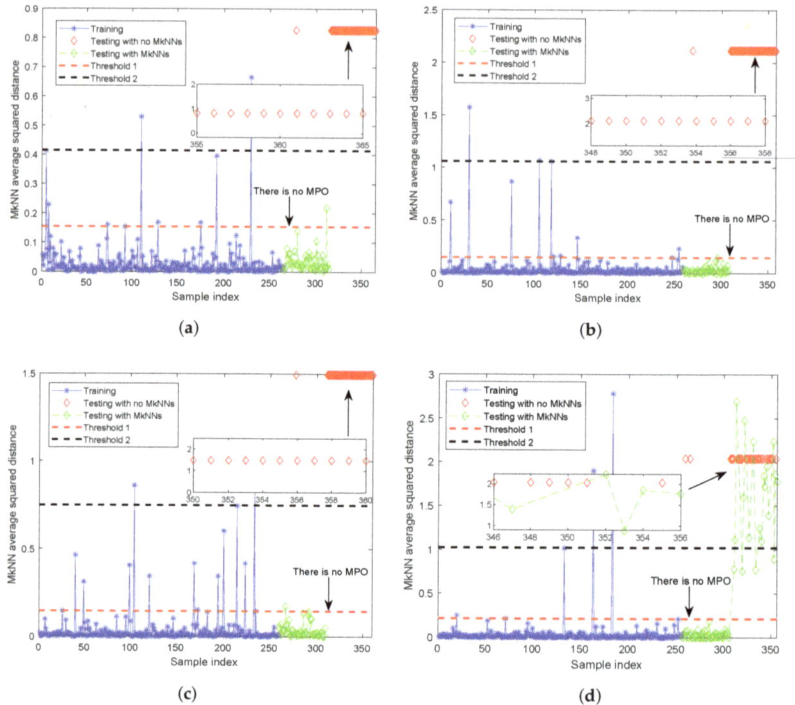

Figure 7. Fault detection performance of FD-MkNN of the numerical example that eliminates outliers with mutual protection phenomenon. (**a**) 2%; (**b**) 3%; (**c**) 4%; (**d**) 5%.

Table 4. FDR (%) and FAR (%) of FD-MkNN of the numerical example that eliminates outliers with mutual protection phenomenon.

	2% Outliers	3% Outliers	4% Outliers	5% Outliers
FDR	100.00	100.00	100.00	100.00
FAR	4.00	2.00	4.00	4.00

3.2. The Tennessee Eastman Process

When comparing the performance or effectiveness of process monitoring methods, the TEP [27] is a benchmark choice. In [28,29], Downs and Vogel proposed the simulation platform. There are five major operating units in the TEP, namely, a product stripper, a recycle compressor, a vapor–liquid separator, a product condenser, and a reactor. The process has four kinds of reactants (A, C, D, E), two products (G, H), contains catalyst (B), and byproducts (F). There are 11 manipulated variables (No.42–No.52), 22 process measurements (No.1–No.22), and 19 composition variables (No.23–No.41). For detailed information on the 52 monitoring variables and 21 fault patterns, see ref. [27]. The flowchart of the process is given in Figure 8.

The number of training samples and the number of validation samples are 960 and 480, respectively. In addition, there are 960 testing samples where the fault is introduced from the 161st sample. To simulate the situation that the training data contains outliers, outliers whose magnitude is twice the normal data are randomly added to the training data. The thresholds of different methods are all calculated at a confidence level of 99%.

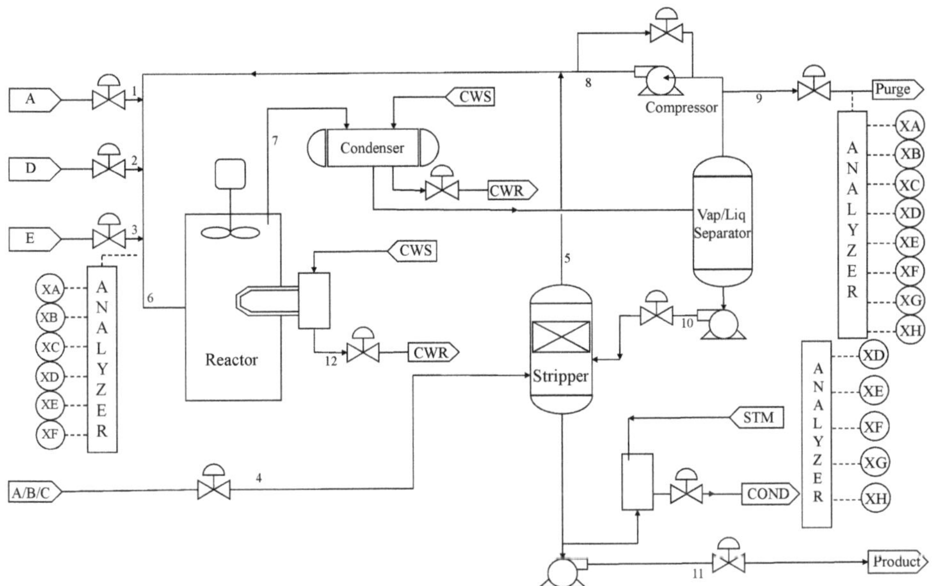

Figure 8. Flowchart of Tennessee Eastman process.

These three faults are chosen to demonstrate the effectiveness of the proposed method. The parameter k of FD-kNN is 3. The parameters k_1 and k_2 of FD-MkNN are 42 and 45, respectively. For the FD-MkNN method, the outliers with mutual protection phenomenon are first eliminated by the elbow method, as shown in Figure 9.

According to [29,30], fault 1 is a step fault with a significant amplitude change. When fault 1 occurs, eight process variables are affected.

Figures 10 and 11 are the monitoring results of fault 1 by FD-kNN and FD-MkNN, respectively. As the proportion of outliers increases, the detection results of kNN and MkNN for fault 1 are not significantly affected. For example, the FDR of MkNN for fault 1 remains at 99.00%, as shown in Table 5. Because fault 1 is a step fault with a significant amplitude change, the outliers introduced in this experiment are insignificant in the face of this fault. Although these outliers also deviate the control limits from normal levels, they do not have much impact on the fault detection phase. The fault false alarm rate of FD-kNN and FD-MkNN is shown in Table 6.

The fault 7 is also a step fault, but its magnitude changes are small, and only one process variable (i.e., variable 45) is affected.

Figures 12 and 13 are the monitoring results of fault 7 by FD-kNN and FD-MkNN, respectively. As shown in Table 7, as the proportion of outliers increases, the FDR of FD-kNN drops from 100.00% to 18.75%, while the FDR of FD-MkNN does not decrease significantly and remains above 90.00%. The fault false alarm rate of FD-kNN and FD-MkNN is shown in Table 8.

According to the detection results of fault 1 and fault 7, it can be seen that FD-MkNN is suitable for the processing of incipient faults. Because outliers will significantly increase the threshold, the detection statistic of incipient faults is lower than the threshold. The proposed method eliminates outliers by judging whether the samples have MkNNs, thereby improving the fault detection performance.

Fault 13 is a slow drift in the reaction kinetics. Figures 14 and 15 are the monitoring results of fault 13 by FD-kNN and FD-MkNN, respectively. In Tables 9 and 10, as the proportion of outliers increases, the FDR of the FD-MkNN is always better than that of FD-kNN, while the FAR is higher than that of kNN. Due to the appearance of outliers, the threshold of the kNN is increased so the FAR of FD-kNN is always 0.00%.

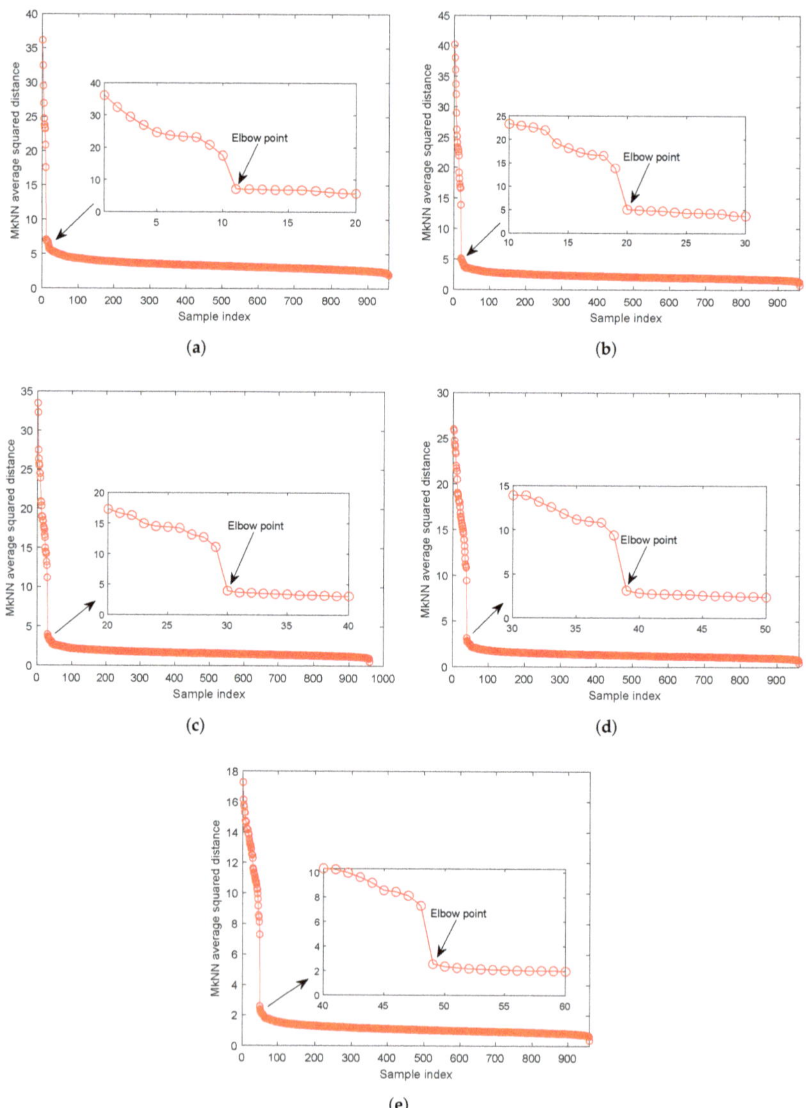

Figure 9. The descent curve of distance statistic for the TEP with different proportions of outliers. (**a**) 1%; (**b**) 2%; (**c**) 3%; (**d**) 4%; (**e**) 5%.

Table 5. FDR (%) of FD-kNN and FD-MkNN for fault 1 of TEP.

Method	No Outlier	1% Outliers	2% Outliers	3% Outliers	4% Outliers	5% Outliers
FD-kNN	99.50	98.75	98.50	98.75	98.50	98.50
FD-MkNN	99.50	99.00	99.00	99.00	99.00	99.00

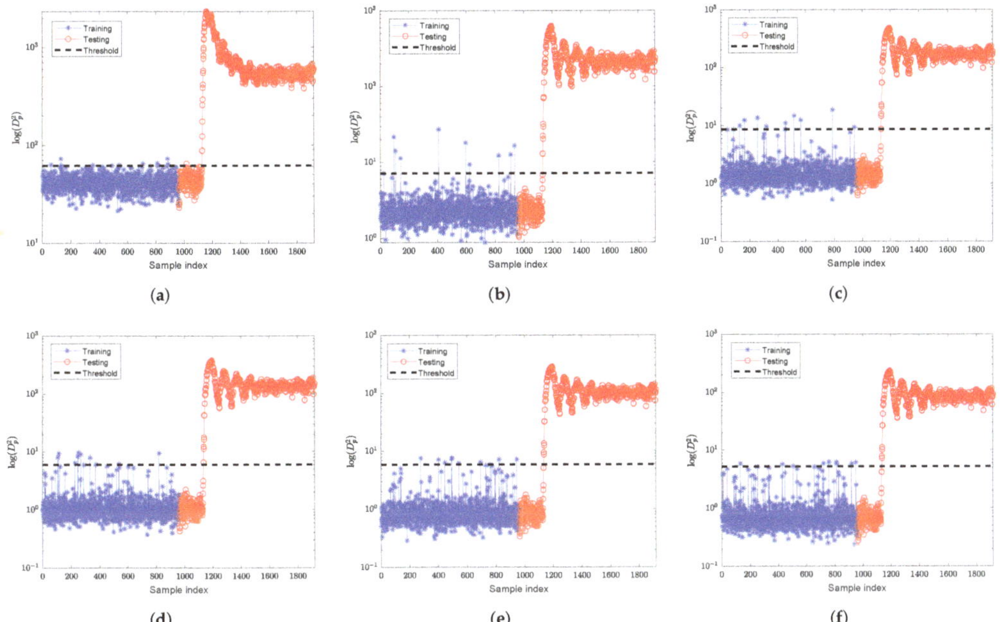

Figure 10. Fault detection results of FD-kNN for fault 1 of TEP. (**a**) no outlier; (**b**) 1%; (**c**) 2%; (**d**) 3%; (**e**) 4%; (**f**) 5%.

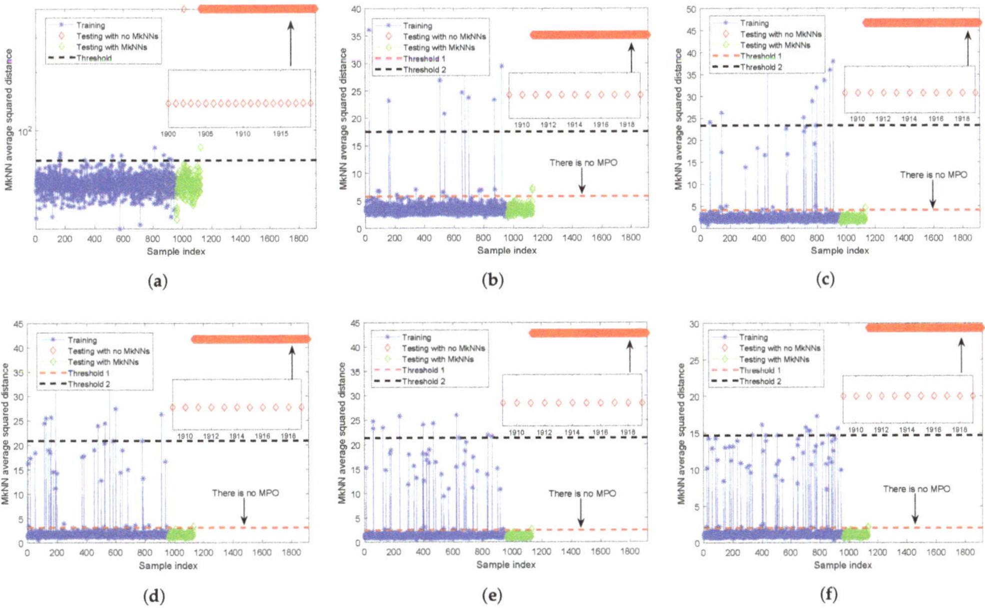

Figure 11. Fault detection results of FD-MkNN for fault 1 of TEP. (**a**) no outlier; (**b**) 1%; (**c**) 2%; (**d**) 3%; (**e**) 4%; (**f**) 5%.

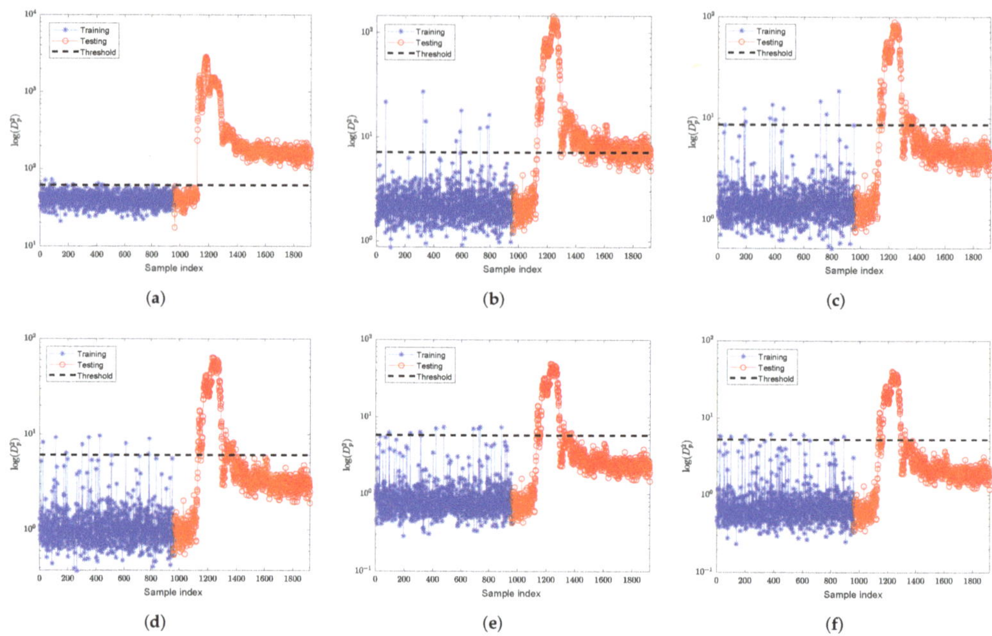

Figure 12. Fault detection results of FD-kNN for fault 7 of TEP. (**a**) no outlier; (**b**) 1%; (**c**) 2%; (**d**) 3%; (**e**) 4%; (**f**) 5%.

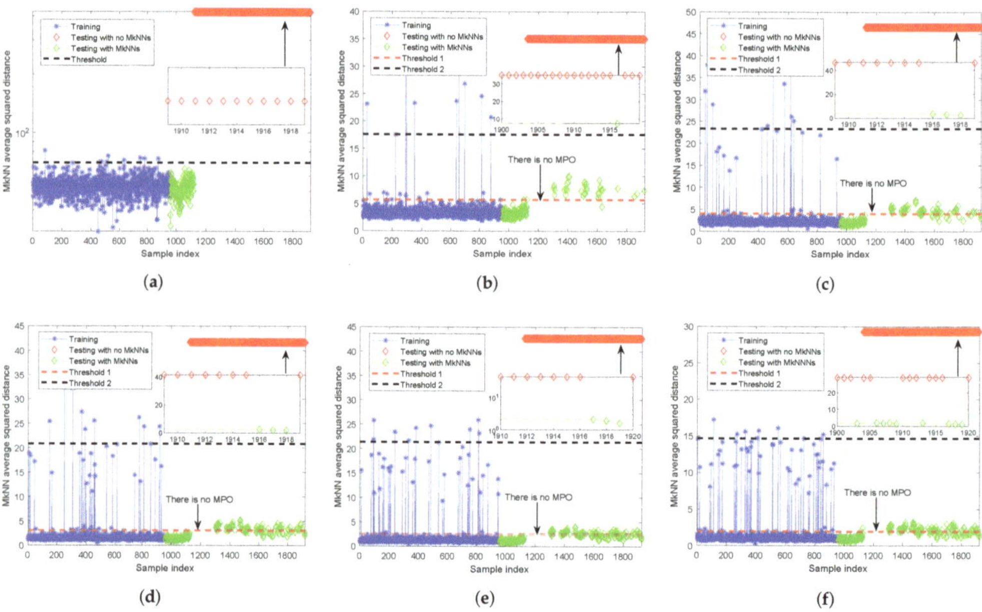

Figure 13. Fault detection results of FD-MkNN for fault 7 of TEP. (**a**) no outlier; (**b**) 1%; (**c**) 2%; (**d**) 3%; (**e**) 4%; (**f**) 5%.

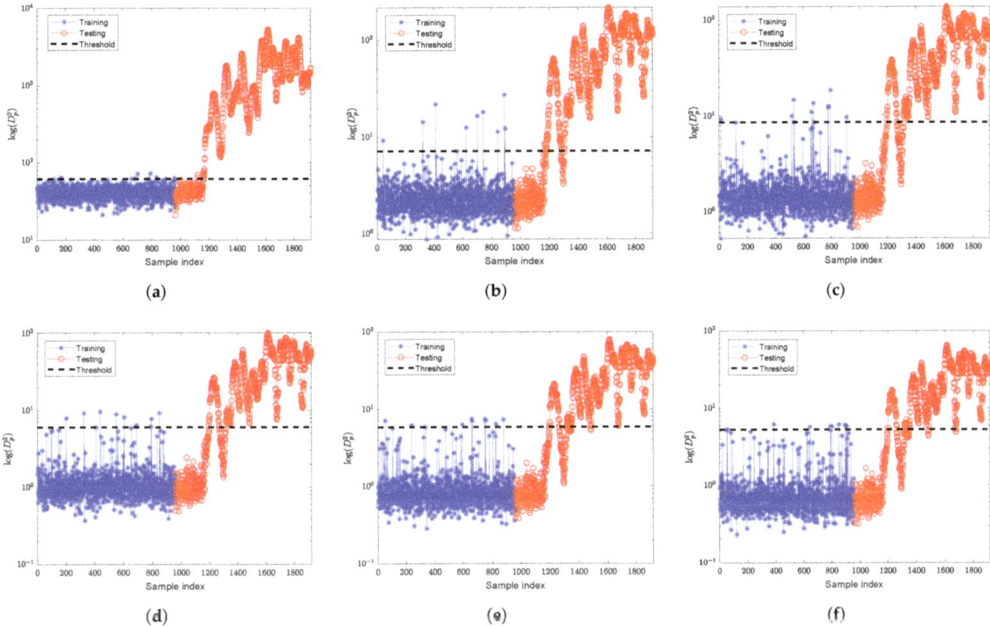

Figure 14. Fault detection results of FD-kNN for fault 13 of TEP. (**a**) no outlier; (**b**) 1%; (**c**) 2%; (**d**) 3%; (**e**) 4%; (**f**) 5%.

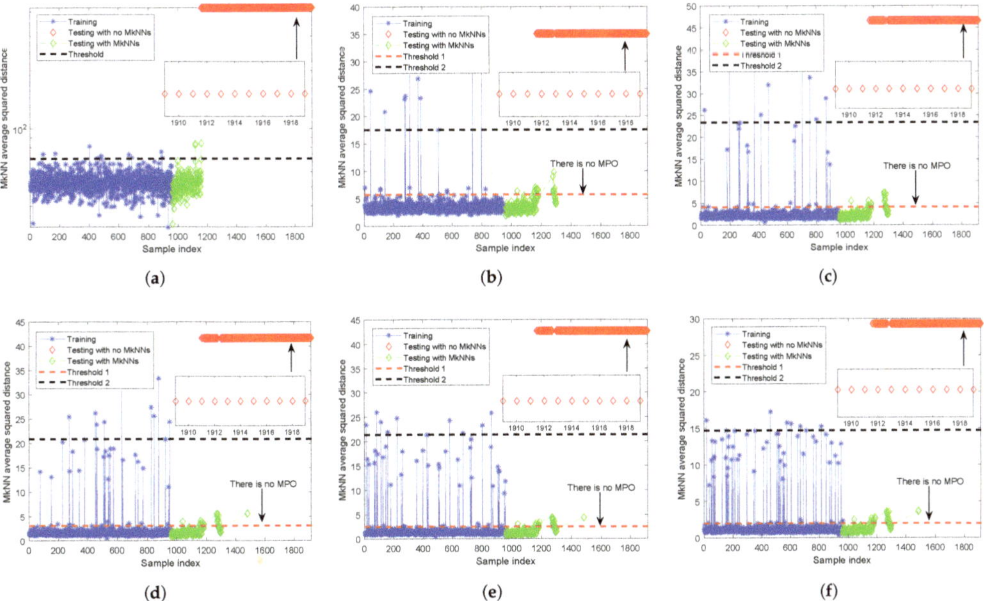

Figure 15. Fault detection results of FD-MkNN for fault 13 of TEP. (**a**) no outlier; (**b**) 1%; (**c**) 2%; (**d**) 3%; (**e**) 4%; (**f**) 5%.

Table 6. FAR (%) of FD-kNN and FD-MkNN for fault 1 of TEP.

Method	No Outlier	1% Outliers	2% Outliers	3% Outliers	4% Outliers	5% Outliers
FD-kNN	0.63	0.00	0.00	0.00	0.00	0.00
FD-MkNN	0.63	0.00	0.00	0.00	0.00	0.00

Table 7. FDR (%) of FD-kNN and FD-MkNN for fault 7 of TEP.

Method	No Outlier	1% Outliers	2% Outliers	3% Outliers	4% Outliers	5% Outliers
FD-kNN	100.00	79.75	25.25	25.25	20.38	18.75
FD-MkNN	100.00	99.63	97.63	94.75	93.63	92.75

Table 8. FAR (%) of FD-kNN and FD-MkNN for fault 7 of TEP.

Method	No Outlier	1% Outliers	2% Outliers	3% Outliers	4% Outliers	5% Outliers
FD-kNN	0.00	0.00	0.00	0.00	0.00	0.00
FD-MkNN	0.00	0.00	0.00	0.00	0.00	0.00

Table 9. FDR (%) of FD-kNN and FD-MkNN for fault 13 of TEP.

Method	No Outlier	1% Outliers	2% Outliers	3% Outliers	4% Outliers	5% Outliers
FD-kNN	95.38	90.50	84.88	85.00	82.25	80.25
FD-MkNN	95.38	92.75	91.88	91.63	91.75	92.00

Table 10. FAR (%) of FD-kNN and FD-MkNN for fault 13 of TEP.

Method	No Outlier	1% Outliers	2% Outliers	3% Outliers	4% Outliers	5% Outliers
FD-kNN	1.25	0.00	0.00	0.00	0.00	0.00
FD-MkNN	1.25	0.63	0.63	0.63	0.63	0.63

4. Discussion

The neighbors of the samples found by kNN on a data set containing outliers may not be true neighbors, but a kind of pseudo neighbor. If such pseudo-nearest neighbors are used to calculate the threshold, the threshold or control limit will deviate significantly from the normal level, thereby degrading the fault detection performance.

The MkNN method determines outliers through checking whether the samples have MkNNs, which simultaneously realizes the elimination of outliers and the fault detection by using the same rule. Through the detection of fault 1 and fault 7 in the TEP, it can be seen that the FD-MkNN has obvious advantages for detecting incipient faults because the incipient faults are more sensitive to outliers.

This work stresses the superiority and promise of the MkNN rule for fault detection, especially for industrial processes with outliers. The MkNN-method-based fault isolation or diagnosis part is currently underway.

5. Conclusions

In this paper, a novel fault detection approach based on the mutual k-nearest neighbor method is proposed. The primary characteristic of our method is that the calculation of the distance statistics for fault detection uses the MkNN rule instead of kNN. The proposed method simultaneously realizes the elimination of outliers and the fault detection using Mutual kNN rule. Specifically, before the training phase, part of the outliers in the training samples are removed so that the outliers will not affect the determination of the control limit in the training phase; in the fault detection phase, MkNN carries more valuable and reliable information than kNN. Furthermore, the effect of PNN is eliminated. Furthermore, the mutual protection problem of outliers is solved using the elbow rule, which improves the performance of fault detection. The experiments on numerical examples and TEP verify the effectiveness of the proposed method.

The proposed FD-MkNN can be seen as an alternative method in monitoring the industrial processes with outliers. In addition, the MkNN method based fault isolation or diagnosis part is currently underway.

Author Contributions: J.W., Z.Z., Z.L. and S.D. conceived and designed the method. J.W. and Z.Z. wrote the paper. J.W. performed the experiments. All authors have read and agreed to the published version of the manuscript.

Funding: This research was funded by Zhejiang Provincial Public Welfare Technology Application Research Project [grant number LGG22F030023]; the Huzhou Municipal Natural Science Foundation of China [grant number 2021YZ03]; the Zhejiang Province Key Laboratory of Smart Management & Application of Modern Agricultural Resources [grant number 2020E10017]; and the Postgraduate Scientific Research and Innovation Projects of HUZHOU UNIVERSITY [grant number 2020KYCX25].

Institutional Review Board Statement: Not applicable.

Informed Consent Statement: Not applicable.

Data Availability Statement: The used data of TEP are available online: http://web.mit.edu/braatzgroup/TE_process.zip (accessed on 24 January 2022).

Conflicts of Interest: The authors declare no conflict of interest. The funders had no role in the design of the study; in the collection, analyses, or interpretation of data; in the writing of the manuscript, or in the decision to publish the results.

Abbreviations

kNN	k Nearest Neighbor
MkNN	Mutual k Nearest Neighbor
PCA	Principal Component Analysis
MSPM	Multivariate Statistical Process Monitoring
SPE	Squared Prediction Error
PNN	Pseudo Nearest Neighbor
KDE	Kernel Density Estimation
FD-kNN	Fault Detection based on kNN
FDMkNN	Fault Detection Scheme based on Mutual kNN Method
FDR	Fault Detection Rates
FAR	False Alarm Rates
TEP	Tennessee Eastman process

References

1. Chiang, L.H.; Russell, E.L.; Braatz, R.D. *Fault Detection and Diagnosis in Industrial Systems*; Springer: London, UK, 2001. [CrossRef]
2. Arunthavanathan, R.; Khan, F.; Ahmed, S.; Imtiaz, S. An analysis of process fault diagnosis methods from safety perspectives. *Comput. Chem. Eng.* **2021**, *145*, 107197. [CrossRef]
3. Zhao, S.; Zhang, J.; Xu, Y. Monitoring of processes with multiple operating modes through multiple principle component analysis models. *Ind. Eng. Chem. Res.* **2004**, *43*, 7025–7035. [CrossRef]
4. Fezai, R.; Mansouri, M.; Okba, T.; Harkat, M.F.; Bouguila, N. Online reduced kernel principal component analysis for process monitoring. *J. Process Control* **2018**, *61*, 1–11. [CrossRef]
5. Peres, F.; Peres, T.N.; Fogliatto, F.S.; Anzanello, M.J. Fault detection in batch processes through variable selection integrated to multiway principal component analysis. *J. Process Control* **2019**, *80*, 223–234. [CrossRef]
6. Deng, X.G.; Lei, W. Modified kernel principal component analysis using double-weighted local outlier factor and its application to nonlinear process monitoring. *ISA Trans.* **2017**, *72*, 218–228. [CrossRef]
7. Wang, J.; He, Q.P. Multivariate statistical process monitoring based on statistics pattern analysis. *Ind. Eng. Chem. Res.* **2010**, *49*, 7858–7869. [CrossRef]
8. Zhou, Z.; Wen, C.L.; Yang, C.J. Fault Isolation Based on k-Nearest Neighbor Rule for Industrial Processes. *IEEE Trans. Ind. Electron.* **2016**, *63*, 2578–2586. [CrossRef]

9. Zhou, Z.; Li, Z.X.; Cai, Z.D.; Wang, P.L. Fault Identification Using Fast k-Nearest Neighbor Reconstruction. *Processes* **2019**, *7*, 340. [CrossRef]
10. Qin, S.J. Survey on data-driven industrial process monitoring and diagnosis. *Annu. Rev. Control* **2012**, *36*, 220–234. [CrossRef]
11. He, Q.P.; Wang, J. Fault detection using the k-nearest neighbor rule for semiconductor manufacturing processes. *IEEE Trans. Semicond. Manuf.* **2007**, *20*, 345–354. [CrossRef]
12. Nomikos, P.; Macgregor, J.F. Monitoring Batch Process Using Multiway Principal Componet Analysis. *AIChE J.* **1994**, *40*, 1361–1375. [CrossRef]
13. Ku, W.; Storer, R.H.; Georgakis, C. Disturbance detection and isolation by dynamic principal component analysis. *Chemom. Intell. Lab. Syst.* **1995**, *30*, 179–196. [CrossRef]
14. Lee, J.M.; Yoo, C.K.; Choi, S.W.; Vanrolleghem, P.A.; Lee, I.B. Nonlinear process monitoring using kernel principal component analysis. *Chem. Eng. Sci.* **2004**, *59*, 223–234. [CrossRef]
15. Lee, J.M.; Yoo, C.K.; Lee, I.B. Fault detection of batch processes using multiway kernel principal component analysis. *Comput. Chem. Eng.* **2004**, *28*, 1837–1847. [CrossRef]
16. Sang, W.C.; Lee, C.; Lee, J.M.; Jin, H.P.; Lee, I.B. Fault detection and identification of nonlinear processes based on kernel pca. *Chemom. Intell. Lab. Syst.* **2005**, *75*, 55–67. [CrossRef]
17. Zhang, Y.; Wu, X. Integrating induction and deduction for noisy data mining. *Inf. Sci.* **2010**, *180*, 2663–2673. [CrossRef]
18. Liu, H.; Zhang, S. Noisy data elimination using mutual k-nearest neighbor for classification mining. *J. Syst. Softw.* **2012**, *85*, 1067–1074. [CrossRef]
19. Barnett, V.; Lewis, T. *Outliers in Statistical Data*; Wiley: Hoboken, NJ, USA, 1974. [CrossRef]
20. Zhu, J.L.; Ge, Z.Q.; Song, Z.H.; Gao, F.R. Review and big data perspectives on robust data mining approaches for industrial process modeling with outliers and missing data. *Annu. Rev. Control* **2018**, *46*, 107–133. [CrossRef]
21. Brighton, H.; Mellish, C. Advances in Instance Selection for Instance-Based Learning Algorithms. *Data Min. Knowl. Discov.* **2002**, *6*, 153–172. [CrossRef]
22. He, Q.P.; Wang, J. Statistics Pattern Analysis: A New Process Monitoring Framework and its Application to Semiconductor Batch Processes. *AIChE J.* **2011**, *57*, 107–121. [CrossRef]
23. Verdier, G.; Ferreira, A. Adaptive mahalanobis distance and k-nearest neighbor rule for fault detection in semiconductor manufacturing. *IEEE Trans. Semicond. Manuf.* **2011**, *24*, 59–68. [CrossRef]
24. Ding, C.; Chris, H.Q.; He, X.F. K-nearest-neighbor consistency in data clustering: Incorporating local information into global optimization. In Proceedings of the 2004 ACM Symposium on Applied Computing, Nicosia, Cyprus, 14–17 March 2004. [CrossRef]
25. Zhu, J.L.; Ge, Z.Q.; Song, Z.H. Robust supervised probabilistic principal component analysis model for soft sensing of key process variables. *Chem. Eng. Sci.* **2015**, *122*, 573–584. [CrossRef]
26. Liu, F.; Deng, Y. Determine the number of unknown targets in Open World based on Elbow method. *IEEE Trans. Fuzzy Syst.* **2020**, *29*, 986–995. [CrossRef]
27. Ricker, N.L. Optimal steady-state operation of the tennessee eastman challenge process. *Comput. Chem. Eng.* **1995**, *19*, 949–959. [CrossRef]
28. Downs, J.J.; Vogel, E.F. A plant-wide industrial process control problem. *Comput. Chem. Eng.* **1993**, *17*, 245–255. [CrossRef]
29. Russell, E.L.; Chiang, L.H.; Braatz, R.D. *Data-Driven Techniques for Fault Detection and Diagnosis in Chemical Processes*; Springer: London, UK, 2000. [CrossRef]
30. Liu, J. Fault diagnosis using contribution plots without smearing effect on non-faulty variables. *J. Process Control* **2012**, *22*, 1609–1623. [CrossRef]

processes

Article

Bearing Fault Diagnosis Based on a Novel Adaptive ADSD-gcForest Model

Shuo Zhai, Zhenghua Wang and Dong Gao *

School of Information Science and Technology, Beijing University of Chemical Technology, Beijing 100029, China; shzhaishuo@163.com (S.Z.); 2019210490@mail.buct.edu.cn (Z.W.)
* Correspondence: gaodong@mail.buct.edu.cn

Abstract: With the continuous improvement of industrial production requirements, bearings work significantly under strong noise interference, which makes it difficult to extract fault features. Deep Learning-based approaches are promising for bearing diagnosis. They can extract fault information efficiently and conduct accurate diagnosis. However, the structure of deep learning is often determined by trial and error, which is time-consuming and lacks theoretical support. To address the above problems, an adaptive (Adaptive Depthwise Separable Dilated Convolution and multi-grained cascade forest) ADSD-gcForest fault diagnosis model is proposed in this paper. Multiscale convolution combined with convolutional attention mechanism (CBAM) concentrates on effectively extracting fault information under strong noise, and the Meta-Activate or Not (Meta-ACON) activation function is integrated to adaptively optimize the model structure according to the characteristics of input samples, then gcForest outputs the final diagnosis result as the classifier. The experiment compares the effects of three bearings failure diagnoses under various noise and load conditions. The experimental results show the effectiveness and practicability of the proposed method.

Keywords: fault diagnosis; Meta-ACON; ADSD-gcForest; SDPimage

Citation: Zhai, S.; Wang, Z.; Gao, D. Bearing Fault Diagnosis Based on a Novel Adaptive ADSD-gcForest Model. *Processes* **2022**, *10*, 209. https://doi.org/10.3390/pr10020209

Academic Editors: Jie Zhang and Meihong Wang

Received: 3 January 2022
Accepted: 21 January 2022
Published: 22 January 2022

Publisher's Note: MDPI stays neutral with regard to jurisdictional claims in published maps and institutional affiliations.

1. Introduction

With the development of the manufacturing industry, rolling bearings, as one of the core components of mechanical equipment, play an increasingly irreplaceable role. However, under the condition of strong noise and multiple loads for a long time, the bearings are prone to wear or breakage. An expected failure, such as a crack in the bearings, may cause the breakdown of the entire machine, resulting in magnificent economic loss and severe safety accidents [1]. Therefore, it is of great significance to realize the high efficiency and accuracy of bearing fault diagnosis.

The location of the bearing failure is generally located in the inner ring, outer ring, and rolling element, and the bearing fault usually produces periodic vibrations when machinery is running, so analysis of the vibration signal during bearing operation is the key to achieving the diagnosis of the fault [2]. Traditional fault diagnosis methods are divided mainly into linear and non-linear methods. Linear diagnosis methods mainly contain time domain analysis, frequency domain analysis, and time-frequency analysis [3]. Nonlinear analysis is less adopted in fault diagnosis than linear analysis, chaos theory, fractal dimension, and entropy value theory, are commonly applied nonlinear analysis methods, among others. However, due to the increase in bearing fault datasets and the increasing complexity of production environments, traditional fault diagnosis methods that rely on traditional manual fault sign extraction have become no longer applicable [4]. Therefore, constructing novel fault diagnosis models based on approaches of deep learning have become a research hotspot.

Frequently used deep learning models include the Deep Autoencoder (DAE), the Deep Belief Network (DBN), the Convolutional Neural Network (CNN) and the Recurrent

Neural Network (RNN). Among them, the improved Stack Denoising Autoencoder (SDAE) diagnostic method was proposed by Hou et al. [5], in which the hyperparameters of the DAE network were adaptively selected by the particle swarm algorithm to determine the structure of the SDAE network. On this basis, the characteristic representation of the fault state was obtained, which was input into the Softmax classifier for fault classification and recognition; this method has achieved accurate fault diagnosis under the circumstance of variable operating conditions. In-depth research on DAE was conducted by Shao et al. [6], in which DAE and shrinking auto-encoding were introduced to improve the fault extraction capabilities of faulty features, and local preservation of projection fusion features was applied to optimize feature quality. Liang T et al. [7] presented a method for the diagnosis of rolling bearing failures, which consisted mainly of three steps: a series of DBNs with different hyperparameters were constructed and trained, after which the improved ensemble method was applied to acquire the weight matrix for each DBN and then each DBN voted together according to its respective weight matrix to obtain the final result of the diagnosis. The method of DBN-based degradation assessment under accelerated life testing of bearings was adopted by Ma et al. [8]. Shao et al. proposed a DBN for the diagnosis of induction motors faults, in which vibration signals were introduced directly as input [9], and the t-SNE algorithm was adopted to visualize the learning representation. Han Tao et al. [10] used CNN training to obtain the corresponding characteristic diagram of the multi-wavelet coefficient branching process through the wavelet transform to realize the intelligent diagnosis of rolling bearing composite faults. Liang et al. [11] have constructed two different CNNs, one for extracting time-domain features and the other was applied to extract time-frequency domain features, and then fused them with the time-frequency features and time-domain features extracted by continuous wavelet transform diagnose faults of rolling bearings in a characteristic way. Bearing fault diagnosis based on LSTM (Long Short-Term Network) and CNN models was established by Pan et al. [12], a fault diagnosis method was proposed by Zhang et al. [13], in which self-encoding of convolutional noise reduction was performed to achieve feature extraction and CNN was introduced for pattern recognition. The long- and short-term memory stacking network was designed by Yu et al. [14], where 12 different bearing health conditions were classified using augmented data, including the type and severity of bearing failure. A convolutional bidirectional long- and short-term memory network was designed by Zhao et al. [15] for bearing fault diagnosis. In this method, a convolutional neural network was applied as a feature extractor of the original signal, and then the bearing faults were classified through a bidirectional long and short-term memory network.

Based on the brief review of the existing diagnosis approaches, the challenges can be summarized as follows: first of all, numerous methods only conduct comparative experiments for a single type of noise and other types of noise are not considered. Second, deep learning structures are often determined by trial and error, which means this structure is randomly defined; as a result, the model with the best performance is adopted after many experiments [16]. To solve the above problem, an adaptive ADSD-gcForest fault diagnosis model is proposed in this paper, and based on the basis of the existing traditional network, the core fault features at different scales are effectively extracted by using dilated convolution with different dilation rates and CBAM fusion under strong noise interference. On this basis, deep separable convolution is incorporated into the dilated convolution mechanism to improve the efficiency of the calculation [17]. In recent years, many adaptive optimization methods have been developed for network structure, but most of these approaches require the assistant of an intelligent optimization algorithm or migration learning [18,19]. In contrast, the network structure can be simply optimized by the Meta-ACON activation based on the input samples without the need for additional complex algorithms and can not only optimize the model structure but also make the model better deal with different sample data. Then, through the multigranular scanning of the deep forest classifier and the cascade forest algorithm, the hidden fault features in the feature vector are analyzed and extracted, and the final classification results are output. The main contributions of this article are as follows:

1. An adaptive ADSD-gcForest diagnostic model is proposed for the diagnosis of rolling bearing fault diagnosis, allowing the extraction of features under the high-noise and complex working conditions that could be realized. The structure of the diagnosis model achieves adaptive optimization based on the characteristics of the sample data.
2. Combining the multiscale depth-separable dilated convolution with CBAM can effectively extract fault features under strong noise interference. On the basis of the lack of adjust the original structure of the model, the Meta-ACON activation function is introduced into the convolution layer of the model to achieve adaptive optimization of the model structure according to the fault data of different bearings.
3. The comparative experiment shows that the ADSD-gcForest model proposed in this paper has strong generalization ability and robustness with certain practical value.

The rest of this paper is organized in the following way: the introduction of the related theories is mainly in Section 2, the specific structure of the adaptive ADSD-gcForest model is described in Section 3, Section 4 is the experimental comparison part, and the conclusion is drawn in Section 5.

2. Related Works

2.1. SDP Image

Through the normalization method, the time domain signal can be described in the polar coordinate system. Thus, the vibration signal can be converted to SDP images [20], and the relationship between the amplitude and the frequency of the vibration signal can be described simply and directly through the geometric shape. The specific mapping relationship is as follows:

$$r(i) = \frac{x_i - x_{min}}{x_{max} - x_{mian}} \tag{1}$$

$$\theta(i) = \theta_l + \frac{x_{i+a} - x_{min}}{x_{max} - x_{min}} \delta \tag{2}$$

$$\varphi(i) = \theta_l - \frac{x_{i+a} - x_{min}}{x_{max} - x_{min}} \delta \tag{3}$$

where the input vibration signal is represented as x_i, i represents the sequence number of the discrete sampling point of the signal in the time domain, the maximum and minimum values of the vibration signal are, respectively, described as x_{max} and x_{min} and the amplitude of x_i corresponding to the time lag coefficient a is shown as x_{i+a}, the radius of the polar coordinates is indicated as $r(i)$, δ is the magnification angle, θ_l is the angle of the l-th mirror symmetry plane, $\theta(i)$ and $\varphi(i)$ are the deflection angles of the mirror symmetry plane, where $\delta \leq \theta_l$ and $\theta_l = \frac{360l}{N}$ ($l \in (0, N-1)$), and N is the number of symmetry planes. SDP images of different fault characteristics will present various geometric characteristics, which are manifested mainly in the curvature, thickness, geometric center and concentrated area of the image arm of the SDP image [21]. The SDP images of different fault characteristics are shown in Figure 1. Among them, IR, OR and B, respectively, represent the inner ring, outer ring and rolling element, while 007, 014, and 021 indicate that the fault diameter is 0.1778 mm, 0.3556 mm and 0.5334 mm separately.

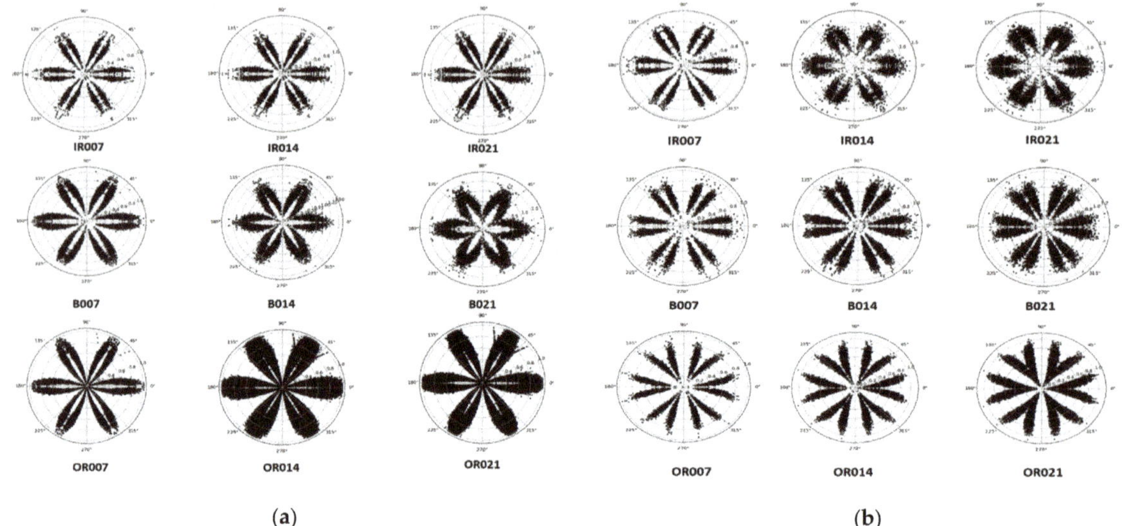

(a)　　　　　　　　　　　　　　　　　　　　**(b)**

Figure 1. SDP images. (**a**) Drive end bearing, (**b**) Fan end bearing.

2.2. Dilated Convolutional

Unlike ordinary convolution, the dilated convolution is one of the convolutional neural networks which increases the receptive field of the output unit without increasing the parameters, which is mainly implemented by introducing the dilation rate parameter. Specifically, it mainly depends on the way of interval sampling, which means that the spacing of each value is defined by the dilation rate when the convolution kernel processes the data. Thus, the receptive field can be increased without reducing the image resolution. For convolution kernels of the same size, the larger the dilation rate, the greater the receptive field of the convolution kernel [22]. The calculation formula for the receptive field of the dilated convolution is as follows:

$$r_n = r_{n-1} - 1 + k \tag{4}$$

where the receptive field of the current layer is presented as r_n, r_{n-1} is the upper receptive field and k is the size of the convolution kernel. The sampling process is shown in Figure 2.

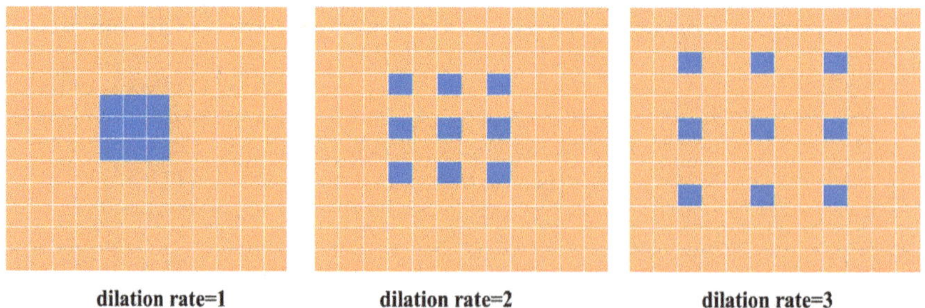

Figure 2. Dilated convolution.

2.3. Depth-Separable Convolution

The standard convolution is decomposed into deep convolution and point-wise convolution by depth-separable convolution [23–25]. First, each channel of the input sample is convolved one by one by the deep convolution; thus, the number of feature maps generated is the same as the number of channels of the input sample, and after that, the feature map is reconstructed with weights which are assigned on the basis of the designed algorithm according to point-by-point convolution. In this way, the amount of calculation and parameters of the model can be reduced in both time and space. For example, the dimension of the input samples is set as D_{in1}, D_{in2}, C_{in}, the output dimension is arranged accordingly as $D_{out1}, D_{out2}, C_{out}$ and the sizes of the convolution kernel are D_{k1} and D_{k2}, where C_{out}, C_{in} are the number of channels. The formula to calculate the total number of parameters for ordinary convolution and deeply separable convolution is shown in Formulas (5) and (6), respectively, where C_{conv} represents the total number of parameters for ordinary convolution and the total number of parameters for deeply separable convolution is $C_{separableconv}$. The specific schematic diagram is shown in Figure 3, where channel 1, channel 2 and channel 3, respectively, indicate the three dimensions of the input image.

$$C_{conv} = D_{out1} \times D_{out2} \times D_{k1} \times D_{k2} \times C_{out} \times C_{in} \tag{5}$$

$$C_{separableconv} = D_{out1} \times D_{out2} \times D_{k1} \times D_{k2} \times C_{in} + D_{out1} \times D_{out2} \times C_{out} \times C_{in} \tag{6}$$

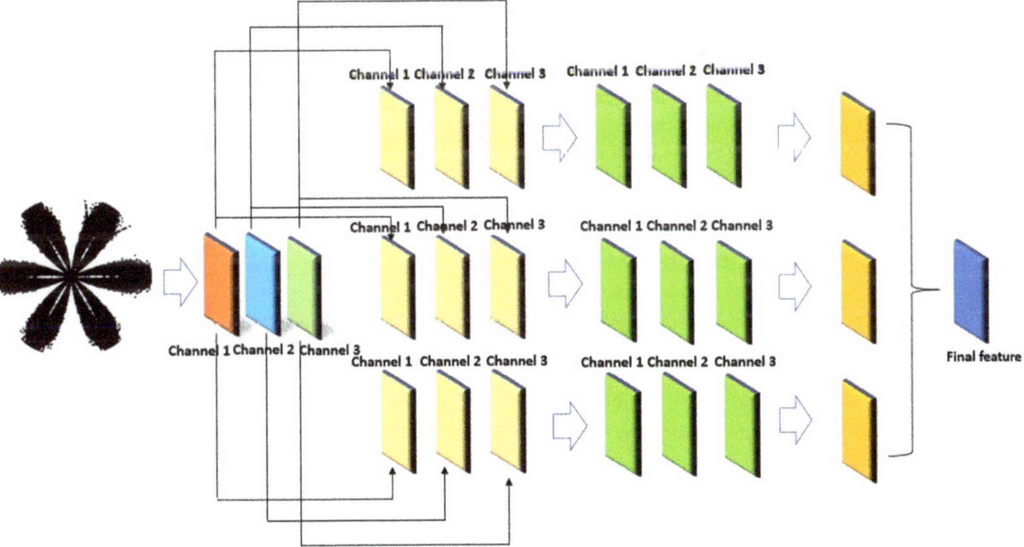

Figure 3. Depth-separable convolution.

2.4. CBAM

The attention mechanism is derived from the human visual mechanism, which is widely used in image processing. CBAM has been widely used in target detection by skillfully integrating spatial attention mechanism and channel attention mechanism [26–28]. Primarily, the channel attention mechanism selects which features are the key features, and then uses the spatial attention mechanism to learn the location of the key features, strengthening the extraction of the core features of the input sample; from this, in addition, the model can achieve adaptive refinement of the core features in the images. The specific composition structure is shown in Figure 4, where Avgpooling and Maxpooling, respectively, represent the average pooling and maximum pooling, while shared FC means the shared full connectivity layer.

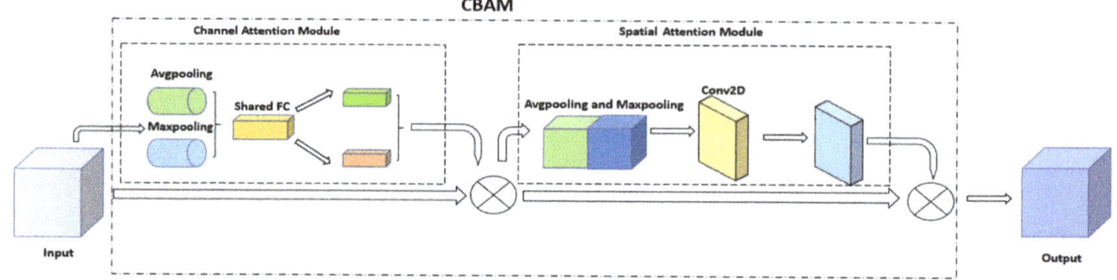

Figure 4. CBAM.

2.5. gcForest

Different from the traditional Softmax classifier, the hidden features of the input feature vector can be learned by gcForest through the superposition of multi-layer random forests, which then output the final classification results [29,30]. It has been proven that the accuracy of the deep forest classifier is about 1–4% higher than that of the Softmax classifier. The deep forest classifier mainly consists of two parts: multi-granularity scanning and cascade forest. The feature vector is sampled in a sliding window to form a new feature vector, which will be input into the cascade forest. After passing through the multi-layer random forest, the final output class probability distribution vector is taken as the final classification result. The specific structure of the deep forest classifier is shown in Figure 5, where K is the dimension of the input vector table type, n is the dimension of the sliding window, m as the category of the classification number and P is the final output vector whose dimension. Furthermore, in Figure 5, the scanned vector is input into the cascade forest, the blue and green two color Forests, respectively, represent random forest and completely random forest, each layer contains two random forests and two completely random forests and each forest after the completion of a training will be an output vector, of which the dimension is C. The output vectors of the four forests are stacked with the output vectors of multi-granularity scanning, and the vector with dimension (C × 4 + P) is the output; moreover, after multi-layer learning, the output vectors of the last layer are averaged to obtain the final probability category vector with dimension C and the maximum probability of the vector is taken as the classification result.

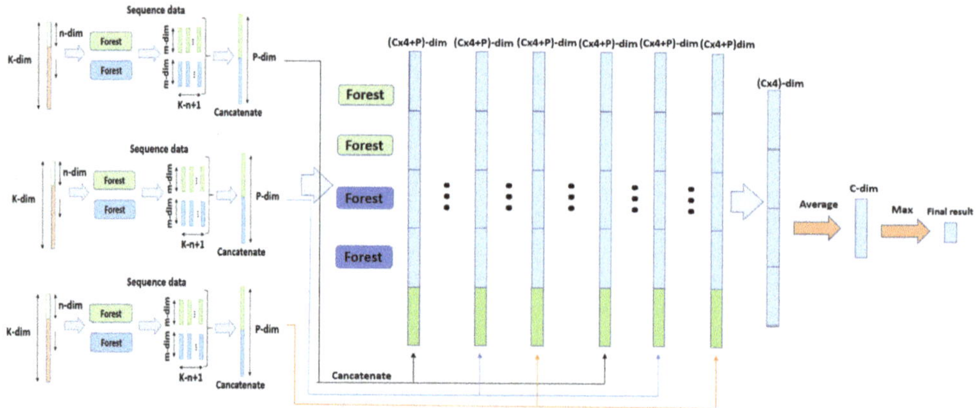

Figure 5. gcForest.

3. Method

The ADSD-gcForest model will be described in detail in this section. The detailed implementations of the method are described in the following three steps.

Step 1: A sliding window is used to sample vibration signals, then the noise of different intensity is added and the signals are converted into SDP image, and then the sample data are divided into a training set and a test set.

Step 2: The training set is entered into the adaptive ADSD-gcForest model for training and the Meta-ACON activation function is applied to adaptively adjust the network structure, according to different types of sample data to obtain the current optimal model structure, after which the trained model is saved.

Step 3: The trained model is used to directly extract fault features from new images, which results in the final diagnosis. The overall flowchart of the fault diagnosis is drawn in Figure 6.

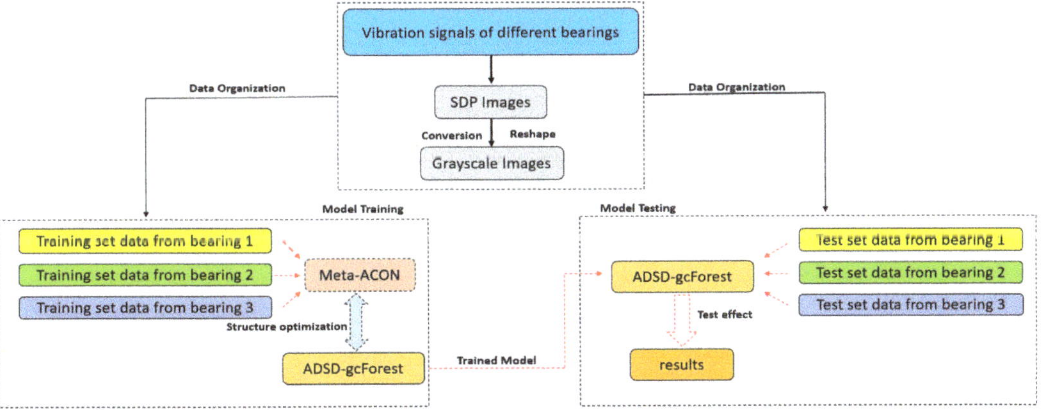

Figure 6. The overall flowchart of the fault diagnosis.

3.1. Meta-ACON

In order to achieve more effective fault diagnosis based on different bearing fault data, it may be necessary to continuously adjust the existing structure to achieve higher accuracy. In order to solve the above problems, a relatively simple way to achieve adaptive adjustment of the network model is proposed in this paper: by setting a single conversion factor β, the Meta-ACON activation function can simply select whether to activate the neurons in this layer according to different sample data (activation represents nonlinear output, while on the contrary, non-activation represents linear output). The design of the Meta-ACON activation function is derived from the smooth maximum function, and its formula is as follows:

$$S_\beta(x_1, \ldots x_n) = \frac{\sum_{i=1}^{n} x_i e^{\beta x_i}}{\sum_{i=1}^{n} e^{\beta x_i}} \tag{7}$$

where x_i represents the input signal sequence, β is the conversion factor, when $\beta \to \infty$, $S_\beta \to max$ and $\beta \to 0$, and S_β is the arithmetic mean value. Many common activation functions have the form $max\,(\eta a\,(x), \eta b\,(x))$. $\eta a\,(x)$ and $\eta b\,(x)$ are two freely configurable functions. For example, in the ReLU function, $\eta a\,(x) - x$ and $\eta b\,(x) - 0$, many activation functions can be expressed in the form of $max\,(\eta a\,(x), \eta b\,(x))$. To simplify the design, only two variables are considered here, and the sigmoid function is simplified as σ. At this time, the approximate relationship is represented as:

$$S_\beta(\eta_a(x), \eta_b(x)) = \eta_a(x) * \frac{e^{\beta \eta_a(x)}}{e^{\beta \eta_a(x)} + e^{\beta \eta_b(x)}} + \eta_b(x) * \frac{e^{\beta \eta_b(x)}}{e^{\beta \eta_a(x)} + e^{\beta \eta_b(x)}}$$
$$= \eta_a(x) * \frac{1}{1 + e^{-\beta(\eta_a(x) - \eta_b(x))}} + \eta_b(x) * \frac{1}{1 + e^{-\beta(\eta_b(x) - \eta_a(x))}} \qquad (8)$$
$$= (\eta_a(x) - \eta_b(x)) * \sigma[\beta(\eta_a(x) - \eta_b(x))] + \eta_b(x)$$

Furthermore, $\eta_a(x) = p_1 x$, $\eta_b(x) = p_2 x$ and $p_1 x \neq p_2 x$. The Meta-ACON activation function is as follows:

$$S_\beta = (p_1 - p_2)x * \sigma[\beta(p_1 - p_2)x] + p_2 x \qquad (9)$$

Among them, p_1 and p_2 are two random trainable parameters; therefore, the activation of neurons in this layer can be easily controlled by means of conversion factor β, where $\beta = \sigma W_1 W_2 \sum_{h=1}^{H} \sum_{w=1}^{W} x_{c,h,w}$, the input sample data is represented as $x_{c,h,w}$ and c, h and w, respectively, describe the number of channels, width and height of the input sample data. W_1 is the convolution of the sample data with the number of input channels as the width of the sample, the number of output channels as the width/r (r is a constant, generally taken as 16) and the convolution core size of 1×1. Similarly, W_2 is also obtained by the convolution with the convolution core size of 1×1, except that the number of output channels and input channels of convolution are opposite to the setting of W_1. Since the β value is directly determined by the structural characteristics of the sample data, different sample data will produce different β values, Therefore, after many times of training, with the continuous updating of Meta-ACON parameters, the structure of the model can be continuously optimized. The specific calculation process is shown in Figure 7.

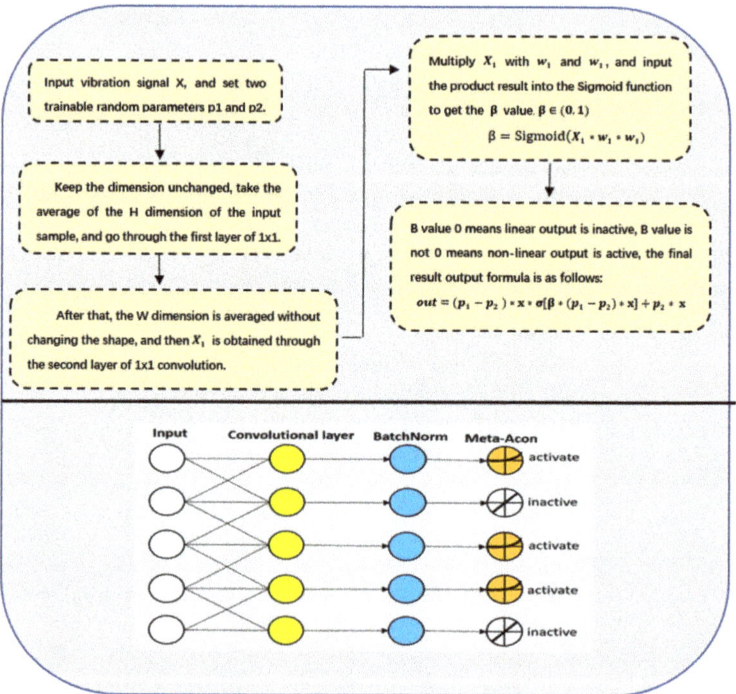

Figure 7. The specific calculation process of Meta-ACON.

3.2. ADSD-gcForest

Compared with time-domain signals, SDP images can represent different fault types in a more intuitive and simple way by presenting different geometric features. Therefore, the key to achieve an accurate fault diagnosis is to design a diagnosis model that can effectively extract geometric features from images. The visual geometry group 16 (VGG16) is one of the commonly used models in image processing. Feature extraction is effectively realized by stacking multilayer convolution, and network parameters are reduced by pooling layer. The model in this paper takes VGG16 as the basic framework. However, the structure of VGG16 network is relatively simple. Firstly, although the network is deep, ordinary convolution is widely used in convolution layers, which cannot extract the sample feature information in multiple scales, which limits the feature extraction ability of the network under the intervention of strong noise. Second, most of the activation functions in the convolution layer simply make the input signal become non-linear, so the network does not have good migration learning ability; thus, in the face of different sample data, the performance of the network will become unstable. Moreover, most diagnostic models use Softmax as the final classifier, However, Softmax is not an advanced classifier and cannot learn the feature information that has not been extracted, so as to reduce the final accuracy. In response to the above problems, the ADSD-gcForest model proposed in this paper makes the following improvements.

Due to the large number of input sample types, in order to increase the model feature extraction range and enrich the feature information, the characteristics of the receptive field are expanded by using the dilated convolution and combined with the residual network to build three branches. Therefore, the construction of three kinds of dilated convolutions with different dilation rates is connected through the residual network, the dilation rate is set to 1, 2 and 3 and the size of the convolution kernel is 3 $\times$ 3. Thus, multi-scale feature extraction is realized. After the dilated convolution with different expansion rates, it is combined with the CBAM, and the channel attention mechanism is used to measure the importance of different kinds of channel feature information in the feature map at different scales, so as to determine the key points under different channels in the feature map features. Then, the spatial attention mechanism is introduced to locate these key features and extract the key feature information from the feature map to obtain key features at different scales. Next, three feature maps are obtained and integrated using the residual network and input into the next layer. Due to the use of more dilated convolution and attention mechanisms in the network, it may lead to a longer network training time. Since the convolution operations for different channels of the input image can be simultaneously performed by the depth-separable convolution, the depth-separable convolution mechanism is led into the dilated convolution layer, after which the weight ratio of each feature map is determined through quasi-point convolution, and the the feature maps are integrated according to the weights. Thus, computational efficiency could be improved in this way. In order to realize that the network model can be adaptively adjusted according to the sample data of different fault types, the original ReLU activation function in the convolutional layer is replaced with the Meta-ACON activation function. The Meta-ACON activation function can be based on the size characteristics of the input image. By setting the conversion factor β, the value of β determines whether to activate the neurons in this layer after multiple trainings, and a flexible and efficient network structure can be adopted for the training model according to different input samples. Softmax is replaced by gcForest, which learns the hidden fault characteristics and gives the final results of the diagnosis results. The structure of the model is shown in Figure 8. SD convolution stands for dilated convolution with a deep separable mechanism. Detailed parameters of the optimized network are shown in Table 1.

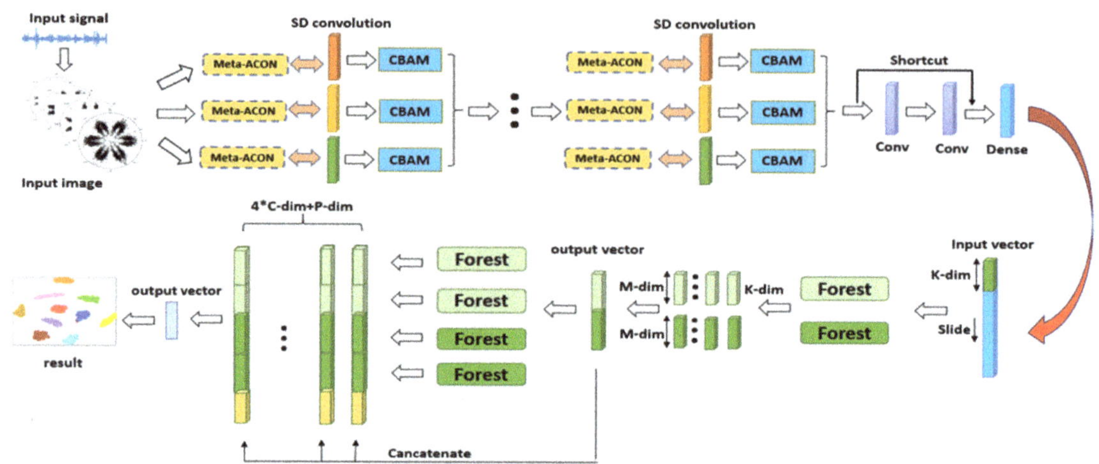

Figure 8. The model structure of ADSD-gcForest.

Table 1. Detailed parameters of the optimized network.

Layers	Filters	Kernel_Size/ Dilation Rate	Trainable Parameters	Input_Shape	Output_Shape
separable_conv_1	64	3/1	137	$28 \times 28 \times 1$	$28 \times 28 \times 64$
CBAM_1			677	$28 \times 28 \times 64$	$28 \times 28 \times 64$
separable_conv_2	64	3/2	137	$28 \times 28 \times 1$	$28 \times 28 \times 64$
CBAM_2			677	$28 \times 28 \times 64$	$28 \times 28 \times 64$
separable_conv_3	64	3/3	137	$28 \times 28 \times 1$	$28 \times 28 \times 64$
CBAM_3			677	$28 \times 28 \times 64$	$28 \times 28 \times 64$
Add_1			0	$28 \times 28 \times 64, 28 \times 28 \times 64, 28 \times 28 \times 64$	$28 \times 28 \times 64$
separable_conv_4	128	3/1	8896	$28 \times 28 \times 64$	$28 \times 28 \times 128$
CBAM_4			2277	$28 \times 28 \times 128$	$28 \times 28 \times 128$
separable_conv_5	128	3/2	8896	$28 \times 28 \times 64$	$28 \times 28 \times 128$
CBAM_5			2277	$28 \times 28 \times 128$	$28 \times 28 \times 128$
separable_conv_6	128	3/3	8896	$28 \times 28 \times 64$	$28 \times 28 \times 128$
CBAM_6			2277	$28 \times 28 \times 128$	$28 \times 28 \times 128$
Add_2			0	$28 \times 28 \times 128, 28 \times 28 \times 128, 28 \times 28 \times 128$	$28 \times 28 \times 128$
separable_conv_7	256	3/1	34,176	$28 \times 28 \times 128$	$28 \times 28 \times 256$
CBAM_7			8949	$25 \times 28 \times 256$	$28 \times 28 \times 256$
separable_conv_8	256	3/2	34,176	$28 \times 28 \times 128$	$28 \times 28 \times 256$
CBAM_8			8949	$25 \times 28 \times 256$	$28 \times 28 \times 256$
separable_conv_9	256	3/3	34,176	$28 \times 28 \times 128$	$28 \times 28 \times 256$
CBAM_9			8949	$25 \times 28 \times 256$	$28 \times 28 \times 256$
Add_3			0	$28 \times 28 \times 256, 28 \times 28 \times 256, 28 \times 28 \times 256$	$28 \times 28 \times 256$
separable_conv_10	256	3/1	68,096	$28 \times 28 \times 256$	$28 \times 28 \times 256$
separable_conv_11	256	3/1	68,096	$28 \times 28 \times 256$	$28 \times 28 \times 256$
Add_4			0	$28 \times 28 \times 256$	$28 \times 28 \times 256$
Flatten			0	$28 \times 28 \times 256$	$200,704 \times 1$
dense_1(1000)			200,705,000	$200,704 \times 1$	1000×1
dense_2 (256)			256,256	1000×1	256×1

4. Experiments

4.1. Introduction of Datasets

The datasets used in the experiment were the Case Western Reserve University bearing dataset and the Canadian University of Ottawa bearing dataset. Two different bearings are contained in the Western Reserve University bearing dataset: drive end bearing SKF6205 and fan end bearing SKF6203. The drive end bearing included the two different sampling frequencies of 12 KHZ and 48 KHZ, while the sampling frequency of the fan end bearing was only 12 KHZ. Ten types of states are contained in each bearing dataset, which are normal state, inner ring failure, outer ring failure, and rolling element failure. Each fault state contains three different fault levels, represented by a fault diameter. A total of four different load conditions were applied when measuring the bearing data. A total of 8 normal samples, 53 outer ring damage samples, 23 inner ring damage samples and 11 rolling element damage samples were obtained. The Canadian Ottawa dataset is the bearing vibration signal of different health conditions measured under time-varying speed conditions, which had 36 datasets. The bearing conditions include: normal, inner ring failure and outer ring failure. The working speed conditions are speed increase, speed deceleration, deceleration after speed increase and speed increase after deceleration. Each dataset contains two channels, and channel 1 represents the vibration data measured by the accelerometer, channel 2 signifies the speed data measured by the encoder, the sampling frequency is 200 KHZ and the sampling duration is 10 s.

The drive end and the fan end bearing data of Western Reserve University used in this paper are at the sampling frequency of 12 KHZ, and for part of the dataset in Channel 1 of the University of Ottawa in Canada, the sample data used were randomly selected from the dataset, where B, IR and OR indicate that the fault location is located in the rolling element, inner ring and outer ring of the bearing, respectively. Moreover, 007, 014 and 021, respectively, indicate that the fault diameter is 0.1778 mm, 0.3556 mm and 0.5334 mm, and the number at the end indicates the size of the load. For example, "−1" means that the load is 1 horsepower. A total of 1000 samples were sampled for each fault category, and the sample ratio of training set to test set was 7:3. The details are shown in Table 2.

Table 2. Sample distribution table.

Bearing Number	Dataset 1	Dataset 2	Dataset 3	Dataset 4	Dataset 5	Dataset 6
	Normal-1	Normal-2	Normal-3	Normal-1	Normal-2	Normal-3
	B007-1	B007-3	B007-3	B007-2	B007-1	B007-2
	B014-1	B014-3	B014-3	B014-2	B014-3	B014-3
	B021-1	B021-3	B021-2	B021-3	B021-2	B021-1
SKF6205	IR007-1	IR007-1	IR007-2	IR007-3	IR007-2	IR007-1
	IR014-2	IR014-2	IR014-2	IR014-3	IR014-3	IR014-2
	IR021-3	IR021-2	IR021-2	IR021-2	IR021-3	IR021-3
	OR007-1	OR007-2	OR007-1	OR007-1	OR007-3	OR007-1
	OR014-2	OR014-2	OR014-1	OR014-1	OR014-2	OR014-1
	OR021-3	OR021-1	OR021-2	0R021-2	OR021-1	OR021-3
	Normal-2	Normal-2	Normal-1	Normal-3	Normal-1	Normal-2
	B007-1	B007-1	B007-3	B007-2	B007-1	B007-3
	B014-3	B014-2	B014-2	B014-3	B014-2	B014-3
	B021-3	B021-1	B021-2	B021-3	B021-2	B021-1
SKF6203	IR007-3	IR007-2	IR007-2	IR007-3	IR007-2	IR007-1
	IR014-3	IR014-1	IR014-3	IR014-1	IR014-1	IR014-1
	IR021-3	IR021-3	IR021-1	IR021-1	IR021-3	IR021-3
	OR007-2	OR007-2	OR007-1	OR007-1	OR007-1	OR007-2
	OR014-2	OR014-2	OR014-3	OR014-2	OR014-3	OR014-2
	OR021-3	OR021-3	OR021-1	OR021-2	OR021-3	OR021-1

Six noises of different intensities were added to the sample dataset, namely, Noise 1, Noise 2, Noise 3, Noise 4, Noise 5 and Noise 6. Each type of noise contains three different types of noise. The proportions of the three noises in the six noises were Gaussian noise with signal-to-noise ratios of -4, -2, 0, 2, 4 and 6, salt and pepper noise, with ratios of 0.3, 0.25, 0.2, 0.15, 0.1 and 0.05, and Cauchy noise with position parameter 0 and scale parameter 1. gcForest was set as the classifier in all comparison methods, and SigDSD-gcforest means that the Sigmoid function is the activation function of the convolutional layers. Similarly, the activation functions of the convolutional layers in ReluDSD-gcforest and PReluDSD-gcforest are Relu and PRelu, respectively. The parameter settings of the ADSD-gcForest model were as follows: the network training parameters were set at a learning rate of 0.00005, the number of batch processing was 580, the number of iterations was 350, Adam was used as the optimization algorithm, the sliding window dimension used in MGS was 240, the number of trees in the random forests of MGS was 35 and the number of trees in a single random forest in the cascade forest was 150. The diagnostic effect is analyzed by comparing the accuracy rate, F1 value and Area Under Curve (AUC) value of different diagnostic models after training. The accuracy rate is generally expressed as $\frac{TP+TN}{TP+TN+FP+FN}$, and FI value is calculated as $\frac{2FP}{2TP+FP+FN}$, where TP refers to True Positives, FP represents True Negatives, FN indicates False Negatives and FP signifies False Positives. AUC is defined as the area under the area under curve. Generally, the higher the AUC value, the better the classification effect of the model.

4.2. Case Study 1: Performance of Drive End Bearing Fault Diagnosis

It can be seen from Figures 9 and 10 that when the noise environment is Noise 1, after the sample is trained by the ADSD-gcForest model, there are three categories of samples with low recognition rates, and there is also a small amount of aliasing in the T-SNE image. In other noise environments, there are only one or two fault categories with a low recognition rate. It can be seen from Table 3 that, compared to other methods, the ADSD-gcForest model achieves the highest fault accuracy rate and F1 value under various noises and different working conditions. Among them, the VGG16-gcForest model obtained the lowest accuracy and the F1 values, which is about 26–35% lower than those of the ADSD-gcForest model, while the accuracy and F1 values of the Res50-gcForest model are about 18% higher than the VGG16-gcForest model. Since the Relu function can better solve the network convergence problem than the Sigmoid function, the accuracy and F1 values of the Relu-gcForest model are about 0.6–0.7% higher than that of the SigDSD-gcForest model, and PRelu updates the weight according to the input data, which makes the network have a certain adaptive optimization capability. The accuracy and F1 values obtained after training is about 1.3% higher than that of the Relu-gcForest model, but its values are still lower than the ADSD-gcForest model. Figure 11 mainly describes the comparison of the AUC values of different methods. From Figure 11, it can be found that the AUC values of ADSD-gcForest under different noises are the highest and all are above 92%, indicating that the ADSD-gcForest model has a good fault diagnosis effect. It can be seen from the experimental results presented above that the ADSD-gcForest model can more accurately diagnose drive-end bearing failures under different working conditions and strong noise interference with a high accuracy rate.

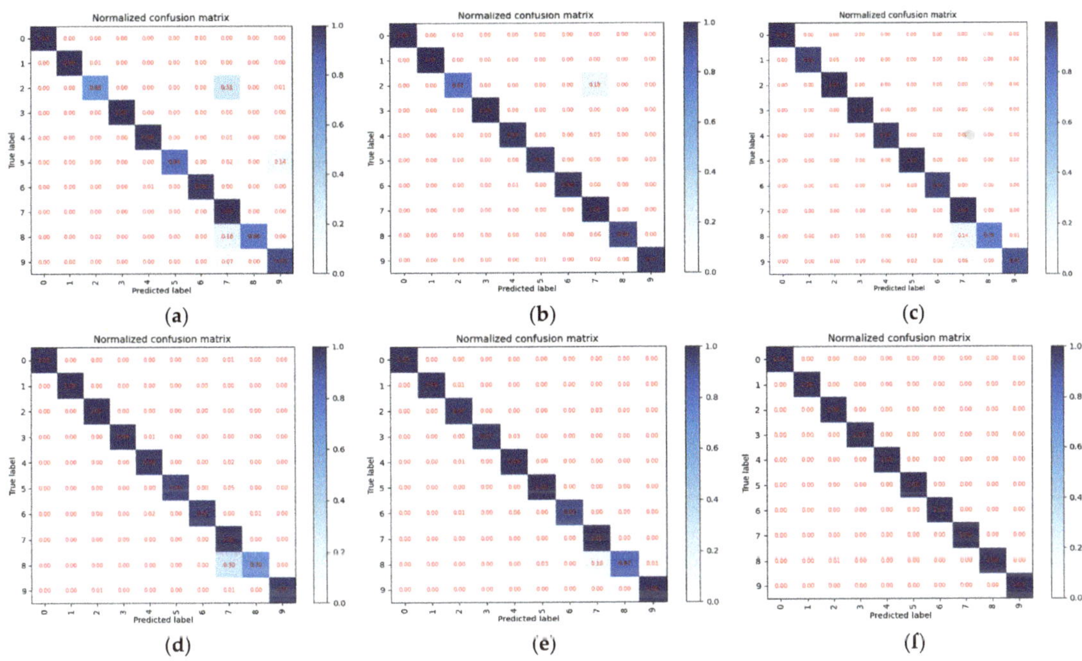

Figure 9. Confusion matrix obtained by ADSD-gcforest model training (drive end). (**a**) Noise 1, (**b**) Noise 2, (**c**) Noise 3, (**d**) Noise 4, (**e**) Noise 5. (**f**) Noise 6.

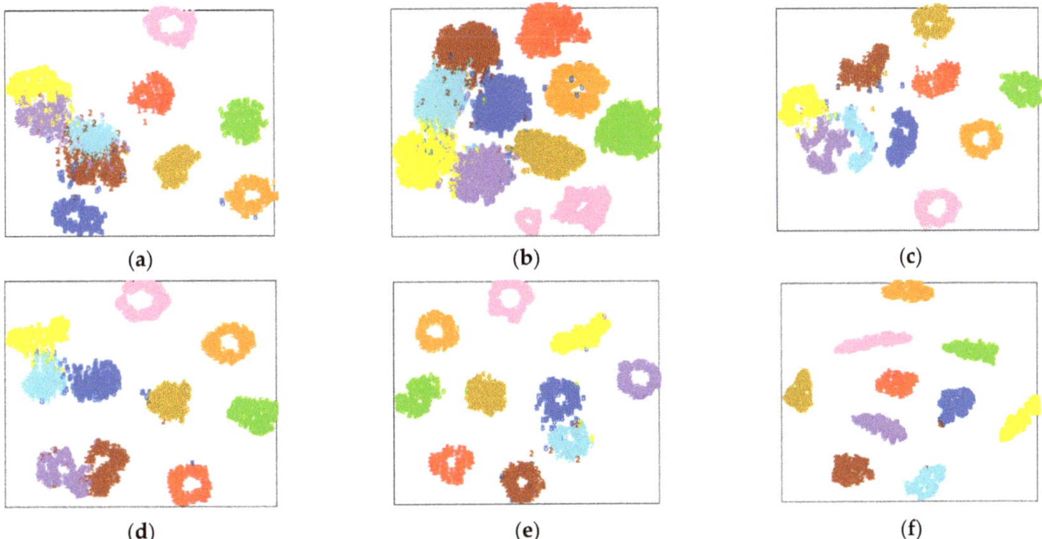

Figure 10. T-SNE images obtained by ADSD-gcforest model training (drive end). (**a**) Noise 1, (**b**) Noise 2, (**c**) Noise 3, (**d**) Noise 4, (**e**) Noise 5. (**f**) Noise 6.

Table 3. Comparison table of the accuracy (AC) and F1 values of driver end data.

Dataset	Methods	N1(AC/F1)	N2(AC/F1)	N3(AC/F1)	N4(AC/F1)	N5(AC/F1)	N6(AC/F1)
Dataset 1	VGG16-gcForest	65.10%/66.20%	69.13%/68.78%	71.20%/72.02%	72.41%/73.10%	75.36%/74.48%	77.54%/76.98%
	Res50- gcForest	83.25%/82.86%	86.42%/85.96%	87.85%/88.12%	89.22%/90.03%	92.02%/92.05%	93.33%/94.40%
	SigDSD- gcForest	92.43%/91.73%	94.17%/93.92%	95.95%/94.65%	96.04%/96.11%	96.35%/96.30%	96.85%/96.80%
	ReluDSD-gcforest	92.85%/93.05%	94.22%/93.88%	96.20%/96.22%	96.47%/96.50%	96.70%/96.73%	97.03%/96.98%
	PReluDSD-gcForest	93.09%/92.86%	94.72%/94.70%	96.62%/96.65%	96.73%/96.80%	96.83%/96.96%	97.45%/97.52%
	ADSD-gcForest	94.32%/94.30%	95.85%/95.78%	97.70%/97.73%	97.83%/98.03%	97.92%/98.15%	98.23%/98.28%
Dataset 2	VGG16-gcForest	64.72%/63.56%	69.20%/68.95%	69.96%/70.14%	72.02%/71.92%	73.46%/73.45%	76.85%/76.88%
	Res50- gcForest	83.50%/83.47%	86.72%/86.75%	87.60%/87.59%	89.27%/89.30%	91.26%/91.33%	92.96%/93.05%
	SigDSD- gcForest	89.98%/90.13%	91.40%/91.48%	92.53%/92..60%	93.06%/93.10%	94.12%/94.08%	95.23%/95.26%
	ReluDSD-gcforest	90.48%/90.43%	91.92%/92.03%	92.90%/92.86%	93.55%/93.57%	94.67%/94.65%	95.73%/95.70%
	PReluDSD-gcForest	91.52%/91.55%	92.03%/92.05%	93.63%/93.71%	94.52%/94.50%	95.03%/95.11%	96.15%/96.18%
	ADSD-gcForest	92.65%/92.66%	93.79%/93.82%	94.42%/94.45%	95.91%/96.02%	96.51%/96.47%	97.83%/97.85%
Dataset 3	VGG16-gcForest	66.25%/66.31%	67.43%/67.40%	69.84%/69.86%	72.45%/72.50%	74.62%/75.06%	77.11%/78.23%
	Res50- gcForest	83.85%/83.80%	85.62%/86.15%	86.87%/86.93%	88.38%/88.43%	91.67%/91.72%	92.63%/92.73%
	SigDSD- gcForest	91.61%/91.78%	92.96%/93.32%	95.41%/95.60%	96.08%/96.05%	96.21%/96.34%	96.91%/97.01%
	ReluDSD-gcforest	91.95%/91.86%	93.60%/93.74%	95.92%/96.02%	96.52%/96.46%	96.87%/96.94%	97.32%/97.28%
	PReluDSD-gcForest	92.07%/91.96%	94.30%/94.52%	95.42%/95.40%	96.90%/96.86%	97.24%/97.35%	97.75%/97.92%
	ADSD-gcForest	93.22%/93.25%	95.43%/95.48%	96.48%/96.56%	97.11%/97.16%	97.84%/98.02%	98.33%/98.30%
Dataset 4	VGG16-gcForest	65.89%/65.92%	68.34%/69.16%	70.54%/71.17%	73.52%/73.58%	75.66%/76.16%	77.35%/77.49%
	Res50- gcForest	77.82%/77.80%	80.35%/81.42%	84.48%/85.53%	86.76%/87.36%	90.81%/91.28%	92.28%/93.16%
	SigDSD- gcForest	92.25%/91.89%	92.72%/92.70%	93.78%/93.89%	94.75%/94.82%	95.52%/95.64%	96.44%/97.13%
	ReluDSD-gcforest	92.71%/93.04%	93.21%/93.19%	94.36%/94.28%	95.46%/95.51%	96.01%/96.33%	97.08%/97.26%
	PReluDSD-gcForest	93.07%/93.26%	94.96%/95.14%	95.58%/96.27%	96.82%/97.14%	97.42%/97.59%	97.64%/98.01%
	ADSD-gcForest	94.18%/95.64%	95.78%/95.74%	96.28%/96.32%	97.34%/97.64%	97.92%/98.12%	98.17%/98.34%
Dataset 5	VGG16-gcForest	66.14%/66.10%	67.94%/68.23%	71.13%/71.06%	72.86%/72.93%	75.43%/76.15%	76.29%/77.35%
	Res50- gcForest	79.85%/80.15%	81.62%/81.64%	84.96%/84.86%	88.62%/88.75%	91.52%/91.67%	93.53%/93.78%
	SigDSD- gcForest	92.02%/92.35%	94.40%/94.67%	95.22%/96.37%	95.72%/96.89%	96.51%/97.02%	97.05%/97.28%
	ReluDSD-gcforest	92.47%/92.43%	94.86%/94.82%	95.76%/95.84%	96.12%/96.15%	97.02%/97.46%	97.53%/97.88%
	PReluDSD-gcForest	93.08%/93.47%	95.48%/94.86%	96.27%/96.20%	96.76%/97.16%	97.67%/98.16%	98.08%/98.24%
	ADSD-gcForest	94.27%/94.35%	96.27%/96.37%	97.19%/97.10%	97.46%/97.65%	98.15%/98.10%	98.67%/98.76%
Dataset 6	VGG16-gcForest	65.85%/65.78%	66.72%/67.05%	70.32%/71.14%	72.61%/72.76%	74.50%/75.65%	76.76%/77.04%
	Res50- gcForest	78.95%/79.13%	84.06%/84.23%	85.46%/85.63%	87.42%/87.25%	89.73%/89.53%	91.32%/92.03%
	SigDSD- gcForest	91.06%/91.32%	92.92%/93.16%	93.72%/94.07%	95.03%/95.06%	96.40%/96.32%	97.02%/97.14%
	ReluDSD-gcforest	91.62%/91.76%	93.52%/94.31%	94.34%/94.54%	95.43%/95.32%	97.02%/96.53%	97.33%/97.27%
	PReluDSD-gcForest	92.07%/92.15%	94.08%/93.64%	95.17%/95.67%	96.06%/96.05%	97.47%/97.36%	97.78%/98.05%
	ADSD-gcForest	93.76%/93.48%	95.61%/95.53%	96.86%/96.57%	97.59%/97.42%	98.34%/98.28%	98.55%/98.62%

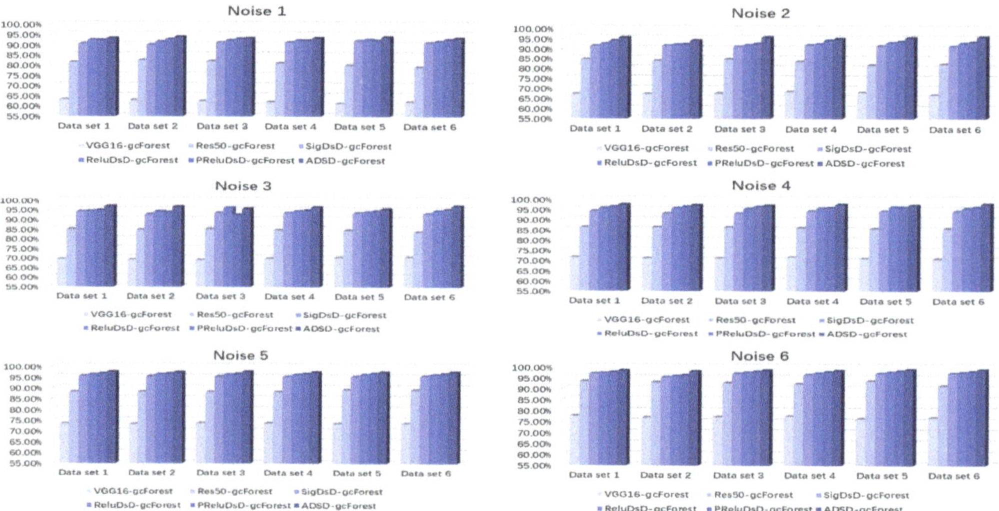

Figure 11. Comparison figures of the AUC of the driver end data under Noise 1, 2, 3, 4, 5 and 6.

4.3. Case Study 2: Performance of Fan End Bearing Fault Diagnosis

It can be seen from Figures 12 and 13 that only when the noise environment is Noise 1, a few fault categories cannot be effectively identified. In other noise environments, the entire fault category can be accurately identified. It can be seen in Table 4 that the accuracy and F1 values obtained from the training of the VGG16-gcForest and Res50-gcForest models have dropped by approximately 1.5–1.6% compared to the driving end values. The overall accuracy and F1 value of the VGG16-gcForest model are between 61–75%. The accuracy and F1 values of the training of the SigDSD-gcForest, ReluDSD-gcForest and PreluDSD-gcForest models has also decreased. Among them, the most obvious decrease is SigDSD-gcForest, with a decrease from 0.3% to 0.4%, while the accuracy and F1 value of the PreluDSD-gcForest model drops by at least about 1.3–1.5%. The accuracy value and F1 values of the ADSD-gcForest model are the highest, and these values are similar to case study 1. Figure 14 depicts the AUC values obtained by different diagnostic methods under different noises. It can be found that the AUC values obtained by the ADSD-gcForest model are still the highest, which are close to those obtained in case study 1. Through the experimental results presented above, it can be found that the ADSD-gcForest model proposed in this paper can basically realize an effective fault diagnosis for different bearings under multiple working conditions.

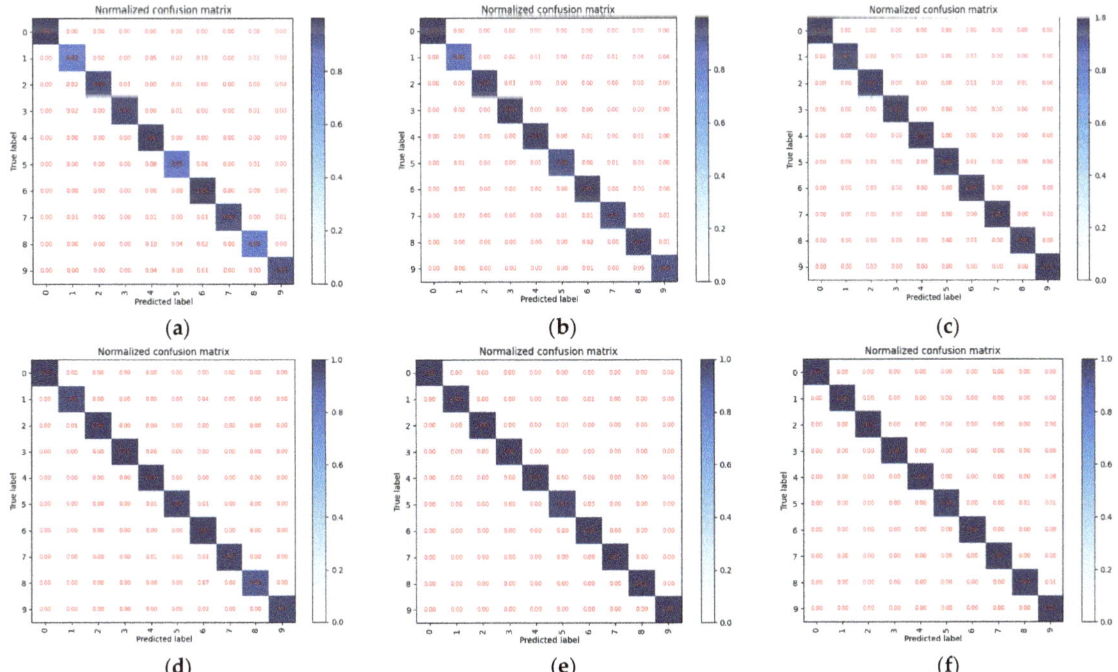

Figure 12. Confusion matrix obtained by ADSD-gcForest model training (fan end). (**a**) Noise 1, (**b**) Noise 2, (**c**) Noise 3, (**d**) Noise 4, (**e**) Noise 5. (**f**) Noise 6.

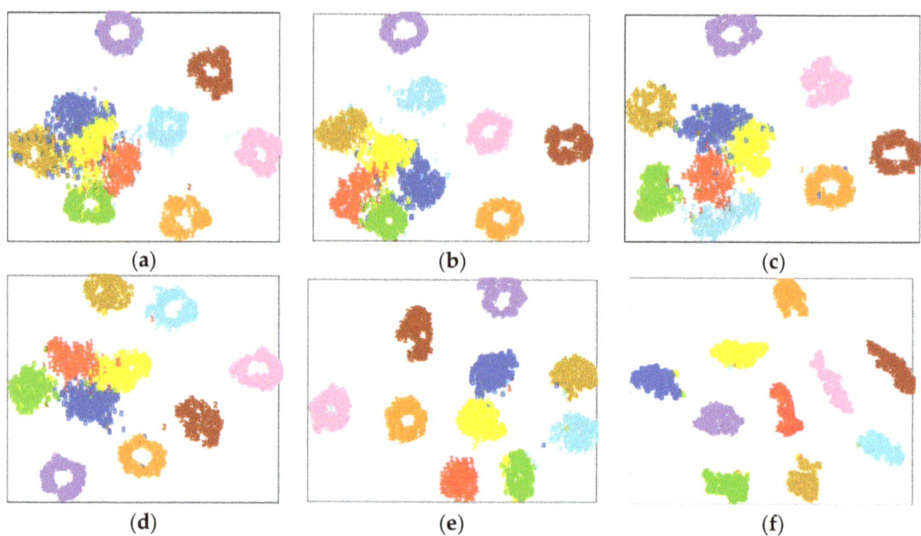

Figure 13. T-SNE images obtained by ADSD-gcForest model training (fan end). (**a**) Noise 1, (**b**) Noise 2, (**c**) Noise 3, (**d**) Noise 4, (**e**) Noise 5. (**f**) Noise 6.

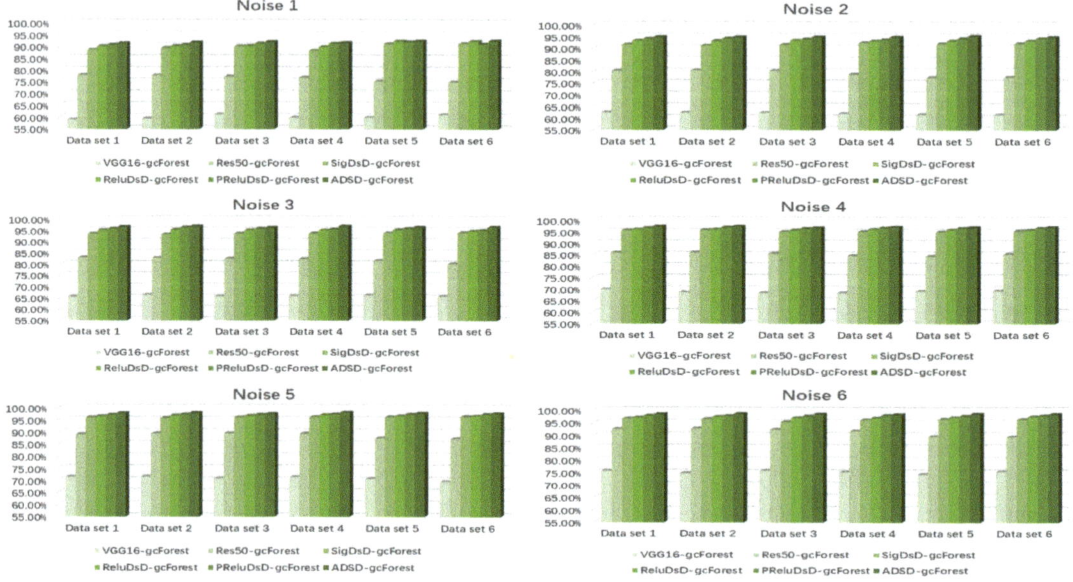

Figure 14. Comparison figures of the AUC of the fan end data under Noise 1, 2, 3, 4, 5 and 6.

Table 4. Comparison table of the accuracy (AC) and F1 values of the fan end data.

Dataset	Methods	N1(AC/F1)	N2(AC/F1)	N3(AC/F1)	N4(AC/F1)	N5(AC/F1)	N6(AC/F1)
Dataset 1	VGG16-gcForest	61.87%/61.80%	64.84%/64.75%	67.45%/68.12%	69.85%/70.23%	73.46%/73.32%	75.75%/75.82%
	Res50- gcForest	79.75%/80.04%	81.28%/81.42%	84.86%/84.90%	87.91%/88.34%	91.03%/91.06%	92.46%/92.50%
	SigDSD- gcForest	91.52%/91.68%	92.53%/93.46%	94.48%/95.62%	95.56%/95.63%	96.01%/96.20%	96.65%/96.60%
	ReluDSD-gcforest	92.03%/92.12%	93.20%/93.18%	95.06%/94.92%	96.01%/95.68%	96.53%/96.48%	97.04%/97.06%
	PReluDSD-gcForest	92.62%/92.65%	93.95%/93.86%	95.65%/95.60%	96.62%/96.63%	97.03%/97.18%	97.75%/97.70%
	ADSD-gcForest	93.32%/93.34%	94.74%/94.70%	96.54%96.46%	97.23%/97.20%	97.72%/97.67%	98.34%/98.25%
Dataset 2	VGG16-gcForest	62.41%/63.26%	64.27%/64.68%	66.75%/67.35%	68.58%/69.14%	71.59%/71.48%	73.84%/73.76%
	Res50- gcForest	79.61%/79.53%	81.45%/81.39%	83.68%/83.61%	86.84%/86.72%	89.83%/89.76%	92.69%/92.53%
	SigDSD- gcForest	90.37%/90.28%	93.12%/93.36%	94.27%/95.26%	95.76%/95.63%	95.67%/95.72%	96.43%/96.84%
	ReluDSD-gcforest	91.06%/91.26%	94.08%/94.34%	95.59%/95.62%	96.22%/96.32%	96.75%/97.14%	97.29%/97.38%
	PReluDSD-gcForest	91.56%/92.36%	94.67%/94.37%	96.24%/95.37%	96.87%/97.08%	97.32%/97.46%	97.64%/97.85%
	ADSD-gcForest	92.68%/93.06%	95.26%/95.37%	96.81%/96.68%	97.26%/97.39%	97.83%/97.80%	98.44%/98.40%
Dataset 3	VGG16-gcForest	63.21%/63.46%	65.32%/64.89%	67.63%/67.90%	68.42%/68.36%	70.94%/71.26%	72.84%/73.58%
	Res50- gcForest	77.12%/77.42%	80.23%/80.68%	82.47%/82.86%	86.58%/86.69%	89.52%/90.15%	92.27%/92.38%
	SigDSD- gcForest	90.46%/91.68%	93.48%/93.56%	94.59%/94.75%.	95.03%/96.15%	95.67%/96.82%	95.33%/95.59%
	ReluDSD-gcforest	90.72%/91.08%	93.73%/94.28%	95.07%/95.36%	95.68%/96.06%	96.28%/96.33%	96.87%/97.11%
	PReluDSD-gcForest	91.25%/91.20%	94.36%/94.38%	95.53%/95.49%	96.36%/96.42%	96.82%/96.74%	97.35%/97.19%
	ADSD-gcForest	91.85%/92.09%	94.75%/95.13%	96.15%/96.18%	96.98%/97.26%	97.23%/97.46%	98.31%/98.48%
Dataset 4	VGG16-gcForest	61.74%/62.31%	63.87%/64.26%	66.88%/67.18%	68.36%/69.45%	70.42%/71.26%	71.29%/72.21%
	Res50- gcForest	76.42%/76.48%	78.73%/78.62%	82.23%/82.26%	84.64%/85.04%	88.37%/88.49%	91.44%/91.57%
	SigDSD- gcForest	91.23%/91.34%	92.34%/92.86%	94.68%/95.71%	95.82%/96.02%	96.42%/96.74%	97.08%/97.06%
	ReluDSD-gcforest	91.75%/91.60%	93.03%/93.09%	95.02%/95.16%	96.35%/96.42%	96.97%/97.05%	97.34%/97.48%
	PReluDSD-gcForest	92.20%/92.34%	93.58%/93.64%	95.42%/95.56%	96.72%/96.83%	97.41%/97.56%	97.86%/98.01%
	ADSD-gcForest	93.45%/93.40%	94.71%/94.65%	96.75%/96.63%	97.32%/97.46%	98.03%/98.14%	98.29%/98.24%
Dataset 5	VGG16-gcForest	62.76%/62.64%	64.52%/65.38%	67.12%/67.48%	69.29%/69.70%	70.68%/71.17%	71.36%/72.47%
	Res50- gcForest	75.12%/75.49%	77.37%/77.25%	79.52%/79.26%	82.67%/82.54%	84.46%/84.69%	87.53%/88.14%
	SigDSD- gcForest	91.40%/91.42%	93.01%/92.89%	94.81%/95.17%	95.02%/95.43%	96.24%/96.39%	96.82%/97.16%
	ReluDSD-gcforest	92.22%/92.47%	93.50%/93.49%	95.22%/95.24%	95.75%/96.78%	96.64%/96.51%	97.27%/97.38%
	PReluDSD-gcForest	92.86%/92.92%	94.08%/94.06%	95.82%/95.86%	96.22%/96.36%	97.25%/97.36%	97.76%/97.82%
	ADSD-gcForest	93.24%/93.28%	95.78%/95.83%	96.28%/97.36%	96.87%/97.12%	97.73%/97.68%	98.38%/98.06%
Dataset 6	VGG16-gcForest	61.98%/61.86%	63.45%/64.58%	65.62%/65.52%	68.29%/68.34%	69.56%/69.96%	71.42%/71.86%
	Res50- gcForest	74.86%/74.92%	77.53%/77.50%	80.22%/80.36%	82.23%/82.18%	84.29%/84.27%	86.36%/86.34%
	SigDSD- gcForest	92.25%/92.37%	93.45%/93.57%	94.09%/94.16%	95.42%/95.26%	96.24%/96.31%	96.83%/96.80%
	ReluDSD-gcforest	92.86%/92.79%	93.73%/93.76%	94.42%/94.39%	95.73%/95.70%	96.89%/96.92%	97.36%/97.34%
	PReluDSD-gcForest	93.03%/93.17%	94.24%/94.28%	95.02%/94.88%	96.36%/97.32%	97.25%/97.05%	97.76%/97.65%
	ADSD-gcForest	93.55%/93.64%	94.61%/94.78%	95.42%/95.57%	96.92%/97.18%	97.83%/98.07%	98.34%/98.64%

4.4. Case Study 3: Performance of the Ottawa Bearing Dataset

In order to further test the generalization and robustness of the ADSD-gcForest model, case study 3 focused on the University of Ottawa dataset, which was specifically divided into six datasets. The setting method of adding noise was the same as case study 1. The specific sample types are shown in Table 5. There were three operation conditions of the bearings in the datasets, i.e., normal (H), inner race fault (I) and out race fault (O), and also contained four speed transformation conditions, i.e., speed up (A), slow down (B), speed up and slow down (C) and slow down and speed up (D). The noise setting used in case study 3 is the same as the case study 1. The training parameter settings of the ADSD-gcForest model are as follows: the network training parameters were set to a learning rate of 0.00005, the number of batch processing was 550, the number of iterations was 350, Adam was used as the optimization algorithm, the sliding window dimension used in MGS was 240, the number of trees in the MGS random forest was 35 and the number of trees in a single random forest in the cascade forest was 150.

Table 5. Sample distribution table.

The Name of Dataset	Dataset 1	Dataset 2	Dataset 3	Dataset 4	Dataset 5	Dataset 6
	HD-1	HA-1	HB-1	HA-1	HC-1	HD-1
	HA-1	HB-1	HC-1	HB-1	HA-1	HB-1
	HB-1	HD-1	HA-1	HD-1	HB-1	HC-1
	IC-1	IA-1	IC-1	IB-1	IB-1	ID-1
University of Ottawa bearing data	ID-1	IB-1	IA-1	IA-1	IC-1	IA-1
	IB-1	ID-1	IB-1	ID-1	IA-1	IB-1
	OB-1	OD-1	OB-1	OB-1	OA-1	OD-1
	OD-1	OC-1	OA-1	OD-1	OC-1	OB-1
	OA-1	OB-1	OD-1	OC-1	OD-1	OC-1
	HC-1	IA-1	HA-1	IB-1	OD-1	OB-1

It can be seen from Figure 15 that, compared to case study 1 and case study 2, when the noise environment is Noise 1 and Noise 2, the degree of discrimination of some fault categories is lower, but in other noise environments, the fault categories can be accurately classified. It can be seen from Figures 16 and 17 that the training accuracy of the ADSD-gcForest model is the highest and the value is relatively stable, while the fluctuation is small, which is consistent with the values in Table 6. At the same time, it can be found from Table 6 that the accuracy and F1 values obtained by training the VGG16-gcForest and Res50-gcForest models are significantly lower than case study 1 and case study 2. In Figures 16 and 17, the accuracy of the two models also fluctuates significantly, and the accuracy of the other three models is more accurate. The rate values have also decreased, but the value fluctuations are relatively small. Figure 18 reflects the AUC values of different diagnostic models, from which it can be found that the AUC values of the ADSD-gcForest model are basically similar to the first two cases, but other diagnostic models have decreased. Through comparative experiments of three groups of different bearings, it can be seen that under different noise conditions and for bearing data under different working conditions, in one way, the ADSD-gcForest model can achieve effective fault feature extraction, while in another way, the use of the Meta-ACON activation function can easily and efficiently complete the self-adaptive optimization of the model structure and realize more accurate fault diagnosis.

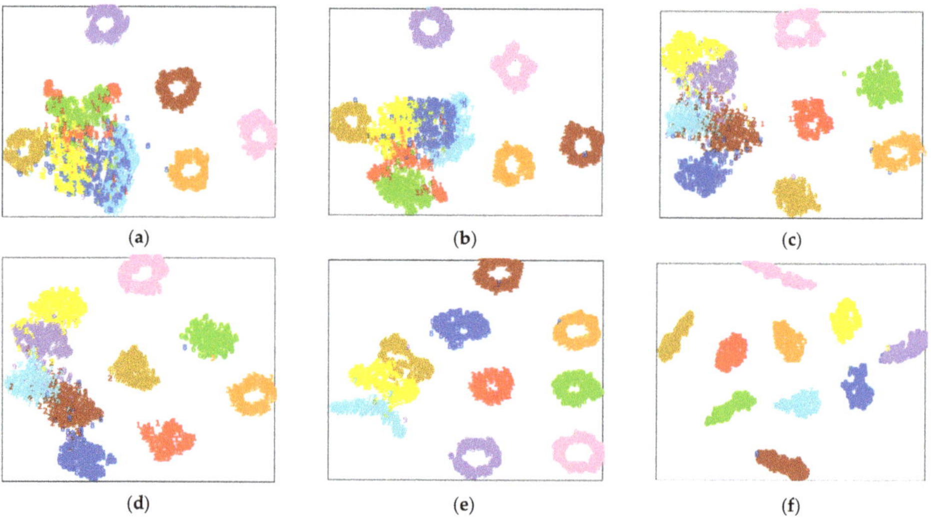

Figure 15. T-SNE images obtained by ADSD-gcforest model training (Ottawa Bearing). (**a**) Noise 1, (**b**) Noise 2, (**c**) Noise 3, (**d**) Noise 4, (**e**) Noise 5. (**f**) Noise 6.

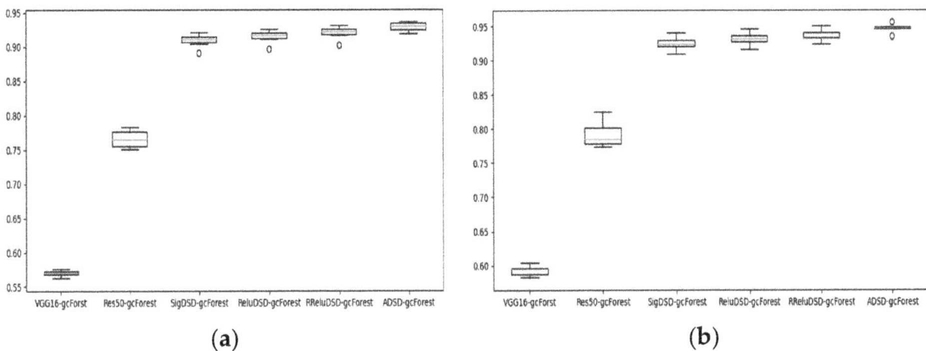

Figure 16. Box plots of accuracy values under different noise conditions. (**a**) Noise 1, (**b**) Noise 2.

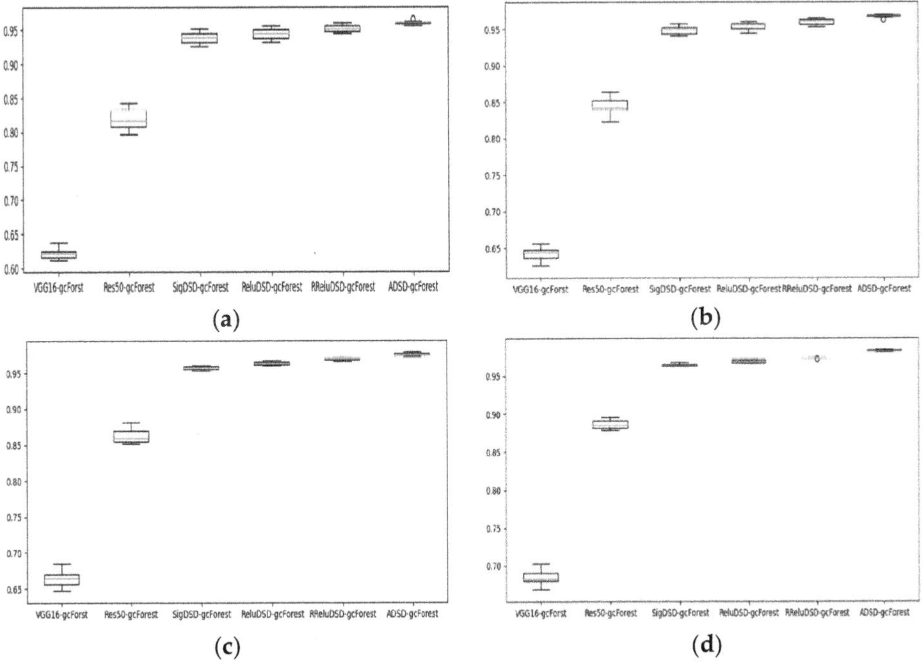

Figure 17. Box plots of accuracy values under different noise conditions. (**a**) Noise 3, (**b**) Noise 4, (**c**) Noise 5, (**d**) Noise 6.

Table 6. Comparison table of the accuracy (AC) and F1 values of the Ottawa dataset.

Dataset	Methods	N1(AC/F1)	N2(AC/F1)	N3(AC/F1)	N4(AC/F1)	N5(AC/F1)	N6(AC/F1)
Dataset 1	VGG16-gcForest	56.23%/56.20%	60.39%/60.34%	63.76%/63.72%	65.63%/65.59%	68.42%/68.45%	70.28%/70.36%
	Res50- gcForest	78.29%/78.26%	82.53%/82.57%	84.31%/84.59%	85.42%/85.31%	87.12%/87.18%	88.03%/87.93%
	SigDSD- gcForest	91.23%/91.36%	92.39%/92.58%	94.52%/95.67%	95.26%/95.68%	96.01%/96.19%	96.75%/97.70%
	ReluDSD-gcforest	91.79%/91.66%	93.42%/93.35%	95.01%/95.16%	95.76%/95.72%	96.42%/96.48%	97.03%/97.18%
	PReluDSD-gcForest	92.34%/92.46%	93.44%/93.67%	95.47%/95.55%	96.12%/96.39%	97.03%/96.98%	97.35%/97.36%
	ADSD-gcForest	93.42%/93.39	94.76%/94.58%	96.08%/96.06%	96.89%/96.92%	97.62%/97.58%	98.42%/98.48%
Dataset 2	VGG16-gcForest	57.62%/57.55%	59.45%/59.52%	62.45%/62.38%	64.81%/64.80%	67.02%/67.28%	69.31%/69.59%
	Res50- gcForest	77.94%/77.80%	80.66%/81.56%	83.52%/84.28%	86.44%/86.50%	88.18%/88.26%	89.50%/89.53%
	SigDSD- gcForest	90.46%/90.63%	92.02%/91.26%	93.25%/93.18%	94.11%/94.07%	95.47%/95.48%	96.26%/96.31%
	ReluDSD-gcforest	91.04%/91.16%	92.68%/92.76%	93.82%/93.96%	94.45%/94.48%	96.06./95.89%	96.63%/96.54%
	PReluDSD-gcForest	91.65%/91.77%	93.24%/93.20%	94.75%/94.79%	95.36%/95.42%	96.62%/96.59%	97.30%/97.28%
	ADSD-gcForest	92.42%/92.38%	94.66%/94.76%	96.14%/97.28%	96.76%/97.02%	97.32%/97.48%	98.15%/98.24%
Dataset 3	VGG16-gcForest	56.80%/56.68%	58.55%/59.04%	62.53%/62.68%	64.33%/65.22%	66.86%/66.78%	68.42%/68.40%
	Res50- gcForest	76.73%/76.55%	78.42%/78.61%	81.18%/81.26%	83.92%/83.90%	86.35%/86.42%	88.26%/88.37%
	SigDSD- gcForest	89.08%/88.89%	90.91%/91.26%	92.58%/92.87%	94.27%/94.38%	95.31%/95.36%	96.26%/96.28%
	ReluDSD-gcforest	89.62%/89.52%	91.54%/91.69%	93.18%/92.89%	94.82%/94.80%	96.05%/96.18%	96.86%/96.76%
	PReluDSD-gcForest	90.16%/91.02%	92.32%/92.21%	94.84%/94.56%	95.47%/95.32%	96.76%/96.79%	97.32%/97.42%
	ADSD-gcForest	91.85%/91.29%	93.49%/93.74%	95.68%/95.88%	96.31%/96.28%	97.45%/97.52%	98.27%/98.38%
Dataset 4	VGG16-gcForest	57.34%/57.49%	59.56%/59.63%	61.16%/61.19%	63.47%/63.30%	65.52%/65.71%	67.92%/67.94%
	Res50- gcForest	75.26%/75.36%	77.56%/77.49%	80.64%/80.79%	83.94%/84.06%	85.16%/85.67%	87.85%/88.06%
	SigDSD- gcForest	91.53%/91.68%	92.20%/92.34%	93.23%/94.36%	94.78%/94.82%	95.63%/95.56%	96.28%/96.34%
	ReluDSD-gcforest	92.06%/92.64%	92.76%/92.84%	93.74%/93.70%	95.58%/95.88%	96.17%/96.24%	96.65%/96.69%
	PReluDSD-gcForest	92.60%/91.65%	93.26%/93.36%	94.43%/94.58%	96.03%/96.15%	96.76%/96.72%	97.22%/97.36%
	ADSD-gcForest	93.42%/93.62%	94.60%/94.26%	95.81%/95.68%	96.63%/96.58%	97.45%/97.26%	98.32%/98.30%
Dataset 5	VGG16-gcForest	56.85%/56.79%	58.25%/58.16%	61.83%/62.05%	62.55%/63.18%	64.67%/65.25%	66.86%/66.79%
	Res50- gcForest	76.40%/76.59%	78.62%/78.59%	82.35%/83.59%	84.54%/85.69%	85.46%/85.96%	89.14%/89.19%
	SigDSD- gcForest	91.01%/91.08%	93.06%/93.28%	95.15%/95.28%	95.70%/95.76%	96.02%/96.34%	96.55%/96.64%
	ReluDSD-gcforest	91.57%/91.68%	93.62%/94.68%	95.65%/95.78%	96.04%/96.29%	96.66%/96.64%	97.05%/97.14%
	PReluDSD-gcForest	92.02%/91.89%	94.22%/94.26%	96.03%/96.38%	96.48%/96.58%	97.21%/97.36%	97.43%/97.58%
	ADSD-gcForest	92.52%/93.76%	94.76%/94.88%	96.56%/96.49%	97.03%/96.69%	97.85%/97.68%	98.42%/98.64%
Dataset 6	VGG16-gcForest	57.25%/57.64%	59.60%/60.12%	61.33%/61.24%	64.58%/64.88%	66.02%/65.89%	68.20%/67.96%
	Res50- gcForest	75.05%/76.18%	77.33%/78.38%	79.62%/80.19%	82.32%/83.49%	85.58%/84.99%	88.75%/89.06%
	SigDSD- gcForest	92.08%/93.16%	94.01%/93.86%	94.52%/95.67%	95.27%/95.86%	96.06%/96.18%	96.47%/96.34%
	ReluDSD-gcforest	92.52%/92.36%	94.63%/94.59%	95.02%/94.98%	95.72%/95.64%	96.58%/96.59%	96.95%/96.85%
	PReluDSD-gcForest	93.05%/92.96%	95.06%/94.89%	95.62%/95.57%	96.34%/96.78%	97.10%/96.68%	97.36%/97.29%
	ADSD-gcForest	93.60%/93.84%	95.52%/95.67%	96.06%/95.98%	96.85%/97.93%	97.67%/97.58%	98.29%/98.36%

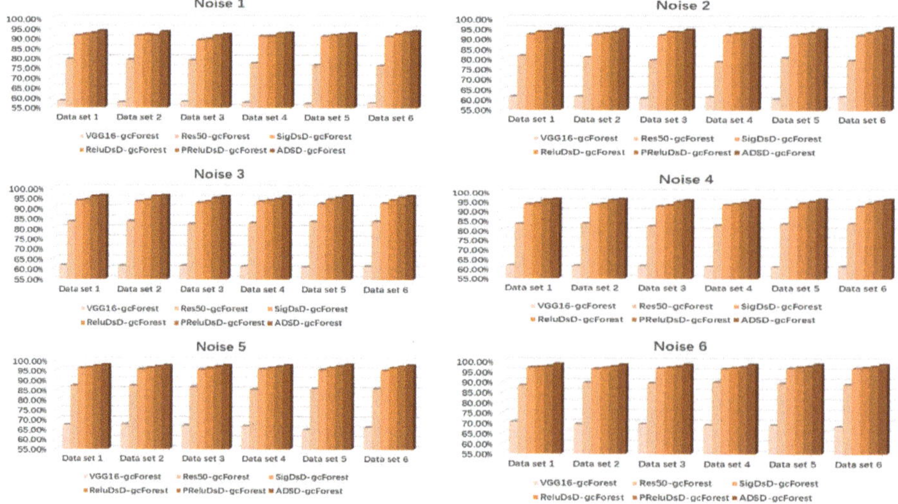

Figure 18. Comparison figures of the AUC of the Ottawa dataset under Noise 1, 2, 3, 4, 5 and 6.

5. Conclusions

This paper proposes an adaptive ADSD-gcForest model. The model uses the VGG network as the basic framework. Multi-scale features of input samples can be extracted through deep separable dilated convolution, and then the CBAM to focus the core features is combined at different scales, the Meta-ACON activation function is integrated into all convolution layers in the network, so that the model can be optimized adaptively according to different input data, and the gcForest as the final classifier can provide the final result. In the experimental part of this paper, datasets of Western Reserve University and University of Ottawa are used, including three bearing data, and it can be seen that faults of different types of bearings under strong noise and multiple load conditions can be effectively diagnosed by the ADSD-gcForest model. This shows that the model proposed in this paper has good robustness. It can also be found that the method proposed in this paper has better improved the migration ability of the model, simplified the design process of the diagnostic model and effectively avoided the problem of repeatedly modifying the model structure.

In modern industrial production, multiple bearings are often required to work together; thus, the effective fault diagnosis of multiple bearings is a hot research topic. The ADSD-gcForest model proposed in this paper can simply optimize the model structure according to different bearing data with the help of the Meta-ACON activation function. It has a certain industrial application value, but the addition of the Meta ACON activation function also increases the number of parameters of the model, which leads to a longer training time. Therefore, how to reduce the training parameters of the Meta-ACON activation function under the premise of ensuring high accuracy will become the focus of future research.

Author Contributions: Conceptualization: S.Z.; Methodology: S.Z. and Z.W.; Formal analysis and investigation: S.Z.; Writing—original draft preparation: S.Z.; Writing—review and editing: D.G. and S.Z. All authors have read and agreed to the published version of the manuscript.

Funding: This research received no external funding.

Institutional Review Board Statement: Not applicable.

Informed Consent Statement: Not applicable.

Data Availability Statement: The used data of bearing fault can be found in CWRU Data Center and University of Ottawa datasets are available online: https://csegroups.case.edu/bearingdatacenter/pages/welcome-case-western-reserve-university-bearing-data-center-website (accessed 18 Octorber 2021), https://data.mendeley.com/datasets/v43hmbwxpm/1 (accessed 20 Octorber 2021).

Conflicts of Interest: The authors declare no conflict of interest.

References

1. Fan, J.; Qi, Y.; Liu, L.; Gao, X.; Li, Y. Application of an information fusion scheme for rolling element bearing fault diagnosis. *Meas. Sci. Technol.* **2021**, *32*, 075013. [CrossRef]
2. Niu, G.; Wang, X.; Golda, M.; Mastro, S.; Zhang, B. An optimized adaptive PReLU-DBN for rolling element bearing fault diagnosis. *Neurocomputing* **2021**, *445*, 26–34. [CrossRef]
3. Jie, D.; Zheng, G.; Zhang, Y.; Ding, X.; Wang, L. Spectral kurtosis based on evolutionary digital filter in the application of rolling element bearing fault diagnosis. *Int. J. Hydromechatronics* **2021**, *4*, 27–42. [CrossRef]
4. Zhao, X.; Qin, Y.; Fu, H.; Jia, L.; Zhang, X. Blind source extraction based on EMD and temporal correlation for rolling element bearing fault diagnosis. *Smart Resilient Transp.* **2021**, *3*, 52–65. [CrossRef]
5. Hou, W.; Ye, M.; Li, W. Rolling bearing fault classification based on improved stack noise reduction self-encoding. *Chin. J. Mech. Eng.* **2018**, *54*, 87–96. [CrossRef]
6. Shao, H.; Jiang, H.; Zhang, X.; Niu, M. Rolling bearing fault diagnosis using an optimization deep belief network. *Meas. Sci. Technol.* **2015**, *26*, 115002. [CrossRef]
7. Liang, T.; Wu, S.; Duan, W.; Zhang, R. Bearing fault diagnosis based on improved ensemble learning and deep belief network. *J. Phys. Conf. Ser.* **2018**, *1074*, 012154. [CrossRef]
8. Ma, M.; Chen, X.; Wang, S.; Liu, Y.; Li, W. Bearing degradation assessment based on weibull distribution and deep belief network. In Proceedings of the IEEE International Symposium on Flexible Automation, Cleveland, OH, USA, 1–3 August 2016; pp. 382–385.

9. Shao, S.; Sun, W.; Wang, P.; Gao, R.X.; Yan, R. Learning features from vibration signals for induction motor fault diagnosis. In Proceedings of the IEEE International Symposium on Flexible Automation, Cleveland, OH, USA, 1–3 August 2016; pp. 71–76.
10. Han, T.; Yuan, J.H.; Tang, J.; An, L.Z. Intelligent composite fault diagnosis method of rolling bearing based on MWT and CNN. *Mech. Transm.* **2016**, *40*, 139–143.
11. Liang, M.; Cao, P.; Tang, J. Tang. Rolling bearing fault diagnosis based on feature fusion with parallel convolutional neural network. *Int. J. Adv. Manuf. Technol.* **2020**, *112*, 819–831. [CrossRef]
12. Pan, H.; He, X.; Tang, S.; Meng, F. An improved bearing fault diagnosis method using one-dimensional CNN and LSTM. *J. Mech. Eng.* **2018**, *64*, 443–452.
13. Zhang, L.; Jing, L.; Xu, W.; Tan, J. Rolling bearing fault diagnosis based on convolutional noise reduction autoencoder and CNN. *Modul. Mach. Tool Autom. Manuf. Technol.* **2019**, *6*, 58–62.
14. Yu, L.; Qu, J.; Gao, F.; Tian, Y. A novel hierarchical algorithm for bearing fault diagnosis based on stacked LSTM. *Shock. Vib.* **2019**, *2019*, 2756284. [CrossRef] [PubMed]
15. Zhao, R.; Yan, R.; Wang, J.; Mao, K. Learing to monitor machine health with convolutional bi-directional lstm networks. *Sensors* **2017**, *17*, 273. [CrossRef] [PubMed]
16. Yan, X.; Xu, Y.; Jia, M. Intelligent Fault Diagnosis of Rolling-Element Bearings Using a Self-Adaptive Hierarchical Multiscale Fuzzy Entropy. *Entropy* **2021**, *23*, 1128. [CrossRef] [PubMed]
17. Yong, Z.; Zhang, X.; Da, N. Research on 3D Object Detection Method Based on Convolutional Attention Mechanism. *J. Phys. Conf. Ser.* **2021**, *1848*, 012097. [CrossRef]
18. Cao, Q.; Yu, L.; Wang, Z.; Zhan, S.; Quan, H.; Yu, Y.; Khan, Z.; Koubaa, A. Wild Animal Information Collection Based on Depthwise Separable Convolution in Software Defined IoT Networks. *Electronics* **2021**, *10*, 2091. [CrossRef]
19. Ma, N.; Zhang, X.; Liu, M.; Sun, J. Activate or Not: Learning Customized Activation. In Proceedings of the IEEE/CVF Conference on Computer Vision and Pattern Recognition, Seattle, WA, USA, 13–19 June 2020.
20. Zhang, N.; Cui, F.; Jiang, B.; He, X. Rotating machinery gearbox fault diagnosis method integrating SDP and CNN. *Ind. Control. Comput.* **2021**, *34*, 89–91.
21. Zhao, L.; Xu, L.; Liu, Y.; Liu, J.; Huang, X. Transformer mechanical fault diagnosis method based on point symmetry transformation and image matching. *Trans. Chin. Soc. Electr. Eng.* **2021**, *36*, 3614–3626.
22. Yin, Q.; Yang, W.; Ran, M.; Wang, S. FD-SSD: An improved SSD object detection algorithm based on feature fusion and dilated convolution. *Signal Processing Image Commun.* **2021**, *98*, 116402. [CrossRef]
23. Yuhui, Z.; Mengyao, C.; Yuefen, C.; Zhaoqian, L.; Yao, L.; Kedi, L. An Automatic Recognition Method of Fruits and Vegetables Based on Depthwise Separable Convolution Neural Network. *J. Phys. Conf. Ser.* **2021**, *1871*, 012075. [CrossRef]
24. Teng, Y.; Gao, P. Generative Robotic Grasping Using Depthwise Separable Convolution. *Comput. Electr. Eng.* **2021**, *94*, 107318. [CrossRef]
25. Liu, T.; Pang, B.; Zhang, L.; Yang, W.; Sun, X. Sea Surface Object Detection Algorithm Based on YOLO v4 Fused with Reverse Depthwise Separable Convolution (RDSC) for USV. *J. Mar. Sci. Eng.* **2021**, *9*, 753. [CrossRef]
26. Chen, Y.; Zhang, X.; Chen, W.; Li, Y.; Wang, J. Research on Recognition of Fly Species Based on Improved RetinaNet and CBAM. *IEEE Access* **2020**, *8*, 102907–102919. [CrossRef]
27. Canayaz, M. C+EffxNet: A novel hybrid approach for COVID-19 diagnosis on CT images based on CBAM and EfficientNet. *Chaos Solitons Fractals* **2021**, *151*, 111310. [CrossRef] [PubMed]
28. Niu, C.; Nan, F.; Wang, X. A super resolution frontal face generation model based on 3DDFA and CBAM. *Displays* **2021**, *69*, 102043. [CrossRef]
29. Sun, Z.; Li, M.; Zhang, J.; Hu, B.; Qi, G.; Zhu, Y. Transient Voltage Stability Assessment Method based on gcForest. *J. Phys. Conf. Ser.* **2021**, *1914*, 012025. [CrossRef]
30. Liu, H.; Zhang, N.; Jin, S.; Xu, D.; Gao, W. Small sample color fundus image quality assessment based on gcforest. *Multimed. Tools Appl.* **2020**, *80*, 17441–17459. [CrossRef]

processes

Article

A Novel Radial Basis Function Neural Network with High Generalization Performance for Nonlinear Process Modelling

Yanxia Yang [1,2,*], **Pu Wang** [1,2] and **Xuejin Gao** [1,2,*]

[1] Faculty of Information Technology, Beijing University of Technology, Beijing 100124, China; wangpu@bjut.edu.cn

[2] Engineering Research Center of Digital Community Ministry of Education, Beijing 100124, China

* Correspondence: yangyx@emails.bjut.edu.cn (Y.Y.); gaoxuejin@bjut.edu.cn (X.G.)

Abstract: A radial basis function neural network (RBFNN), with a strong function approximation ability, was proven to be an effective tool for nonlinear process modeling. However, in many instances, the sample set is limited and the model evaluation error is fixed, which makes it very difficult to construct an optimal network structure to ensure the generalization ability of the established nonlinear process model. To solve this problem, a novel RBFNN with a high generation performance (RBFNN-GP), is proposed in this paper. The proposed RBFNN-GP consists of three contributions. First, a local generalization error bound, introducing the sample mean and variance, is developed to acquire a small error bound to reduce the range of error. Second, the self-organizing structure method, based on a generalization error bound and network sensitivity, is established to obtain a suitable number of neurons to improve the generalization ability. Third, the convergence of this proposed RBFNN-GP is proved theoretically in the case of structure fixation and structure adjustment. Finally, the performance of the proposed RBFNN-GP is compared with some popular algorithms, using two numerical simulations and a practical application. The comparison results verified the effectiveness of RBFNN-GP.

Keywords: radial basis function neural network (RBFNN); generation performance; local generalization error bound; self-organizing structure method; convergence analysis

Citation: Yang, Y.; Wang, P.; Gao, X. A Novel Radial Basis Function Neural Network with High Generalization Performance for Nonlinear Process Modelling. *Processes* **2022**, *10*, 140. https://doi.org/10.3390/pr10010140

Academic Editors: Jie Zhang and Meihong Wang

Received: 16 December 2021
Accepted: 3 January 2022
Published: 10 January 2022

Publisher's Note: MDPI stays neutral with regard to jurisdictional claims in published maps and institutional affiliations.

1. Introduction

In recent years, with the continuous development of artificial intelligence and intelligent algorithms, data-driven methods have been widely used as an effective modeling method because they do not require complex mathematical models and high maintenance costs. Among them, the radial basis function neural network (RBFNN) is widely used due to its simple structure and strong nonlinear function approximation ability, especially in the fields of pattern classification, industrial control, nonlinear system modeling and so on [1–4]. However, there are still some problems to be solved in practice, for example, how to extend the network performance from limited training set to invisible data set, that is, how to design RBFNN with a good generalization ability [5,6]. The generalization performance of RBFNN is usually measured by generalization error, which mainly includes the approximation error caused by the insufficient representation ability of network and estimation errors caused by a limited number of samples. In order to make the RBFNN learnable, the generalization error should be zero as the data tends to infinity.

Due to the limitation of sample data, the network model will produce an estimation error. In order to make the total generalization error close to zero, the number of parameters and samples should tend to infinity to ensure the learnability of the model. Among them, it is worth mentioning that references [5,7–10] deal with the problem of estimation error according to different assumptions. On this basis, the sample complexity of finite networks is studied to demonstrate that, when the data tends to infinity, the estimation error tends to zero. In addition, due to the limited number of samples, even if the optimal parameter setting is obtained, it will produce functions far from the target, resulting in errors and a poor generalization performance [9]. To solve this problem, Barron et al. [10] introduced the concept of an approximation and estimation bound of artificial neural network, pointing out that, for a kind of common artificial neural network, the integral square error between the estimation network model and the objective function is bounded, and discussing the comprehensive influence of approximation error and estimation error as the objective function on the network accuracy. In addition, Yeung et al. [11] developed a new RBFNN local generalization error model to identify the classifier. By predefining the neighborhood of training samples in the local generalization error model, the upper bound of generalization error of invisible samples was derived. Sarraf [12] proposed a tight upper bound of the generalization error under the assumption of twice continuously differentiable, which was composed of the estimation error under the sample space mean and the expected sensitivity of the error to the input change, and showed how the given upper bound could be used to analyze the generalization error of a feedforward neural network. Although the above methods achieved good results through the generalization error bound based on sensitivity, they still faced challenges due to the computational complexity of partial derivatives, and the generalization error should not only be a function of the number of parameters; it is also important to find a better structure. In addition, Wu et.al. proposed a self-adaptive structural optimal algorithm-based fuzzy neural network (SASOA-FNN) in [13]. This network can improve the generalization ability of the network by minimizing the structural risk model with the number of samples. Terada et al. [14] derived the fast generalization error bound of deep learning under the framework developed in [15]. In the derivation process, they only focused on the minimization of empirical risk and eliminated the scale invariance assumption of activation function. The common feature of the above references is that they are based on risk minimization, and accelerate the convergence speed of the network, while ignoring the influence of the properties of different networks on the generalization error. For example, RBFNN is essentially a local learning method. Each hidden neuron captures the local information of a specific region in the input space by the center and width of its Gaussian activation function [16]. However, the training samples far away from the center of hidden neurons have no effect on the learning of hidden neurons. Therefore, for this local learning method, finding the optimal compromise between model accuracy and generalization error is an effective way to improve the generalization ability of the network.

Different from the estimation error caused by the insufficient samples mentioned above, the approximation error of the network is greatly affected by the network structure. Thus, how to obtain a suitable network structure has always been a hot topic. For instance, Zhou et al. [17] proposed a self-organizing fuzzy neural network with hierarchical pruning scheme (SOFNN-HPS). In SOFNN-HPS, the adaptive ability and robustness of the prediction model were improved through the effective combination of a hierarchical pruning strategy and adaptive allocation strategy. Finally, the accurate prediction of ammonia nitrogen, the key variable in the wastewater treatment process, is realized. To predict the outlet ferrous ion concentration on-line, Xie et al. [18] developed a self-adjusting structure radial basis function neural network (SAS-RBFNN). This algorithm uses the supervised clustering algorithm to initialize the RBFNN, and combines or segments the hidden neurons according to the clustering distance to realize the structural self-organization of RBFNN. In addition, Huang et al. [19] proposed a growing and pruning RBF (GAP-RBF) method based on the significance of a neuron. For the GAP-RBF, the number of neurons can be self-designed to realize a compact RBFNN by linking the significance of neurons and a desired accuracy. On this basis, an improved GAP-RBF algorithm (IGAP-RBF) for any arbitrary distribution of input training data was proposed in reference [20]. This algorithm only adjusts the parameters of the nearest neuron to reduce the computational complexity while ensuring the learning performance. A common feature of GAP-RBF and IGAP-RBF is that the self-organizing strategy is based on the contribution of hidden neurons, and the training samples need to be known in advance. In reference [21], an adaptive gradient multi-objective particle swarm optimization algorithm was proposed to predict the biochemical oxygen demand, a key water quality parameter in the wastewater treatment process. This method adopts a multi-objective gradient method and adaptive flight parameter mechanism, which not only greatly reduces computational complexity, but also improves generalization performance; other structure self-organization methods are outlined in [22–25]. The advantage of the above algorithms is that they can adjust the network parameters while adjusting the network structure, and the disadvantage is that different parameter adjustment methods make the learning speed of the network different, which may affect the accuracy of the network.

Most of the existing methods focus on using self-organizing strategies to obtain appropriate structures, or using effective learning algorithms to obtain a higher accuracy. However, good training accuracy is not equal to good generalization performance. Therefore, designing an effective learning model to improve the generalization performance of RBFNN is still an urgent problem that needs to be solved. Based on the above analysis, a self-organizing RBFNN based on network sensitivity is proposed to improve the generalization performance. The main contributions of this method are as follows.

The generalization ability is quantified by network sensitivity. Then, an RBFNN-GP algorithm is constructed to improve the generalization performance:

1. The convergence of the RBFNN-GP is verified in theory, which ensures its successful application;
2. The effectiveness and feasibility of the RBFNN-GP are verified by predicting the key water quality parameters in wastewater treatment process.

The remainder of this paper is organized as follows. Section 2 briefly introduces the basic RBFNN and the local generalization error bound of the network. Then, the details of RBFNN-GP are given in Section 3. The convergence of RBFNN-GP is discussed in Section 4. Section 5 presents the experimental results of RBFNN-GP to demonstrate its advantages. The application field and future work direction of the proposed method are shown in Section 6. Finally, the conclusions are given in Section 7.

2. Materials and Methods

2.1. Radial Basis Function Neural Network (RBFNN)

In general, RBFNN consists of three layers: the input layer, the hidden layer and the output layer. A typical multiple-input, single-output RBFNN (MISO-RBFNN) is shown in Figure 1. The MISO-RBFNN is a k-m-1 network, and each neuron in the RBFNN hidden layer is constructed in the form of Gaussian function. The mathematical description of RBFNN output is as follows:

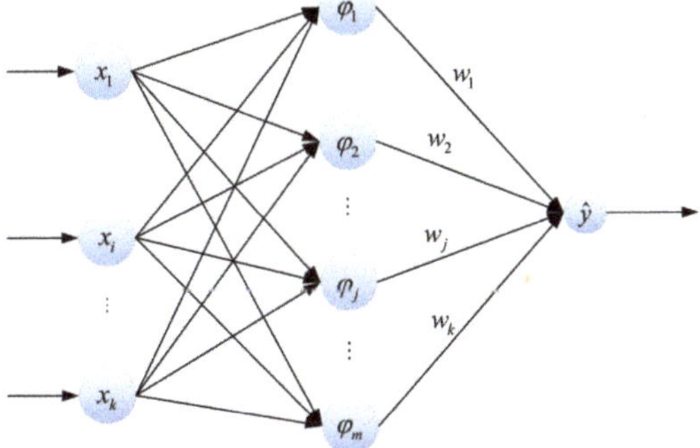

Figure 1. The structure of MISO-RBFNN.

$$\hat{y}(t) = \sum_{j=1}^{m} w_j(t)\theta_j(t), \tag{1}$$

where m is the number of hidden neurons, $w_j(t)$ is the weight between the jth hidden layer and the output layer at time t, and $\theta_j(t)$ is the output of the jth hidden layer neuron, described as:

$$\theta_j(t) = e^{-\frac{\|\mathbf{x}(t)-\mathbf{c}_j(t)\|^2}{2\sigma_j^2(t)}}, \tag{2}$$

$$\mathbf{c}_j(t) = \left[c_{1,j}(t), c_{2,j}(t), \cdots, c_{n,j}(t)\right]^T, \tag{3}$$

where $\mathbf{x}(t) = [x_1(t), x_2(t), \ldots, x_n(t)]^T$ is the input vector, $\mathbf{c}_j(t)$ is the center vector of the jth hidden neuron at time t, and n is the dimension of the input vector; $\|\mathbf{x}(t)-\mathbf{c}_j(t)\|$ is the Euclidean distance between $\mathbf{x}(t)$ and $\mathbf{c}_j(t)$, and $\sigma_j(t)$ is the width of the jth hidden neuron at time t.

2.2. Local Generalization Error Bound

The generalization error of the whole input space is defined as [11]:

$$E_{gen}(t) = \int_D [f(\mathbf{x}(t)) - F(\mathbf{x}(t))]^2 p(\mathbf{x}(t))d\mathbf{x}(t), \tag{4}$$

where $\mathbf{x}(t)$ is the input vector in the input space and $p(\mathbf{x}(t))$ is the unknown probability density function of $\mathbf{x}(t)$. Given a training data set $D = (\mathbf{x}_a(t), F_a(t)), a = 1, 2, \cdots, N, N$ is the number of pairs of input and output, namely the number of samples of training set. The empirical error of network can be defined as:

$$E_{emp}(t) = \frac{1}{N} \sum_{a=1}^{N} \left[\hat{f}(\mathbf{x}_a(t)) - F(\mathbf{x}_a(t))\right]^2, \tag{5}$$

where $\hat{f}(\mathbf{x}_a(t))$ and $F(\mathbf{x}_a(t))$ represent the approximate and real mapping functions between the ath input and output in the input space, respectively. The ultimate goal of improving the generalization ability is minimize the approximation error, and the network can directly predict the unseen data.

Since RBFNN is a local method, for each sample $\mathbf{x}_a(t) \in D$, we can find a sample set:

$$\mathbf{X}_{S,\mathbf{x}_a}(t) = \{\mathbf{x}(t) | \mathbf{x}(t) = \mathbf{x}_a(t) + \Delta\mathbf{x}(t)\}, \tag{6}$$

where $\mathbf{X}_{S,\mathbf{x}_a}(t)$ defines an S-neighborhood of the training sample $\mathbf{x}_a(t)$, $\Delta\mathbf{x}(t) = [\Delta x_1(t), \cdots, \Delta x_n(t)]$ is regarded as perturbations, n denotes the number of input features, and S is a given number. The samples in $\mathbf{X}_{S,\mathbf{x}_a}(t)$ (except $\mathbf{x}_a(t)$) are regarded as unseen samples. For $0 \leq S_1 \leq \cdots \leq S_k \leq \infty$, the relationship holds $D \subseteq \mathbf{X}_{S_1}(t) \subseteq \cdots \subseteq \mathbf{X}_{S_k}(t) \subseteq I$, where I is the entire input space. By Hoeffding's inequality, we can derive the definition of local generalization error bound as follows [11,26]:

$$E_{gen,S}(t) = \sqrt{E_{emp}(t)} + \sqrt{E_{\mathbf{X}_S,\Delta y^2}(t)}, \tag{7}$$

with:

$$\Delta y(t) = \hat{f}(\mathbf{x}_a(t)) - F(\mathbf{x}_a(t))$$
$$E_{\mathbf{X}_S,\Delta y^2}(t) = \frac{1}{N} \sum_{a=1}^{N} \int_{\mathbf{X}_S(\mathbf{x}_a(t))} (\Delta y^2(t)) \frac{1}{(2S)^a} dx(t), \tag{8}$$

where $\Delta y(t)$ is the difference between the network output and the real value; the term $E_{\mathbf{X}_S,\Delta y^2}(t)$ is introduced in Section 3.1.

Remark 1. *Different from previous methods, to the best of our knowledge, this work is a new attempt to obtain looser generalization error bounds by eliminating high-order terms and reducing partial accuracy in exchange for better generalization ability.*

3. RBFNN with High Generation Performance (RBFNN-GP)

The proposed RBFNN-GP, which can improve the network generalization ability, is introduced in this section. It contains the following two parts: (1) sensitivity measurement (SM) method and (2) structural self-organization optimization (SSO) strategy.

3.1. Sensitivity Measurement (SM) Method

Sensitivity analysis provides an effective method to evaluate the impact of different inputs on output results, and it can accurately express the causal response between input changes and corresponding outputs [27,28]. Different from existing studies that focus on improving modeling accuracy or looking for indicator variables, in this study, SM is introduced to quantify the impact of network input changes on output changes, and to intuitively represent the sensitivity of network output to input changes. Thus, the network structure can be adjusted accordingly. Suppose that the inputs are independent and not identically distributed; then, from this, each input feature has its own expectation μ_{x_i} and variance, $\delta^2_{x_i}$:

$$\phi_j(t) = w_j^2(t)e^{\delta_{d_j}(t)/2\sigma_j^4(t) - \mu_{d_j}(t)/\sigma_j^2(t)}, \tag{9}$$

with:

$$\begin{cases} d_j(t) = \|\mathbf{x}(t) - c_j(t)\|^2 \\ \delta_{d_j}(t) = \sum_{i=1}^n \mu\left[(x_i(t) - \mu_{x_i}(t))^4\right] - \left(\delta^2_{x_i}(t)\right)^2 + 4\delta^2_{x_i}(t)(\mu_{x_i}(t) - c_{ji}(t))^2 + \\ 4\mu\left[(x_i(t) - \mu_{x_i}(t))^3\right](\mu_{x_i}(t) - c_{ji}(t)) \end{cases} \tag{10}$$

and:

$$\begin{cases} \mu_{d_j}(t) = \sum_{i=1}^n \left[\delta^2_{x_i}(t) + (\mu_{x_i}(t) - c_{ji}(t))^2\right] \\ \sigma_j(t) = \phi_j(t)\sum_{i=1}^n \left(\delta^2_{x_i}(t) + (\mu_{x_i}(t) - c_{ji}(t))^2/\sigma_j^4(t)\right) \end{cases}. \tag{11}$$

In theory, as long as the input variation is finite, we do not strictly limit its data distribution. In this instance, we assume that the unseen samples $\mathbf{S}$-neighborhood of the training samples obey the uniform distribution, and thus we obtain $\delta^2_{\Delta x_i}(t) = S^2/3$. By the law of large numbers, the sensitivity of RBFNN is:

$$\begin{aligned} Ex_{S,\Delta y^2}(t) &= \frac{1}{N}\left\{\sum_{a=1}^N \int_{\mathbf{X}_{S,x_a}} [f(\mathbf{x}_a(t) + \Delta\mathbf{x}(t)) - f(\mathbf{x}_a(t))]^2 p(\Delta\mathbf{x}(t))d\Delta\mathbf{x}(t)\right\} \\ &\approx \sum_{j=1}^m \phi_j(t)\left\{\sum_{i=1}^n \left[\delta^2_{\Delta x_i}(t)\left(\delta^2_{x_i}(t) + (\mu_{x_i}(t) - c_{ji}(t))^2 + 0.2\delta^2_{\Delta x_i}(t)\right)\right]/\sigma_j^4(t)\right\} \end{aligned} \tag{12}$$

with:

$$\phi_j(t) = \sigma_j^4(t)\xi_j(t). \tag{13}$$

therefore, we can obtain:

$$Ex_{S,\Delta y^2}(t) \approx \frac{1}{45}S^4 n\sum_{j=1}^m \xi_j(t) + \frac{1}{3}S^2\sum_{j=1}^m \sigma_j(t). \tag{14}$$

Remark 2. *Limiting cases of* $E_{gen,S}(t)$*. Clearly, when* $S \to \infty$*,* $\mathbf{X}_S(t) \to I$*, that is* $S \to 0$*,* $\mathbf{X}_S(t) \to D$*. If* $0 \le S_1 \le \cdots \le S_k \le S_\infty$*, the relationship* $D \subseteq \mathbf{X}_{S_1}(t) \subseteq \cdots \subseteq \mathbf{X}_{S_k}(t) \subseteq I$ *holds. Therefore, in the case of* $S \to \infty$*,* $E_{gen}(t) < E_{gen,S}(t)$*.*

Remark 3. *Statistical performance. Compared with the regression error bound, which only uses the effective parameters and the number of training samples, the proposed RBFNN-GP algorithm has clear advantages, because it considers statistical characteristics, such as the mean and variance of the training data set.*

3.2. Structural Self-Organization Optimization (SSO) Strategy

In order to construct a RBFNN with a high generalization performance, an SSO strategy is designed based on the sensitivity measurement. This SSO strategy can adjust the structure and parameters (including center, width and weight) of RBFNN at the same time. The self-organization strategies are shown as follows:

$$
\begin{cases}
\mathbf{c}(t+1) = \mathbf{c}(t) - \eta\Delta\mathbf{c}(t) \\
\boldsymbol{\sigma}(t+1) = \boldsymbol{\sigma}(t) - \eta\Delta\boldsymbol{\sigma}(t) \\
\mathbf{w}(t+1) = \mathbf{w}(t) - \eta\Delta\mathbf{w}(t) \\
m(t+1) = \begin{cases} m(t) + 1, \sigma_j < \lambda_1 \\ m(t), \lambda_1 < \sigma_j < \lambda_2 \\ m(t) - 1, \sigma_j > \lambda_2 \end{cases}
\end{cases}
\tag{15}
$$

where $\sigma_j \geq 1$, $\mathbf{c}(t+1) = [\mathbf{c}_1(t+1), \mathbf{c}_2(t+1), \ldots, \mathbf{c}_m(t+1)]$ and $\boldsymbol{\sigma}(t+1) = [\sigma_1(t+1), \sigma_2(t+1), \ldots, \sigma_m(t+1)]$ are the center and width vectors of the hidden neuron at time $t+1$, $\mathbf{w}(t+1)$ is the weight of output layer at time $t+1$, $m(t)$ represents the number of hidden layer neurons at time t, and $m(t) \geq 1$, η is the learning rate, $\lambda_1 \leq -2/3S^2 n \zeta_m$, ζ_m is the ratio of statistical output and width of the mth neuron (λ_1 is a dynamic threshold and the statistical residual is negative), and $\lambda_2 \geq 0$ are used to ensure the convergence performance of the network (here the value is twice that of the input dimension). It should be noted that, only when λ_1 and λ_2 acquire the equals sign at the same time, will the number of neurons remain unchanged, that is, the structure of the neural network holds. The variables of $\Delta\mathbf{c}(t)$, $\Delta\boldsymbol{\sigma}(t)$, $\Delta\mathbf{w}(t)$ present the changes of the centers, widths and weights at time t, respectively. We obtain:

$$
\begin{cases}
\Delta\mathbf{c}(t) = [\Delta\mathbf{c}_1(t), \Delta\mathbf{c}_2(t), \cdots, \Delta\mathbf{c}_m(t)] \\
\Delta\boldsymbol{\sigma}(t) = [\Delta\sigma_1(t), \Delta\sigma_2(t), \cdots, \Delta\sigma_m(t)] \\
\Delta\mathbf{w}(t) = [\Delta w_1(t), \Delta w_2(t), \cdots, \Delta w_m(t)]
\end{cases}
\tag{16}
$$

where $\Delta\mathbf{c}_m(t) = [\Delta c_{m,1}(t), \cdots, \Delta c_{mn}(t)]$ is the change of the center of the mth neuron at time t, and $\Delta\sigma_m(t)$ and $\Delta w_m(t)$ are the changes of width and weight of the mth neuron at time t, respectively. Moreover, the updates details of the parameters are:

$$
\begin{cases}
\Delta c_j(t) = \partial E_{emp}(t)/\partial c_j(t) = \left(\mathbf{x}(t) - c_j(t)/\sigma_j^2\right)w_j\theta_j(1-\theta_j)e(t) \\
\Delta \sigma_j(t) = \partial E_{emp}(t)/\partial \sigma_j(t) = \left[(\mathbf{x}(t) - c_j(t))^2/\sigma_j^3\right]w_j\theta_j(1-\theta_j)e(t) \cdot \\
\Delta w_j(t) = \partial E_{emp}(t)/\partial w_j(t) = \theta_j \mathbf{e}(t)
\end{cases}
\tag{17}
$$

Based on the above analysis, the detailed steps of neuron growth and pruning are given below.

3.2.1. Growth Stage

If $\sigma_j < \lambda_1$, new neurons are added to the neural network to reduce the approximation error and improve the generalization performance. At this time, the number of neurons becomes, and the parameter of new neurons is:

$$
\begin{cases}
c_{new} = \mathbf{x}(t) \\
\sigma_{new} = \frac{1}{m}\sum_{j=1}^{m} \sigma_i(t) \\
w_{new} = (y(t) - \hat{y}(t))e^{\sum_{i=1}^{n} \frac{(x_i(t)-c_{i,new}(t))^2}{2\sigma_{i,new}^2(t)}}
\end{cases}
\tag{18}
$$

where $\mathbf{c}_{new}$, σ_{new} and w_{new} represent the center, width and weight of the new neuron, $x_i(t)$ represents the ith element in the input vector at time t, $c_{i,new}(t)$ and $\sigma_{i,new}(t)$ are the ith elements of the center and width of the new neuron at time t, respectively. After structural adjustment, the parameters are updated as:

$$\begin{cases} \mathbf{c}(t+1) = [\mathbf{c}(t); \mathbf{c}_{new}] \\ \sigma(t+1) = [\sigma(t); \sigma_{new}] \\ \mathbf{w}(t+1) = [\mathbf{w}(t); w_{new}] \end{cases} , \tag{19}$$

where $\mathbf{c}(t+1)$, $\sigma(t+1)$ and $\mathbf{w}(t+1)$ are the center, width and weight of RBFNN at time $t+1$, respectively.

3.2.2. Prune Stage

If $\sigma_j > \lambda_2$, the jth neuron with the least information in the hidden layer is deleted. In this way, the network complexity is reduced under the premise of ensuring generalization ability. At this time, the number of neurons becomes $m(t-1)$, and the parameters of the new neurons are:

$$\begin{cases} c_j(t+1) = 0 \\ \sigma_j(t+1) = 0 \\ w_j(t+1) = 0 \end{cases} , \tag{20}$$

where $c_j(t+1)$, $\sigma_j(t+1)$ and $w_j(t+1)$ represent the center, width and weight of the jth neuron at time $t+1$, respectively. After structural adjustment, the parameters of the ith neuron were as follows:

$$\begin{cases} c_i(t+1) = c_i(t) \\ \sigma_i(t+1) = \sigma_i(t) \\ w_i(t+1) = w_i(t) + \frac{w_j \theta_j(t)}{\theta_i(t)} \end{cases} . \tag{21}$$

Among them, the ith neuron is the neuron closest to the Euclidean distance from the jth neuron, $c_i(t+1)$, $\sigma_i(t+1)$ and $w_i(t+1)$ are the center, width and weight of the ith neuron at time $t+1$.

Remark 4. *Because there is only one hidden layer in the model, the network structure is simple, which greatly reduces the error accumulation in the process of back propagation. Furthermore, no extra parameters are added during network training, which reduces the amount of computation. Therefore, the stability of the network is well guaranteed.*

4. Convergence Analysis

Another important problem of neural network structure is convergence analysis, which not only affects the application in practical engineering, but also reflects the generalization ability of neural network. If the neural network cannot guarantee convergence or meet convergence conditions, it is difficult to realize the successful application of the neural network. In addition, for RBFNN-GP, its convergence is not only related to the parameter optimization algorithm, but also related to structural changes. Therefore, this paper analyzes the convergence from three aspects: convergence in the stable stage, growth stage and deletion stage.

Hypothesis 1 (H1). *The center c of the hidden layer, the width σ of the hidden layer and the input–output weight w satisfy the boundedness, that is $\|\mathbf{c}\| \leq \mu_c$, $\|\sigma\| \leq \mu_\sigma$, $\|\mathbf{w}\| \leq \mu_w$, where μ_c, μ_σ, μ_w are positive real numbers.*

Hypothesis 2 (H2). *There is a set of "optimal" network parameters, $\mathbf{c}^*, \sigma^*$ and $\mathbf{w}^*$, that is, the optimal center, width and weight.*

4.1. Convergence Analysis of RBFNN with Fixed Structure

For the convenience of discussing its convergence, the differential equation of e is expressed as follows [29].

$$\dot{e}(t) = -e(t) + \mathbf{w}^{*T}(t)\boldsymbol{\theta}^*(t) - \mathbf{w}^T(t)\boldsymbol{\theta}(t), \tag{22}$$

where $\Delta\boldsymbol{\theta} = \boldsymbol{\theta}^* - \boldsymbol{\theta}, \Delta\mathbf{w} = \mathbf{w}^* - \mathbf{w}$, $\Delta\boldsymbol{\theta}$ is the change in the hidden layer neuron output, $\Delta\mathbf{w}$ is the change of weight. Equation (22) is reformulated as:

$$\dot{e}(t) = -e(t) + \mathbf{w}^{*T}(t)\Delta\boldsymbol{\theta}(t) + \Delta\mathbf{w}^T(t)\boldsymbol{\theta}(t). \tag{23}$$

In order to transform the nonlinear output of the network into a partially linear form, the Taylor expansion of $\Delta\boldsymbol{\theta}$ is:

$$\Delta\boldsymbol{\theta} = \varphi_c^T(\mathbf{c}^* - \mathbf{c}) + \varphi_\sigma^T(\boldsymbol{\sigma}^* - \boldsymbol{\sigma}) + \Omega, \tag{24}$$

where Ω is the higher order infinitesimal of Taylor expansion.

Theorem 1. *Suppose the number of hidden-layer neurons of RBFNN is fixed, and the network parameters are updated according to Equtions (14)–(16); when $t \to \infty, e(t) \to 0$, the convergence of the network is guaranteed.*

Proof of Theorem 1. The Lyapunov function is defined as:

$$V(e, \mathbf{c}, \boldsymbol{\sigma}, \mathbf{w}) = \frac{1}{2}\left(e^2 + \Delta\mathbf{c}^T\Delta\mathbf{c} + \Delta\boldsymbol{\sigma}^T\Delta\boldsymbol{\sigma} + \Delta\mathbf{w}^T\Delta\mathbf{w} + \Delta\mathbf{v}^T\Delta\mathbf{v}\right) \tag{25}$$

with:

$$\begin{cases} \Delta\mathbf{c} = \mathbf{c}^* - \mathbf{c} \\ \Delta\boldsymbol{\sigma} = \boldsymbol{\sigma}^* - \boldsymbol{\sigma} \\ \Delta\mathbf{w} = \mathbf{w}^* - \mathbf{w} \\ \dot{\mathbf{v}} = e \\ \Delta\mathbf{v} = \mathbf{v}^* - \mathbf{v} = \mathbf{w}^{*T}\left(\varphi_c^T\mathbf{c}^* + \varphi_\sigma^T\boldsymbol{\sigma}^* + \Omega\right) - \mathbf{w}^T\left(\varphi_c^T\mathbf{c}^* + \varphi_\sigma^T\boldsymbol{\sigma}^*\right) \end{cases} \tag{26}$$

where $\mathbf{v}$ is the compensator and $\mathbf{v}^*$ is the optimal compensator. The partial derivative of the Lyapunov function is:

$$V'(e, \mathbf{c}, \boldsymbol{\sigma}, \mathbf{w}) = ee' + \Delta\mathbf{c}'^T\Delta\mathbf{c} + \Delta\boldsymbol{\sigma}'^T\Delta\boldsymbol{\sigma} + \Delta\mathbf{w}'^T\Delta\mathbf{w} + \Delta\mathbf{v}'^T\Delta\mathbf{v}. \tag{27}$$

According to Equations (22)–(26), we can obtain:

$$\begin{aligned} V'&(e, \mathbf{c}, \boldsymbol{\sigma}, \mathbf{w}) \\ &= ee' + (\mathbf{c}^* - \mathbf{c})'^T\Delta\mathbf{c} + (\boldsymbol{\sigma}^* - \boldsymbol{\sigma})'^T\Delta\boldsymbol{\sigma} + (\mathbf{w}^* - \mathbf{w})'^T\Delta\mathbf{w} + (\mathbf{v}^* - \mathbf{v})'^T\Delta\mathbf{v} \\ &= -e^2 + e\left\{\begin{array}{c} \mathbf{w}^{*T}\left[\varphi_c^T(\mathbf{c}^* - \mathbf{c}) + \varphi_\sigma^T(\boldsymbol{\sigma}^* - \boldsymbol{\sigma}) + \Omega\right] + \Delta\mathbf{w}^T\boldsymbol{\theta} - \\ \mathbf{w}^T\left[\varphi_c^T(\mathbf{c}^* - \mathbf{c}) + \varphi_\sigma^T(\boldsymbol{\sigma}^* - \boldsymbol{\sigma})\right] - \varphi_\mathbf{w}^T\Delta\mathbf{w} \end{array}\right\} - e(\mathbf{v}^* - \mathbf{v}) \\ &= -e^2 + e\left[\mathbf{v}^* - \mathbf{v} + \Delta\mathbf{w}^T\left(-\varphi_c^T\mathbf{c} - \varphi_\sigma^T\boldsymbol{\sigma} + \boldsymbol{\theta}\right) - \varphi_\mathbf{w}^T\Delta\mathbf{w}\right] - e(\mathbf{v}^* - \mathbf{v}) \\ &= -e^2 \end{aligned} \tag{28}$$

and:

$$V'(e, \mathbf{c}, \boldsymbol{\sigma}, \mathbf{w}) \le 0. \tag{29}$$

Thus, V' is the seminegative definite in the above space. In light of the Lyapunov theorem, we can obtain:

$$\lim_{t \to \infty} e(t) = 0. \tag{30}$$

So far, the convergence of a fixed-structure RBFNN is proved. $\square$

4.2. Convergence Analysis of RBFNN with Changeable Structure

Theorem 2. *If the network structure is self-organizing in the learning process, then the parameters are adjusted according to Equations (14)–(21). When $t \to \infty, e(t) \to 0$, the convergence of RBFNN based on sensitivity can be guaranteed.*

Proof Theorem 2. The structure self-organization process of RBFNN is divided into two parts: structure growth and structure pruning stages. □

4.2.1. Growth Stage

At time t, there are m neurons in the hidden layer of self-organizing RBFNN based on sensitivity, and the network error is $e_m(t)$. When the growth condition is satisfied, the number of hidden layer neurons is increased by 1. Then, the number of hidden layer neurons is $m + 1$, and the output error of the RBFNN is:

$$e_{m+1}(t) = \frac{1}{2}(\hat{y}(t) - \hat{y}_{m+1}(t))^2, \tag{31}$$

where $e_{m+1}(t)$ is the network error of $m + 1$ hidden neurons, $\hat{y}_m(t)$ and $\hat{y}_{m+1}(t)$ represent the network output before and after the addition of hidden layer neurons, respectively. According to Equations (17)–(19), we can obtain:

$$e_{m+1}(t) = \frac{1}{2}\left[\hat{y}(t) - \left(\sum_{j-1}^{m} w_j(t)\theta_j(t) + w_{m+1}(t)\theta_{m+1}(t)\right)\right]^2 = 0. \tag{32}$$

so:

$$e_{m+1}(t) = 0. \tag{33}$$

It can be seen that when a new neuron is added to the hidden layer, the convergence speed of the network is accelerated based on the parameter setting of the newly added neurons.

4.2.2. Prune Stage

When the pruning condition is satisfied, the jth neuron in the hidden layer is deleted, and the error of the network is $e_{m-1}(t)$. Thus, the error of the network can be rewritten as:

$$e_{m-1}(t) = \hat{y}_m(t) - \left(\sum_{l=1}^{m} w_l\theta_l(t) - w_j\theta_j(t)\right), \tag{34}$$

where $\hat{y}_m(t)$ represents the network output when the number of hidden layer neurons is m:

$$\begin{aligned} e_{m-1}(t) &= \hat{y}_m(t) - \left(\sum_{l=1,l\neq i}^{m} w_l\theta_l(t) - w_j\theta_j(t) + \left(w_i + w_j\frac{\theta_j(t)}{\theta_i(t)}\right)\theta_i(t)\right) \\ &= \hat{y}_m(t) - \left(\sum_{l=1,l\neq i}^{m} w_l\theta_l(t) + w_i\theta_i(t)\right) \\ &= 0 \end{aligned} \tag{35}$$

In summary, the error of the neural network remains unchanged before and after the jth neuron is deleted, that is, the process of deleting neurons does not destroy the convergence of the original neural network.

5. Experimental Studies

This section describes the experiments conducted to assess the effectiveness of the generalization performance of the RBFNN-GP algorithm. The experiment includes one benchmark and two practical problems, namely, the approximation of the Mexican straw hat function and the prediction of key water quality parameters, ammonia nitrogen and membrane permeability in the wastewater treatment process. In addition, the good generalization performance of the proposed RBFNN-GP is further illustrated by comparisons with the existing six algorithms.

5.1. Benchmark Example A

In this case, the RBFNN-GP algorithm is applied to approximate the Mexican straw hat function, which is a benchmark problem used in [30,31] to checkout many prevalent algorithms. The Mexican straw hat function is:

$$y = \sin(\sqrt{x_1^2 + x_2^2}) / \sqrt{x_1^2 + x_2^2}. \tag{36}$$

In the training phase, the training samples are $\mathbf{x} = \{x_1, x_2; y\}^N$, where x_1, x_2 are stochastic and generated within the scope of $(-2\pi, 2\pi)$, $N = 2100$ is the number of training samples; the testing samples are $\{x_1, x_2, y\}^M$, where $M = 700$ is the number of testing samples, and the learning rate is set to $\eta = 0.003$.

The experimental results are shown in Figure 2, where Figure 2a shows the number of hidden neurons in the training process. It can be seen that the proposed RBFNN-GP can adjust the network structure by pruning or increasing the number of neurons in the learning process. Figure 2b shows the change trajectory of the root-mean-square error (RMSE) in the learning process. Meanwhile, the prediction error results and output error surface are depicted in Figure 2c,d. It is clear that the proposed RBFNN-GP can approximate the Mexican straw hat function with small predicting errors. In order to further prove the excellent generalization ability of the proposed method, the prediction results of the RBFNN-GP are compared with those of the other dynamic neural networks based on structural adjustment, such as SASOA-FNN [12], SOFNN-HPS [17], SAS-RBFNN [18], AG-MOPSO [21], ASOL-SORBFNN [23] and the RBFNN with a fixed structure (fixed-RBFNN). In order to make the comparison more meaningful, all algorithms in this experiment use the same data set, including training samples and test samples, and ensure that the initial number of neurons is the same. In addition, all of the algorithms run 10 times, and then take the average value to make the results more convincing. The results are shown in Table 1, where Max. is the maximum and Dev. is the deviation.

As can be seen from Table 1, the proposed RBFNN-GP requires fewer hidden nodes and output errors and has a better generalization ability than the self-organizing network based on information minimization and structural risk minimization. This mainly depends on the fact that the method considers not only the number of effective parameters in the network, but also the mean and variance of input data.

Table 1. Comparison results with other self-organizing methods.

Methods	No. of NNs	CPU Time(s)		Testing RMSE			Training RMSE
		Mean	Dev.	Mean	Dev.	Max.	Mean
Fixed-RBFNN	8	100.10	0.096	0.031	0.0037	0.040	0.042
SASOA-FNN [12]	8	108.29	0.031	0.029	0.0043	0.033	0.041
SOFNN-HPS [17]	12	131.02	0.076	0.047	0.0053	0.063	0.057
SAS-RBFNN [18]	9	119.35	0.024	0.039	0.0051	0.059	0.052
AGMOPSO [21]	8	106.29	0.028	0.024	0.0037	0.039	0.041
ASOL-SORBFNN [23]	8	135.46	0.031	0.035	0.0046	0.041	0.040
RBFNN-GP	7	100.36	0.012	0.027	0.0033	0.029	0.039

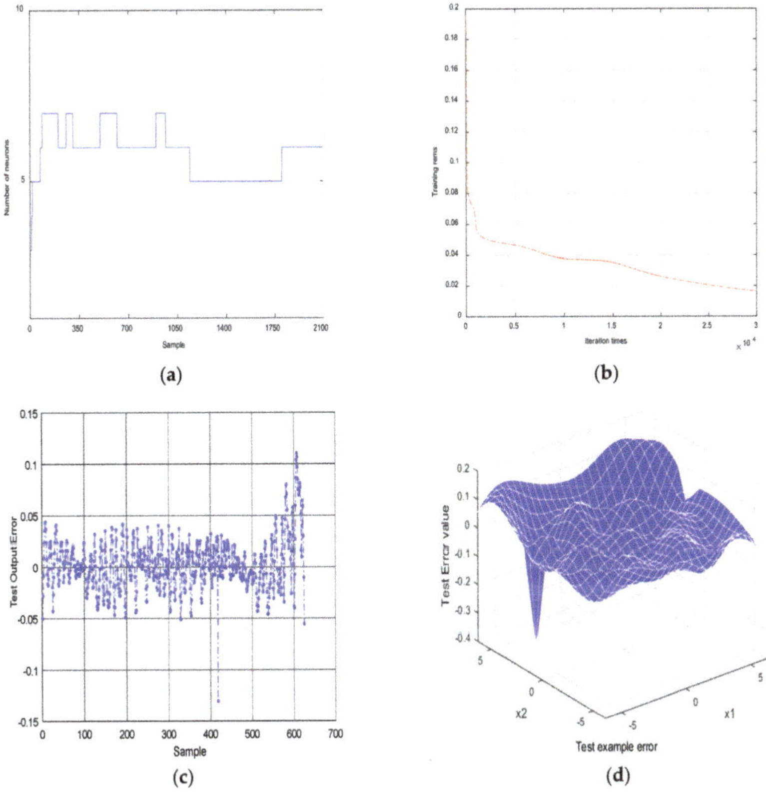

Figure 2. Approximation results of the RBFNN-GP in example A. (**a**) Number of neurons; (**b**) The training RMSE value in example A; (**c**) Test examples output error in example A; (**d**) Test examples output error surface.

5.2. Benchmark Example B

In this example, the effectiveness of the RBFNN-GP is applied for the Mackey–Glass chaotic time series prediction problem, which is a famous benchmark [32,33]. The discrete expression of the time series is given by:

$$x(t+1) = (1 - a_1)x(t) + a_2 x(t - \tau) / \left[1 + x^{10}(t - \tau) \right]$$
$$x(t+1) = f[x(t), x(t-6), x(t-12), x(t-18)] \tag{37}$$

where a_1, a_2, τ are constants, $x(0)$ represents the initial value, and $a_1 = 0.1$, $a_2 = 0.2$, $\tau = 17$, $x(0) = 1.2$. In the training phase, 1000 data samples are extracted from $t = 21$ to 1021. Additionally, 500 samples are used as training samples and 500 samples as test samples. The preset training error is 0.001, the initial learning rate $\eta = 0.02$, and the number of neurons is 5. It should be noted that the other parameters in all comparison methods are the same, and the running results of the experiment are shown in Figure 3.

From Figure 3a, it can be seen that there are several increasing and decreasing stages in the learning process of RBFNN-GP. In the early period of training, the number of neurons changes frequently, and the network structure is unstable. Figure 3b shows the comparison between the predicted output of RBFNN-GP and the real value. Figures 3c,d show the prediction error value of the network and the RSME value in the training stage, respectively. It is worth mentioning that the reason why the RSME curve is so smooth is closely related to the sensitivity of the network. Since the mean and variance of the training samples are fully considered, the stability of the network is improved.

Similarly, Table 2 shows the comparative experimental results with the other four methods: SASOA-FNN [12], SOFNN-HPS [17], SAS-RBFNN [18], AGMOPSO [21], ASOL-SORBFNN [23] and fixed-RBFNN. As shown in Table 2, the optimal number of hidden nodes obtained by using the RBFNN-GP is only six, and the mean and deviation of the test error are minimal in the comparison method. Since the parameters are updated at the same time in the process of structural adjustment to avoid repeated calculations, the computational complexity is greatly reduced compared with the dynamic structural adjustment method based on information strength. Nevertheless, compared with other self-organizing networks based on structural risk, the proposed RBFNN-GP takes a little longer time to calculate the mean and variance information of input samples. However, we still have reason to believe that RBFNN-GP demonstrates a good compromise between generalization ability and training accuracy.

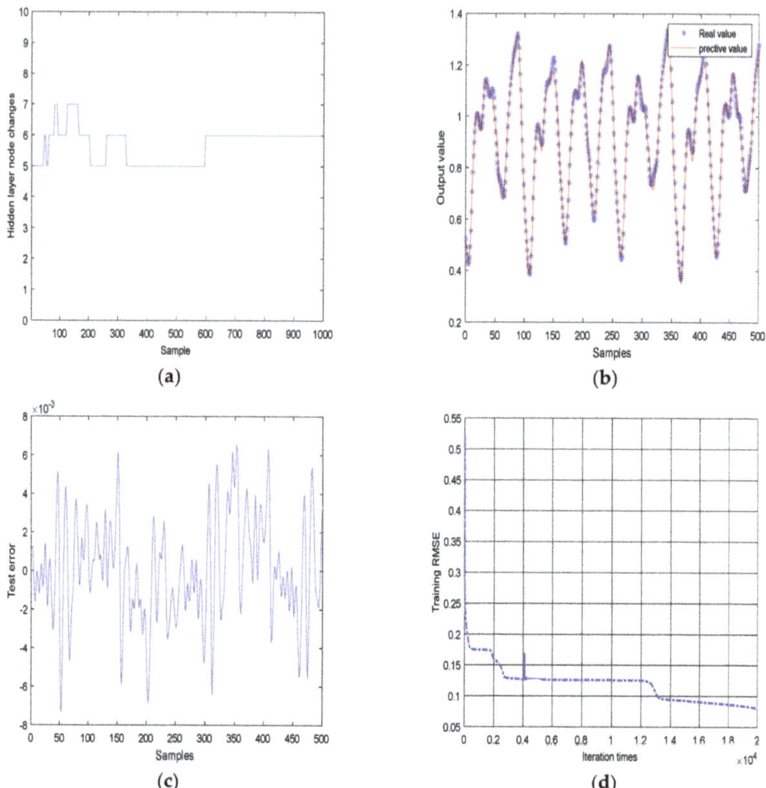

Figure 3. Prediction results of the RBFNN-GP in example B: (**a**) Number of neurons; (**b**) Fitting results of the RBFNN-GP in example B; (**c**) Test output error value; (**d**) The training RMSE value in example B.

Table 2. Comparison results with other self-organizing methods.

Methods	No. of NNs	CPU Time(s)		Testing RMSE			Training RMSE
		Mean	Dev.	Mean	Dev.	Max.	Mean
Fixed-RBFNN	8	104.10	0.069	0.036	0.0039	0.035	0.312
SASOA-FNN [12]	7	126.29	0.030	0.029	0.0041	0.028	0.124
SOFNN-HPS [17]	11	116.3	0.076	0.054	0.0066	0.068	0.261
SAS-RBFNN [18]	10	123.97	0.026	0.034	0.0045	0.031	0.219
AGMOPSO [21]	8	119.89	0.029	0.027	0.0039	0.031	0.217
ASOL-SORBFNN [23]	8	133.91	0.065	0.043	0.0044	0.040	0.251
RBFNN-GP	6	105.36	0.022	0.026	0.0032	0.019	0.215

5.3. Permeability Prediction of Membrane Bio-Reactor

Membrane bio-reactor (MBR) is a new wastewater treatment technology combining membrane separation technology and biotechnology, which is widely used in wastewater treatment process (WWTP). However, in the process of MBR wastewater treatment, membrane pollution will shorten the service life of the membrane and cause an unnecessary loss of process energy consumption. It is of great practical significance to correctly predict the permeability of the membrane and increase its working efficiency [34–36]. Therefore, the proposed RBFNN-GP is applied to predict the permeability of MBR in WWTP. The real data of the experiment come from a wastewater treatment plant in Beijing, China. After disposing of the abnormal data, 500 samples were obtained and normalized.

In this experiment, 50 training samples and 50 test samples are selected to test the performance of RBFNN-GP. In order to remove the correlation between variables, partial least squares method was used to select nine variables from twenty-two variables as the input of RBFNN-GP. Due to the wide range of membrane bioreactor permeability, the number of iterations is $T = 200$, the learning rate is $\eta = 0.003$, and the time length is $h = 10$. The intuitive prediction results are shown in Figure 4. Figure 4a shows the comparison results between the actual and predicted values of membrane bioreactor permeability, and the prediction error is shown in Figure 4b. It can be seen from Figure 4 that the proposed RBFNN-GP has a good prediction performance for the permeability of the MBR, and the prediction error within the range $[-4, 3]$.

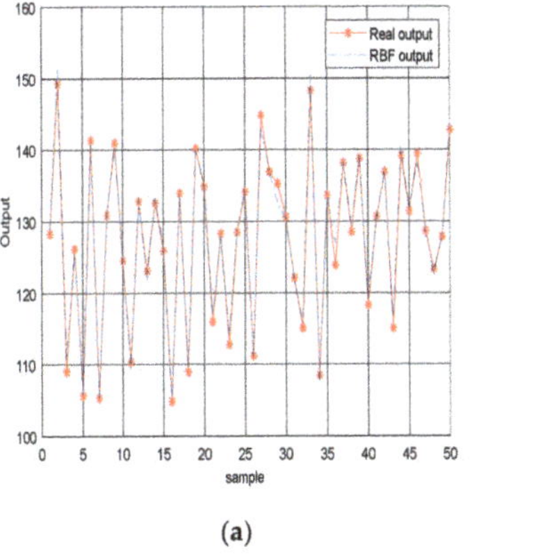

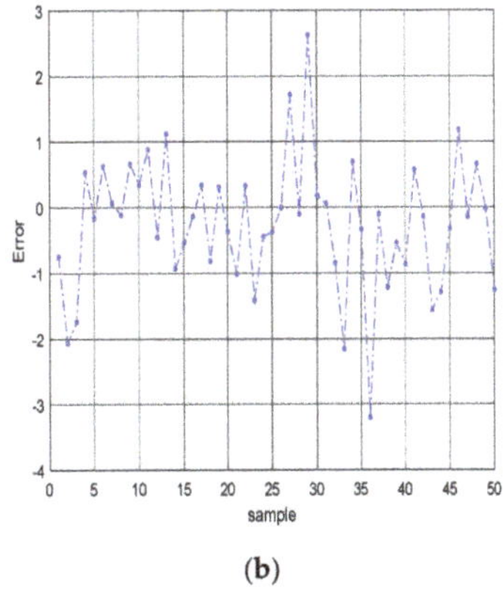

Figure 4. The permeability prediction results for MBR in WWTP: (**a**) Fitting results of RBFNN-GP for MBR; (**b**) Test output error for MBR.

Moreover, Table 3 shows the comparison results of SASOA-FNN [12], SOFNN-HPS [17], SAS-RBFNN [18], AGMOPSO [21], ASOL-SORBFNN [23] and Fixed-RBFNN in predicting the permeability of MBR. It can be seen from Table 3 that, under the same iteration times, the learning time of SASOA-FNN [12] and RBFNN-GP is almost the same. However, the accuracy of the latter is better than that of the former. Thus, we have sufficient reasons to believe that the RBFNN-GP method shows a great improvement in training accuracy, calculation speed and generalization ability compared with the above methods.

Table 3. Comparison results with other self-organizing methods.

Methods	No. of NNs	CPU Time(s)		Testing RMSE			Training RMSE
		Mean	Dev.	Mean	Dev.	Max.	Mean
Fixed-RBFNN	8	105.65	0.037	0.033	0.0034	0.037	0.031
SASOA-FNN [12]	8	109.21	0.035	0.028	0.0033	0.021	0.024
SOFNN-HPS [17]	11	126.34	0.032	0.042	0.0049	0.025	0.032
SAS-RBFNN [18]	9	121.62	0.042	0.025	0.0038	0.029	0.035
AGMOPSO [21]	8	114.25	0.031	0.030	0.0042	0.031	0.114
ASOL-SORBFNN [23]	11	125.31	0.040	0.028	0.0041	0.040	0.127
RBFNN-GP	6	108.24	0.032	0.025	0.0040	0.022	0.022

6. Discussion

6.1. Computational Complexity

Computational complexity is an important indicator for evaluating the model. For the proposed RBFNN-GP, the calculation involved is closely related to the training process of and N. Suppose $[\mathbf{x}(t), y(t); t = 1, \cdots, N]$ is a set of training samples. When the structure of RBFNN-GP is k-m-l (k represents the input variable, m represents the number of hidden layer neurons, and l represents the output variable), the computational complexity of RBFNN-GP is $O[m(t)]$. It can be seen that the computational burden of RBFNN-GP is not heavy.

6.2. Future Trends

In this paper, which aims to realize the online prediction of key water quality parameter membrane pollution in wastewater treatment process, a prediction model based on self-organizing RBFNN is established that meets the needs of accurate prediction of key water quality parameters in wastewater treatment process. At the same time, the self-organizing network modeling method is developed. Due to its good generalization performance and theoretical support, the proposed method can also be extended to other types of networks, such as multi-input, single-output fuzzy neural networks and multi-input, multiple-output RBF neural network.

7. Conclusions

In this paper, an RBFNN-GP algorithm is proposed to improve the model generalization ability. Firstly, the upper bound of the local generalization error is found, and the network structure and parameters are adjusted within the allowable error range. Then, a generalization error formula based on network sensitivity and approximate error is introduced to improve the generalization performance, while ensuring its accuracy. Finally, experimental validation is carried out on two different benchmark data sets and a real application of wastewater treatment process. The results show that the RBFNN-GP algorithm can learn robust networks with good generalization performance and compact scale. In summary, RBFNN-GP has the following advantages:

1. With the help of sensitivity measurements and locally generalized error bounds, the network has a statistical performance and can reasonably achieve structure self-organization without a high dependence on sample numbers.
2. The convergence of RBFNN-GP for fixed and variable structures is guaranteed by the thresholds λ_1 and λ_2. Therefore, the proposed RBFNN-GP can not only reduce the number of additional parameters, but also decreases the computational burden.
3. Compared with existing algorithms, the proposed RBFNN-GP shows a good generalization ability in the prediction of key water quality parameters in wastewater

treatment processes. Furthermore, this approach can be extended to other types of networks and industrial domains.

Author Contributions: Conceptualization, X.G. and Y.Y.; methodology, P.W.; software, P.W.; validation, Y.Y.; formal analysis, Y.Y.; investigation, Y.Y.; resources, X.G.; data curation, P.W.; writing—original draft preparation, Y.Y.; writing—review and editing, Y.Y.; visualization, Y.Y.; supervision, X.G.; project administration, P.W.; funding acquisition, P.W. All authors have read and agreed to the published version of the manuscript.

Funding: This study was funded by the National Natural Science Foundation of China, grant number is 61640312; the Natural Science Foundation of Beijing Municipality, grant numbers are 4172007 and 4192011.

Institutional Review Board Statement: Not applicable.

Conflicts of Interest: The authors declare that they have no conflicts of interest to report regarding the present study.

References

1. Xu, L.; Qian, F.; Li, Y.; Li, Q.; Yang, Y.-W.; Xu, J. Resource allocation based on quantum particle swarm optimization and RBF neural network for overlay cognitive OFDM System. *Neurocomputing* **2016**, *173*, 1250–1256. [CrossRef]
2. Xiong, T.; Bao, Y.; Hu, Z.; Chiong, R. Forecasting interval time series using a fully complex-valued RBF neural network with DPSO and PSO algorithms. *Inf. Sci.* **2015**, *305*, 77–92. [CrossRef]
3. Han, H.G.; Zhang, L.; Hou, Y.; Qiao, J.F. Nonlinear Model Predictive Control Based on a Self-Organizing Recurrent Neural Network. *IEEE Trans. Neural Netw. Learn. Syst.* **2015**, *27*, 402–415. [CrossRef] [PubMed]
4. Yang, Y.-K.; Sun, T.-Y.; Huo, C.-L.; Yu, Y.-H.; Liu, C.-C.; Tsai, C.-H. A novel self-constructing Radial Basis Function Neural-Fuzzy System. *Appl. Soft Comput.* **2013**, *13*, 2390–2404. [CrossRef]
5. Niyogi, P.; Girosi, F. On the Relationship between Generalization Error, Hypothesis Complexity, and Sample Complexity for Radial Basis Functions. *Neural Comput.* **1996**, *8*, 819–842. [CrossRef]
6. Qiao, J.-F.; Han, H.-G. Optimal Structure Design for RBFNN Structure. *Acta Autom. Sin.* **2010**, *36*, 865–872. [CrossRef]
7. Vapnik, V.N. *Estimation of Dependences Based on Empirical Data*; Springer: Berlin, Germany, 1982.
8. Pollard, D. *Convergence of Stochastic Processes*; Springer: Berlin, Germany, 1984.
9. Haussler, D. Decision-theoretic generalizations of the PAC model for neural net and other learning applications. *Inf. Comput.* **1992**, *100*, 78–150. [CrossRef]
10. Barron, A.R. Approximation and estimation bounds for artificial neural networks. *Mach. Learn.* **1994**, *14*, 115–133. [CrossRef]
11. Yeung, D.S.; Ng, W.W.Y.; Wang, D.; Tsang, E.C.C.; Wang, X.-Z. Localized Generalization Error Model and Its Application to Architecture Selection for Radial Basis Function Neural Network. *IEEE Trans. Neural Netw.* **2007**, *18*, 1294–1305. [CrossRef]
12. Sarraf, A. A tight upper bound on the generalization error of feedforward neural networks. *Neural Netw.* **2020**, *127*, 1–6. [CrossRef]
13. Han, H.; Wu, X.; Liu, H.; Qiao, J. An Efficient Optimization Method for Improving Generalization Performance of Fuzzy Neural Networks. *IEEE Trans. Fuzzy Syst.* **2019**, *27*, 1347–1361. [CrossRef]
14. Terada, Y.; Hirose, R. Fast generalization error bound of deep learning without scale invariance of activation functions. *Neural Netw.* **2020**, *129*, 344–358. [CrossRef]
15. Suzuki, T. Fast generalization error bound of deep learning from a kernel perspective. In Proceedings of the International Conference on Artificial Intelligence and Statistics, Playa Blanca, Lanzarote, 9–11 April 2018; pp. 1397–1406.
16. Ng, W.; Yeung, D.; Wang, X.-Z.; Cloete, I. A study of the difference between partial derivative and stochastic neural network sensitivity analysis for applications in supervised pattern classification problems. In Proceedings of the 2004 International Conference on Machine Learning and Cybernetics (IEEE Cat. No. 04EX826), Shanghai, China, 26–29 August 2004.
17. Zhou, H.; Zhang, Y.; Duan, W.; Zhao, H. Nonlinear systems modelling based on self-organizing fuzzy neural network with hierarchical pruning scheme. *Appl. Soft Comput.* **2020**, *95*, 106516. [CrossRef]
18. Xie, Y.; Yu, J.; Xie, S.; Huang, T.; Gui, W. On-line prediction of ferrous ion concentration in goethite process based on self-adjusting structure RBF neural network. *Neural Netw.* **2019**, *116*, 1–10. [CrossRef] [PubMed]
19. Huang, G.-B.; Saratchandran, P.; Sundararajan, N. An Efficient Sequential Learning Algorithm for Growing and Pruning RBF (GAP RBF). *IEEE Trans. Syst. Man Cybern. Part B-Cybern.* **2004**, *34*, 2284–2292. [CrossRef]
20. Huang, G.-B.; Saratchandran, P.; Sundararajan, N. A Generalized Growing and Pruning RBF (GGAP-RBF) Neural Network for Function Approximation. *IEEE Trans. Neural Netw.* **2005**, *16*, 57–67. [CrossRef]
21. Han, H.G.; Wu, X.L.; Zhang, L.; Tian, Y.; Qiao, J.F. Self-Organizing RBF neural network using an adaptive gradient multi-objective particle swarm optimization. *IEEE Trans. Cybern.* **2019**, *49*, 69–82. [CrossRef] [PubMed]
22. Han, H.-G.; Chen, Q.-L.; Qiao, J.-F. An efficient self-organizing RBF neural network for water quality prediction. *Neural Netw. Off. J. Int. Neural Netw. Soc.* **2011**, *24*, 717–725. [CrossRef]

23. Han, H.-G.; Ma, M.-L.; Yang, H.-Y.; Qiao, J.-F. Self-organizing radial basis function neural network using accelerated second-order learning algorithm. *Neurocomputing* **2021**, *469*, 1–12. [CrossRef]
24. Li, Z.-Q.; Zhao, Y.-P.; Cai, Z.-Y.; Xi, P.-P.; Pan, Y.-T.; Huang, G.; Zhang, T.-H. A proposed self-organizing radial basis function network for aero-engine thrust estimation. *Aerosp. Sci. Technol.* **2019**, *87*, 167–177. [CrossRef]
25. Shi, J.R.; Wang, D.; Shang, F.H.; Zhang, H.Y. Research advances on stochastic gradient descent algorithms. *Acta Autom. Sin.* **2021**, *47*, 2103–2119. [CrossRef]
26. Hoeffding, W. Probability Inequalities for Sums of Bounded Random Variables. *J. Am. Stat. Assoc.* **1963**, *58*, 13–30. [CrossRef]
27. Jiang, Q.; Gao, D.-C.; Zhong, L.; Guo, S.; Xiao, A. Quantitative sensitivity and reliability analysis of sensor networks for well kick detection based on dynamic Bayesian networks and Markov chain. *J. Loss Prev. Process. Ind.* **2020**, *66*, 104180. [CrossRef]
28. Belouz, K.; Nourani, A.; Zereg, S.; Bencheikh, A. Prediction of greenhouse tomato yield using artificial neural networks combined with sensitivity analysis. *Sci. Hortic.* **2021**, *293*, 110666. [CrossRef]
29. Han, H.-G.; Qiao, J.-F. Adaptive computation algorithm for RBF neural network. *IEEE Trans. Neural Netw. Learn. Syst.* **2011**, *23*, 342–347. [CrossRef]
30. Xie, T.; Yu, H.; Hewlett, J.; Rózycki, P.; Wilamowski, B. Fast and efficient second-order method for training radial basis function networks. *IEEE Trans. Neural Networks Learn. Syst.* **2012**, *23*, 609–619. [CrossRef]
31. Han, H.-G.; Ma, M.-L.; Qiao, J.-F. Accelerated gradient algorithm for RBF neural network. *Neurocomputing* **2021**, *441*, 237–247. [CrossRef]
32. Zemouri, R.; Racoceanu, D.; Zerhouni, N. Recurrent radial basis function network for time-series prediction. *Eng. Appl. Artif. Intell.* **2003**, *16*, 453–463. [CrossRef]
33. Han, H.-G.; Lu, W.; Hou, Y.; Qiao, J.-F. An Adaptive-PSO-Based Self-Organizing RBF Neural Network. *IEEE Trans. Neural Netw. Learn. Syst.* **2016**, *29*, 104–117. [CrossRef]
34. Teychene, B.; Guigui, C.; Cabassud, C. Engineering of an MBR supernatant fouling layer by fine particles addition: A possible way to control cake compressibility. *Water Res.* **2011**, *45*, 2060–2072. [CrossRef]
35. Huyskens, C.; Brauns, E.; Vanhoof, E.; Dewever, H. A new method for the evaluation of the reversible and irreversible fouling propensity of MBR mixed liquor. *J. Membr. Sci.* **2008**, *323*, 185–192. [CrossRef]
36. Feng, H.-M. Self-generation RBFNs using evolutional PSO learning. *Neurocomputing* **2006**, *70*, 241–251. [CrossRef]

Article

Adaptive PID Control and Its Application Based on a Double-Layer BP Neural Network

Ming-Li Zhang [1,2,3], Yi-Jie Zhang [1,*], Xiao-Long He [1] and Zheng-Jie Gao [1]

[1] School of Economics and Management, Yanshan University, Qinhuangdao 066000, China; zhangml@ysu.edu.cn (M.-L.Z.); hxl@ysu.edu.cn (X.-L.H.); shanav@126.com (Z.-J.G.)
[2] School of Mechanical Engineering, Hebei University of Chinese Medicine, Shijiazhuang 050200, China
[3] Key Laboratory for Health Care with Chinese Medicine of Hebei Province, Shijiazhuang 050200, China
* Correspondence: zhangyijie@ysu.edu.cn

Abstract: In this paper, focusing on the inconvenience of variable value PID based on manual parameter adjustment for the hydraulic drive unit (HDU) of a legged robot, a method employing double-layer back propagation (BP) neural networks for learning the law of PID control parameters is proposed. The first layer is used to learn the relationship between different control parameters and the control performance of the system under various working conditions. The second layer is used to study the relationship between the parameters of the working conditions and the optimizing control parameters under various working conditions. The effectiveness of the proposed control method was verified by simulation and experiment. The results showed that the proposed method can provide a theoretical and experimental basis for the selection of control parameters, and can be extended to similar controllers, therefore possessing engineering application value.

Keywords: legged robot; hydraulic drive unit (HDU); BP neural network; PID control

Citation: Zhang, M.-L.; Zhang, Y.-J.; He, X.-L.; Gao, Z.-J. Adaptive PID Control and Its Application Based on a Double-Layer BP Neural Network. *Processes* **2021**, *9*, 1475. https://doi.org/10.3390/pr9081475

Academic Editor: Jie Zhang

Received: 24 July 2021
Accepted: 17 August 2021
Published: 23 August 2021

1. Introduction

Robots can walk in a variety of ways. At present, the movement forms can be roughly divided into wheeled [1], tracked [2], wheel-foot compound [3], snake-like [4], bionic legged [5], and so on. Compared with other types of robots, bionic-legged robots have the characteristic of discontinuous support because they have a similar leg structure to tetrapods. Especially when combined with hydraulic drive, which has a high power-to-weight ratio, it not only has good adaptability to unknown and unstructured environments but can also pass through the barrier. Therefore, this type of robot is particularly suitable for use in complex environments in the wild.

The leg controller serves as the bottom-level controller of this kind of robot, and each leg of the robot has several degrees of freedom controlled by highly integrated valve cylinders, also known as the hydraulic drive unit (HDU) [6,7]. While the HDU serves as the bottom-level controller of each leg, its control performance directly affects the control strategy and performance of the robot. Commonly, HDU bottom-level control methods can be divided into position control and force control. Based on bottom-level control, control methods of the leg can be extended to compliance control, contact force control, and so on. The above methods are not only applied in electrically driven robots such as Scara [8] and Stewart [9], but they can also be applied to robots such as Bigdog [10], Hydraulic quadrupedal (HyQ) [11], Light Weight Robot (LWR) [12], and Atlas [13].

This paper mainly researched the performance of the HDU in position control. The position control system in the HDU is a kind of high-order nonlinear system. Designing a superior control method requires a very detailed understanding of the characteristics of the controlled system. The establishment of a mathematical model involves analysis of the controlled system, and an accurate mathematical model can truly reflect the dynamic characteristics of the system, fully simulate the actual system in simulation research, and shorten the design cycle of the control method. High-performance intelligent control methods suitable for low-order nonlinear systems can also be used in it. However, in order to ensure the control stability and reliability of the whole machine, such a control method is not often used in engineering practice. The traditional control method is simple to implement and the effect is obvious. Furthermore, the change in the control parameters can truly reflect the system characteristics, which can be used to conduct a preliminary analysis of the system performance. Thus, the HDU position control system is still based on traditional PID control.

A neural network is a computational model that comprehensively simulates the human brain neural network in terms of structure, mechanism, and function [14–16]. By virtue of its complex nonlinear network structure and efficient iterative learning performance, it has obvious advantages compared with other nonlinear optimization methods. Some research works have shown that neural networks can fit arbitrary nonlinear functions. Swic presented an original machine learning-based automated approach for controlling the process of machining of low-rigidity shafts using artificial intelligence methods. Three models of hybrid controllers based on different types of neural networks and genetic algorithms were developed [17]. Rego deals with the problem of finding the control Lyapunov function that keeps the system stable. To find the Lyapunov function, this paper proposes the use of reinforcement learning with two neural networks based on the Lyapunov stability theory [18]. Nobahari focuses on developing a nonlinear controller based on the convolutional neural networks to control different plants. It is assumed that prior knowledge of the plants is very limited and there are only sensory input–output data history of the plants [19]. Wang studied the hysteresis nonlinear characteristics of piezoelectric actuators, a novel hybrid modeling method based on long short-term memory (LSTM) and nonlinear autoregressive with external input (NARX) neural networks is proposed [20].

The neural network is used to learn the relationship between parameters and control performance under different working conditions, and to find out the optimal control parameters under the current working conditions, which can improve the control accuracy of the system under various working conditions and eliminate the work of manual adjustment of parameters. Compared with variable value PID based on manual parameter adjustment, the method based on neural networks can output parameters with continuous variation according to different working conditions, thereby improving the accuracy of control. In addition, the latter method is not restricted by a specific number of conditions in the expert table. Thus, the applicable scope of the improved expert table holds great significance for the application of engineering.

The structure and the contribution of this paper is organized as follows: in Section 2, a mathematical model is established for the HDU position control system. In the model, many factors are carefully considered, such as servo valve nonlinearity, flow-pressure nonlinearity, and load characteristics. In Section 3, aiming at the inconvenience of variable value PID based on manual parameter adjustment in engineering practice, a method of employing double-layer back propagation (BP) neural networks for learning the law of PID control parameters is proposed, and the simulation results are shown, this is the main contribution of our paper. In Section 4, experimental research is carried out on the HDU performance test platform.

2. Introduction to the Sampling System

The HDU is a highly integrated system that includes a servo valve-controlled cylinder, which is the legged robot joint actuator. Figure 1 shows the quadruped robot prototype, the single leg hydraulic drive system, and the HDU.

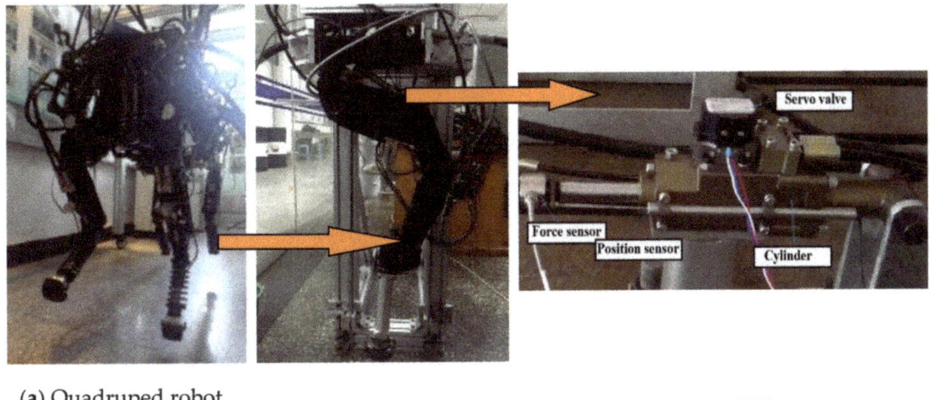

(a) Quadruped robot prototype   **(b)** Single leg   **(c)** HDU

Figure 1. Photos of the quadruped robot prototype, the single leg, and the HDU.

Figure 2 shows a block diagram of the closed-loop position control system for the HDU in Figure 1c. The block diagram takes into account such factors as the flow pressure nonlinearity of the servo valve, the asymmetry of the servo cylinder, the complex variability of the load, and so on. The detailed derivation process is shown in Appendix A.1.

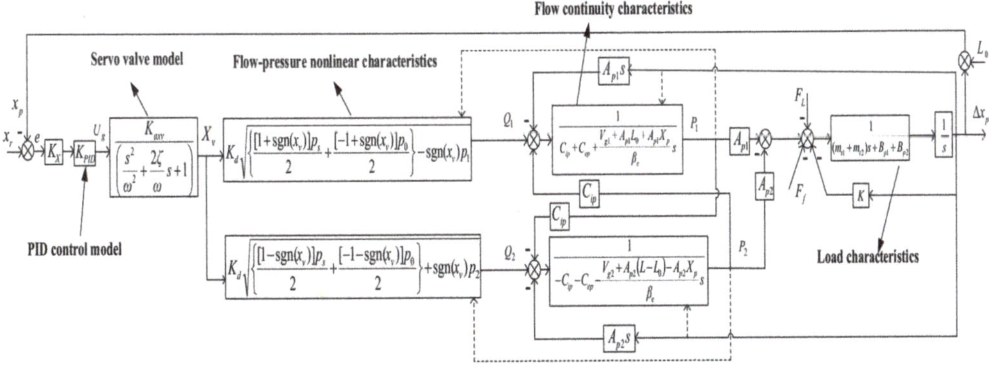

Figure 2. Block diagram of the HDU position closed-loop control system.

The parameters definition and simulation values of the above system are shown in Table 1. The purpose of this paper is to present a new PID controller based on neural networks instead of the PID control parameter in Figure 2.

Table 1. Parameters definition and simulation values of HDU position control system.

Parameter/Input	Value	Unit	Parameter/Input	Value	Unit
Servo valve gain K_{axv}	4.5×10^{-4}	m/V	Density of hydraulic oil ρ	0.867×10^3	kg/m^3
Natural frequency of servo valve ω_{sv}	628	rad/s	Position sensor gain K_X	200	V/m
Damping ratio of servo valve ζ_{sv}	0.82	-	Internal leakage coefficient of servo cylinder C_{ip}	2.38×10^{-13}	m^3/(s·Pa)
Area of the cavity without rod A_1	5.98×10^{-4}	m^2	Conversion mass m_t	1.1315	kg
Area of the cavity with rod A_2	3.97×10^{-4}	m^2	Effective bulk modulus β_e	8×10^8	Pa
Volume of inlet oil cavity pipe V_{g1}	1.1×10^{-6}	m^3	Load stiffness K	0	N/m
Volume of return oil cavity pipe V_{g2}	2.0×10^{-7}	m^3	Load damping B_p	2000	N·s/m
Total stroke of the servo cylinder piston L	0.07	m	Conversion coefficient K_d	1.248×10^{-4}	m^2/s
Initial position of the servo cylinder piston L_0	0.035	m	Input position X_r	-	m
System supply oil pressure p_s	5×10^6	Pa	Output position X_p	-	m
System return oil pressure p_0	0.5×10^6	Pa	Control deviation e	-	m
External leakage coefficient of servo cylinder C_{ep}	0	m^3/(s·Pa)	Load Force F_L	0	N

3. Adaptive PID Parameter Control Method Based on a Double-Layer BP Neural Network

3.1. Learning Strategy Design

Neurons are the basic unit of neural networks and their main function is to simulate the functional characteristics of biological neurons [21–23]. Considering that the input of the neural network in this paper comes from the sensor data of the control system, a Tanh activation function in the Sigmoid activation functions (the latter is generally referred to as a Sigmoid activation function) was selected as the activation function of neurons.

In order to make the system automatically output the optimal control parameters according to the working conditions, it is necessary to design the appropriate neural network structure first. If the neural network is too simple, the fitting accuracy will be reduced; if the neural network is too complex, the convergence will be slow, and even the generalization ability of the neural network will be reduced. Therefore, it is very important to design a neural network with an appropriate structure. Then, designing learning strategies to enable the neural network to learn effectively are needed, including the learning objects of the neural network, the selection of samples, the initial processing of samples, and iterative learning methods. In this section, a parameters learner based on a double-layer BP neural network is designed, which can realize automatic parameter learning. The overall learning strategy is shown in Figure 3, and the details are explained in the following sections.

3.2. Generation of Learning Samples

The sample is a very important part of neural network learning problems and is the source of learning for effective information. The sample data in this paper were driven by position control system simulation or experimental collection in the HDU. The data contained random interference generated by the system itself, and the range of each variable data was also different, so it was necessary to process the data before it was used for learning. The sample data used in this section had to meet the following conditions:

(1) The samples should cover a wider range of working conditions and control parameters as much as possible, and the performance indexes under the corresponding working conditions should be obtained through experiment or simulation, so that the neural network can learn the characteristics of the control system and improve the adaptive ability of the control method.

(2) The sample should be universal. The hydraulic system is a highly nonlinear time-varying system, and the dynamic characteristics of the system change with the differ-

ent external conditions. Collection of data should be carried out after the hydraulic system has been started up and run stably under good heat dissipation conditions.

(3) The data interval of each variable in the sample should be as consistent as possible, which is beneficial for improving the convergence speed and stability of neural networks.

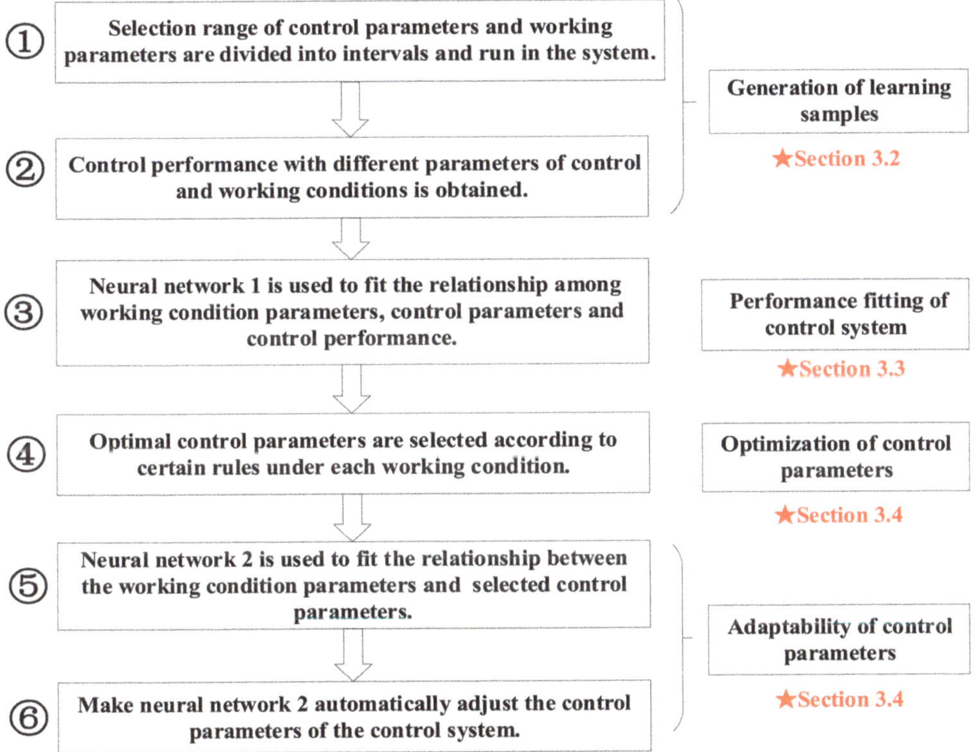

Figure 3. Learning steps of the adaptive parameters in the neural network.

According to the above conditions and principles, a plan of learning data for the PID position control system of the HDU was designed in this section. By generating the input signals and change signals of the control parameters, then importing them into the control model, automatic data acquisition was realized.

In order to prove the effectiveness of the proposed learning strategy, part of the overall working conditions of the HUD were selected for verification to reduce unnecessary work, and then the control parameter range was simplified based on the simulation results of the PID control system shown in Section 2. The working conditions and control parameters finally determined in this section are shown in Table 2.

Table 2. Collection range of simulation learning samples.

Parameters		Range of Value
Working conditions	Sinusoidal frequency	0.5~4 Hz, with a step of 0.5 Hz
	Sinusoidal amplitude	1~10 mm, with a step of 1 mm
Control parameters	P gain	7~14, with a step of 0.5
	I gain	2
	D gain	0

The final working conditions are generated by the permutation and combination of sinusoidal frequency, sinusoidal amplitude, and P gain in Table 2, and there are eight groups of sinusoidal frequency, 15 groups of P gain, 10 groups of sinusoidal amplitude, and 1200 working conditions in total. In order to avoid the mutual influence between two adjacent working conditions, each working condition runs for two cycles, with an overall sampling time of approximately 1632 s. Moreover, the mean of the control deviation absolute value at each moment of the last cycle is taken as the basis for evaluating the control performance.

The desired input signals in the simulation are shown in Figure 4. Due to the long sampling time, sinusoidal curves at different frequencies are relatively dense, as shown in the Figure 4 below.

The working conditions parameters include sinusoidal frequency and amplitude of input signal, the control parameters are P gain of the PID control method, and the performance index in the system is the mean of control deviation absolute value. It can be seen that in Table 2, there is an order of magnitude difference in the size of these three variables, which is not beneficial to the learning of the neural network. Therefore, the above three variables should be appropriately transformed to make their interval roughly between 0 and 1. So, the concept "data after processing" in the following section is the data after normalization.

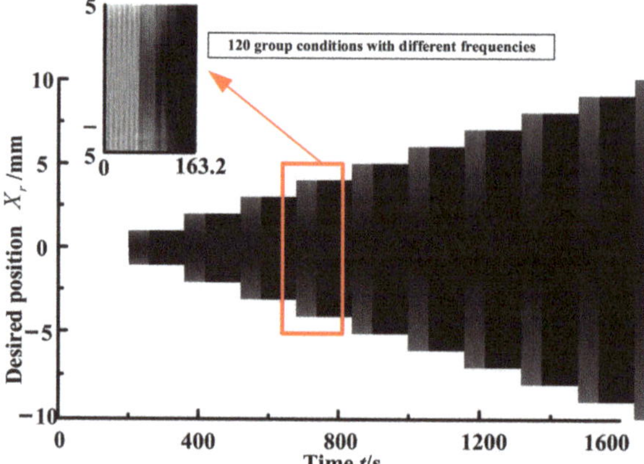

Figure 4. Desired position signals of input in the simulation.

The P gain of controller in the simulation is shown in Figure 5.

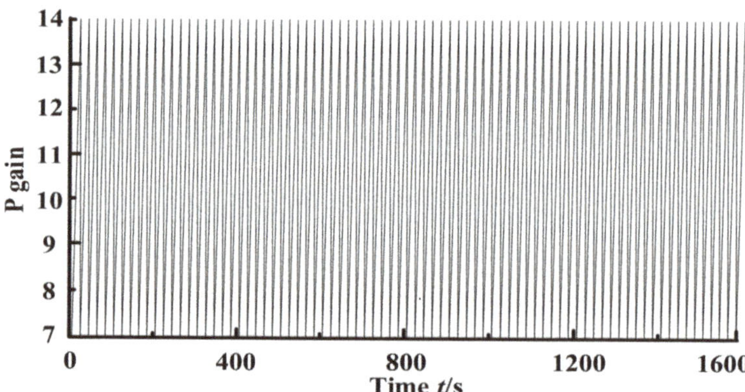

Figure 5. P gain of controller in the simulation.

3.3. Performance Fitting of Control System

In Section 3.2, the mean of control deviation e under different working conditions and control parameters are obtained through simulation. In this section, neural network 1 is used to fit the relationship among the working condition parameters, control parameters, and the mean of control deviation e. Then, neural network 1 can be used to calculate the mean of control deviation e with different control parameters under each working condition. The parameters with the minimum of mean of control deviation e under each working condition are selected, so as to complete the optimization process of the control parameters.

(1)   Input and output of the neural network

Neural network 1 was designed. The input of the neural network is a three-dimensional vector, which represents the sinusoidal frequency and amplitude of the input signal and P gain, respectively, and the output is the mean of control deviation e of the corresponding set of parameters.

$$u1 = \left[\frac{iam}{5}; \frac{freq}{2}; \frac{pgain - 5}{15}\right] \tag{1}$$

(2)   Selection of the loss function

The loss function is the index used to evaluate the model fitting effect, and the goal of neural network learning is to make the loss function as small as possible. The input and output variables of the neural network are continuous values, and the mean square error function is adopted. Its expression is as follows:

$$J(\theta) = \sum_{i=1}^{m} \left(h_\theta(x_i) - y_i\right)^2 \tag{2}$$

(3)   Determination of the neural network structural parameters

The total number of neural network layers is three, including the input layer, the output layer, and a hidden layer. The number of neurons in the hidden layer is 13, and the activation function is Sigmoid, the overall structure of neural network 1 is shown in Figure 6. The sinusoidal input signals and control parameters are shown in Table 2, the output of the neural network (mean of control deviation e) indicates the mean of the control deviation e between the input signals and output signals of the HDU position control system.

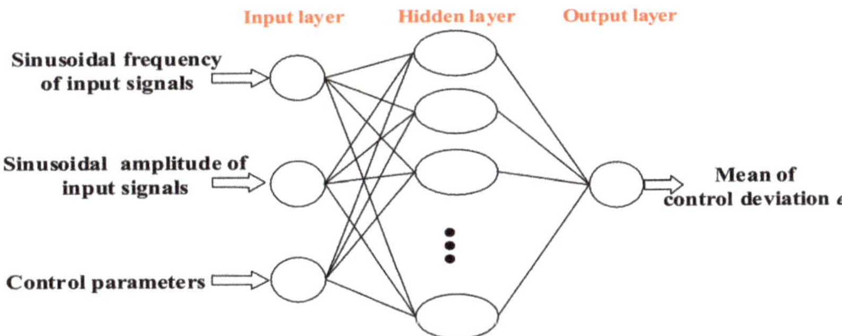

Figure 6. The overall structure of neural network 1.

(4)   Training of neural network 1

The input of neural network 1 after data processing is shown in Figure 7.
The output of neural network after data processing is shown in Figure 8.

The processed data are fed into the neural network for learning until the gradient is less than 10^{-6} or the mean square deviation is less than 10^{-4}.

3.4. Optimization of the Control Parameters

The sinusoidal frequency and amplitude of the input signals can be determined for a specific working condition. Taking the control parameters as independent variables, mapping the relationship established through neural network 1 as a function and the mean of control deviation e as the dependent variable, the relationship between the control performance and control parameters can be obtained under this working condition. There is an obvious rule between the control performance and the control parameters, so the control parameters with better control performance can be obtained through the curves. The optimal control parameters under the working conditions are selected according to certain rules and neural network 2 is used to learn the relationship between the working condition parameters and the selected control parameters. After learning, the neural network is used to adaptively change the control parameters according to the working conditions, so as to realize the adaptive control. The specific learning model was designed as follows:

(1) Selection of the neural network input and output

The purpose of neural network 2 is to calculate the control parameters that meet the rules under different working conditions. Therefore, the input of the neural network are the sinusoidal frequency and amplitude of the input signals, which are generated through permutation and combination with a sinusoidal frequency of 0.4~2 Hz and a sinusoidal amplitude of 1~5 mm, forming at the intervals of 0.01 Hz and 0.05 mm, respectively. The neural network output are the selected control parameters which could control the model in Figure 2 instead of the PID. The overall structure of neural network 2 is shown in Figure 9.

(2) Rules of parameter selection

The control parameters with the minimum of control deviation e are selected to form the output sample of the neural network.

(3) Training of neural network 2

Neural network 2 consists of three layers, including a hidden layer and 10 neurons in this hidden layer. The activation function is Sigmoid, and the loss function is the mean square error.

3.5. Simulation

Neural network 2, after training, was applied to the HDU position control system. Then, the updated schematic diagram of the HDU position control system is shown in Figure 10.

While the working conditions parameters changed, the neural network 2 automatically adjusted the control parameters according to the working condition to realize the adaptive control. Based on the MATLAB/Simulink model of the system established in Section 2, this section introduces a MATLAB function module for the neural network 2 calculation, and the results were output to the PID control model.

In the simulation, the initial position of the hydraulic cylinder piston was 25 mm, the P gain was the output of neural network 2, the I gain was 2, and the D gain was 0. The simulation working conditions are shown in Table 3.

The ideal control deviation (reference signal) was 0 which means that there is no control deviation between the input and the output. The comparison curves with constant and variable value PID are shown in Figure 11 (adaptive PID control based on a neural network is neural network PID for short, control deviation e is deviation e for short).

The control deviation of the adaptive PID control system based on the neural network (the blue curves in Figure 11) is shown in Table 4 (maximal relative deviation is equal to the ratio of the maximum deviation to the sinusoidal amplitude).

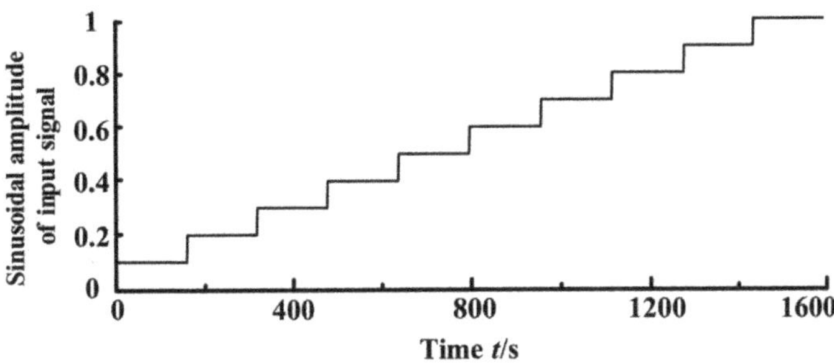

(**a**) Sinusoidal amplitude of the input signal after data processing.

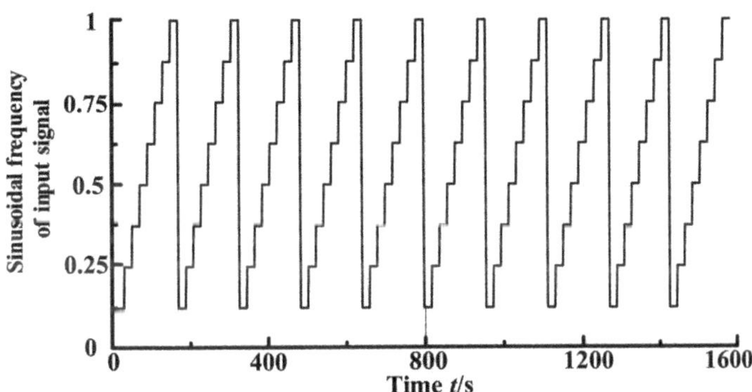

(**b**) Sinusoidal frequency of the input signal after data processing.

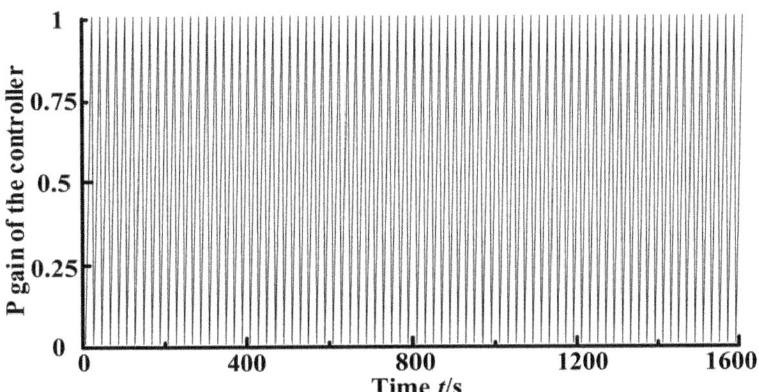

(**c**) P gain of the controller after data processing.

Figure 7. Input of neural network 1 after data processing.

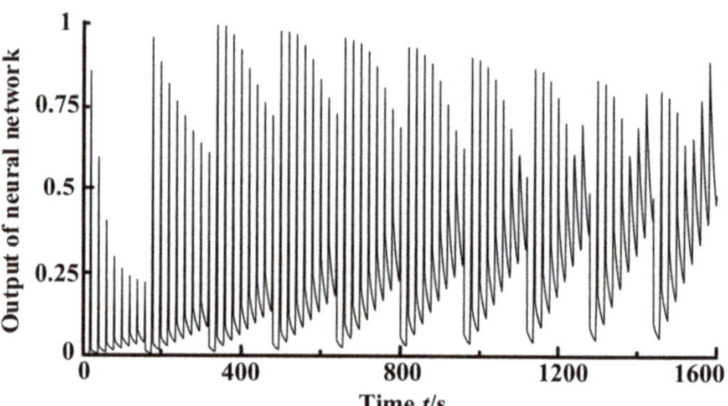

Figure 8. Output of the neural network after data processing.

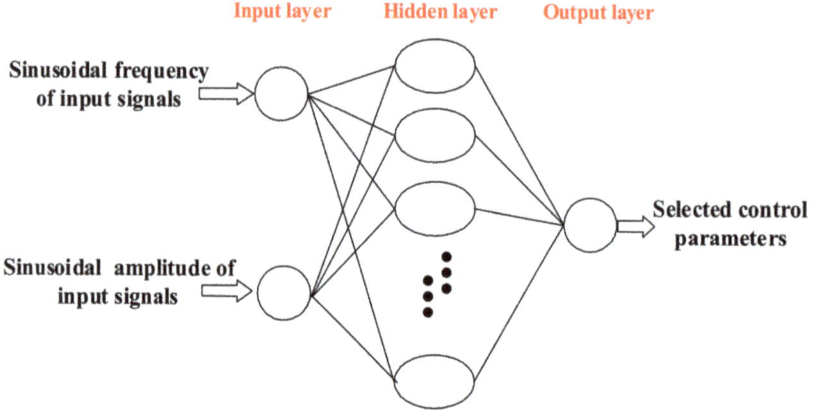

Figure 9. Overall structure of neural network 2.

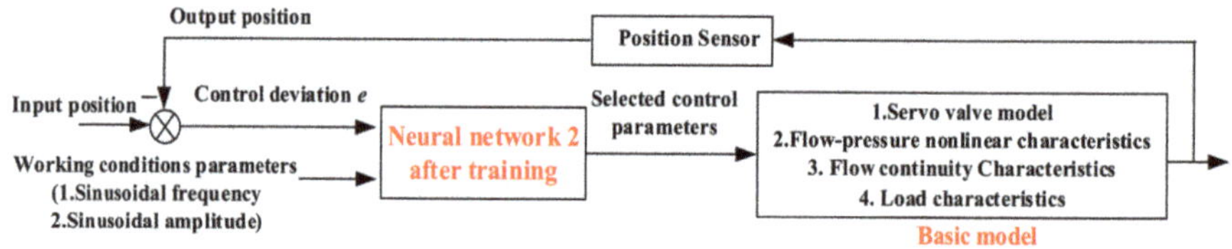

Figure 10. Updated schematic diagram of the HDU position control system.

Table 3. Working conditions in simulation.

Amplitude \ Frequency	0.5 Hz	1 Hz	2 Hz
2 mm	Working condition 1	-	-
4 mm	-	Working condition 2	-
6 mm	-	-	Working condition 3

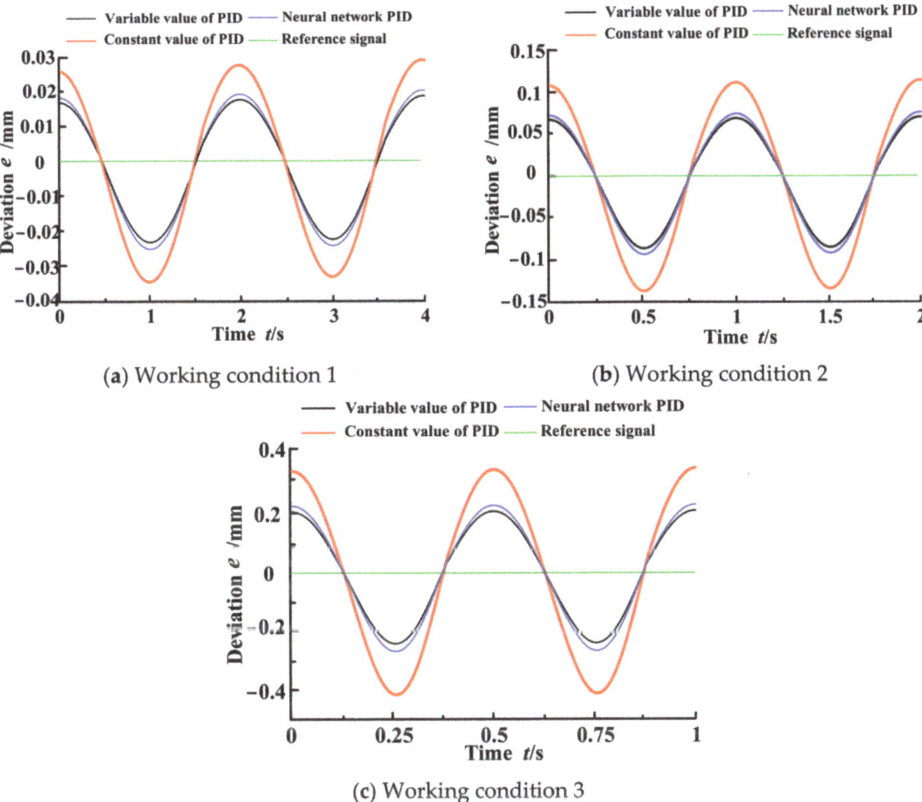

(**a**) Working condition 1

(**b**) Working condition 2

(**c**) Working condition 3

Figure 11. Comparative deviation curves of different PID control methods.

Table 4. Control deviation of the adaptive PID control system based on a neural network.

Working Conditions	Working Condition 1	Working Condition 2	Working Condition 3
Maximal deviation/mm	0.026	0.096	0.28
Maximal relative deviation	1.3%	2.4%	4.67%

According to the simulation results, under the three working conditions, the maximum relative deviation of the adaptive PID method based on a neural network decreased by an average of 31.3% compared with the maximum relative deviation of the constant value PID and increased by 7.87% compared with the maximum relative deviation of the variable value PID. The deviation of the adaptive PID method based on a neural network was greatly reduced compared with the constant value PID, which approached the effect of the manually adjusted PID control parameters and maintained a good control performance under multiple working conditions. Due to space limitations, additional simulation results are not included in this paper.

4. Experiments

4.1. Introduction to the Experimental System

The experiment of this study was carried out on the performance test platform of the HDU. The platform is mainly composed of two HDUs, which are installed in the top. The HDU on the left adopts the position of closed-loop control, while the HDU on the right adopts the force closed-loop control position. In the experiment, the HDU on the left carried out the performance test of the relevant control algorithm, and the HDU on the right carried out the zero-force servo control. In each experiment, the working conditions of the left and right HDUs were the same. The photo of the experimental platform is shown in Figure 12a.

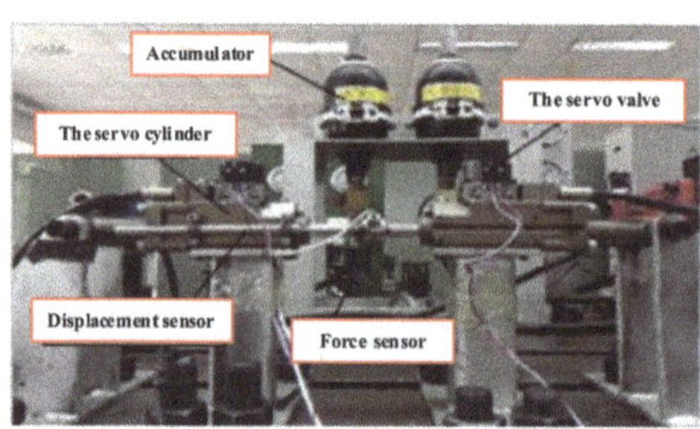

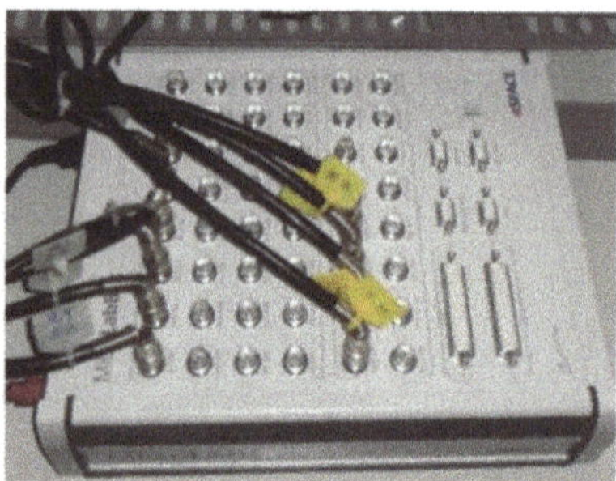

(**a**) Performance test bench for HDU (**b**) dSPACE controller

Figure 12. Photos of the experimental platform.

The controller used in the experiments is a semi-physical simulation experiment platform dSPACE-MicroLabBox shown in Figure 12b. MicroLabBox is supported by a comprehensive dSPACE software package, real-time interface (RTI) for Simulink (Math-Works, Natick, America) for model-based I/O integration and the experiment software ControlDesk, which provides access to the real-time application during run time by means of graphical instruments.

After the control algorithm in MATLAB/simulink, we used the code to automatically generate the target C code that could then be identified by the controller. Compared with manual C coding, combining MATLAB/simulink with the encoder can quickly design and test the control algorithms, avoid the complexity of the underlying C code writing, and improve the speed of the controller implementation stage. In the experiment, the data sampling frequency was 1 KHz. Figure 13 is the schematic diagram of the experimental signal input and data acquisition.

4.2. Collection of Learning Samples

As a joint actuator of robots, the HDU is the key to determining the motion performance of robots. According to the movement of the robot during trotting, pacing, and other gaits, the proposed sampling range of experimental learning samples is shown in Table 5.

The final working conditions were obtained by permutation and combination in the table, with a total of 324 groups of working conditions, and each group of working conditions ran for three cycles. In order to avoid mutual influence between adjacent conditions, the mean of control deviation for the last two working conditions was taken as the evaluation of the performance index. The generated system input signal sequence is shown in Figures 14 and 15, and the signal acquisition interface is shown in Figure 16.

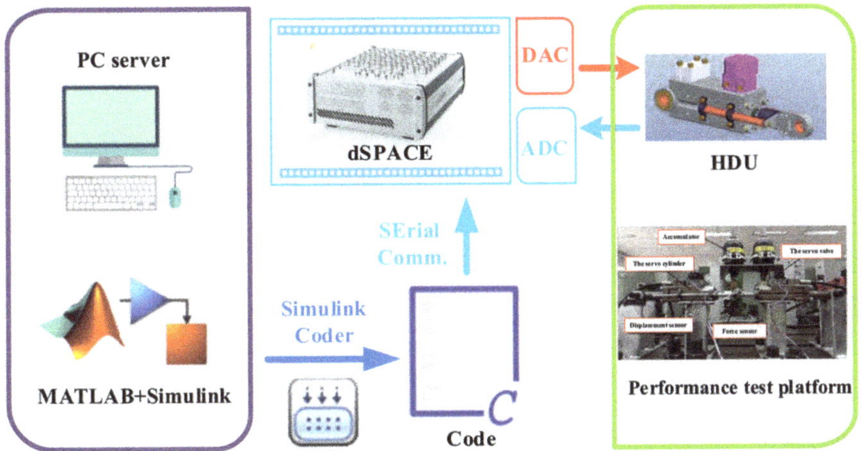

Figure 13. Schematic diagram of the experimental signal input and data acquisition.

Table 5. Collection range of the learning samples.

Parameters		Range of Value
Working conditions	Sinusoidal frequency	0.5~2 Hz, with a step of 0.3 Hz
	Sinusoidal amplitude	2~5 mm, with a step of 1 mm
Control parameters	P gain	10~50, with a step of 5
	I gain	2
	D gain	0

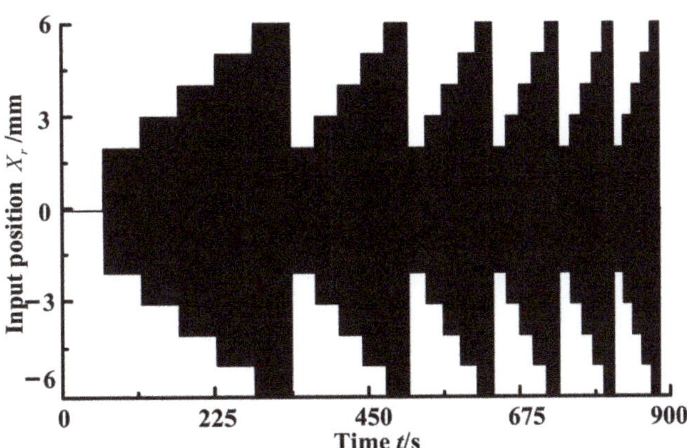

Figure 14. Input position signals of the input in the experiment.

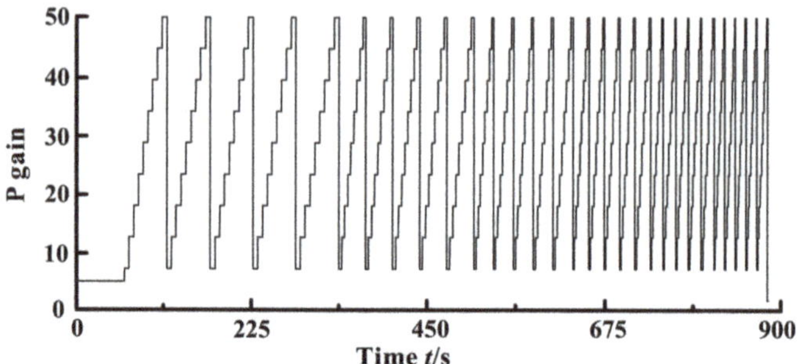

Figure 15. Control parameter signals in the experiment.

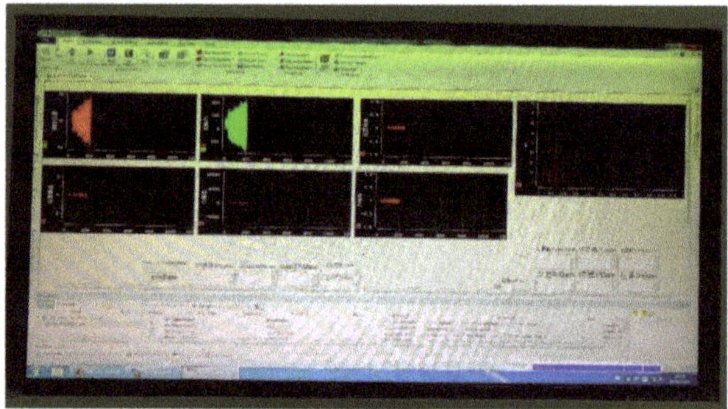

Figure 16. Signal acquisition interface.

4.3. Optimization of the Control Parameters

The samples obtained in Section 4.2 were used to learn the relationship among the working conditions parameters, the control parameters, and the control performance, and the neural network structure and data processing methods used were the same as those in Section 3.3. The training performance of the neural network is shown in Figure 17.

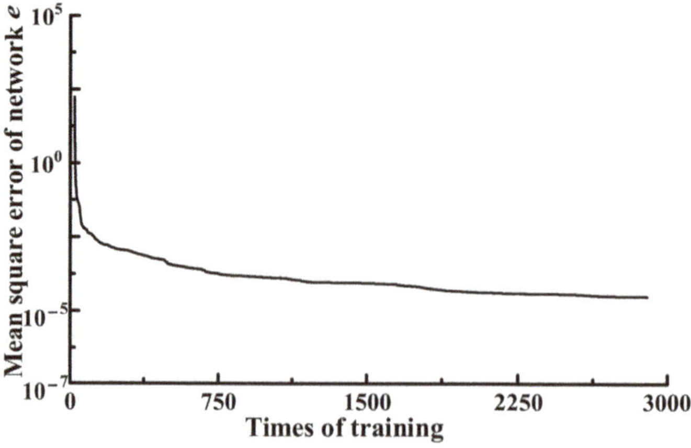

Figure 17. Training performance of the neural network.

It can be seen that after the completion of neural network learning, the value of the mean square error reached the magnitude 10^{-4}, which well estimated the control performance and laid a foundation for the next calculation of control parameters.

The control performance index of the HDU was set as follows: the maximum of control deviation e should not exceed 5% of the sinusoidal amplitude. Based on the obtained neural network, the corresponding system performance under different working conditions and the control parameters were calculated, and the control parameters required to meet the control performance requirements were selected. The working condition parameters were taken as the input of neural network 2, and the selected control parameters were taken as the desired output of neural network 2. The sinusoidal frequency of the input signal was 0.5–2 Hz and the amplitude was 5–15 mm, and the input signals were generated by permutation and combination at intervals of 0.01 Hz and 0.05 mm, respectively.

The neural network structure and data processing methods used were the same as those used in Section 3.4. The learning performance of neural network 2 is shown in Figure 18.

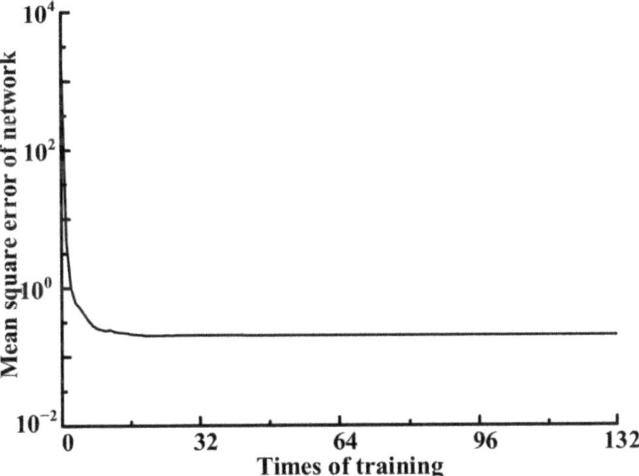

Figure 18. Training performance of the neural network.

It can be seen that the neural network converged rapidly, and the value of the mean square error reached an order of magnitude 10^{-1} after learning, which meets the requirements of controlling parameter adjustment accuracy.

4.4. Experiment of Adaptive PID Control Based on a Neural Network

In order to verify the performance of the adaptive PID control based on a neural network, an experiment was carried out on the performance test platform of the HDU under the working conditions shown in Table 3, and the control performance of the system under different working conditions was tested.

The initial position of the piston of the HDU was 25 mm, and the oil source pressure of the system was 5 MPa. The working conditions were input into the adaptive PID control system based on the neural network, and a deviation curve was obtained, which was compared with the deviation curve of the PID control with constant and variable values, as shown in Figure 19.

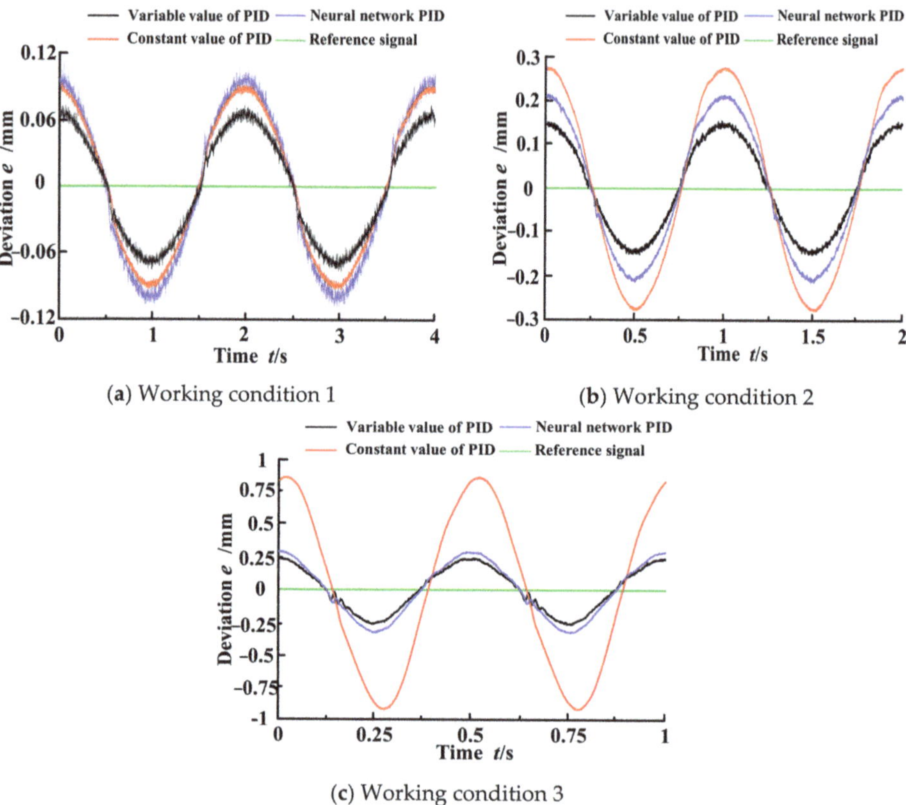

(**a**) Working condition 1

(**b**) Working condition 2

(**c**) Working condition 3

Figure 19. Comparative deviation curves of the different PID control methods.

The control deviation of the adaptive PID method based on the neural network (the blue curves in Figure 19) is shown in Table 5.

As shown in Figure 19 and Table 6, due to the setting of the parameter selection rules, the control deviation was slightly larger than that of the constant value PID under working condition 1. It greatly improved over that of the constant value PID method under the other two working conditions. The maximum relative deviation of the three working conditions reduced by 22.13% on average compared with that of the constant value PID method, which is close to the deviation level of the variable value PID method. On the whole, the control accuracy of the adaptive PID method based on a neural network was between the constant value PID method and the variable value PID method, which is slightly worse than the variable value PID method. However, its control accuracy was better than that of the constant value PID method, which has good adaptability and can maintain better control accuracy under various working conditions.

Table 6. Control deviation of the adaptive PID control system based on a neural network.

Working Conditions	Working Condition 1	Working Condition 2	Working Condition 3
Maximal deviation/mm	0.1	0.21	0.3
Maximal relative deviation	5%	5.25%	5%

According to the proposed method in this paper, more parameter information corresponding to working conditions can be learned, and the same research idea can be extended to other control systems with similar structures. Moreover, based on this double-layer BP neural network, other "machine learning" methods such as deep deterministic policy gradient (DDPG) could be researched.

5. Conclusions

In this paper, an adaptive PID control method using a double-layer BP was designed. Neural network 1 is used to fit the relationship among the working parameters, the control parameters, and the control performance. Neural network 2 is used to fit the relationship between the working condition parameters and the selected control parameters, and to realize the adaptive adjustment of the PID control parameters according to the working condition parameters. The results showed that the designed method can automatically adjust the control parameters in the learning range and the working conditions near it, and it has a certain adaptability. It basically achieved the desired control precision. Compared with the constant value PID method, the deviation was reduced by 31.3%, and the performance was close to that of the variable value PID. Avoiding the disadvantage of the variable value PID requiring repeated manual adjustment of parameters, it provides practical value in engineering.

Author Contributions: Conceptualization, M.-L.Z.; Y.-J.Z.; X.-L.H.; Z.-J.G.; methodology, M.-L.Z. and Y.-J.Z.; software, X.-L.H. and Z.-J.G.; validation, M.-L.Z. and Y.-J.Z.; formal analysis, X.-L.H.; Z.-J.G.; investigation, M.-L.Z.; Y.-J.Z.; resources, M.-L.Z.; data curation, X.-L.H.; Z.-J.G.; writing—original draft preparation, Z.-J.G.; writing—review and editing, M.-L.Z.; Y.-J.Z.; visualization, X.-L.H.; supervision, M.-L.Z.; project administration, M.-L.Z.; funding acquisition, M.-L.Z. and Y.-J.Z. All authors have read and agreed to the published version of the manuscript.

Funding: This research was funded by Soft Science Research Project of Inovation Competence Enhancement Plan of Hebei Province (no. 21556105D and no. 21552501D) and Graduate Innovation Funding Project of Hebei Province (no. CXZZB2021120).

Institutional Review Board Statement: Not applicable.

Informed Consent Statement: Not aplicable.

Data Availability Statement: Data are contained within the article.

Conflicts of Interest: The authors declare no conflict of interest.

Appendix A.

Appendix A.1. Mathamatical Model of HDU Position Closed-Loop Control System

Whether the electro-hydraulic servo valve can output the corresponding flow and pressure under the control of the electrical analog signals is the core of the electro-hydraulic servo control system, and whether the model is accurate or not has a great influence on the overall modeling accuracy. The general modeling method for the electro-hydraulic servo valve is linearization at a specific working point (usually at zero position of the servo valve). However, this method cannot accurately reconstruct the characteristics of the servo valve in all working areas. In order to improve the accuracy of the model, the nonlinear factors of pressure and flow for the electro-hydraulic servo valve are considered in this paper, and the flow equations of the electro-hydraulic servo valve were obtained as follows:

The inlet oil flow of the servo valve is:

$$q_1 = \begin{cases} K_d x_v \sqrt{p_s - p_1} & x_v \geq 0 \\ K_d x_v \sqrt{p_1 - p_0} & x_v < 0 \end{cases} \tag{A1}$$

The return oil flow of the servo valve is:

$$q_2 = \begin{cases} K_d x_v \sqrt{p_2 - p_0} & x_v \geq 0 \\ K_d x_v \sqrt{p_s - p_2} & x_v < 0 \end{cases} \tag{A2}$$

The equivalent flow coefficient K_d is expressed as:

$$K_d = C_d W \sqrt{\frac{2}{\rho}} \tag{A3}$$

For the convenience of expression and research, let:

$$K_1 = \begin{cases} K_d \sqrt{p_s - p_1} & x_v \geq 0 \\ K_d \sqrt{p_1 - p_0} & x_v < 0 \end{cases} \tag{A4}$$

$$K_2 = \begin{cases} K_d \sqrt{p_2 - p_0} & x_v \geq 0 \\ K_d \sqrt{p_s - p_2} & x_v < 0 \end{cases} \tag{A5}$$

According to Equations (A1), (A2), (A4), and (A5), the flow equations of the servo valve can be written.

The inlet oil flow of the servo valve is:

$$q_1 = K_1 \cdot x_v \tag{A6}$$

The return oil flow of the servo valve is:

$$q_2 = K_2 \cdot x_v \tag{A7}$$

The response of the servo valve is often much higher than that of the hydraulic power components. In order to simplify the analysis and design of the dynamic characteristics in the system, the transfer function of the electro-hydraulic servo valve is equivalent to a second-order oscillation link, and the transfer function of the spool position and the input voltage of servo valve is obtained in Equation (A8).

$$\frac{X_v}{U_g} = \frac{K_a K_{xv}}{\frac{s^2}{\omega^2} + \frac{2\zeta}{\omega}s + 1} \tag{A8}$$

The hydraulic cylinder is a component of the hydraulic actuator, the final carrier of the output power of the hydraulic system, and the control object of the electro-hydraulic servo valve. Its dynamic characteristics largely determine the performance of the system. Assuming that the connecting pipe diameter between the servo valve and the hydraulic cylinder is large enough, the pressure loss, fluid quality influence, and pipeline dynamic characteristics are all ignored; the hydraulic cylinder pressure in the working cavity is equal, the oil bulk modulus and the oil temperature are constant, and the internal and external leakage of the hydraulic cylinder are laminar flow, then the flow equations of the two working cavities for the asymmetric hydraulic cylinder can be obtained.

The rodless cavity flow of the asymmetric hydraulic cylinder and the volume of the servo valve to the rodless cavity are:

$$\begin{cases} q_1 = A_1 \frac{dx_p}{dt} + C_{ip}(p_1 - p_2) + \frac{V_1}{\beta_e} \frac{dp_1}{dt} \\ V_1 = V_{01} + A_1 x_p \end{cases} \tag{A9}$$

The rod cavity flow of the asymmetric hydraulic cylinder and the volume of the servo valve to the rod cavity are:

$$\begin{cases} q_2 = A_2 \frac{dx_p}{dt} + C_{ip}(p_1 - p_2) - C_{ep}p_2 - \frac{V_2}{\beta_e} \frac{dp_2}{dt} \\ V_2 = V_{02} - A_2 x_p \end{cases} \tag{A10}$$

The inlet or return oil cavities of the HDU are all set inside the servo cylinder body. Considering the difference of the initial position for the piston in the servo cylinder, the initial volume of the rodless and rod cavities can be obtained as:

$$\begin{cases} V_{01} = V_{g1} + A_1 L_0 \\ V_{02} = V_{g2} + A_2(L - L_0) \end{cases} \tag{A11}$$

Considering the coulomb friction force of the hydraulic cylinder is very small relative to the load force, the coulomb friction force is included in the load force and not considered separately. According to Newton's second law, the dynamic equilibrium equation on the piston is:

$$A_1 p_1 - A_2 p_2 = m_t \frac{dx_p^2}{dt} + B_p \frac{dx_p}{dt} + Kx_p + F_L \tag{A12}$$

The transfer function between the feedback voltage of the position sensor and the position of the piston rod in the servo cylinder is:

$$\frac{U_p}{x_p} = K_X \tag{A13}$$

A block diagram of the position closed-loop control system of the HDU can be obtained by combining Equations (A1)–(A13), which is shown in Figure 2 in Section 2.

Appendix A.2. Neuron Model

Let the input of neurons be an n-dimensional vector, $u_1, u_2, \cdots, u_n$, and let the vector $\mathbf{u} = [u_1; u_2; \cdots u_n]$ represent the input of neurons. Neurons assign different weights to each input element, and the final input is obtained after summing, which is called the net input:

$$z = \sum_{i=1}^{n} w_i u_i + b = \boldsymbol{w}^T \boldsymbol{u} + \boldsymbol{b} \tag{A14}$$

where $\boldsymbol{w} = [w_1; w_2; \cdots w_n]$ is the n-dimensional weight vector and b is the offset value.

In the human brain, different input signals cause the neurons to produce different electrical signals. Artificial neurons use a nonlinear function to simulate this function and finally obtain the output value of the neuron x:

$$x = f(z) \tag{A15}$$

where $f(\cdot)$ is referred to as an activation function.

The introduction of the activation function improves the ability of expression and learning in neural networks. Derivable activation functions can use numerical optimization methods to update the network parameters, and self-defined activation functions can limit the scope of input and output in neural networks, keeping the overall calculation domain within a reasonable range, then improving the stability of the learning.

Sigmoid activation functions are S-shaped on the whole, closing to linear near 0 and tending to saturate at both ends [24,25]. The commonly used Sigmoid activation functions can be divided into logistic activation functions and Tanh activation functions.

Logistic activation functions are expressed as:

$$\sigma(x) = \frac{1}{1 + e^{-x}} \tag{A16}$$

It can be seen that the standard logistic activation functions can map the data from the real interval to the scope of 0 and 1. After a certain transformation, the input can cover the whole range of data for the sensors in the control system, and the output can be limited to a certain effective interval, which can be continuously derivable.

Tanh activation functions are expressed as:

$$\text{Tanh}(x) = \frac{e^x - e^{-x}}{e^x + e^{-x}} \tag{A17}$$

The standard Tanh activation functions can map data from the real interval to the scope of -1 and 1, which can be used to control the output control value in the system.

References

1. Savaee, E.; Hanzaki, A.R. A new algorithm for aalibration of an omni-directional wheeled mobile robot based on effective kinematic parameters estimation. *J. Intell. Robot. Syst.* **2021**, *101*, 28. [CrossRef]
2. Li, Z.Q.; Chen, L.Q.; Zheng, Q.; Dou, X.Y.; Yang, L. Control of a path following caterpillar robot based on a sliding mode variable structure algorithm. *Biosyst. Eng.* **2019**, *186*, 293–306. [CrossRef]
3. Chen, Z.H.; Wang, S.K.; Wang, J.Z.; Xu, K.; Lei, T.; Zhang, H.; Wang, X.W.; Liu, D.H.; Si, J.G. Control strategy of stable walking for a hexapod wheel-legged robot. *ISA Trans.* **2020**, *108*, 367–380. [CrossRef] [PubMed]
4. Luo, M.; Wan, Z.Y.; Sun, Y.N.; Skorina, E.H.; Tao, W.J.; Chen, F.C.; Gopalka, L.; Yang, H.; Onal, C.D. Motion planning and iterative learning control of a modular soft robotic snake. *Front. Robot. AI* **2020**, *7*, 299242. [CrossRef] [PubMed]
5. Rodino, S.; Curcio, E.M.; Bella, A.D.; Persampieri, M.; Funaro, M.; Carbone, G. Design, simulation, and preliminary validation of a four-legged robot. *Machines* **2020**, *8*, 82. [CrossRef]
6. Ba, K.X.; Song, Y.H.; Yu, B.; He, X.L.; Huang, Z.P.; Li, C.H.; Yuan, L.P.; Kong, X.D. Dynamics compensation of impedance-based motion control for LHDS of legged robot. *Robot. Auton. Syst.* **2021**, *139*, 103704. [CrossRef]
7. Li, M.T.; Jiang, Z.Y.; Wang, P.F.; Sun, L.N.; Ge, S.S. Control of a quadruped robot with bionic springy legs in trotting gait. *J. Bionic Eng.* **2014**, *11*, 188–198. [CrossRef]
8. Souzanchi, K.M.; Arab, A.; Akbarzadeh, T.M.R.; Fate, M.M. Robust impedance control of uncertain mobile manipulators using time-delay compensation. *IEEE Trans. Control. Syst. Technol.* **2017**, *26*, 1942–1953. [CrossRef]
9. Chen, Y.H.; Zhao, J.B.; Wang, J.Z.; Li, D.Y. Fractional-order impedance control for a wheel-legged robot. In Proceedings of the 2017 29th Chinese Control and Decision Conference (CCDC), Chongqing, China, 28–30 May 2017.
10. Playter, R.; Buehler, M.; Raibert, M. BigDog. In Proceedings of the Conference on Unmanned Systems Technology VIII, Kissimmee, FL, USA, 17–20 April 2006.
11. Semini, C.; Barasuol, V.; Goldsmith, J.; Frigerio, M.; Focchi, M.; Gao, Y.F.; Caldwell, D.G. Design of the hydraulically actuated, torque-controlled quadruped robot HyQ2Max. *IEEE/ASME Trans. Mechatron.* **2016**, *22*, 635–646. [CrossRef]
12. Focchi, M.; Barasuol, V.; Havoutis, I.; Buchili, J.; Semini, C.; Caldwell, D.G. Local reflex generation for obstacle negotiation in quadrupedal locomotion. In Proceedings of the Conference on Climbing and Walking Robots (CLAWAR), Sydney, Australia, 14–17 July 2015.
13. Wiedebach, G.; Bertrand, S.; Wu, T.F.; Fiorio, L.; Mccrory, S.; Griffin, R.; Nori, F.; Pratt, J. Walking on partial footholds including line contacts with the humanoid robot atlas. In Proceedings of the 16th IEEE-RAS International Conference on Humanoid Robots (Humanoids), Cancun, Mexico, 15–17 November 2016.
14. Chen, Y.F.; Shen, L.G.; Li, R.J.; Xu, X.C.; Hong, H.C.; Lin, H.J.; Chen, J.R. Quantification of interfacial energies associated with membrane fouling in a membrane bioreactor by using BP and GRNN artificial neural networks. *J. Colloid Interface Sci.* **2020**, *565*, 1–10. [CrossRef] [PubMed]
15. Tang, M.C.; Zhou, C.C.; Zhang, N.N.; Liu, C.; Pan, J.H.; Cao, S.S. Prediction of the ash content of flotation concentrate based on froth image processing and BP neural network modeling. *Int. J. Coal Prep. Util.* **2021**, *3*, 191–202. [CrossRef]
16. Lee, C.Y.; Chen, Y.H. Motor Fault Detection Using wavelet transform and improved PSO-BP neural network. *Processes* **2020**, *8*, 1322. [CrossRef]
17. Swic, A.; Wolos, D.; Klosowski, G. The Use of Neural Networks and Genetic Algorithms to Control Low Rigidity Shafts Machining. *Sensors* **2020**, *20*, 4683. [CrossRef]
18. Rego, R.C.B.; de Araujo, F.M.U. Nonlinear Control System with Reinforcement Learning and Neural Networks Based Lyapunov Functions. *IEEE Lat. Am. Trans.* **2021**, *19*, 1253–1260. [CrossRef]
19. Nobahari, H.; Seifouripour, Y. A Nonlinear Controller Based on the Convolutional Neural Networks. In Proceedings of the 7th International Conference on Robotics and Mechatronics (ICRoM), Tehran, Iran, 20–21 November 2019.
20. Wang, G.; Yao, X.M.; Zhao, W. A novel piezoelectric hysteresis modeling method combining LSTM and NARX neural networks. *Mod. Phys. Lett. B* **2020**, *34*, 2050306. [CrossRef]
21. Nair, V.; Hinton, G.E. Rectified linear units improve restricted boltzmann machines. In Proceedings of the 27th International Conference on Machine Learning, Haifa, Israel, 21–24 June 2010.
22. Jinsakul, N.; Tsai, C.F.; Tsai, C.E.; Wu, P. Enhancement of deep learning in image classification performance using xception with the swish activation function for colorectal polyp preliminary screening. *Mathematics* **2019**, *7*, 1170. [CrossRef]

23. Goodfelow, I.; Warde-Farley, D.; Mirza, M.; Courville, A.; Bengio, Y. Maxout networks. In Proceedings of the 30th International Conference on Machine Learning, Atlanta, GA, USA, 16–21 June 2013.
24. Langer, S. Approximating smooth functions by deep neural networks with sigmoid activation function. *J. Multivar. Anal.* **2021**, *182*, 104696. [CrossRef]
25. Uteuliyeva, M.; Zhumekenov, A.; Kabdolov, O. Fourier neural networks: A comparative study. *Intell. Data Anal.* **2020**, *24*, 1107–1120. [CrossRef]

processes

Article

Group Acceptance Sampling Plan Using Marshall–Olkin Kumaraswamy Exponential (*MOKw-E*) Distribution

Abdullah M. Almarashi [1], **Khushnoor Khan** [1], **Christophe Chesneau** [2] **and Farrukh Jamal** [3,*]

[1] Department of Statistics, Faculty of Science, King Abdulaziz University, Jeddah 21589, Saudi Arabia; am_1235@hotmail.com (A.M.A.); khushnoorkhan64@gmail.com (K.K.)

[2] Department of Mathematics, Université de Caen, LMNO, Campus II, Science 3, 14032 Caen, France; christophe.chesneau@gmail.com

[3] Department of Statistics, The Islamia University of Bahawalpur, Punjab 63100, Pakistan

* Correspondence: farrukh.jamal@iub.edu.pk

Abstract: The current research concerns the group acceptance sampling plan in the case where (i) the lifetime of the items follows the Marshall–Olkin Kumaraswamy exponential distribution (*MOKw-E*) and (ii) a large number of items, considered as a group, can be tested at the same time. When the consumer's risk and the test terminsation period are defined, the key design parameters are extracted. The values of the operating characteristic function are determined for different quality levels. At the specified producer's risk, the minimum ratios of the true average life to the specified average life are also calculated. The results of the present study will set the platform for future research on various nano quality level topics when the items follow different probability distributions under the Marshall–Olkin Kumaraswamy scheme. Real-world data are used to explain the technique.

Keywords: Marshall–Olkin Kumaraswamy; consumer's risk; group acceptance plan

Citation: Almarashi, A.M.; Khan, K.; Chesneau, C.; Jamal, F. Group Acceptance Sampling Plan Using Marshall–Olkin Kumaraswamy Exponential (*MOKw-E*) Distribution. *Processes* **2021**, *9*, 1066. https:// doi.org/10.3390/pr9061066

Academic Editors: Jie Zhang and Meihong Wang

Received: 18 May 2021
Accepted: 15 June 2021
Published: 18 June 2021

Publisher's Note: MDPI stays neutral with regard to jurisdictional claims in published maps and institutional affiliations.

1. Introduction

It is now an established fact that nanotechnology affects our daily lives like never before. Advancement in nanotechnology might never have been possible without the use of appropriate statistical methods. Ref. [1] presented a detailed review of the use of statistical methods in nanoscale applications. Wherever we rely on destructive tests for testing the quality of products, sampling is the only way out. For selecting a representative sample, traditional sampling techniques (simple random sampling, systematic sampling, etc.), and their hybrids are used. For the nano process, however, new sampling techniques have been created [1]. With a smaller sample size and good use of the sampling plan, we can obtain a more precise inference and save money on sampling. In the manufacturing area, sampling plans are employed to decide on the acceptance or rejection of the incoming or outgoing batches based on some pre-specified quality, which is commonly known as lot sentencing. The size of the sample and the duration of the experiment are the two most critical factors for design engineers to consider, and both must be optimized. Acceptance sampling plans can help to achieve this optimization. Simple acceptance plans give us the minimum sample size to be used for testing. In this case, it is presumed that a single item is evaluated in a tester at a time, in order to maximize both cost and time. Ref. [2] posited the use of group to cut back on the amount of time and money invested on research. When more than one item is checked in a tester, the set of items is considered a group, justifying the name Group Acceptance Sampling Plan (GASP). As GASP is combined with truncated life testing, the result is known as a GASP based on truncated life test, which assumes that a product's lifespan matches a certain probability distribution. The attributes group acceptance sampling plan was originally established by [3] for the truncated life test, assuming that the lifetime of each item followed the Weibull distribution. For the given values of producer and consumer risks, the number of groups and acceptance numbers are obtained simultaneously in such a sampling plan.

When the product followed various forms of probability distributions, some authors worked on a GASP based on a truncated life test. For instance, ref. [3] considered the inverse Rayleigh and log-logistic distributions, ref. [4] the extended Lomax distribution, ref. [5] the Marshall–Olkin (MO) extended Weibull distribution, ref. [6] the generalized exponential distribution and finally, and [7] the odd generalized exponential log-logistic distribution.

Since 1997, there has been a surge of interest in designing new distributions based on baseline distributions and composite approaches, with the prospect of incorporating new parameters. Indeed, the inclusion of parameters has proven to be helpful in terms of exploring skewness and tail properties, as well as enhancing the goodness-of-fit of the created family. For an exhaustive list of references on extended family of probability distributions, see [7]. Additionally, ref. [8] offered a detailed explanation of how new families of univariate continuous distributions are formed by using additional parameters. The most relevant source for the present study's theoretical foundation is [9], where the authors developed the Kumaraswamy MO (*KwMO*) family of distributions, by using the MO [10] and the cumulative distribution function (cdf) as a baseline distribution in the Kumaraswamy-G family by [11], and studied its many properties. The second most relevant study is [12], which developed the MO Kumaraswamy-G (*MOKw-G*) family of distributions. It was based on the Kumaraswamy-G family cdf as the baseline distribution in the MO extended family, and studied its various properties at length. The primary goal of this paper is to further improve the GASPs for the MO Kumaraswamy exponential distribution (*MOKw-E*). As sketched in [12], the basic interests of considering the *MOKw-E* in this context are as follows: (i) the *MOKw-E* extends the modeling capabilities of the exponential distribution, and some of its powerful exponentiated versions, via a simple ratio scheme with several strategically well placed tuning parameters, (ii) the *MOKw-E* has a strong physical interpretation in terms of order statistics; it corresponds to the distribution of the time to the first failure of a component in a series system with N independent components, where N can be modeled by a random variable following a geometric distribution and the lifetime of a component that can be modeled by a random variable with the Kumaraswamy exponential distribution *(Kw-E)*, (iii) thanks to the *MO* scheme, the *MOKw-E* directly benefits from strong stochastic ordering properties, and (iv) diverse sub-distributions of the *MOKw-E* have been proved to be particularly efficient to analyze lifetime data of various kinds (see [13,14]). As a result, we suggest that the *MOKwE* is an ideal candidate distribution for GASP.

For the current study, the median is taken as the quality parameter. We can refer to [15], which states that the median is a better-quality parameter for a skewed distribution than the mean. Since the *MOKw-E* is a skewed distribution, percentile point shall be employed as the quality parameter. Hence, the main purpose of the present study is to offer a GASP based on truncated life test assuming that the lifetimes of a product follow the *MOKw-E* developed by [12] with known shape parameters. Scrolling through the literature, the authors were not able to find GASP in the context of the *MOKw-E*. The GASP for the *MOKw-E* is constructed, satisfying specific consumer's and producer's risks at some specific quality level. Furthermore, the minimum number of groups and approval number needed for a given customer risk and test termination time are calculated for a given group size. The results of the present study will set a platform for future research based on the proposed sampling plan for studying nano quality level (NQL) when products follow different probability distributions under the MO Kumaraswamy family scheme.

Format of the Paper

The remainder of this paper is structured as follows. In Section 2, we establish the theoretical background of *MOKw-G* and how the probability distribution function (pdf), cdf, and quantile function of *MOKw-E* are worked out. Section 3 addresses the design of GASP for the lifetime percentiles under a truncated life test. Section 4 provides a summary of the proposed approach as well as real-world data examples. Finally, Section 5 summarizes the observations and addresses several possible future consequences.

2. Marshall–Olkin Kumaraswamy Exponential (MOk_w-E) Distribution

First, let us recall the pdf, cdf, and quantile function of *MOKw-G* from which the pdf, cdf, and *quantile* function of *MOKw-E* are derived. For in-depth mathematical derivations, see [12]. The pdf *MOKw-G* is listed by:

$$f^{MOKwG}(t) = \frac{\alpha a b g(t) G(t)^{a-1}\left[1 - G(t)^a\right]^{b-1}}{\left[1 - \bar{\alpha}\left[1 - G(t)^a\right]^b\right]^2} \tag{1}$$

where a and b constitute shape parameters with $a, b > 0$, α is the tilt parameter for the extended family of distributions with "$\alpha > 0$", $\bar{\alpha} = 1 - \alpha$, and $G(t)$ is the cdf of a baseline distribution with pdf $g(t)$. Then, the cdf of *MOKw-G* is given by:

$$F^{MOKwG}(t) = \frac{1 - \left[1 - G(t)^a\right]^b}{1 - \bar{\alpha}\left[1 - G(t)^a\right]^b} \tag{2}$$

and the pth quantile function t_p of *MOKw-G* is taken form [12] as:

$$t_p = G^{-1}\left[1 - \left\{1 - \frac{\alpha p}{1 - \bar{\alpha}p}\right\}^{\frac{1}{b}}\right]^{\frac{1}{a}} \tag{3}$$

Using the exponential distribution as a baseline, the pdf, cdf, and the quantile function for *MOKw-E* can be worked out by substituting the pdf, cdf, and the quantile function of the exponential distribution in Equations (1)–(3). That is, we consider $g(t) = \lambda e^{-\lambda t}$, and $G(t) = 1 - e^{-\lambda t}$, $\lambda > 0$, $t > 0$. Thus, the pdf of *MOKw-E* is given by:

$$f^{MOKwE}(t) = \frac{\alpha a b \lambda e^{-\lambda t}\left[1 - e^{-\lambda t}\right]^{a-1}\left[1 - \left(1 - e^{-\lambda t}\right)^a\right]^{b-1}}{\left[1 - \bar{\alpha}\left[1 - \left(1 - e^{-\lambda t}\right)^a\right]^b\right]^2} \tag{4}$$

and the cdf of *MOKw-E* is given by:

$$F^{MOKwE}(t) = \frac{1 - \left[1 - \left(1 - e^{-\lambda t}\right)^a\right]^b}{1 - \bar{\alpha}\left[1 - \left(1 - e^{-\lambda t}\right)^a\right]^b} \tag{5}$$

Since the pth quantile of the exponential distribution is obtained as $t_p = -\frac{1}{\lambda}\log[1 - p]$, the pth quantile function t_p of *MOKw-E* using Equation (3) is:

$$t_p = -\frac{1}{\lambda}\log\left[1 - \left\{1 - \left(1 - \frac{\alpha p}{1 - \bar{\alpha}p}\right)^{\frac{1}{b}}\right\}^{\frac{1}{a}}\right]. \tag{6}$$

3. Description of the Gasp

The design parameters of a GASP are now obtained in the context of *MOKw-E*. The steps for implementing the group acceptance plan and obtaining the design parameters were followed from [3,16], and consist of:

- Selecting 'g' number of groups, and allocating predefining r items to each group. Thus, the sample size for a lot is obtained as 'n' = g × r.

- Selecting 'c' with reference to the acceptance number for a group with experiment time t_0.
- Simultaneously performing the experiment for 'g' groups and recording the number of failures for each group.
- Accepting the lot if no more than 'c' failures occur in all groups.
- Truncating the experiment and refusing the lot if more than 'c' failures occur in any group.

Thus, the proposed GASP is defined by two design parameters (g, c) for a given r. From Equation (5), it can be seen that the *cdf* of *MOKw-E* depends on α, t, a, and b and the median life of *MOKw−E* is given by Equation (6). It would be appropriate to calculate the termination time t_0 as $t_0 = a_1 m_0$, where a_1 denotes a certain constant and m_0 refers to the specified life. For instance, if $a_1 = 0.5$, the experiment time is half that of the specified life, or, if $a_1 = 3$, the experiment time is three times that of the specified life. In this setting, the probability of accepting a lot is:

$$P_{a(p)} = \left[\sum_{i=0}^{c} \binom{r}{i} p^i (1-p)^{r-i} \right]^g ,\qquad(7)$$

where 'p' refers to the probability that an item in a group fails before t_0, and this probability of failure is derived by inserting Equation (6) in Equation (5).

Based on Equation (6), we set

$$m = -\tfrac{1}{\lambda} \log \left[1 - \left\{ 1 - \left(1 - \tfrac{\alpha p}{1-\bar{\alpha} p} \right)^{\frac{1}{b}} \right\}^{\frac{1}{a}} \right].$$

$$\text{Let } \eta = \log \left[1 - \left\{ 1 - \left(1 - \tfrac{\alpha p}{1-\bar{\alpha} p} \right)^{\frac{1}{b}} \right\}^{\frac{1}{a}} \right]$$

Now, substituting $\lambda = -\tfrac{\eta}{m}$ and $t = a_1 m_0$ in Equation (5), the probability of failure is given by:

$$p = \frac{1 - \left[1 - (1 - e^{-\lambda t})^a \right]^b}{1 - \bar{\alpha} \left[1 - (1 - e^{-\lambda t})^a \right]^b}$$

which can be expressed as:

$$p = \frac{1 - \left[1 - \left\{ 1 - e^{\eta a_1 (m/m_0)^{-1}} \right\}^a \right]^b}{1 - \bar{\alpha} \left[1 - \left\{ 1 - e^{\eta a_1 (m/m_0)^{-1}} \right\}^a \right]^b}$$

For chosen a and b, p can be determined when a_1 and $r_2 = \tfrac{m}{m_0}$ are specified. The ratio of a product's mean lifetime to the specified lifetime $\tfrac{m}{m_0}$ can be used to express the product's quality level.

All that is required now is to minimize ASN $= n = g \times r$, subject to the following constraints:

$$P_{a(p_1 | \frac{m}{m_0} = r_1)} = \left[\sum_{i=0}^{c} \binom{r}{i} p_1^i (1-p_1)^{r-i} \right]^g \leq \beta \qquad(8)$$

and

$$P_{a(p_2 | \frac{m}{m_0} = r_2)} = \left[\sum_{i=0}^{c} \binom{r}{i} p_2^i (1-p_2)^{r-i} \right]^g \geq 1 - \alpha_1 \qquad(9)$$

where r_1 and r_2 denote the means ratio at the consumer's risk and at the producer's risk, respectively. Here, α_1 and α should not be confused; as stated earlier, α is the tilt parameter for the extended family of distributions. The probabilities of failure to be used in Equations (8) and (9) are as follows:

$$p_1 = \frac{1 - \left[1 - \{1 - e^{\eta a_1}\}^a\right]^b}{1 - \bar{\alpha}\left[1 - \{1 - e^{\eta a_1}\}^a\right]^b} \tag{10}$$

and

$$p_2 = \frac{1 - \left[1 - \left\{1 - e^{\eta a_1 (m/m_o)^{-1}}\right\}^a\right]^b}{1 - \bar{\alpha}\left[1 - \left\{1 - e^{\eta a_1 (m/m_o)^{-1}}\right\}^a\right]^b} \tag{11}$$

Both Equations (10) and (11) above are extracted from Equation (9).

4. Discussion and Example

4.1. Discussion

The design parameters under GASP for different values of the α (1.25 and 1.50) are presented in Tables 1 and 2. The code in R is provided in Appendix A. The values of $r = 5$ and 10 are considered. Then, it is noticed that a reduction in consumer's risk, β, leads to a rise in the number of groups. Furthermore, as r_2 increases, the number of groups rapidly decreases. However, after a certain point, even though the number of groups and acceptance numbers remain constant, the probability of accepting a lot begins to rise. The table also shows the impact of a_1. As an example, observe that, with $\beta = 0.25$, $a_1 = 0.5$, $r_2 = 6$, $\alpha = 1.25$, and, for $r = 5$, a total of eight groups, i.e., $8 \times 5 = 40$ number of units, are necessary on the life test. Additionally, when $r = 10$, then only two groups, i.e., $2 \times 5 = 10$ number of units, are necessary for the life test. As a result, in this case, 10 groups would be preferable. Table 2 reports $\alpha = 1.50$. According to the reported values, increasing the shape parameter value results in a smaller group size for the associated plan. For the considered GASP, under the *MOKw-E* and using median lifetime as the quality parameter, the number of groups decreases and the OC values ($P(a)$) increase when the true median life increases. This is presented in Table 1 for various values of the parameters ($\alpha = 1.25$, $\beta = 0.01$, $r = 10$, and $a_1 = 1.0$)

$m/m_0 = r_2$	4	6	8
G	11	6	3
C	5	4	3
$P(a)$	0.9653	0.9758	0.9793

To cross check the results shown in Tables 1 and 2, an example from [17] is considered. Suppose that the lifetime of ball bearings placed on a test follow *MOKw-E*, with the shape parameter $\alpha = 1.25$, and the mean specified life of the ball bearings is 2000 cycles. When the true mean life is 2000 cycles, the consumer faces a 25% risk, while the producer faces a 5% risk when the true mean life is 4000 cycles. Now, an experimenter wants to run a 1000-cycle experiment with 10 units in each group to see if the ball bearings' mean life is longer than the specified life. In this context, we have $\alpha = 1.50$, $m_0 = 2000$ cycles, $a_1 = 0.5$, $r = 10$, $\beta = 0.25$, $r_1 = 1$, producer's risk $= 0.05$, and $r_2 = 4$. In addition, from Table 2, we have $g = 39$ and $c = 3$. This implies that 195 units ($n = g \times r$) must be drawn, with five units assigned to each of the 39 groups. If no more than three units fail in each of these groups before 1000 cycles, the mean life of the ball bearings will be statistically assured to be greater than the specified life. If a quality control engineer wants to test the hypothesis that ball bearings have a life span of 4000 cycles but a true average life of four times that, he or she can test 39 groups of five units each; if fewer than three units fail in 1000 cycles; as $a_1 = 0.5$ and the mean life length is in thousands of cycles, the engineer will infer that the life is more than 4000 cycles with 95 percent confidence. Therefore, the lot under investigation should be accepted.

Table 1. GASP for a = 1, b = 1, and α = 1.25, showing minimum g and c.

| | | $r = 5$ | | | | | | $r = 10$ | | | | | |
| | | $a_1 = 0.5$ | | | $a_1 = 1$ | | | $a_1 = 0.5$ | | | $a_1 = 1$ | | |
β	r_2	g	c	$P(a)$	g	c	$P(a)$	g	c	$P(a)$	g	c	$P(a)$
	2	-	-	-	-	-	-	-	-	-	-	-	-
0.25	4	41	3	0.9852	8	3	0.9678	3	3	0.9693	2	4	0.9620
	6	8	2	0.9809	3	2	0.9552	2	2	0.9552	1	3	0.9743
	8	8	2	0.9914	3	2	0.9786	2	2	0.9972	1	3	0.9897
	2	-	-	-	-	-	-	-	-	-	-	-	-
0.10	4	67	3	0.9760	83	4	0.9863	13	4	0.9840	6	5	0.9809
	6	13	2	0.9691	13	3	0.9869	5	3	0.9870	3	4	0.9878
	8	13	2	0.9861	4	2	0.9715	3	2	0.9681	2	3	0.9794
	2	-	-	-	-	-	-	-	-	-	-	-	-
0.05	4	88	3	0.9686	107	4	0.9823	17	4	0.9792	8	5	0.9746
	6	16	2	0.9621	16	3	0.9839	7	3	0.9818	4	4	0.9838
	8	16	1	0.9819	5	2	0.9645	3	2	0.9681	2	3	0.9794
	2	-	-	-	-	-	-	-	-	-	-	-	-
0.01	4	134	3	0.9526	165	4	0.9729	26	4	0.9683	11	5	0.9653
	6	134	3	0.9892	25	3	0.9749	10	3	0.9741	6	4	0.9758
	8	25	2	0.9734	25	3	0.9910	10	3	0.9907	3	3	0.9793

Remark: The cells with hyphens (-) indicate that a very large sample size is needed.

Table 2. GASP for a = 1, b = 1, and α = 1.50, showing minimum g and c.

| | | $r = 5$ | | | | | | $r = 10$ | | | | | |
| | | $a_1 = 0.5$ | | | $a_1 = 1$ | | | $a_1 = 0.5$ | | | $a_1 = 1$ | | |
β	r_2	g	c	$P(a)$	g	c	$P(a)$	g	c	$P(a)$	g	c	$P(a)$
	2	-	-	-	-	-	-	-	-	-	-	-	-
0.25	4	39	3	0.9803	9	3	0.9553	3	3	0.9853	5	5	0.9786
	6	8	2	0.9747	9	3	0.9877	3	3	0.9888	1	3	0.9668
	8	8	2	0.9886	9	3	0.9726	2	2	0.9716	1	3	0.9860
	2	-	-	-	-	-	-	-	-	-	-	-	-
0.10	4	65	3	0.9674	109	4	0.9762	13	4	0.9763	7	5	0.9701
	6	12	2	0.9623	15	3	0.9797	5	3	0.9814	4	4	0.9773
	8	12	2	0.9826	5	3	0.9547	3	2	0.9576	2	3	0.9721
	2	-	-	-	-	-	-	-	-	-	-	-	-
0.05	4	85	3	0.9575	142	4	0.9691	17	4	0.9691	9	5	0.9617
	6	16	2	0.9501	20	3	0.9730	7	3	0.9741	5	4	0.9717
	8	16	2	0.9768	20	3	0.9898	3	2	0.9576	3	3	0.9585
	2	-	-	-	-	-	-	-	-	-	-	-	-
0.01	4	132	3	0.9876	218	4	0.9530	25	4	0.9549	10	5	0.9607
	6	125	3	0.9846	30	3	0.9597	10	3	0.9632	7	4	0.9606
	8	24	2	0.9654	30	3	0.9848	10	3	0.9862	5	3	0.9872

4.2. Example

We consider now a data set which consists of a sample of 50 observed values of breaking stress of carbon fibers given by [18]. The unit is Gba. The data set can be expanded as follows: {1.12,0.17,0.64,4.32,1.22,0.37,1.16,1.42,0.09,1.67,0.13,0.25,0.08,0.04,2.35,0.20,0.78,0.34, 1.02,0.17,1.76,2.39,0.50,1.35,3.36,0.45,0.90,2.92,6.53,1.62,7.46,3.19,2.49,1.40,7.49,0.57,0.14,0.63, 5.23,0.71,0.68,0.12,0.09,3.47,5.93,1.82,4.20,7.29,3.13,3.41}.

The maximum likelihood estimates with standard errors (in parentheses) of the four parameters of $MOKw-E$ for the data are $\hat{\lambda} = 0.2978$ (0.5117), $\hat{a} - 0.9356$ (0.2371), $\hat{b} = 1.2805$ (0.6596), and $\hat{\alpha} = 0.6361$ (0.5377). The 'maximum distance' between the data and the fitted $MOKw-E$ is 0.0681 with a p-value of 0.9743, according to the Kolmogorov–Smirnov (K–S) test. Figure 1 depicts the histogram of the data with the estimated pdf, the empirical cdf with the estimated cdf, the probability–probability (P–P) plot, and the quantile–quantile (Q–Q) plot. Figure 2 completes Figure 1 by considering the total time on test (TTT) plot to have some information regarding the underlying hazard rate function (hrf) and the estimated hazard rate function.

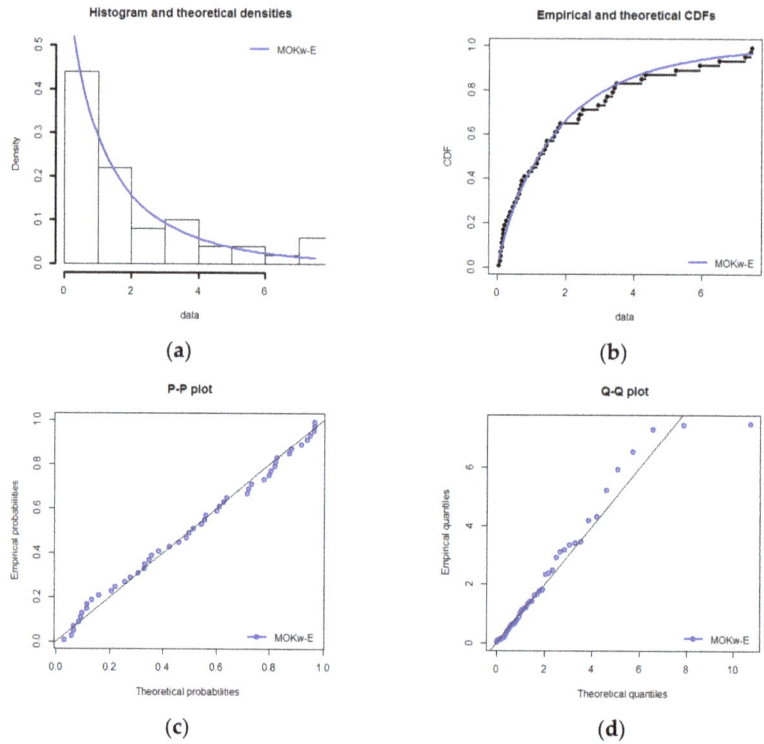

Figure 1. Examples of fits of *MOKw-E* for the carbon fibers dataset: (**a**) histogram fitted by the estimated pdf, (**b**) empirical cdf fitted by the estimated cdf, (**c**) P–P plot, and (**d**) Q–Q plot.

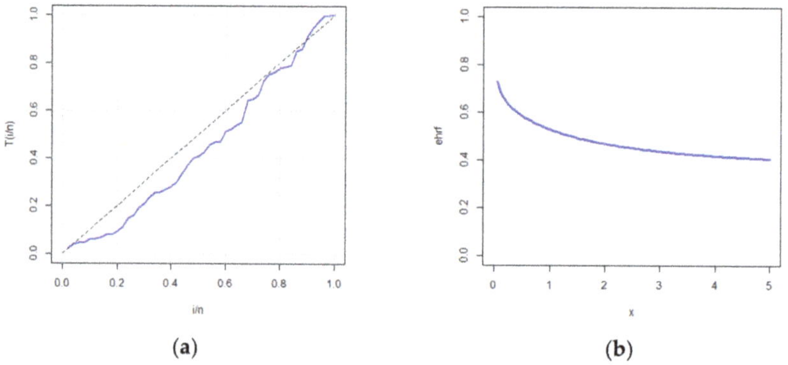

Figure 2. Plot of (**a**) TTT, and (**b**) the estimated hrf (ehrf) for the carbon fibers data set.

Figure 1 shows that *MOKw-E* has a good fit for the carbon fibers data set, whereas Figure 2 displays the total time on test (TTT) plot and the estimated hrf that the given data set has a decreasing hazard rate. Thus, *MOKw-E* provides a reasonable fit of the data. The plan parameters for the 50th percentile are also calculated using fitted parametric values and are shown in Table 3. The behavior of the plan parameters in Table 3 matches the values of the plan parameters in Tables 1 and 2.

Table 3. GASP for MLE $\hat{a} = 0.94$, $\hat{b} = 1.14$, and $\hat{\alpha} = 0.63$, showing minimum g and c.

| | | $r = 5$ | | | | | | $r = 10$ | | | | | |
| | | $a_1 = 0.5$ | | | $a_1 = 1$ | | | $a_1 = 0.5$ | | | $a_1 = 1$ | | |
β	r_2	g	c	$P(a)$	g	c	$P(a)$	g	c	$P(a)$	g	c	$P(a)$
	2	-	-	-	-	-	-	-	-	-	17	5	9575
0.25	4	33	2	0.9785	6	2	0.9715	6	2	0.9650	3	3	0.9813
	6	7	1	0.9576	6	2	0.9917	6	2	0.9896	2	2	0.9728
	8	7	1	0.9764	2	1	0.9717	2	1	0.9714	2	2	0.9882
	2	-	-	-	-	-	-	-	-	-	-	-	-
0.10	4	63	2	0.9646	10	2	0.9530	31	3	0.9873	4	3	0.9752
	6	63	2	0.9900	10	2	0.9861	9	2	0.9845	2	2	0.9728
	8	11	1	0.9632	3	1	0.9578	9	2	0.9935	2	2	0.9882
	2	-	-	-	-	-	-	-	-	-	-	-	-
0.05	4	81	2	0.9547	62	3	0.9872	41	5	0.9832	5	3	0.9691
	6	81	2	0.9872	13	2	0.9820	12	2	0.9793	3	2	0.9595
	8	11	1	0.9534	13	2	0.9926	12	2	0.9914	3	2	0.9824
	2	-	-	-	-	-	-	-	-	-	-	-	-
0.01	4	138	2	0.9635	94	3	0.9806	62	5	0.9747	8	3	0.9510
	6	125	2	0.9803	19	2	0.9738	18	2	0.9692	8	3	0.9896
	8	125	2	0.9920	19	2	0.9892	18	2	0.9871	4	2	0.9765

5. Conclusions

This study emphasizes a GASP assuming that the lifetime of the product follows *MOKw-E*. All the attention is focused on certain key plan parametric quantities. The number of categories, 'g', and the acceptance number, 'c', are calculated by balancing the risks of the manufacturer and the customer. For all the parametric combinations considered in this paper, it is observed in the proposed plan that as the percentile ratio increases, g decreases, and as the number of items in each group increases, the number of groups decreases, which is consistent with the results given in [7].

Author Contributions: Conceptualization, A.M.A., K.K., C.C. and F.J.; methodology, A.M.A., K.K., C.C. and F.J.; validation, A.M.A., K.K., C.C. and F.J.; formal analysis, A.M.A., K.K., C.C. and F.J.; investigation, A.M.A., K.K., C.C. and F.J.; writing—original draft preparation, A.M.A., K.K., C.C and F.J.; writing—review and editing, A.M.A., K.K., C.C. and F.J. All authors have read and agreed to the published version of the manuscript.

Funding: This research received no external funding.

Informed Consent Statement: Not applicable.

Data Availability Statement: The data are fully available in the article or the mentioned references.

Acknowledgments: The authors wish to express their gratitude to the Department of Statistics, King Abdulaziz University for providing the wherewithal coupled with the conducive research environment for carrying out the present study. Moreover, the authors would also like to extend thanks to all the authors whose work has been cited herein for providing the requisite stimulus for research on the current topic.

Conflicts of Interest: There exists no conflict of interest amongst the authors regarding the contents and publication of this research article.

Appendix A

R code for the considered sampling plan.

```
g=sEquation(1,1000,1);c=c(0,1,2,3,4,5);lp2=double(length(g));
lp1=double(length(g));lp21=double(length(c)); lp22=double(length(c));
lp23=double(length(c));lp24=double(length(c));G1=double(length(c));
G2=double(length(c));G3=double(length(c));G4=double(length(c));
p=function(alp,p,ratio,a,b ){
   nu=log(1-(((1-(1-((alp*p)/(1-(1-alp)*p)))^(1/b)))^(1/a)))
   d=(1-((1-exp(nu*((ratio)^-1)*a1))^a))
   y=(((1-d)^b)/(1-((1-alp)*(1-d)^b)));return(y)
}
p2=round(p(2,0.5,c(2,4,6,8,10),1,1,0.5),4);
p2; p1=round(p(2,0.5,1,1,1,0.5),4);p1
for(i in 1:length(c)){
   for(j in 1:length(g)){
      lp2[j]=(pbinom(c[i],10,p2[2]))^j
      lp1[j]=(pbinom(c[i],10,p1))^j
}
   G1[i]=min(which(lp2>=0.95 & lp1<0.25));lp21[i]=round(lp2[G1[i]],4);
   G2[i]=min(which(lp2>=0.95 & lp1<0.10));lp22[i]=round(lp2[G2[i]],4);
   G3[i]=min(which(lp2>=0.95 & lp1<0.05));lp23[i]=round(lp2[G3[i]],4);
   G4[i]=min(which(lp2>=0.95 & lp1<0.01));lp24[i]=round(lp2[G4[i]],4);
}
cbind(c,G1,lp21,G2,lp22,G3,lp23,G4,lp24).
```


References

1. Lu, J.C.; Jeng, S.L.; Wang, K. A review of statistical methods for quality improvement and control in nanotechnology. *J. Qual. Technol.* **2009**, *41*, 148–164. [CrossRef]
2. Jun, C.H.; Balamurali, S.; Lee, S.H. Variables sampling plans for Weibull distributed lifetimes under sudden death testing. *IEEE Trans. Reliab.* **2006**, *55*, 53–58. [CrossRef]
3. Aslam, M.; Jun, C.H. A Group Acceptance Sampling Plans for Truncated Life Tests based on The Inverse Rayleigh And Log-Logistic Distributions. *Pak. J. Stat.* **2009**, *25*, 107–119.
4. Rao, G.S. A group acceptance sampling plans based on truncated life tests for Marshall-Olkin extended Lomax distribution. *Electron. J. Appl. Stat. Anal.* **2009**, *3*, 18–27.
5. Rao, G.S. A group acceptance sampling plans for lifetimes following a Marshall–Olkin extended Weibull distribution. *Stat. Appl.* **2010**, *8*, 135–144.
6. Aslam, M.; Kundu, D.; Jun, C.H.; Ahmad, M. Time truncated group acceptance sampling plans for generalized exponential distribution. *J. Test. Eval.* **2011**, *39*, 671–677.
7. Sivakumar, D.C.U.; Kanaparthi, R.; Rao, G.S.; Kalyani, K. The Odd generalized exponential log-logistic distribution group acceptance sampling plan. *Stat. Transit. New Ser.* **2019**, *20*, 103–116. [CrossRef]
8. Tahir, M.H.; Nadarajah, S. Parameter induction in continuous univariate distributions: Well-established G families. *An. Acad. Bras. Ciências* **2015**, *87*, 539–568. [CrossRef] [PubMed]
9. Alizadeh, M.; Tahir, M.H.; Cordeiro, G.M.; Zubair, M.; Hamedani, G.G. The Kumaraswamy Marshal-Olkin family of distributions. *J. Egypt. Math. Soc.* **2015**, *23*, 546–557. [CrossRef]
10. Marshall, A.W.; Olkin, I. A new method for adding a parameter to a family of distributions with application to the exponential and Weibull families. *Biometrika* **1997**, *84*, 641–652. [CrossRef]
11. Cordeiro, G.M.; de Castro, M. A new family of generalized distributions. *J. Stat. Comput. Simul.* **2011**, *81*, 883–898. [CrossRef]
12. Handique, L.; Chakraborty, S. The Marshall-Olkin-Kumaraswamy-G family of distributions. *arXiv* **2015**, arXiv:1509.08108. [CrossRef]
13. Ghitany, M.E.; Al-Hussaini, E.K.; Al-Jarallah, R.A. Marshall–Olkin extended Weibull distribution and its application to censored data. *J. Appl. Stat.* **2005**, *32*, 1025–1034. [CrossRef]
14. Ristić, M.M.; Kundu, D. Marshall-Olkin generalized exponential distribution. *Metron* **2015**, *73*, 317–333. [CrossRef]
15. Gupta, S.S. Life test sampling plans for normal and lognormal distributions. *Technometrics* **1962**, *4*, 151–175. [CrossRef]
16. Khan, K.; Alqarni, A. A group acceptance sampling plan using mean lifetime as a quality parameter for inverse Weibull distribution. *Adv. Appl. Stat.* **2020**, *649*, 237–249. [CrossRef]
17. Singh, S.; Tripathi, Y.M. Acceptance sampling plans for inverse Weibull distribution based on truncated life test. *Life Cycle Reliab. Saf. Eng.* **2017**, *6*, 169–178. [CrossRef]
18. Nichols, M.D.; Padgett, W.J. A bootstrap control chart for Weibull percentiles. *Qual. Reliab. Eng. Int.* **2006**, *22*, 141–151. [CrossRef]

processes

MDPI

Article

A TITO Control Strategy to Increase Productivity in Uncertain Exothermic Continuous Chemical Reactors

Ricardo Aguilar-López [1,*,†], **Juan Luis Mata-Machuca** [2,*,†] and **Valeria Godinez-Cantillo** [3,†]

1   CINVESTAV-IPN, Department of Biotechnology and Bioengineering, San Pedro Zacatenco, Mexico City 07360, Mexico
2   Department of Advanced Technologies, Instituto Politecnico Nacional, UPIITA, Mexico City 07340, Mexico
3   Instituto Politecnico Nacional, UPIITA, Mechatronics Engineering, Mexico City 07340, Mexico; vgodinezc1300@alumno.ipn.mx
*   Correspondence: raguilar@cinvestav.mx (R.A.-L.); jmatam@ipn.mx (J.L.M.-M.); Tel.: +52-5557-4738-00 (ext. 4307) (R.A.-L.); +52-5557 2960 00 (ext. 56880) (J.L.M.-M.)
†   These authors contributed equally to this work.

Abstract: In this manuscript, a two-input two-output (TITO) control strategy for an exothermic continuous chemical reactor is presented. The control tasks of the continuous chemical reactor are related to temperature regulation by a standard proportional-integral (PI) controller. The selected set point increases reactor productivity due to the temperature effect and prevents potential thermal runaway, and the temperature increases until it reaches isothermal operating conditions. Then, an optimal controller is activated to increase the mass reactor productivity. The optimal control strategy is based on a Euler-Lagrange framework, in which the corresponding Lagrangian is based on the model equations of the reactor, and the optimal controller is coupled with an uncertainty estimator to infer the unknown terms required by the proposed controller. As a benchmark, a continuous stirred tank reactor (CSTR) with a Van de Vusse chemical reaction is considered as an application case study. Notably, the proposed methodology is generally applicable to any continuous stirred tank reactor. The results of numerical experiments verify the satisfactory performance of the proposed control strategy.

Keywords: exothermic chemical reactors; temperature stabilization; optimal control; optimal reactor productivity

Citation: Aguilar-López, R.; Mata-Machuca, J.L.; Godinez-Cantillo, V. A TITO Control Strategy to Increase Productivity in Uncertain Exothermic Continuous Chemical Reactors. *Processes* **2021**, *9*, 873. https://doi.org/10.3390/pr9050873

Academic Editors: Jie Zhang, Siman Upreti and Meihong Wang

Received: 26 March 2021
Accepted: 12 May 2021
Published: 16 May 2021

1. Introduction

The chemical reactor is widely regarded as the most important equipment in the transformation industry since it houses chemical reactions that produce highly valuable compounds or, conversely, degrade toxic pollutants. Given its significance, the design, optimization, and control of chemical reactors have been an important focus for process engineers. However, highly nonlinear behavior related to steady-state multiplicity, input multiplicity, instabilities, and sustained oscillation, among other factors, presents difficulties in the operation of these instruments [1–3]. Therefore, research on these topics remains challenging for scientists and engineers. In particular, the control of continuous chemical reactors has been studied for several years. One of the main control strategies for these devices is related to temperature regulation; this is an important issue because an appropriate temperature control strategy leads to the adequate yield of chemical products. Furthermore, the operation of an isotherm reactor is generally conditionally stable, and temperature regulation is essential for process security to prevent hot points during reactor operations [4]. Conventional proportional–integral–derivative (PID) controllers are widely used because of their simple structures and tuning methods. Although these devices are simple, they fail to perform well for nonlinear processes. The fundamental requirement for these controllers is the use of a tuning algorithm to maintain the desired levels of outputs. In general, PID controllers are locally robust to parametric uncertainty since their design is independent of the phenomenological model of the system, but they are strongly dependent on tuning [5,6].

A feedback linearization controller for an exothermic reactor with a single reaction is introduced in [7,8], but the corresponding design is based on the state-space model of the reactors, which can be a significant drawback. However, this can be avoided by employing observer-based uncertainty estimators, and stabilization by state feedback has been proven effective for a well-defined domain [9,10]. Another control technique applied to temperature regulation is input–output linearizing control, which aims to reduce the original nonlinear control problem to a linear control problem under a differential geometry framework, but technical difficulties may arise in constructing the required diffeomorphisms [11,12]. Various solutions have been proposed to overcome the robustness problem. In particular, an integral action has been added to the controller obtained from input–output linearization to develop generic control. However, the main drawback of this approach is the over-parameterization of the controller [13,14]. Thus, several control strategies have been presented in the open literature, including linear and nonlinear back-stepping control, mass concentration regulation controllers, etc. [15–18].

The necessity of high process performance has led to efforts to improve the operation of reactors by optimizing operational trajectories, which include operation security, maximum productivity, and optimal cost, among others, leading to the tracking trajectory control problem, where optimal control designs have been successful. Model-based Hamiltonian techniques have been applied to nonlinear systems as optimal control approaches. In such cases, Hamiltonian equations must be developed, and then an adequate functional related to the objective function and corresponding restrictions must be applied to obtain an optimal controller for the required task. In this case, Pontryagin's principle is applied to determine the best possible control strategy under constraints for the state or input controls. Although Lagrangian-based optimal control approaches have been studied as well, they are mostly oriented to the control of mechanical systems [19].

Chemical processes frequently involve structured uncertainties and output disturbances. Some examples are variations in feed quality, uncertain initial and ambient conditions, and uncertainty in model parameters [20,21]. Failing to account for uncertainties in the optimal design may lead to a nonoptimal and potentially high-risk solution. Therefore, methodologies that compensate for uncertainty in chemical processes are essential for realizing a robust process [22–26].

To compensate for uncertainty, a probabilistic approach based on the polynomial chaos expansion (PCE) was proposed by [27]. PCE is used to calculate an approximation of the expected values and variances, such as the first two statistical moments of the objective function and nonlinear inequality constraints. A similar approach is presented in [28], which also used PCE to optimize biological networks under model parameter uncertainty. Additionally, in [29], a multi-model approach was applied to manage uncertainties for the optimization of semi-batch processes; in this method, multiple worst-case parameter scenarios are selected based on a heuristic approach, but the corresponding real-time implementation is complex due to the large algorithms and the control effort. Another approach that is increasing in popularity is the application of the unscented transformation (UT) for optimization under uncertainty. The UT is a method for the approximation of the statistical moments of nonlinearly transformed probability distributions in the context of nonlinear filtering [30]. For implementation purposes, the UT shares similarities with certain numerical integration techniques, such as cubature rules, which are well known in the numerical integration literature. Cubature rules are used to approximate multi-dimensional integrals and have been applied for optimization under uncertainty (see [31,32]).

However, the above procedures can lead to complex control structures, often with high computing requirements. Heuristic optimization is another control approach, which does not usually involve assumptions about the problem to be optimized. The heuristic approach can search large spaces of candidate solutions to identify optimal or near-optimal solutions at a reasonable computational cost, but it is unable to guarantee either feasibility or optimization, and, in many cases, it does not indicate how close a certain feasible solution is to the optimum [33–35]. A wide range of direct search methods have been developed from heuristic optimization, such as genetic algorithms, evolutionary programming, differential evolution, genetic programming, evolutionary strategy, particle swarm optimization, and artificial bee colonies [36,37]; however, a theoretical analysis of convergence is not available, which is a major shortcoming [16,38]. A variety of approaches considering uncertainties in the optimization of chemical processes have been reported in the open literature [39,40].

Therefore, this work proposes a method to increase reactor productivity by employing standard temperature regulation with an appropriate set point via a PI controller. The proposed method increases reactor productivity via thermal effects until reaching an isothermal operating condition. After that, an optimal control law is applied in the Euler–Lagrange framework, in which the corresponding Lagrangian is based on the state equations of the reactor. This allows the construction of a controller for optimal reactor productivity as an objective function. However, this controller is based on the kinetic reaction rate, which is unavailable. To overcome this drawback, a reduced-order uncertainty estimator is coupled with the proposed optimal controller. The proposed method results in the satisfactory operating performance of the reactor and increases the corresponding productivity of the desired chemical product.

2. Mathematical Model of a Continuous Stirred Tank Reactor(CSTR)

In general, let us consider an exothermic CSTR model. According to mass and energy conservation principles, the reactor model represents the following system:

$$\dot{x}_1 = ER(x_1, x_2) + (x_{1,in} - x_1)u_c \tag{1}$$

$$\dot{x}_2 = \theta(x_{2,in} - x_2) + \Delta H^T R(x_1, x_2) + \gamma(u_T - x_2) \tag{2}$$

where $x_1 \in \mathbb{R}^n$ denotes the concentration vector of the chemical species; $x_2 \in \mathbb{R}$ is the reactor temperature; $E \in \mathbb{R}^{n \times m}$ denotes the stoichiometric matrix; $R(x_1, x_2) := R_1(x_1)R_2(x_2) \in \mathbb{R}^n$ represents the vector of reaction rates, with $R_1(x_1) := diag(R_{1,i}(x_1) \in \mathbb{R}^{m \times m})$ and $R_2(x_2) \in \mathbb{R}^m$; $\Delta H \in \mathbb{R}^m$ defines the vector of reaction enthalpies; the positive defined real scalar $u_c := F/V$ denotes the quotient between the inlet flow F and reactor volume V; γ represents the heat transfer coefficient; and u_T is the coolant temperature (the manipulable control variable). The system (1)–(2) is a standard model that satisfies general conditions for the design of controllers for chemical reactors (see, for instance, the classical contribution by [41]).

As an application case study, let us consider the following chemical pathway from the Van de Vusse reaction [42]:

$$A \xrightarrow{k_1} B \xrightarrow{k_2} C \tag{3}$$

$$2A \xrightarrow{k_3} D \tag{4}$$

This chemical reaction pathway contains series and parallel reactions. The above chemical reactions are considered elemental chemical reactions, that is, those that occur in a single stage, where the order of the reaction coincides with the corresponding stoichiometric coefficient of the chemical reaction, as is assumed for the Van de Vusse kinetic model [43].

From the above, a continuous stirred tank reactor mathematical model can be constructed via the mass conservation principle under the following assumptions: the reacting mixture is perfectly mixed to avoid temperature and concentration gradients; the reacting mixture volume remains constant; the inlet mass flow is equal to the outlet mass flow; and physical properties such as mixture density, heat capacity, transport coefficients, inlet concentration, and inlet temperature to the reactor are constant. In addition, the cooling jacket temperature is assumed to be the same as the temperature control input, and the mass input flow to the reactor is considered to be the same as the other control input.

Therefore, the mass and energy balance equations are as follows:

- Mass balance equations:

$$\frac{dC_A}{dt} = \frac{u_c}{V}(C_{Ain} - C_A) - k_1 C_A - k_3 C_A^2 \tag{5}$$

$$\frac{dC_B}{dt} = -\frac{u_c}{V}C_B + k_1 C_A - k_2 C_B \tag{6}$$

$$\frac{dC_C}{dt} = -\frac{u_c}{V}C_C + k_2 C_B \tag{7}$$

$$\frac{dC_D}{dt} = -\frac{u_c}{V}C_D + \frac{1}{2}k_3 C_A^2 \tag{8}$$

- Energy balance equation:

$$\frac{dT}{dt} = \frac{u_c}{V}(T_{in} - T) + \frac{\Delta H_T}{\rho C_p} + \frac{UA}{V\rho C_p}(u_T - T) \tag{9}$$

In (5)–(9), k_i and ΔH_T are defined as

$$k_i(T) = k_{i0}exp\left(-\frac{E_i}{RT}\right), \text{ for } i = 1, 2, 3 \tag{10}$$

$$\Delta H_T = \Delta h_1 k_1 C_A + \Delta h_2 k_2 C_B + \Delta h_3 k_3 C_A^2 \tag{11}$$

The corresponding nomenclature and parameter values are included in Table 1 [44].

Table 1. Reactor parameters.

Description	Parameter	Value	Units
Heat transfer area	A	0.215	m^2
Temperature initial condition	T_0	387.05	K
Heat transfer coefficient	U	67.2	$kJ{\cdot}min^{-1}m^{-2}K^{-1}$
Heat capacity	C_p	3.01	$kJ{\cdot}kg^{-1}\,K^{-1}$
Nominal cooling jacket temperature	u_T	125	$^\circ C$
Reacting mixture density	ρ	934.2	$kg{\cdot}m^3$
Reactor volume	V	0.01	m^{-1}
Concentration initial conditions	C_{A0}	2.1	$kmol \cdot m^{-3}$
	C_{B0}	0	$kmol \cdot m^{-3}$
	C_{C0}	0	$kmol \cdot m^{-3}$
	C_{D0}	0	$kmol \cdot m^{-3}$
Pre-exponential kinetic factors	k_{10}	2.145×10^{10}	min^{-1}
	k_{20}	2.145×10^{10}	min^{-1}
	k_{30}	1.5072×10^8	min^{-1}
Activation energies	E_1/R	9758.3	K
	E_2/R	9758.3	K
	E_3/R	8560	K
Reaction heat	Δh_1	-4200	$kJ \cdot kmol^{-1}$
	Δh_2	11000	$kJ \cdot kmol^{-1}$
	Δh_3	41850	$kJ \cdot kmol^{-1}$

3. Control Strategy Design

3.1. PI Temperature Control

Most temperature controllers in industrial chemistry are classical PI controllers [41]. There are many reasons for this, including their proven operating performance and the fact that their operation is well understood by technicians, industrial operators, and maintenance personnel. Furthermore, the fact that a properly designed and well-tuned PID controller achieves control objectives makes it attractive for many applications. The general structure of PI controllers is defined by the following well-known equation:

$$u = k_p\big(x(t) - x_{sp}\big) + k_i \int_0^t \big(x(\sigma) - x_{sp}\big) d\sigma \tag{12}$$

Studies on the design and analysis of PI controllers for the stabilization and regulation of CSTR reactors are abundant in the literature, which includes numerous successful applications [41,45,46].

3.2. Optimal Control Design

The general framework of optimal control designs relies on the calculus of variations, which is involved in the trajectory optimization problem, where a functional $\mathcal{F}(\ell(\cdot)) : \mathbb{R}^q \to \mathbb{R}$ is a scalar, namely, the cost index, cost function, or performance index, which is minimizing or maximizing. The corresponding objective can be attained by solving the well-known Euler–Lagrange equation [47]:

$$\frac{\partial \ell}{\partial x_1} - \frac{d}{dx_2}\left(\frac{\partial \ell}{\partial \dot{x}_1}\right) = 0 \tag{13}$$

The term ℓ denotes the Lagrangian of the system under study.

In general, the cost functional $\mathcal{F}(\ell(\cdot))$ can be represented as follows:

$$\mathcal{F}(\ell(\cdot)) = \Omega\big(x_f, t_f\big) + \int_{t_0}^{t_f} \ell(t, x, u)dt \tag{14}$$

where $\Omega\big(x_f, t_f\big)$ is an algebraic term to be minimized (or maximized) in final conditions, subject to the following constraints:

- The state equation:

$$\dot{x} := \frac{dx}{dt} = f(x) + g(x)\,u \tag{15}$$

- The terminal constraints:

$$\Omega\big(x_f, t_f\big) = 0 \tag{16}$$

- The initial conditions:

$$x(t_0) = x_0 \tag{17}$$

In Equation (15), $x \in \mathbb{R}^n$ is the state vector; $f(x) : \mathbb{R}^n \to \mathbb{R}^n$ is a nonlinear function, where $f(x) \subset \Sigma \in C^\infty$ and Σ is a compact set; $g(x)$ is a smooth and invertible bounded function; and $u \in \mathbb{R}^m$, with $m \leq n$, is the exogenous control input.

Next, consider the following functional form:

$$P(x, \dot{x}, u) = \int_0^T \ell(x, \dot{x}, u)dt \tag{18}$$

The problem is to minimize the functional (18); therefore,

$$\delta P(x, \dot{x}, u) = \int_0^T \delta \ell(x, \dot{x}, u)dt \tag{19}$$

Here, the differential of the Lagrangian ℓ is

$$\delta \ell(x, \dot{x}, u) = \frac{\partial \ell}{\partial x}\delta x + \frac{\partial \ell}{\partial \dot{x}}\delta \dot{x} + \frac{\partial \ell}{\partial u}\delta u \tag{20}$$

Equation (20) is substituted into Equation (19):

$$\delta P(x, \dot{x}, u) = \int_0^T \left(\frac{\partial \ell}{\partial x} \delta x + \frac{\partial \ell}{\partial \dot{x}} \delta \dot{x} + \frac{\partial \ell}{\partial u} \delta u \right) dt \tag{21}$$

The first term in Equation (20) is represented by

$$\frac{\partial \ell}{\partial x} \delta x = \frac{\partial \ell}{\partial x} \delta x \frac{\delta u}{\delta u} = \frac{\partial \ell}{\partial x} \frac{\delta u}{u'} \tag{22}$$

where u' is defined as

$$\frac{\delta u}{\delta x} := u' \tag{23}$$

The second term in Equation (20) is integrated by parts:

$$\int_0^T \frac{\partial \ell}{\partial \dot{x}} \delta \dot{x} dt = \frac{\partial \ell}{\partial \dot{x}} \delta x \Big|_0^T - \int_0^T \frac{d}{dt} \left(\frac{\partial \ell}{\partial \dot{x}} \right) \delta x \, dt \tag{24}$$

Next, let us consider the following nonlinear control affine dynamic system representation of Equation (15):

$$\dot{x} := \frac{dx}{dt} = f(x) + a_1 u + p(\dot{x}, x, u) \tag{25}$$

From Equation (25),

$$p(\dot{x}, x, u) = \dot{x} - f(x) - a_1 u \tag{26}$$

Therefore, the corresponding functional and the Lagrangian for the system (18) are defined as

$$P(x, \dot{x}, u) \equiv p(x, \dot{x}, u) \tag{27}$$

and

$$\ell(x, \dot{x}, u) \equiv \dot{x} - f(x) - a_1 u \tag{28}$$

Note that a useful characteristic of the Lagrangian in Equation (28) is the dependence on the state Equation (25).

Therefore, from Equation (28),

$$\frac{\partial \ell}{\partial x} = -\frac{df(x)}{dx} = -f'(x) \tag{29}$$

$$\frac{\partial \ell}{\partial \dot{x}} = 1 \tag{30}$$

$$\frac{\partial \ell}{\partial u} = -a_1 \tag{31}$$

Then,

$$\int_0^T \frac{d}{dt} \left(\frac{\partial \ell}{\partial \dot{x}} \right) \delta \dot{x} dt = 0 \tag{32}$$

$$\frac{\delta P(x, \dot{x}, u)}{\delta u} = \int_0^T \left(-\frac{f'}{u'} - a_1 \right) dt \tag{33}$$

The optimum productivity $P(x, \dot{x}, u)$ is then determined by the following restriction:

$$\frac{\delta P(x, \dot{x}, u)}{\delta u} = 0 \tag{34}$$

From Equations (33) and (34), the following equality must hold:

$$-\frac{f'}{u'} - a_1 = 0 \tag{35}$$

or equivalently,

$$a_1 u' + f' = 0 \tag{36}$$

By solving Equation (36),

$$
\begin{aligned}
u &= -a_1^{-1} \int_{x_0}^{x} f'(z)dz + u_0 \tag{37} \\
&= -a_1^{-1} f(x) - u_0 \tag{38}
\end{aligned}
$$

In (37), $f(x_0) = 0$.

According to the above, the control law (38) is realizable only if the nonlinear term $f(x)$ is available. However, as is well known, nonlinear terms are challenging to model accurately and serve as a significant source of parametric and/or structured uncertainties [16,36–38].

Therefore, an alternative form of the controller (38) must be considered to avoid this drawback.

If there are non-ideal conditions and uncertain terms, then the control approach in Equation (38) cannot be realized. Thus, a strategy must be proposed to compensate for the uncertain terms and obtain a realizable control design. For this purpose, an uncertainty observer-based controller is considered.

3.3. Uncertainty Estimator Design

Let us consider the system (25) where the nonlinear term $f(x)$ is unknown, which is now viewed as a new unknown state variable. The extended dynamical system is defined as

$$
\begin{aligned}
\dot{x} &= f(x) + a_1 u + p(x, u) \tag{39} \\
\dot{f} &= g(x) \tag{40}
\end{aligned}
$$

coupled with a measured linear output, $y = Cx$.

In Equations (39) and (40), $f(x)$ and $g(x)$ are bounded unknown nonlinear terms.

Let us assume that the state variable can be measured online, i.e., $y = x$. This is a typical assumption in chemical systems, where the mass concentrations can be regarded as the output measurements. Therefore, the following reduced-order observer is applied to estimate the unknown term $f(x)$ [48,49].

$$\dot{\hat{f}} = -\lambda\left(\hat{f} - f\right) \tag{41}$$

In (41), $\hat{f}$ is the estimated value of f, and λ is the observer gain.

The corresponding output injection for this reduced-order observer is f, which is the unknown term to be estimated. To circumvent this issue, the nonlinear term f is obtained from Equation (39) as follows:

$$f(x) = \dot{x} - a_1 u_c - p(x, u_c) \tag{42}$$

By substituting (42) into Equation (41) and considering the above assumption of $y = x$, the following is obtained:

$$\dot{\hat{f}} = -\lambda\left(\hat{f} - \dot{y} - a_1 u_c - p(y, u_c)\right) \tag{43}$$

With a change in the variable,

$$\eta = \hat{f} - \lambda y \tag{44}$$

In consequence,

$$\dot{\eta} = \dot{\hat{f}} - \lambda \dot{y} \tag{45}$$

Finally, the reduced-order observer (41) can be expressed as

$$\dot{\eta} = -\lambda(\eta - a_1 u_c - p(y, u_c)) \tag{46}$$

Note that the uncertainty estimator (46) contains only known variables and parameters, where k is the observer gain.

From Equation (44), the uncertain term can be estimated by

$$\hat{f} = \eta + \lambda y \tag{47}$$

Finally, the designed optimal controller is

$$u_c = -a_1^{-1}\hat{f} - u_0 \tag{48}$$

An important characteristic of the Lagrangian in (28) is that its inclusion in the corresponding functional (Equation (27)) results in an analytic and explicit form of the corresponding controller, which avoids the high computational effort required to numerically calculate the controller system, which is a common issue in other optimization strategies.

Finally, the following closed-loop stability analysis is considered.

The closed-loop stability of the reactor is evaluated via zero-dynamic analysis. Let us consider the following representation of Equation (39):

$$\dot{x} = f(x) + g(x)u \rightarrow \begin{cases} \dot{x}_C = f_C(x) + g(x)u \\ \dot{x}_D = f_D(x) + g(x)u \end{cases} \tag{49}$$

where x_C denotes controller state variables, and x_D represents uncontrolled state variables. The dynamical system (49) is closed-loop stable for $t > 0$ if and only if (50)–(52) are fully satisfied.

$$x_C = x_C^{sp} \tag{50}$$
$$\dot{x}_C^{sp} = 0 \tag{51}$$
$$\dot{x}_D = f_D(x) + g_D(x)u^{sp} \leq 0 \tag{52}$$

where

$$u^{sp} = -g_C^{-1}(x)f_C(x) \tag{53}$$

This analysis is based on a Lyapunov framework, and the proof of this proposition is in [50].

4. Numerical Results and Discussion

The process in the application case study is described by the exothermic CSTR modeled by Equations (5)–(9), which were solved by employing the ordinary differential equation (ODE) 23s library from Matlab, v2020a, on a PC with an Intel i7 processor. The open-loop behavior of the reactor is based on the conditions in Table 1.

The proposed control strategy is as follows. The reactor operation is initiated together with the implementation of the temperature regulation via a standard PI controller. A temperature set point is selected to increase the chemical conversion of the corresponding compounds. The abovementioned process increases the reactor temperature, and the thermal effects increase the chemical conversion of reactant A. Because the reaction rate is related to temperature via the Arrhenius model, reactant A can react to generate chemical products B and D, in accordance with the kinetic pathway shown above. In fact, this presents an interesting selectivity problem that must be solved if an intermediate chemical product is desired. Once the temperature is stabilized, the reactor operates under isothermal conditions, at which point, the optimal controller is activated, and the selected control input is the mass input flow. This controller acts to optimize the mass productivity of chemical compound B.

Upon initiation of the proposed closed-loop operation, the temperature of the reactor is regulated via the temperature of the cooling jacket, which is considered the manipulated variable u_T, under a standard PI control structure. The controller gains are $k_p = 80$ min^{-1} and $k_i = 35$ min^{-1}, which are calculated by an identification process via a step disturbance in the temperature control input and the application of internal model control (IMC) tuning rules [51]. Figure 1 shows a generalized scheme of the closed-loop operation of an exothermic CSTR.

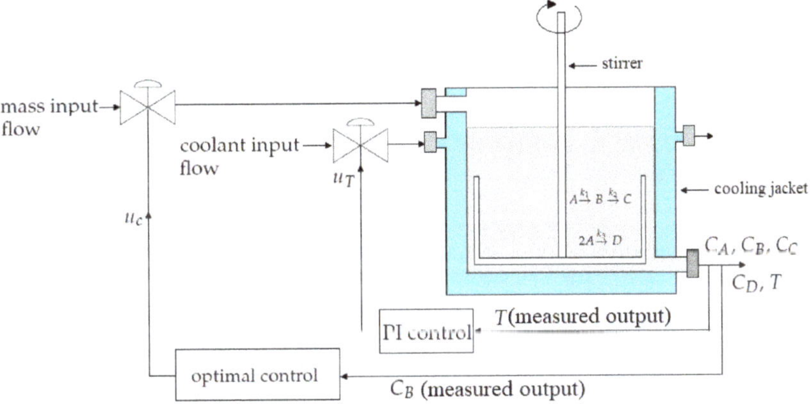

Figure 1. Exothermic CSTR.

As can be observed in Figure 2, the temperature response of the closed-loop reactor is increasingly close to the required set point ($T_{sp} = 185$ °C), maintaining a small offset of 2.3 °C. Furthermore, when the reactor temperature is in the steady state (isothermal operation), at $t = 0.1$ min, the optimal controller is activated. This is a significant disturbance in the thermal behavior of the reactor, but the temperature PI controller is able to resist it with an acceptable margin and maintains a temperature of 187 °C, again with a small offset of 2.5 °C from the required set point. Thus, the PI controller is able to maintain the isothermal operation of the reactor, as illustrated in Figure 3, which shows that the temperature control input has a brief temperature decrease to 50 °C but, almost immediately, it behaves as a first order-type response and reaches a steady state of 97 °C. The temperature of the reactor is maintained within a small range around the required set point. When the optimal controller has been activated, an increase in the mass input flow is required to increase the concentration of chemical reactant A. Then, the temperature controller increases the temperature of the cooling jacket to 116 °C, initiating the chemical pathway that forms product B, which is the desired product. This reaction pathway has a high activation energy (see Table 1), so sufficient energy must be maintained to continue generating product B. In this sense, the temperature controller facilitates the supply of the necessary energy to generate product B.

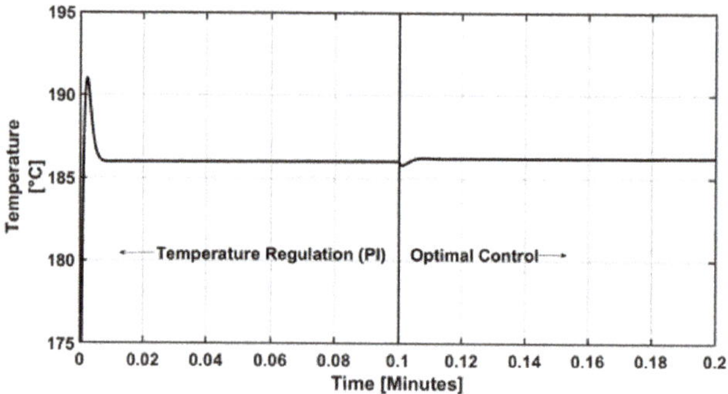

Figure 2. Closed-loop dynamic behavior of the reactor temperature.

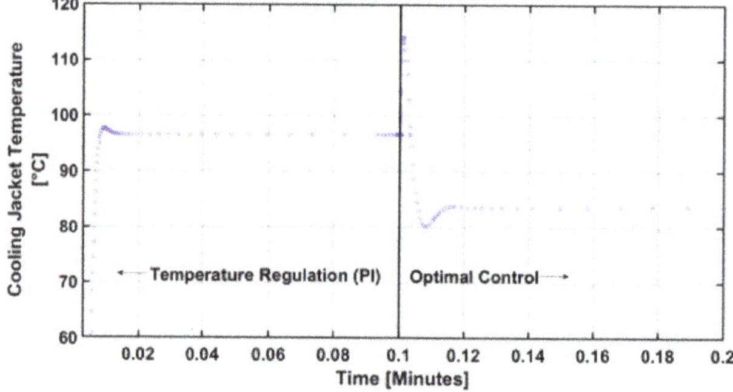

Figure 3. Control effort for temperature regulation.

As previously mentioned, the objective is to optimize the productivity of the generation of chemical product B, which is an intermediate compound in the reaction network. Compound B is simultaneously a product and a reactant, so the optimal control must solve this chemical selectivity problem. As observed in Figure 4, after the operation is initiated, the concentration of the main reactant A is consumed from the initial condition via temperature effects (recall that compound A reacts in a parallel reaction). Then, the concentration of compound B increases due to temperature effects and reaches a steady-state value of $C_B = 0.4$ mol/L, and when the optimal controller is activated, this concentration increases to $C_B = 0.48$ mol/L. However, as can be seen below, the optimal controller increases the outlet mass flow and significantly increases the productivity. Furthermore, the dynamic behavior of the uncontrolled concentration is stable. Therefore, the results of numerical experiments show that the zero or inner dynamic of the reactor is stable and, as a consequence, the reactor is stabilizable.

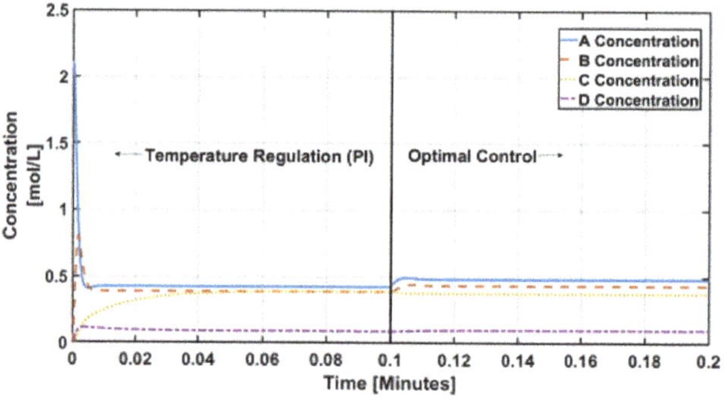

Figure 4. Closed-loop dynamic behavior of concentrations.

Figure 5 shows the control effort of the optimal controller. The corresponding mass control input is adjusted to a nominal value ($u_c = 72.5$ mol/L) when the process is operating under the temperature regulation regimen. When the optimal controller is activated at $t = 0.1$ min, the corresponding input flow drastically and rapidly increases to $u_c = 84$ mol/L. This is an important characteristic of the behavior of the control input. Because of this substantial flow increase, the productivity of the reactor increases, as shown below (recall that the reactor productivity P is defined as $P = u_c C_B$).

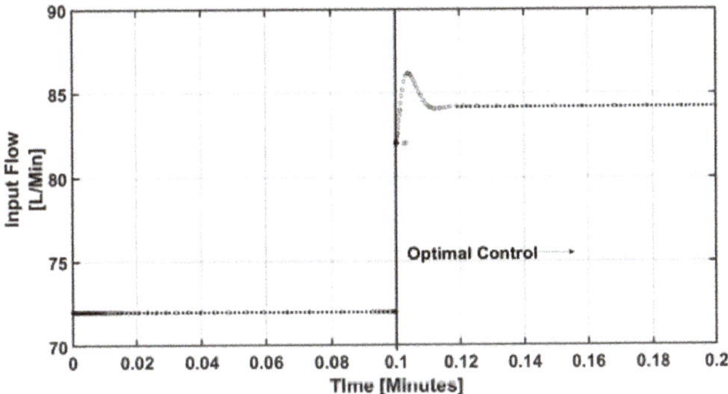

Figure 5. Control effort of the optimal controller.

As mentioned previously, the optimal controller requires online information about the reaction rate term of chemical compound B. To provide this information, an observer-based uncertainty estimator (see Equation (41)) is coupled with the control algorithm. Figure 6 shows the performance of the proposed observer, where the observer gain is $\lambda = 500$. The results of numerical experiments show that the controller has adequate performance, despite the significant disturbances related to the activation of the temperature and productivity controllers.

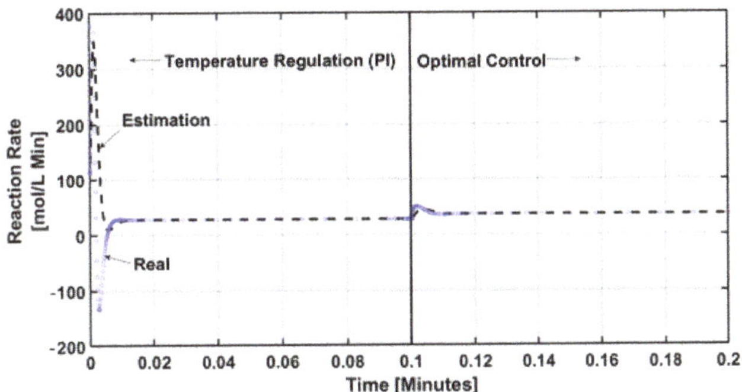

Figure 6. Dynamic performance of the uncertainty estimator.

Finally, as one of the main results, Figure 7 shows the productivity performance of compound B in the reactor. When the reactor operates in the open-loop regimen, the steady-state productivity is reached at $P = 20$ mol/L min, and when the proposed control strategy is applied, the productivity is increased to $P = 27$ mol/L min in the first step of control temperature operation. When the optimal controller is activated, the corresponding productivity of compound B is successfully increased to $P = 38$ mol/L min.

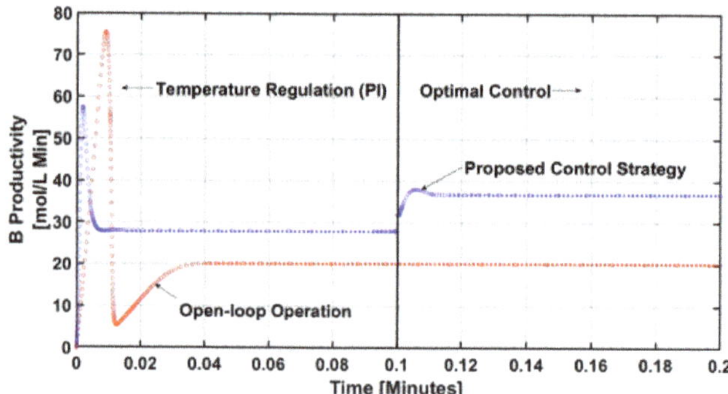

Figure 7. Open-loop and closed-loop dynamic behaviors of reactor productivity for compound B.

Figure 8 depicts the behavior of the relative stability values, i.e., the time derivatives of the uncontrolled concentrations, from the analysis of closed-loop stability in accordance with inequality (52). After the startup of the reactor under the temperature closed-loop regimen, compounds B and C are unstable because their relative values are positive, but their corresponding relative stability values asymptotically approach zero, becoming stable state variables. Compounds A and D show dynamic trajectories that are close to zero, indicating that they are acting as stable state variables. When the optimal controller is activated, the reactor dynamics undergo a considerable disturbance, and all relative stability values become positive, with the exception of compound D, whose relative value is always close to zero. As observed in the figure, all the zero dynamics of the reactor asymptotically reach a stable operating condition. These results agree with the dynamic behavior of the uncontrolled concentrations shown in Figure 4.

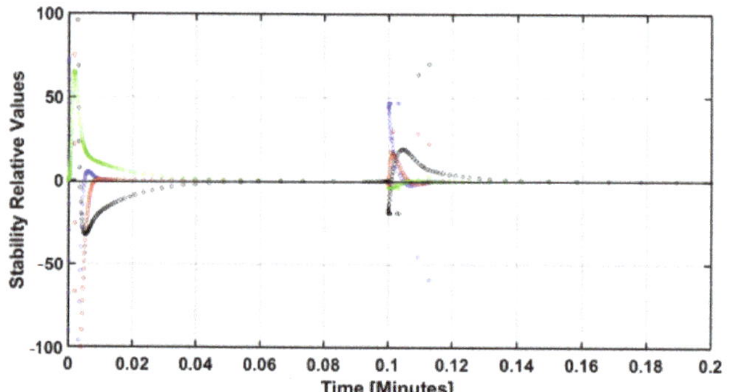

Figure 8. Relative stability values for uncontrolled concentrations (red—compound A; blue—compound B; green—compound C; black—compound D).

5. Conclusions

This work presents a two-input two-output (TITO) control strategy in which the control pairs are the reactor temperature/jacket temperature and productivity/input mass flow. At the first stage, the proposed closed-loop strategy aims to regulate the reactor temperature via a standard PI control law to increase the concentration of chemical compound B via a temperature increase until the reactor operation reaches isothermal conditions. Furthermore, an optimal controller designed by using the Euler–Lagrange approach is proposed. An important characteristic is that the Lagrangian is based on the model equation of the reactor, and when it is included in the corresponding functional, an analytic and explicit form of the corresponding controller can be obtained. This avoids the high computational effort required to numerically calculate the controller system, which is a common issue in other optimization strategies. The reactor productivity, which is the objective function, is optimized for compound *B*. The optimal controller design depends on online measurement of the reaction rate, which is unavailable. However, a reduced-order uncertainty observer is coupled with the controller to provide the required feedback term. The numerical results show the efficiency of the proposed TITO control methodology, and the temperature regulation regimen is confirmed to increase the productivity of the reactor by 40% in comparison with the open-loop operation. The increase in productivity after temperature regulation achieves optimal control is 32.15%, and the global productivity increase with the optimal control operation is 85.4% compared with the open-loop operation. To build on the theoretical results for the optimal controller design and the initial positive results of numerical experiments, future work will be oriented to real-time implementation to validate the performance of the proposed control strategy.

Author Contributions: Conceptualization, R.A.-L. and J.L.M.-M.; methodology, R.A.-L.; software, J.L.M.-M. and V.G.-C.; validation, R.A.-L., J.L.M.-M. and V.G.-C.; formal analysis, R.A.-L. and J.L.M.-M.; investigation, R.A.-L.; writing—original draft preparation, R.A.-L.; writing—review and editing, J.L.M.-M. and V.G.-C. All authors have read and agreed to the published version of the manuscript.

Funding: The APC was funded by the Comisión de Operación y Fomento de Actividades Académicas del Instituto Politécnico Nacional.

Institutional Review Board Statement: Not applicable.

Informed Consent Statement: Not applicable.

Data Availability Statement. Not applicable.

Conflicts of Interest: The authors declare no conflict of interest.

References

1. Roberge, D.; Noti, C.; Irle, E.; Eyholzer, M.; Rittiner, B.; Penn, G.; Schenkel, B. Control of Hazardous Processes in Flow: Synthesis of 2-Nitroethanol. *J. Flow Chem.* **2014**, *4*, 26–34. [CrossRef]
2. Chehade, G.; Dincer, I. Advanced Kinetic Modelling and Simulation of a New Small Modular Ammonia Converter. *Chem. Eng. Sci.* **2021**, *236*, 116512. [CrossRef]
3. Contreras, L.A.; Franco, H.A.; Alvarez, J. Saturated output-feedback nonlinear control of a 3-continuous exothermic reactor train. *IFAC-Pap. Online* **2018**, *51*, 425–430. [CrossRef]
4. Gavalas, G.R. Uniform Systems with Chemical Change. In *Nonlinear Differential Equations of Chemically Reacting Systems*; Springer Tracts in Natural Philosophy; Springer: Berlin/Heidelberg, Germany, 1968; Volume 17.
5. Ackermann, J.; Kaesbauer, D. Design of robust PID controllers. In Proceedings of the European Control Conference (ECC), Porto, Portugal, 4–7 September 2001; pp. 522–527.
6. Aguilar, R.; Poznyak, A.; Martínez-Guerra, R.; Maya-Yescas, R. Temperature control in catalytic cracking reactors via a robust PID controller. *J. Process Control* **2002**, *12*, 695–705. [CrossRef]
7. Oriolo, G.; De Luca, A.; Vendittelli, M. WMR control via dynamic feedback linearization: Design, implementation, and experimental validation. *IEEE Trans. Control Syst. Technol.* **2002**, *10*, 835–852. [CrossRef]
8. Barkhordari, M.; Jahed-Motlagh, M.R. Stabilization of a CSTR with two arbitrarily switching modes using modal state feedback linearization. *Chem. Eng. J.* **2009**, *155*, 838–843. [CrossRef]
9. Rubio, J.J. Robust feedback linearization for nonlinear processes control. *ISA Trans.* **2018**, *74*, 155–164. [CrossRef]
10. Zhang, Z.; Wu, Z.; Durand, H.; Albalawi, F.; Christofides, P.D. On integration of feedback control and safety systems: Analyzing two chemical process applications. *Chem. Eng. Res. Des.* **2018**, *132*, 616–626. [CrossRef]
11. Kravaris, C.; Kantor, J.C. Geometric methods for nonlinear process control. 2. Controller synthesis. *Ind. Eng. Chem. Res.* **1990**, *29*, 2310–2323. [CrossRef]
12. Alvarez-Gallegos, J. Nonlinear regulation of a Lorenz system by feedback linearization techniques. *Dyn. Control* **1994**, *4*, 277–298. [CrossRef]
13. Jana, A.K. Nonlinear State Estimation and Generic Model Control of a Continuous Stirred Tank Reactor. *Int. J. Chem. React. Eng.* **2007**, *5*, A42. [CrossRef]
14. Cott, B.J.; Macchietto, S. Temperature control of exothermic batch reactors using generic model control. *Ind. Eng. Chem. Res.* **1989**, *28*, 1177–1184. [CrossRef]
15. Chu, Y.; You, F. Model-based integration of control and operations: Overview, challenges, advances, and opportunities. *Comput. Chem. Eng.* **2015**, *83*, 2–20. [CrossRef]

16. Baruah, S.; Dewan, L. A comparative study of PID based temperature control of CSTR using Genetic Algorithm and Particle Swarm Optimization. In Proceedings of the International Conference on Emerging Trends in Computing and Communication Technologies (ICETCCT), Dehradun, India, 17–18 November 2017; pp. 1–6.

17. Khanduja, N.; Bhushan, B. CSTR Control Using IMC-PID, PSO-PID, and Hybrid BBO-FF-PID Controller. In *Applications of Artificial Intelligence Techniques in Engineering*; Malik, H., Srivastava, S., Sood, Y., Ahmad, A., Eds.; Advances in Intelligent Systems and Computing; Springer: Singapore, 2019; Volume 697.

18. Romero-Bustamante, J.A.; Moguel-Castañeda, J.G.; Puebla, H.; Hernandez-Martinez, E. Robust Cascade Control for Chemical Reactors: An Approach based on Modelling Error Compensation. *Int. J. Chem. React. Eng.* **2017**, *15*. [CrossRef]

19. Murugesan, R.; Solaimalai, J.; Chandran, K. Computer-Aided Controller Design for a Nonlinear Process Using a Lagrangian-Based State Transition Algorithm. *Circuits Syst. Signal Process* **2020**, *39*, 977–996. [CrossRef]

20. Zerari, N.; Chemachema, M. Robust adaptive neural network prescribed performance control for uncertain CSTR system with input nonlinearities and external disturbance. *Neural Comput. Applic.* **2020**, *32*, 10541–10554. [CrossRef]

21. Zhang, R.; Wu, S.; Cao, Z.; Lu, J.; Gao, F. A Systematic Min–Max Optimization Design of Constrained Model Predictive Tracking Control for Industrial Processes against Uncertainty. *IEEE Trans. Control Syst. Technol.* **2018**, *26*, 2157–2164. [CrossRef]

22. Maurer, J.; Freund, H. Efficient calculation of constraint back-offs for optimization under uncertainty: A case study on maleic anhydride synthesis. *Chem. Eng. Sci.* **2018**, *192*, 306–317.

23. Rodríguez-Pérez, B.E.; Flores-Tlacuahuac, A.; Ricardez-Sandoval, L.; Lozano, F.J. Optimal Water Quality Control of Sequencing Batch Reactors Under Uncertainty. *Ind. Eng. Chem. Res.* **2018**, *57*, 9571–9590. [CrossRef]

24. Jia, R.; You, F. Multi-stage economic model predictive control for a gold cyanidation leaching process under uncertainty. *AIChE J.* **2021**, *67*, e17043.

25. Aguilar, R.; Martínez-Guerra, R.; Poznyak, A. Reaction heat estimation in continuous chemical reactors using high gain observers. *Chem. Eng. J.* **2002**, *87*, 351–356. [CrossRef]

26. Aguilar-López, R. State estimation for nonlinear systems under model unobservable uncertainties: Application to continuous reactor. *Chem. Eng. J.* **2005**, *108*, 139–144. [CrossRef]

27. Harinath, E.; Foguth, L.C.; Paulson, J.A.; Braatz, R.D. Nonlinear model predictive control using polynomial optimization methods. In Proceedings of the IEEE American Control Conference (ACC), Boston, MA, USA, 6–8 July 2016; pp. 1–6.

28. Hashem, I.; Telen, D.; Nimmegeers, P.; Logist, F.; Van Impe, J. A novel algorithm for fast representation of a Pareto front with adaptive resolution: Application to multi-objective optimization of a chemical reactor. *Comput. Chem. Eng.* **2017**, *106*, 544–558. [CrossRef]

29. Puschke, J.; Djelassi, H.; Kleinekorte, J.; Hannemann-Tamás, R.; Mitsos, A. Robust dynamic optimization of batch processes under parametric uncertainty: Utilizing approaches from semi-infinite programs. *Comput. Chem. Eng.* **2018**, *116*, 253–267. [CrossRef]

30. Maurer, J.; Freund, H. Optimization under uncertainty in chemical engineering: Comparative evaluation of unscented transformation methods and cubature rules. *Chem. Eng. Sci.* **2018**, *183*, 329–345.

31. Bernardo, F.P.; Pistikopoulos, E.N.; Saraiva, P.M. Integration and computational issues in stochastic design and planning optimization problems. *Ind. Eng. Chem. Res.* **1999**, *38*, 3056–3068. [CrossRef]

32. Bernardo, F.P.; Pistikopoulos, E.N.; Saraiva, P.M. Quality costs and robustness criteria in chemical process design optimization. *Comput. Chem. Eng.* **2001**, *25*, 27–40. [CrossRef]

33. Zhang, C.; Seow, V.Y.; Chen, X.; Too, H.P. Multidimensional heuristic process for high-yield production of astaxanthin and fragrance molecules in Escherichia coli. *Nat. Commun.* **2018**, *9*, 1–12. [CrossRef]

34. Yuan, J.; Liu, C.; Zhang, X.; Xie, J.; Feng, E.; Yin, H.; Xiu, Z. Optimal control of a batch fermentation process with nonlinear time-delay and free terminal time and cost sensitivity constraint. *J. Process Control* **2016**, *44*, 41–52. [CrossRef]

35. Gurubel, K.J.; Osuna-Enciso, V.; Coronado-Mendoza, A.; Cuevas, E. Optimal control strategy based on neural model of nonlinear systems and evolutionary algorithms for renewable energy production as applied to biofuel generation. *J. Renew. Sustain. Energy* **2017**, *9*, 033101. [CrossRef]

36. Goud, H.; Swarnkar, P. Investigations on Metaheuristic Algorithm for Designing Adaptive PID Controller for Continuous Stirred Tank Reactor. *MAPAN* **2019**, *34*, 113–119 [CrossRef]

37. Zhang, Y.; Li, S.; Liao, L. Near-optimal control of nonlinear dynamical systems: A brief survey. *Annu. Rev. Control* **2019**, *47*, 71–80. [CrossRef]

38. Bailo, R.; Bongini, M.; Carrillo, J.A.; Kalise, D. Optimal consensus control of the Cucker-Smale model. *IFAC-Pap. OnLine* **2018**, *51*, 1–6. [CrossRef]

39. Lara-Cisneros, G.; Aguilar-López, R.; Femat, R. On the dynamic optimization of methane production in anaerobic digestion via extremum-seeking control approach. *Comput. Chem. Eng.* **2015**, *75*, 49–59. [CrossRef]

40. Verleysen, K.; Coppitters, D.; Parente, A.; De Paepe, W.; Contino, F. How can power-to-ammonia be robust? Optimization of an ammonia synthesis plant powered by a wind turbine considering operational uncertainties. *Fuel* **2020**, *266*, 117049. [CrossRef]

41. Alvarez-Ramirez, J.; Puebla, H. On classical PI control of chemical reactors. *Chem. Eng. Sci.* **2001**, *56*, 2111–2121. [CrossRef]

42. Amte, V.; Nistala, S.; Malik, R.; Mahajani, S. Attainable regions of reactive distillation—Part III. Complex reaction scheme: Van de Vusse reaction. *Chem. Eng. Sci.* **2011**, *66*, 2285–2297. [CrossRef]

43. Van de Vusse, J.G. Plug-flow type reactor versus tank reactor. *Chem. Eng. Sci.* **1964**, *19*, 994–997. [CrossRef]

44. Trierweiler, J.O. A Systematic Approach to Control Structure Design. Ph.D. Thesis, Universität Dortmund, Dortmund, Germany, 1997.
45. Rubi, V.A.; Deo, A.; Kumar, N. Temperature control of CSTR using PID Controller. *Int. J. Eng. Comput. Sci.* **2015**, *4*, 11902–11905.
46. Kumar, M.; Prasad, D.; Giri, B.S.; Singh, R.S. Temperature control of fermentation bioreactor for ethanol production using IMC-PID controller. *Biotechnol. Rep.* **2019**, *22*, e00319. [CrossRef]
47. Aguilar-López, R. Chaos Suppression via Euler-Lagrange Control Design for a Class of Chemical Reacting System. *Math. Probl. Eng.* **2018**, 3802801. [CrossRef]
48. Flores-Flores, J.P.; Martínez-Guerra, R. PI observer design for a class of nondifferentially flat systems. *Int. J. Appl. Math. Comput. Sci.* **2019**, *29*, 655–665. [CrossRef]
49. Martínez-Guerra, R.; Cruz-Victoria, J.C.; Gonzalez-Galan, R.; Aguilar-Lopez, R. A new reduced-order observer design for the synchronization of Lorenz systems. *Chaos Solitons Fractals* **2006**, *28*, 511–517 [CrossRef]
50. Aguilar-Lopez, R.; Maya-Yescas, R. Inverse dynamics: A problem on transient controllability for industrial plants. *Inverse Probl. Sci. Eng.* **2008**, *16*, 811–827. [CrossRef]
51. Babatunde, A.O.; Ray, W.H. *Process Dynamics Modeling Control*; Oxford University Press: Oxford, UK, 1994.

Article

Ripple Attenuation for Induction Motor Finite Control Set Model Predictive Torque Control Using Novel Fuzzy Adaptive Techniques

Zhihui Zhang [1], Hongyu Wei [1], Wei Zhang [1],* and Jianan Jiang [2]

[1] School of Mechanical and Material Engineering, Xi'an University, Xi'an 710065, China; zhihzhangxa@gmail.com (Z.Z.); hongyuwei67@gmail.com (H.W.)

[2] School of Automation, Northwestern Polytechnical University, Xi'an 710129, China; jianan@mail.nwpu.edu.cn

* Correspondence: zhwei_top@126.com; Tel.: +86-13379269302

Abstract: Finite control set model predictive torque control (FCS-MPTC) strategy has been widely used in induction motor (IM) control due to its fast response characteristic. Although the dynamics of the FCS-MPTC method are highly commended, its steady-state performance—ripple deserves attention in the meantime. To improve the steady-state performance of the IM drives, this paper proposes an improved FCS-MPTC strategy, based on a novel fuzzy adaptive speed controller and an adaptive weighting factor, tuning strategy to reduce the speed, torque and flux ripples caused by different factors. Firstly, a discrete predicting plant model (PPM) with a new flux observer is established, laying the ground for achieving an FCS-MPTC algorithm accurately. Secondly, after analyzing the essential factors in establishing a fuzzy adaptive PI controller, with high ripple suppression capacity, an improved three-dimensional controller is designed. Simultaneously, the implementation procedures of the fuzzy adaptive PI controller-based FCS-MPTC are presented. Considering that a weighting factor must be employed in the cost function of an FCS-MPTC method, system ripples increase if the value of the weighting factor is inappropriate. Then, on that basis, a novel fuzzy adaptive theory-based weighting factor tuning strategy is proposed, with the real-time torque and flux performance balanced. Finally, both simulation and hardware-in-loop (HIL) test are conducted on a 1.1 kW IM drive to verify the proposed ripple reduction algorithms.

Keywords: induction motor; finite control set; model predictive torque control; ripple attenuation; fuzzy adaptive theory

Citation: Zhang, Z.; Wei, H.; Zhang, W.; Jiang, J. Ripple Attenuation for Induction Motor Finite Control Set Model Predictive Torque Control Using Novel Fuzzy Adaptive Techniques. *Processes* **2021**, *9*, 710. https://doi.org/10.3390/pr9040710

Academic Editors: Jie Zhang and Meihong Wang

Received: 25 March 2021
Accepted: 12 April 2021
Published: 16 April 2021

1. Introduction

Induction motors (IM) that are endowed with the advantages of simple and robust structure, low cost, stable operations and a wide speed regulation range have been broadly adopted in industrial applications, such as electric vehicles, mining and textile, etc. [1–4]. To ensure or improve the machine control performance, many advanced control strategies have been developed so far, the most famous being field-oriented control (FOC), which is based on the coordinate transformation principle [5–7]. Generally, the traditional FOC strategy is achieved by using three proportional integrate (PI) controllers in the cascaded control loops to regulate the speed, flux and torque. Although the PI control techniques are now mature, several defects still exist [8,9]. For example, it is tedious to tune the internal parameters (at least six) of those PI controllers. Specifically, the complicated parameter-tuning schemes, not only increase the time cost in practical applications, as well as burden the optimization process for obtaining better control performance.

Due to the merits of the optimal control mechanism, easy parameter tuning process and quick response, model predictive control (MPC) has been broadly adopted in the area of IM control [10]. MPC controllers can be used to achieve different functions, such as model predictive torque control (MPTC) [11,12], model predictive speed control (MPSC) [13] and model predictive current control (MPCC) [14,15]. By considering the modulation modes of the MPC controllers, they can be further categorized into two groups, one of which requires a modulator to generate control signals. As the control voltage set is infinite for this method, it is commonly used for a continuous control set (CCS) MPC (CCS-MPC) [11,15]. The other strategy completely abandons the modulator. Instead, the cost function directly determines the optimal switching states to be applied next. As for this method, there are only finite control voltages, so it is known as a finite control set model predictive control (FCS-MPC) [12–14,16]. Comparatively speaking, the FCS-MPC strategy can reduce computational complexity significantly, benefiting from an offline lookup table. In addition, given that the FCS-MPC algorithm is simple to implement and the users do not need to grasp much in-depth professional knowledge, it is well-suited for industrial applications.

One of the main aims of this paper is to develop an FCS-MPTC controller to replace the torque and flux PI regulators in the traditional FOC scheme for the IM drives. However, when an FCS-MPTC strategy is achieved, in addition to its dynamics, the steady-state performance is one of the top concerns as well. In particular, the system ripples should be low. Whereas, there are many reasons that can cause high speed, torque or flux ripples in the practical applications, and these can be grouped into internal and external factors [17–25]. The internal factors mainly include; (1) the values of the parameters (e.g., stator inductance and resistance, etc.) used for predictions are not consistent with the real ones [17]; (2) weighting factor is not appropriate [18–20]; (3) control period is long in some applications [21]; (4) calculation delay reduces prediction accuracy [22]; and (5) the number of voltage candidates in the control set is not enough. Some of the external factors are summarized as follows. First, external disturbances, such as the electromagnetic interference bring about high-frequency noises [23]. Secondly, the speed and current sensors used for signal measurement are not accurate [24], or worse, they can malfunction. Third, the hardware system, which generally includes a power supply, filters and other circuits is not stable [25]. Usually, if the system inertia (including machine rotor inertia and load inertia) is large, the ripples that arise as a result of these factors remain minor. However, for the low-inertia systems, the ripples are large. There are so many factors that are prone to aggravating the system ripples in the stable states for the low-inertia system. Therefore, scholars have paid significant attention to these factors and have developed various ripple minimization methods to improve the high steady-state performance [26–28]. The previous studies show that the solutions to any of the aforementioned factors are capable of reducing the system ripples effectively, but few of them focus on several factors simultaneously. For example, the authors in [26] propose an improved dual voltage vector-based FCS-MPTC method to reduce the torque ripple of the system, which takes into account only the fifth-order harmonics. Reference [27] develops a weighting factor optimization strategy based on non-dominated sorting genetic algorithm II (NSGA-II) to minimize the ripples caused by the inappropriate weighting factor, and the authors in [28] propose an FCS-MPTC strategy without using weighting factors. In [29], a perturbation observer is designed to resolve the external disturbance problem separately.

This paper proposes novel ripple attenuation methods based on a fuzzy adaptive theory-based speed (PI) controller and fuzzy adaptive weighting factor tuning strategy. This is in order to simultaneously reduce the ripples caused by the inaccurate external sensors and inappropriate weighting factor for the IM FCS-MPTC working over the high-speed range. The fuzzy control theories are employed in this paper as the fuzzy logic has the advantage of providing a solution to the machine control problem that can be cast in terms that human operators can understand [30–35], making it easier to mechanize tasks that are already successfully performed by humans. It deserves to be mentioned the adaptive weighting factor tuning method is able to solve the second internal problem above when it is executed alone. While when the adaptive PI controller and the adaptive weighting factor are implemented together, the second external problem is expected to be tackled. In terms of the fuzzy adaptive weighting factor tuning strategy, the importance degree of the torque and flux are balanced according to their real-time values, ensuring that the flux and torque ripples do not exceed the upper limit. The reason why it contributes in reducing the system ripples caused by the external sensors is that if the weighting factor is not proper, the ripples become larger when the sensors are not accurate, and hence, the adaptive weighting factor can prevent the ripples from growing further in this case. The rationale behind the fuzzy adaptive speed controller is that when the speed ripples grow, a set of PI parameters that can narrow the bandwidth of the system would be employed. Then, the high-frequency disturbances arising from the inaccurate sensors are excluded, thereby ensuring that the magnitude of the ripples is limited. Compared to the existing ripple optimization methods in [17–29], the proposed ripple attenuation strategies are totally different, and the novelties are reflected in the following aspects:

(1) The novel idea that using a fuzzy adaptive PI controller to reduce the system ripples (mainly caused by the external sensors) of the FCS-MPTC method is studied. Traditionally, the fuzzy adaptive speed controller is adopted to improve the dynamics of the system, and there are few studies using it to improve the steady-state performance. On this basis, this study is comparatively pretty novel.

(2) Unlike the traditional two-dimensional fuzzy adaptive PI controller, the proposed one aim to improve the steady-state performance of the system. To achieve this goal, firstly, the input scaling factor of the fuzzy controller is optimized, relying on the pre-setting speed, enlarging the variation range of input and output. Secondly, the structure of the fuzzy controller is rebuilt. In detail, both the current and previous speed errors and the error between the current and the previous speed errors (CPSE-E) are treated as the inputs to build a three-dimensional fuzzy controller. This assists in evaluating the effectiveness of the proportional and integral factors designed in the previous period and ultimately provide guidelines for generating new control parameters. What needs to be mentioned is that although the fuzzy control strategies that are achieved by using the previous states or errors were studied previously in [33–35]. The proposed fuzzy controller is still pretty novel considering its structure and role in suppressing the steady-state ripples of an IM. This contributes to enriching the fuzzy control theories and prompting the relevant applications.

(3) Few studies incorporate the fuzzy adaptive theory into the weighting factor tuning process for an FCS-MPTC strategy. However, this paper proposes a new fuzzy adaptive weighting factor tuning strategy, whereby the inputs are torque and flux errors. By balancing the importance degree of the two variables, the torque and flux ripples are inclined to decrease.

The structure of the rest paper is as follows. Section 2 introduces a system model that is suitable for designing an FCS-MPTC algorithm. For the sake of high flux observation estimation accuracy, a new numerical-solution-based flux observer is introduced. In Section 3, the proposed fuzzy adaptive PI controller-based FCS-MPTC method is developed. Section 4 presents the new adaptive weighting factor tuning strategy. Section 5 discusses the simulation and hardware-in-loop (HIL) testing results of the proposed FCS-MPTC algorithms, and Section 6 outlines the conclusion.

2. Modelling for IM FCS-MPTC

A discrete predicting plant model (PPM) is the prerequisite for achieving an IM FCS-MPTC algorithm. In this section, the mathematical model of an IM is established. To obtain the rotor flux used for feedback regulation, a novel numerical solution-based current-type flux observer is presented.

The commonly-used IM models concerning electrical properties include flux model and current model [36]. Considering that the currents, rather than the flux, can usually be measured to construct different control strategies in real applications, this paper uses the state-space current model to describe the general behaviors of the system. In addition, in comparison with the stator field orientation-based model (dq-axis reference frame), because the IM model based on rotor field orientation (MT-axis reference frame) is totally decoupled, it is employed for analysis. The differential equations in *MT*-axis frame are as follows, where the iron saturation and eddy current are assumed to be negligible,

$$\frac{di_{sM}}{dt} = -\frac{R_s}{L_s}i_{sM} + \omega_e i_{sT} - \frac{L_m}{L_s}\frac{di_{rM}}{dt} + \frac{L_m}{L_s}\omega_e i_{rT} + \frac{u_{sM}}{L_s} \tag{1}$$

$$\frac{di_{sT}}{dt} = -\omega_e i_{sM} - \frac{R_s}{L_s}i_{sT} - \frac{L_m}{L_s}\omega_e i_{rM} - \frac{L_m}{L_s}\frac{di_{rT}}{dt} + \frac{u_{sT}}{L_s} \tag{2}$$

$$\frac{di_{rM}}{dt} = -\frac{L_m}{L_r}\frac{di_{sM}}{dt} + \frac{L_m}{L_r}\Delta\omega i_{sT} - \frac{R_r}{L_r}i_{rM} + \Delta\omega i_{rT} \tag{3}$$

$$\frac{di_{rT}}{dt} = -\frac{\Delta\omega L_m}{L_r}i_{sM} - \frac{L_m}{L_r}\frac{di_{sT}}{dt} - \Delta\omega i_{rM} - \frac{R_r}{L_r}i_{rT} \tag{4}$$

where i_{sM} and i_{sT} are stator currents. u_{sM} and u_{sT} are stator voltage. While, ω_e and $\Delta\omega$ are the synchronous speed, and slip speed, respectively; i_{rM} and i_{rT} are the rotor currents; R_s and L_s represent the stator resistance, and inductance, respectively; R_r and L_r are the rotor resistance, and inductance, respectively; and L_m is the mutual inductance between the stator and rotor windings.

In terms of FCS-MPTC, torque and flux (rotor flux in this paper) are the TCOs, but they cannot be detected directly using sensors. In this case, observers are the available solutions to the problem. To observe the torque of an IM, the electromagnetic torque calculation strategy can be used:

$$T_e = pL_m(i_{sT}i_{rM} - i_{sM}i_{rT}) = p\frac{L_m}{L_r}\psi_r i_{sT} \tag{5}$$

where T_e is the electromagnetic torque. ψ_r is the rotor flux, and p is the number of pole pairs. It can be seen that once the rotor flux is obtained, the torque can be calculated subsequently. Now, there exist two main flux observation methods, that is, the current-type observer [37] and the voltage-type observer [38]. Comparatively, the accuracy of the latter approach is relatively lower due to the internal integrators. Therefore, a current-type observer is adopted in this paper. As in [39], according to (1)–(5) and the flux descriptions, the rotor flux can be derived as,

$$\psi_r = \frac{L_m i_{sM}}{T_r s + 1} \tag{6}$$

where T_r is the electrical time constant of the rotor windings, and it equals $\frac{L_r}{R_r}$. s is a differential operator. Obviously, Equation (6) represents a typical delay element. Traditionally, the delay effect would be ignored because T_r is usually small, and a proportional element can be obtained:

$$\psi_r = L_m i_{sM}. \tag{7}$$

Although the flux estimation process by using (7) becomes simple, the accuracy is less accurate. This is prone to degrade the control performance. In this paper, for the purpose of calculating ψ_r precisely, a method based on numerical calculation is adopted. In detail, by transforming (6) to the differential equation, it can be obtained that:

$$T_r\frac{d\psi_r}{dt} + \psi_r - L_m i_{sM} = 0. \tag{8}$$

Then, to solve (8) and the rotor flux can be derived as,

$$\psi_r(t) = L_m i_{sM}(1 - e^{-\frac{t}{T_r}}) \tag{9}$$

where t is time.

So far, the IM model in the time domain has been established completely. Substitute (3) and (4) into (1) and (2), respectively and then, the Euler forward algorithm can be used to discretize them in a time step of T (sampling time) to calculate the future states at t_{k+1} as follows:

$$i_{sM}^{k+1} = (1 - C_1 L_r R_s)i_{sM}^k + C_1[(L_s L_r \omega_e^k - L_m^2 \Delta\omega^k)i_{sT}^k + (\omega_e^k - \Delta\omega^k)L_r L_m i_{rT}^k + L_m R_r i_{rM}^k + L_r u_{sM}^k] \quad (10)$$

$$i_{sT}^{k+1} = (1 - C_1 L_r R_s)i_{sT}^k + C_1[(L_m^2 \Delta\omega^k - L_s L_r \omega_e^k)i_{sM}^k - (\omega_e^k - \Delta\omega^k)L_r L_m i_{rM}^k + L_m R_r i_{rT}^k + L_r u_{sT}^k] \quad (11)$$

where the superscripts $k+1$ and k represent the future and current states, respectively. C_1 is a constant, and it satisfies that $C_1 = \frac{T}{L_s L_r - L_m^2}$. As for the rotor flux, in the control period between t_k and t_{k+1}, where $0 \le t \le T$, the detected current value i_{sM}^k at t_k and T should be adopted to predict the flux at t_{k+1}, that is:

$$\psi_r^{k+1} = L_m i_{sM}^k (1 - e^{-\frac{T}{T_r}}). \quad (12)$$

Then, the future torque can be calculated by:

$$T_e^{k+1} = p\frac{L_m}{L_r}\psi_r^{k+1}i_{sT}^{k+1}. \quad (13)$$

From (12) and (13), it can be seen that only i_{sM}^k and i_{sT}^{k+1} are needed for calculating flux and torque information, so it is not necessary to predict the M-axis current i_{sM}^{k+1}, which is another advantage of the new flux calculation method. Overall, the discrete PPM used for achieving an IM FCS-MPTC algorithm is (11)–(13).

3. Proposed Fuzzy Adaptive PI Controller-Based FCS-MPTC

This section introduces the traditional fuzzy adaptive PI controller that focuses on the system dynamic performance firstly, explaining the reasons why it contributes less to the ripple reduction in the stable state. Then, the improved fuzzy adaptive controller that aims to enhance the steady-state performance is developed. Finally, the implementation procedures of the fuzzy adaptive PI controller-based FCS-MPTC strategy are detailed.

3.1. Traditional Fuzzy Adaptive PI Controller

In order to improve the dynamic performance of the IM control system, a traditional fuzzy adaptive PI controller, which is shown in Figure 1 is usually used for speed regulation, where ω_r and ω_{ref} are the real and reference rotor speed. $\Delta\omega_r$ is the speed error, K_ω and $K_{d\omega}$ are the input scaling factors. In_1, In_2 and Out_p, Out_i are the inputs and outputs of the fuzzy inference engine, respectively. D_{e_kp} and D_{e_ki} are the outputs of the defuzzification part. K_p^* and K_i^* are the output scaling factors. Δk_p and Δk_i are the compensation values for the PI controller.

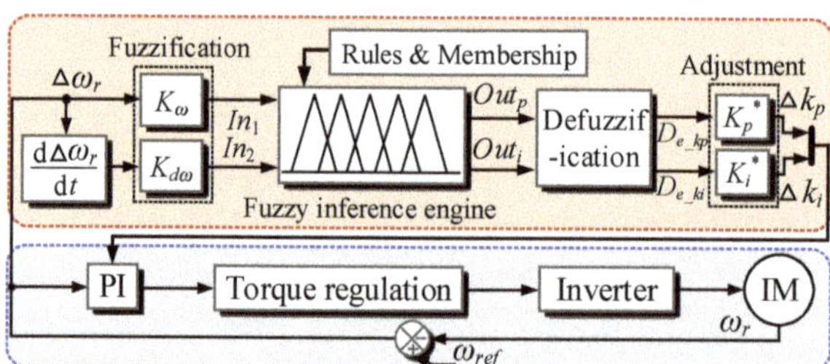

Figure 1. Block diagram of traditional fuzzy adaptive speed PI controller used for IM control.

As for the traditional controller, there are several features that need to be mentioned. First, the fuzzy controller contains five parts, namely, fuzzification, membership function, fuzzy control rules, defuzzification and adjustment. All of them are the fundamental components of a fuzzy controller. Secondly, in terms of the input scaling factors K_ω and

$K_{d\omega}$, they are of fixed magnitude. Usually, to ensure In_1 and In_2 to stay between -1 and 1 [40], K_ω is set as the reciprocal of the maximum speed value ω_{r_max}, that is,

$$K_\omega = \frac{1}{\omega_{r\_max}} \tag{14}$$

and $K_{d\omega}$ is the reciprocal of the maximum acceleration which can be calculated by using the rated torque T_{rated} and overload ratio k,

$$K_{d\omega} = \frac{J}{k \cdot T_{rated}} \tag{15}$$

where J is the rotor inertia. It can be noticed that $K_{d\omega}$ is determined by the largest torque that can be generated by the machine, which is inclined to occur in the starting stage. Thirdly, the derivative of the speed errors (DSE) is selected as one input of the fuzzy inference engine because it represents the change rate of speed (dynamics). The reason why the speed errors and the DSE are able to improve the dynamics can be summarized as follows; (1) when the DSE and speed errors are large, the machine starts to speed up or down. The fuzzy controller will generate large compensation values to ensure that the proportional and integral factors stand at a relatively large position so as to ensure a fast response. (2) When DSE is large but the speed errors are small, load variations occur when the speed nearly levels off. In this case, a large proportional gain is needed to quicken the process of returning to the normal state. It deserves to be mentioned that the traditional fuzzy adaptive controller has an obvious significance in improving the system dynamics regardless of parameter mismatch.

However, as for the increasing ripples caused by the parameter mismatch issue, the traditional fuzzy adaptive controller cannot suppress them effectively, and below are the reasons; (1) in the stable state, both the DSE and the speed errors are small. Namely, the values of $\Delta\omega_r$ and $d\Delta\omega_r/dt$ comply with the trend. After they pass through the fuzzification part, because K_ω and $K_{d\omega}$ are fixed and large, In_1 and In_2 are very close to zero, leading to the fact that the final compensation values generated by the fuzzy controller are too small to influence the initial parameters of the PI controller. Consequently, the ripples will maintain at a large level continuously. (2) $d\Delta\omega_r/dt$ only reflects the speed change rate but it cannot be used to evaluate the control effect (especially the shifting trend of the ripples) of the current PI parameters. For the sake of ripple reduction, these two aspects can be optimized.

3.2. Proposed Fuzzy Adaptive PI Controller Used for Ripple Reduction

To solve the above two issues, in order to make the fuzzy adaptive PI controller suitable for ripple reduction, an improved controller is proposed as in Figure 2, where $\Delta\omega_r'$ and $\Delta\omega_{cp}$ are the previous speed error and the CPSE-E, respectively. K_ω' and K_{cp} are the input scaling factors. In_3 is the input of the fuzzy inference engine and σ represents the maximum allowable speed ripple range. Compared to the traditional fuzzy controller in Figure 1, the novelties of the proposed controller include that firstly, the structure is rebuilt. Specifically, both the current and previous speed errors and the CPSE-E are selected as the inputs of the fuzzy controller. In addition, differing from the two-dimensional fuzzy controller in Figure 1, the improved fuzzy controller is three-dimensional. Secondly, the input scaling factors are modified. Without using fixed parameters, the input scaling factors K_ω, K_ω' and K_{cp} are mutable and determined by the rated speed and the maximum allowable ripple range. This can widen the range of In_1, In_2 and In_3 so the range of the output compensation values is expanded as well. Thirdly, considering that the structure and the range of the input values get changed, the fuzzy control rules and membership functions require to be modified correspondingly. In addition to these differences, the fundamental components of the improved fuzzy adaptive PI controller are the same to those in Figure 1. In this part, the design procedures of each component are discussed, which will reflect the novelties at length.

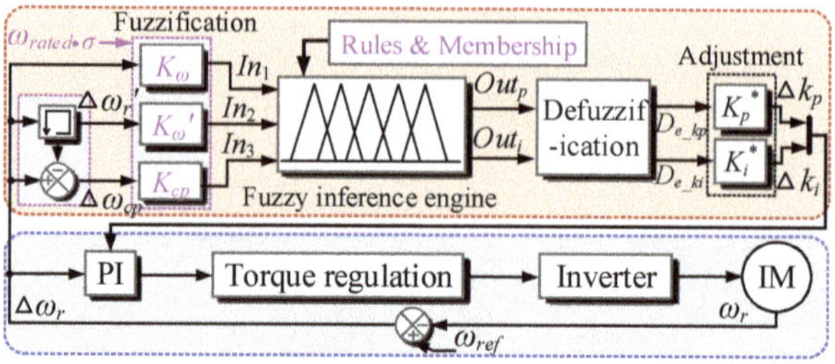

Figure 2. Block diagram of improved fuzzy adaptive speed PI controller used for IM control.

3.2.1. Explanations of Three-Dimensional Fuzzy Controller Concerning Novel Structure

In reference to the traditional fuzzy controller using the DSE, which reflects the system dynamics as one of the inputs, the variables that can reflect the steady-state performance should be selected as the inputs for the new controller considering that this paper focuses on ripple reduction. First of all, when the machine rotates stably, the magnitude of the speed errors directly represents the severity of the ripples. When it is large, the speed tracking capacity of the system is poor, and the bandwidth of the whole system is wide. In this case, a smaller proportional gain or larger integral gain of the PI controller is needed to narrow the system bandwidth so as to improve the steady-state performance. On the contrary, when the speed errors are small, a general conclusion that the current in-service PI controller parameters are suitable can be reached as long as the speed errors are able to level off. Secondly, as the CPSE-E is able to embody the changing trend of the speed ripples, with the support of the speed errors, it can be adopted to evaluate the steady-state performance. In detail, firstly, when the previous and current speed errors are positive, the real speed exceeds the reference value. At the moment, if the CPSE-E is positive, the speed ripples is inclined to increase. This demonstrates that the parameters of the PI controller in the last control period are not proper. It needs to be mentioned that only when the CPSE-E is negative can we reach a conclusion that the control parameters are decent. Secondly, when the previous speed error is positive or negative while the current one is negative or positive, the CPSE-E is undoubtedly negative or positive. In this case, the magnitude of the CPSE-E determines whether the controller parameters are excessively sensitive. Thirdly, when the previous and current speed errors are both negative, the real speed is lower than the pre-setting speed. Then, if the CPSE-E is positive, the speed ripples are decreasing, but if the CPSE-E is negative, the ripples grow continuously. Simultaneously, the magnitude of the CPSE-E determines the response sensitivity of the PI controller parameters.

According to be the above analysis, the previous speed errors, current speed errors and CPSE-E are closely related to the steady-state performance of the system. Hence, they can be employed as the inputs of the fuzzy controller, reducing the IM speed ripples during control. On this ground the proposed three-dimensional fuzzy controller is developed, which is brand-new indeed.

3.2.2. Design of Each Fuzzy controller Component

(a) Fuzzification

With the use of the fuzzification part, the actual input values can be converted to the fuzzy domain. Referring to the standardization control theory, the fuzzy domain is set as $[-1, 1]$. Therefore, the inputs of the proposed fuzzy controller can be described as:

$$In_1 = K_\omega \cdot \Delta\omega_r \tag{16}$$

$$In_2 = K_\omega' \cdot \Delta\omega_r' \tag{17}$$

$$In_3 = K_{cp} \cdot \Delta\omega_{cp}. \tag{18}$$

Now, a new issue is understanding how to design the input scaling factors. In order to enlarge the range of In_1, In_2 and In_3, which can improve the sensitivity of the fuzzy controller in the stable states, K_ω, K_ω' and K_{cp} are designed as,

$$K_\omega = K_\omega' = \frac{1}{\omega_{rated} \cdot \sigma\%}, \quad K_{cp} = \frac{1}{2\omega_{rated} \cdot \sigma\%} \tag{19}$$

where σ represents the maximum allowable speed ripple range (in percentage) in the high-speed situations. The rationale behind the scaling factor design method can be summarized as that, firstly, because the speed ripples need to meet the performance requirement that includes σ in engineering, effective measures must be taken to achieve this goal. In some particular scenarios (e.g., parameter mismatch conditions). while the ripples are inclined to exceed the allowable range, they can be suppressed by a marked parameter adjustment operation before the ripples reach the upper limit point (the goal of this paper), which depends on the fuzzy control rules and membership function. Hence, in real applications, as long as the membership function and fuzzy control rules are well-designed, In_1 and In_2 will range from -1 and 1 when K_ω and K_ω' satisfy (19). Secondly, when the speed ripples are controlled within the limited range, the magnitude of the maximum CPSE-E equals $2\omega_{ref}.\sigma\%$, so K_{cp} should be designed as the value in (19).

(b) Fuzzy control rules

The fuzzy control rules are the core of the fuzzy inference engine, which are used to build bridges between the inputs and the outputs. To formulate the control rules, linguistic expressions need to be defined to describe the fuzzy variables. In this paper, the linguist expressions include positive large (PL), positive medium (PM), positive small (PS), zero (ZO), negative small (NS), negative medium (NM) and negative large (NL). For the outputs, more accurate compensation values could be obtained by using the following subset: {PL, PM, PS, ZO, NS, NM, NL}. However, for the sake of inference burden reduction, only five linguist descriptions are employed for the inputs of the three-dimensional fuzzy controller, that is, {PL, PM, ZO, NM, NL}. Then, as for the proportional gain k_p and integral gain k_i of the PI controller, 85 rules that are extracted from the existing knowledge (the larger k_p/k_i is, the wider the system bandwidth is) and practical experiences can be designed as in series of Tables 1 and 2, respectively. '*' represents that the corresponding situations will never appear. The fuzzy control rules can be generally summarized as follows:

1.   When the signs of In_1 and In_2 are the same, if In_3 indicates that the speed ripples are increasing, a negative value of Δk_p and a positive value of Δk_i should be applied to reduce the bandwidth of the system. In this case, the system response speed drops, so that the speed ripples can be suppressed before it reaches the upper limit.
2.   When the signs of In_1 and In_2 are the same, if In_3 indicates that the speed ripples are decreasing, the in-service PI controller parameters are suitable. Both the values of Δk_p and Δk_i should be only adjusted slightly.
3.   When the signs of In_1 and In_2 are opposite, if the absolute value of In_3 is large, a negative value of Δk_p and a positive value of Δk_i are adopted to restrain the response speed of the system.
4.   When the signs of In_1 and In_2 are opposite, if the absolute value of In_3 is small, the relatively smaller compensation values are generated to adjust the parameters of the PI controller. In these cases, it is not necessary to markedly adjust the controller parameters because the in-service parameter values calculated in the previous period are appropriate.

Table 1. Fuzzy control rules for the proportional gain.

$In_3 = PL$

In_2 \ In_1	PL	PM	ZO	NM	NL
PL	NL	NL	NL	NL	NL
PM	*	NM	NM	NM	NM
ZO	*	*	NS	ZO	NS
NM	*	*	*	PM	PS
NL	*	*	*	*	PM

$In_3 = PM$

In_2 \ In_1	PL	PM	ZO	NM	NL
PL	NL	NL	NL	NL	NL
PM	*	NM	NM	NM	NM
ZO	*	*	NS	ZO	NS
NM	*	*	*	PM	PS
NL	*	*	*	*	PM

$In_3 = ZO$

In_2 \ In_1	PL	PM	ZO	NM	NL
PL	ZO	NS	NM	NL	NL
PM	PS	ZO	NM	NS	NL
ZO	PM	NS	ZO	NS	NM
NM	NS	NM	NM	PS	PS
NL	NM	NM	NM	NM	PM

$In_3 = NM$

In_2 \ In_1	PL	PM	ZO	NM	NL
PL	PM	*	*	*	*
PM	PS	PM	*	*	*
ZO	NS	ZO	NS	*	*
NM	NM	NM	NM	NM	*
NL	NL	NL	NL	NL	NL

$In_3 = NL$

In_2 \ In_1	PL	PM	ZO	NM	NL
PL	PM	*	*	*	*
PM	PS	PM	*	*	*
ZO	NS	ZO	NS	*	*
NM	NM	NM	NM	NM	*
NL	NL	NL	NL	NL	NL

Table 2. Fuzzy control rules for the integral gain.

$In_3 = PL$

In_2 \ In_1	PL	PM	ZO	NM	NL
PL	PL	PL	PL	PL	PL
PM	*	PM	PM	PM	PM
ZO	*	*	PS	ZO	PS
NM	*	*	*	NM	NS
NL	*	*	*	*	NM

$In_3 = PM$

In_2 \ In_1	PL	PM	ZO	NM	NL
PL	PL	PL	PL	PL	PL
PM	*	PM	PM	PM	PM
ZO	*	*	PS	ZO	PS
NM	*	*	*	NM	NS
NL	*	*	*	*	NM

$In_3 = ZO$

In_2 \ In_1	PL	PM	ZO	NM	NL
PL	PL	PM	PM	PL	PL
PM	PM	PM	PS	PS	PL
ZO	ZO	PS	ZO	PS	PM
NM	PM	PM	PS	NS	NS
NL	PM	PM	PM	PM	NM

$In_3 = NM$

In_2 \ In_1	PL	PM	ZO	NM	NL
PL	NM	*	*	*	*
PM	NS	NM	*	*	*
ZO	PS	ZO	PS	*	*
NM	PM	PM	PM	PM	*
NL	PL	PL	PL	PL	PL

$In_3 = NL$

In_2 \ In_1	PL	PM	ZO	NM	NL
PL	NM	*	*	*	*
PM	NS	NM	*	*	*
ZO	PS	ZO	PS	*	*
NM	PM	PM	PM	PM	*
NL	PL	PL	PL	PL	PL

the proposed FCS-MPTC strategy is shown in Figure 4. At the kth period, the proposed FCS-MPTC implementation procedures are:

(a)   Measurement: The phase currents i_a^k, i_c^k and rotor speed ω_r^k are measured by using current and speed sensors (common methods in engineering).

(b)   *abc/MT* transformation: The measured phase currents are transformed to the *MT*-axis currents i_{sM}^k, i_{sT}^k according to the estimated rotor flux position θ^{k+1} in the last period.

(c)   Observation: Use the currents i_{sM}^k, i_{sT}^k to calculate the slip speed $\Delta\omega^k$, synchronous speed ω_e^k and position θ^k.

(d)   Calculation of compensation values: The speed error $\Delta\omega_r^k$ is obtained by subtracting ω_r^k from the reference speed ω_{ref}^k, and then, the fuzzy controller is used to calculate the compensation values Δk_p^k and Δk_i^k.

(e)   Reference torque and flux generation: Use the PI controller of which parameters comply with the results in (22) to calculate the reference torque $T_e{}^*$, while the reference flux $\psi_r{}^*$ is given manually.

(f)   Prediction: Use the speed and current information to estimate the future current states i_{sT}^{k+1}, ψ_r^{k+1} and T_e^{k+1} for the candidate voltage vectors. As for a two-level inverter, a total of seven phase voltage vectors that are denoted as $\mathbf{u}_{000}$, $\mathbf{u}_{100}$, $\mathbf{u}_{110}$, $\mathbf{u}_{010}$, $\mathbf{u}_{011}$, $\mathbf{u}_{001}$, and $\mathbf{u}_{101}$ are among the alternatives,

$$\mathbf{u}_{s_a s_b s_c} = \begin{bmatrix} u_a \\ u_b \\ u_c \end{bmatrix} = \frac{U_{dc}}{3}\begin{bmatrix} 2 & -1 & -1 \\ -1 & 2 & -1 \\ 1 & 1 & 2 \end{bmatrix}\begin{bmatrix} s_a \\ s_b \\ s_c \end{bmatrix} \tag{23}$$

where $[s_a, s_b, s_c]^{\mathrm{T}}$ includes $[0, 0, 0]^{\mathrm{T}}$, $[1, 0, 0]^{\mathrm{T}}$, $[1, 1, 0]^{\mathrm{T}}$, $[0, 1, 0]^{\mathrm{T}}$, $[0, 1, 1]^{\mathrm{T}}$, $[0, 0, 1]^{\mathrm{T}}$, and $[1, 0, 1]^{\mathrm{T}}$, and they are the switching states. U_{dc} is the DC source voltage. $[u_a, u_b, u_c]^{\mathrm{T}}$ are the terminal phase voltages. By the use of *abc/MT* transformation, the control voltage sets used for prediction can be expressed as:

$$\begin{bmatrix} u_{sM}(k) \\ u_{sT}(k) \end{bmatrix} = \sqrt{\frac{2}{3}}\begin{bmatrix} \cos\theta^k & \dfrac{\sqrt{3}\sin\theta^k -\cos\theta^k}{2} & \dfrac{-\sqrt{3}\sin\theta^k -\cos\theta^k}{2} \\ -\sin\theta^k & \dfrac{\sin\theta^k +\sqrt{3}\cos\theta^k}{2} & \dfrac{\sin\theta^k -\sqrt{3}\cos\theta^k}{2} \end{bmatrix} \cdot \mathbf{u}_{s_a s_b s_c} \tag{24}$$

(g)   Evaluation: Substitute the seven the predicted values into a traditional one-step cost function (25) and determine the optimal voltage vector and the corresponding switching states [43].

$$J = \left| \psi_r{}^* - \psi_r^{k+1} \right| + \lambda \left| T_e{}^* - T_e^{k+1} \right| \tag{25}$$

where λ is weighting factor, and in this paper, it is tuned by using the proposed method in the next chapter. $T_e{}^*$ and $\Psi_r{}^*$ are the reference torque, and flux, respectively.

(h)   Actuation: Apply the optimum switching states to the drive system.

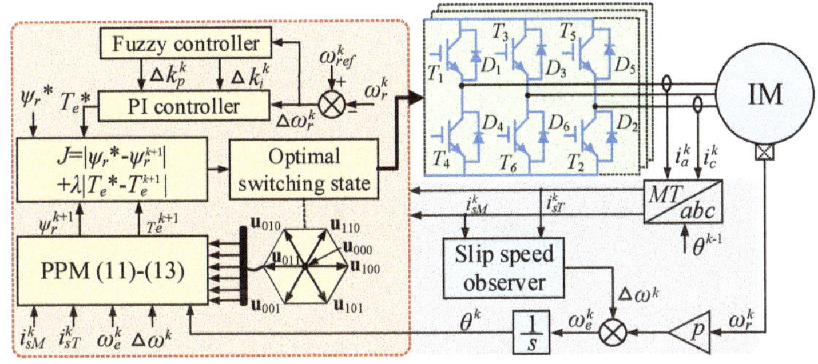

Figure 4. Block diagram of proposed fuzzy adaptive PI controller-based FCS-MPTC.

4. Novel Adaptive Weighting Factor Tuning Strategy

Usually, the weighting factor λ is set as the quotient of the rated flux and rated torque. By doing this, the magnitudes of the torque and flux terms in (25) can be deemed to stand at the same level, representing that the torque performance and the flux performance are of equal importance when using the cost function for evaluation. However, just as we illustrated in [44], this method might not be optimal in some working conditions so as to cause either large flux or torque ripples. To solve this problem, this part introduces a novel fuzzy adaptive weighting factor tuning method, which is shown in Figure 5, where ψ_r^k and T_e^k are the observed flux, and torque at t_k, respectively. ΔT_e and $\Delta \Psi_r$ are the torque and flux errors, respectively. K_T and K_Ψ are the output scaling factors. Out is the output of the fuzzy inference engine. D_{e_λ} is the output of the defuzzification part. $K_w{}^*$ is the output scaling factor. $\Delta\lambda$ and λ_0 are the compensation value, and the initial weighting factor, respectively.

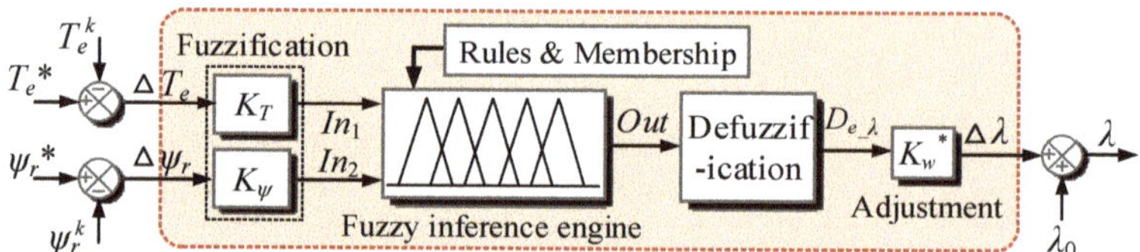

Figure 5. Block diagram of proposed fuzzy adaptive weighting factor tuning method.

It can be seen that, firstly, the input variables of the fuzzy controller are the torque and flux errors. The rationale behind this design method is that when ΔT_e is large, the torque performance needs to be focused on, thereby increasing the value of the weighting factor [44]. On the contrary, if $\Delta \Psi_r$ is large, the importance degree of flux should outweigh that of torque. In this case, λ requires to be decreased. Secondly, the fuzzy controller still contains five parts. Referring to the design procedures and the relevant theories in Chapter 3, each component of the fuzzy controller can be designed as follows.

(a) Fuzzification

To make In_1 and In_2 of the fuzzy controller range from -1 and 1, K_T and K_Ψ are designed as,

$$K_T = \frac{1}{T_{rated} \cdot \sigma_1\%}, \quad K_\psi = \frac{1}{\psi_{rated} \cdot \sigma_2\%} \tag{26}$$

where σ_1 and σ_2 represent the maximum allowable torque and flux ripple range (in percentage) in the high-torque/flux cases. The rationale behind the scaling factor design method is similar to that in Section 3.2.2.

(b) Fuzzy control rules

For the inputs and outputs, they can be either positive or negative, so the fuzzy subset is selected as {PL, PM, PS, ZO, NS, NM, NL}, and 49 fuzzy control rules are shown in Table 3. The fuzzy control rules can be summarized as follows:

Table 3. Fuzzy control rules for weighting factor.

		In_1						
		PL	**PM**	**PS**	**ZO**	**NS**	**NM**	**NL**
	PL	PL	PM	NL	NL	NL	PM	PL
	PM	PL	PM	NM	NM	NM	PM	PL
	PS	PL	PM	NS	NS	NS	PM	PL
In_2	ZO	PM	PS	ZO	ZO	ZO	PS	PM
	NS	PL	PM	NS	NS	NS	PM	PL
	NM	PL	PM	NM	NM	NM	PM	PL
	NL	PL	PM	NL	NL	NL	PM	PL

(1) When In_1 is large or medium (large torque error), a large or medium positive value of $\Delta\lambda$ is generated regardless of the magnitude of In_2. In this case, the priority is to suppress the torque ripples because the torque performance is the direct output of the machine.

(2) When In_1 is small, the output of the fuzzy controller is determined by In_2. If In_2 is large, a large negative compensation value is generated to stress the importance of flux, while if In_2 is small, the weighting factor does not need to be adjusted significantly.

(c) Membership functions

The membership functions for the inputs (In_1 and In_2) and output are depicted in Figure 6. Interestingly, the IDs of NL and NM, PM and PL domains in Figure 6a are weak. This draws clear lines between the large and medium-range for the flux and torque deviations. In this paper, by adopting these kinds of membership functions, when $|In_1| > 0.9$, the torque and flux deviation will be directly identified as the large level. This contributes to ensuring the torque and flux ripples are prevented from exceeding the upper limit as much as possible. In addition, the slopes of the input and output member functions are large when the corresponding values are small or zero, demonstrating that the fuzzy controller is pretty sensitive even if the system ripples are not large.

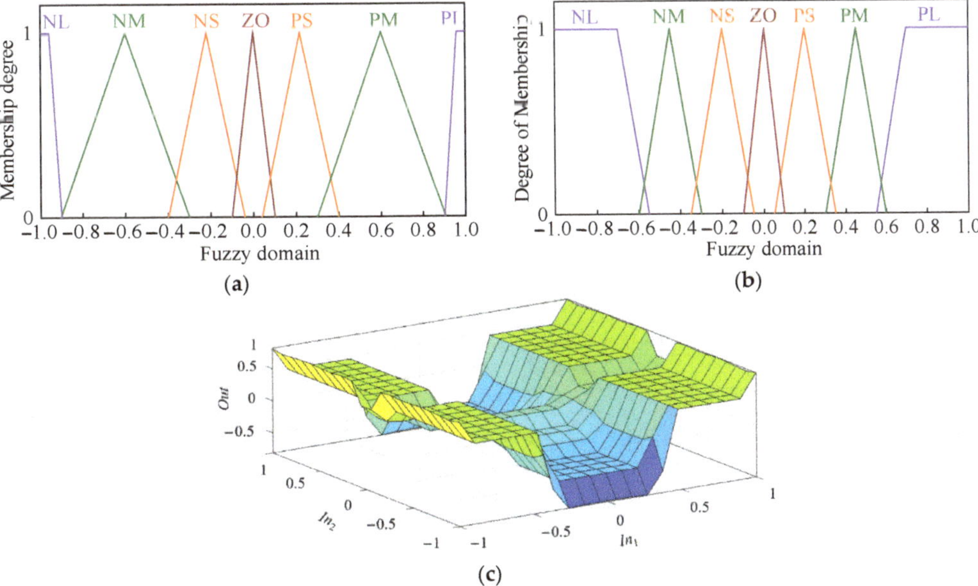

Figure 6. Fuzzy membership functions. (**a**) Membership functions for inputs; (**b**) membership functions for output; (**c**) relationship between inputs and output in the form of surface view.

(d) Defuzzification

By using the centroid defuzzification method, the output of the defuzzification part is,

$$D_{e\_\lambda} = \frac{\sum\limits_{i=1}^{m} z_i \cdot w(z_i)}{\sum\limits_{i=1}^{m} w(z_i)} \tag{27}$$

where $w(z_i)$ is the membership degree, z_i is the value corresponding to its degree of membership.

(e) Adjustment

The compensation value is calculated as follows:

$$\Delta\lambda = K_w{}^* \cdot D_{e\_\lambda} \tag{28}$$

Then, as is illustrated in Figure 5, the real-time weighting factor used for the cost function is,

$$\lambda = \lambda_0 + \Delta\lambda \tag{29}$$

where λ_0 equals the quotient of the rated flux Ψ_{r_rated} and rated torque T_{rated}, that is:

$$\lambda_0 = \frac{\Psi_{r\_rated}}{T_{rated}}. \tag{30}$$

5. Simulation and Hardwar-In-Loop Testing Results

In this part, simulation and HIL tests were conducted on a three-phase IM drive whose parameters are given in Table 4 to verify the ripple suppression effect of the proposed FCS-MPTC strategy in the stable states. The simulation was carried out in MATLAB/Simulink 2018b (see Figure 7a), and the HIL testing results were obtained from an RT Lab–based control board (see Figure 7b). It deserves to be mentioned again that the requirements concerning the allowable speed, torque and flux ripple ranges are only applicable to the high-speed/torque/flux situations because in the low-speed cases, the magnitudes of them (rather than percentage ripples) are usually specified. Hence, in this chapter, the no-load and low-speed situations are not considered.

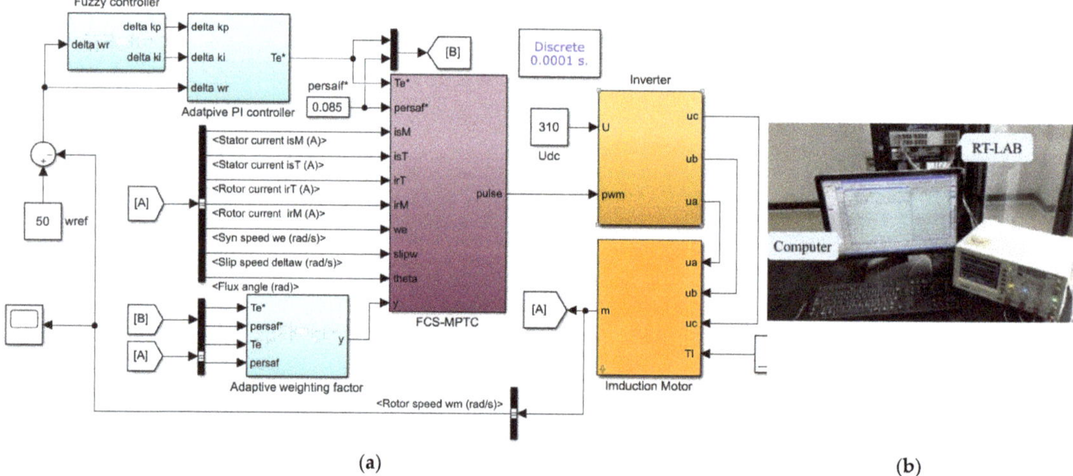

(a)   (b)

Figure 7. Setups for simulation and HIL test. (**a**) Simulation setup for the proposed FCS-MPTC method. (**b**) HIL testing setup for verifying the proposed FCS-MPTC method.

Table 4. Parameters of IM drive system.

Variable	Description	Value	Unit
U_{dc}	DC-link voltage	540	V
L_s	stator inductance	120	mH
R_s	stator resistance	0.065	Ω
L_r	mutual inductance	95	mH
R_r	rotor resistance	0.05	
T	sampling time	0.1	ms
p	number of pole pairs	2	-
ω_{rated}	rated speed	110	rad/s
T_{rated}	rated torque	10	Nm
Ψ_{r_rated}	rated flux	0.085	Wb
$\sigma_1\%$	allowable speed ripple	1%	-
$\sigma_2\%$	allowable torque ripple	25%	
$\sigma_3\%$	allowable flux ripple	20%	
k_{p0}/k_{i0}	initial parameters	0.5/0.75	-
$K_p*/K_i*/K_w*$	scaling factors	0.35/0.5/0.0064	-

5.1. Simulation Results

On the one hand, in order to verify that the proposed adaptive weighting factor tuning strategy was able to suppress the system ripples caused by the internal factors, a comparative study was carried out. In detail, the performance of the traditional FCS-MPTC with fixed PI parameters and fixed weighting factor (λ_0) were firstly given, and then it was compared with the performance of the FCS-MPTC with fixed PI parameters but adaptive weighting factor. The simulation process for the two algorithms was conducted as follows: Between 0 and 5.0 s, the IM motor operated at the rated speed stably under the rated load torque. Then, at 5.0 s, a sudden load of 5 Nm was removed, and then the machine works at 110 rad/s under 5 Nm until 10.0 s. Figure 8 shows the system performance of the traditional FCS-MPTC method. It can be seen that when the weighting factor equals λ_0, the speed ripples reached 0.4% of the rated speed (nearly 0.4 rad/s or 4 rpm) under the rated load situation and they are 0.3% of the rated speed under 5 Nm. As for the current, the ripples of i_{sT} over the high-load range are ±2 A and they are ±1.6 A in the medium-load case. In terms of the torque, when 10 Nm imposed on the rotor, the ripples are ±2.5 Nm (25%), satisfying the requirement in Table 4. However, when the load is was 5 Nm, although the magnitude of the ripples decreases to 2.2 Nm, the ripples in percentage are 44% because the torque base becomes smaller. This proves that σ_2 is only applicable to the high-torque situations. Whereas, it cannot be concluded that the FCS-MPTC algorithm is not suitable for IM control in the medium- and low-torque cases because other kinds of requirements (such as ripple magnitude) are usually specified for these situations in engineering. As far as the flux is concerned, it is interesting that the ripples were too small to be visible. According to (6), this occurs because of the machine rotor parameters (L_r/R_r is large). However, another phenomenon that needs to be mentioned is that when using an FCS-MPTC algorithm to control the motor, steady-state errors existed in the flux. In Figure 8, the steady-state errors are about 0.012 Wb and 0.005 Wb for the high-load, and medium-load situations, respectively. Before leaving Figure 8, it can be seen that the ripples of the control effort of the PI controller reach ±1.2 in the high-speed and high-load cases.

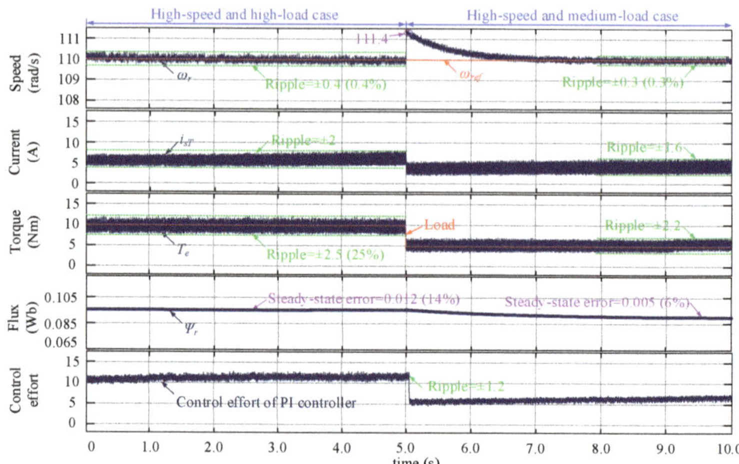

Figure 8. Performance of traditional FCS-MPTC with fixed PI parameters and fixed weighting factor.

Figure 9 illustrates the system performance of the proposed FCS-MPTC method with fixed PI parameters but adaptive weighting factor. In comparison with Figure 8, the speed, current and torque ripples are smaller regardless of load conditions. In detail, in the high-speed and high-load situations, they are ±0.3 rad/s (0.3%), ±1.8 A, and ±1.75 Nm (17.5%), respectively. It can be seen that the torque ripples decrease so significantly that the proposed weighting factor tuning strategy is able to make the system satisfy the requirement. Although the speed ripples just drop slightly, it still proves that the proposed strategy has higher speed reduction capacity compared to the traditional FCS-MPTC. Besides, in the medium-load conditions, the torque ripples the speed, current and torque ripples are just ±0.2 rad/s (0.2%), ±1.3 A, and ±1.6 Nm, respectively, which are also smaller than those in Figure 8. When it comes to the flux, the steady-state error was reduced by 0.002 Wb in the high-load situation, but it stays at the previous level when the load is 5 Nm. Compared to Figure 8, the ripples of the control effort of the PI controller in Figure 9 are much smaller (±0.9), which is one important reason why the proposed algorithm is able to suppress torque ripples.

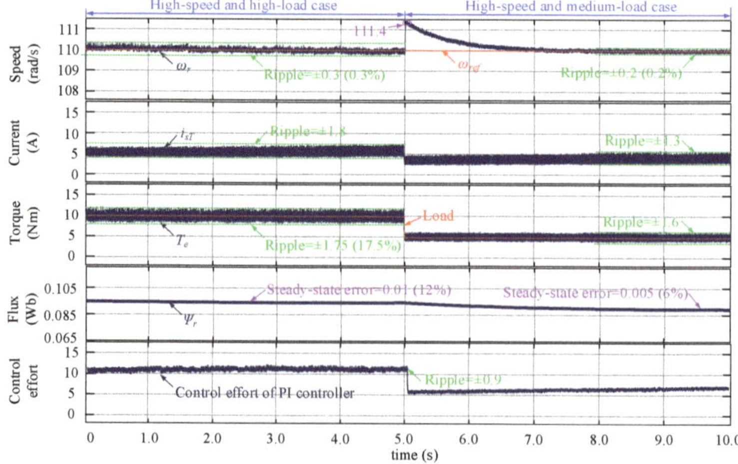

Figure 9. Performance of proposed FCS-MPTC with fixed PI parameters but adaptive weighting factor.

On the other hand, for the purpose of verifying that the proposed FCS-MPTC (with adaptive PI controller parameters and adaptive weighting factor) is able to attenuate the system ripples caused by the external factors, assuming that the speed sensor has random errors, the control performance of the traditional and the proposed FCS-MPTC strategies are compared. In the simulation, random errors were added to the measured speed to form the feedback speed, and the simulation process was conducted in the same way as before. Figures 10 and 11 demonstrate the control performance of the traditional FCS-MPTC, and the proposed FCS-MPTC, respectively. First of all, compared to the results in Figure 8, the speed, current and torque ripples in Figure 10 witness a significant increase when the external disturbances are applied to the system. In detail, in the high-load situations, the speed, current and torque ripples are ± 0.8 rad/s (0.8%), ± 3 A, and ± 4.8 Nm (48%), respectively, and in the medium-load condition, they are ± 0.9 rad/s (0.9%), ± 3.5 A, and ± 5 Nm (50%), respectively. Obviously, the system performance, especially the torque ripples, cannot meet the requirements in Table 4, and the speed ripples are very close to the upper limit. Secondly, it can be noticed that after using the proposed FCS-MPTC method with adaptive PI controller parameters and adaptive weighting factor, the steady-state performance of the system has been improved greatly. Specifically, the speed ripples in the high-torque and medium-torque conditions are ± 0.5 rad/s (0.5%) and ± 0.55 rad/s (0.55%), respectively. As for the current ripples, the magnitude of them decreases by 1.1 A under 10 Nm. Most importantly, the torque ripples are only 2 Nm (20%) at the rated working point, satisfying the requirement, though the ripples in percentage for medium-load condition are still high (44%) due to the lower torque base value. These prove that the proposed FCS-MPTC has great capacity in suppressing the system ripples. Finally, being similar to Figures 8 and 9, the flux ripples in Figure 11 are small, but there still exist steady-state errors.

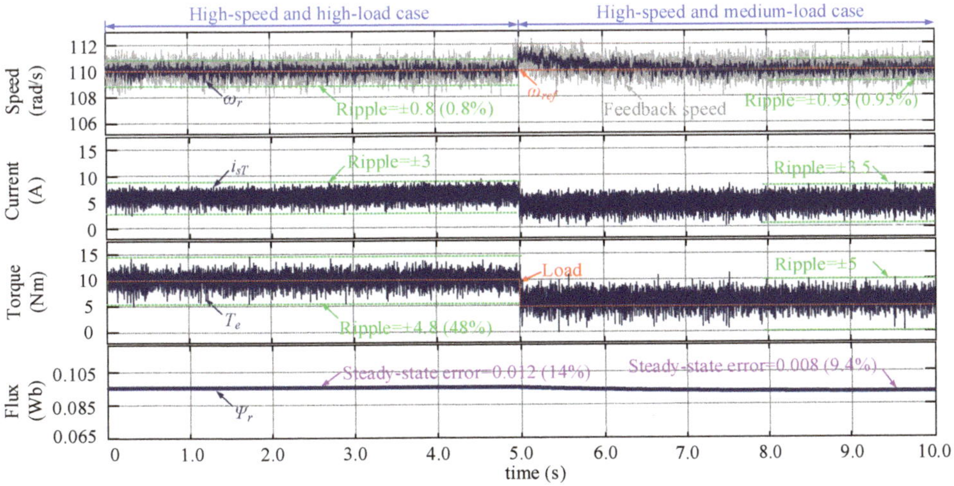

Figure 10. Performance of traditional FCS-MPTC with fixed PI parameters and fixed weighting factor.

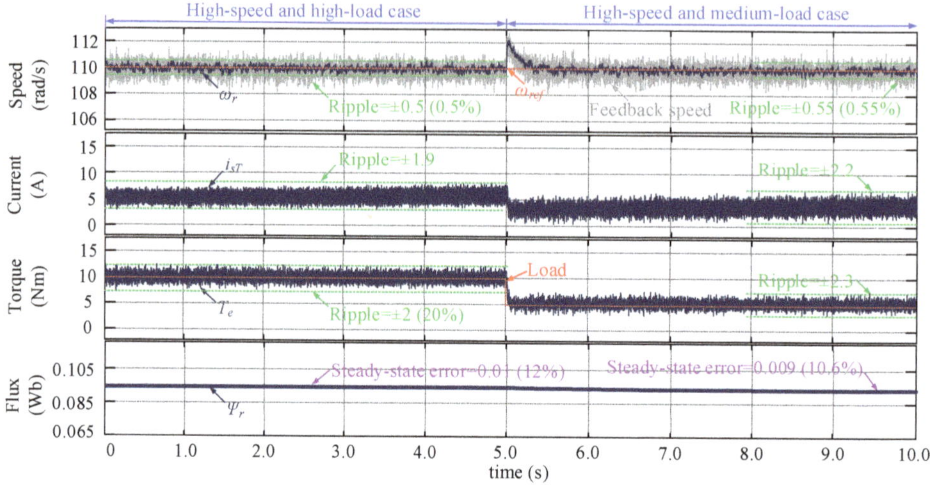

Figure 11. Performance of proposed FCS-MPTC with adaptive PI parameters and adaptive weighting factor.

5.2. HIL Testing Results

The HIL testing verification procedures are consistent with those in simulation. Figure 12 compares the system performance of the traditional FCS-MPTC and the proposed strategy with fixed PI parameters but adaptive weighting factor. As with the simulation results, intuitively, Figure 12b shows slightly better performance than Figure 12a. Figure 13b illustrates that the proposed FCS-MPTC with adaptive PI parameters and adaptive weighting factor has much greater capacity in suppressing the speed, current and torque ripples when the external disturbances exist. Overall, the steady-state control performance of the system can be improved by using the proposed FCS-MPTC strategy, and hence, it can be of great significance in industrial applications.

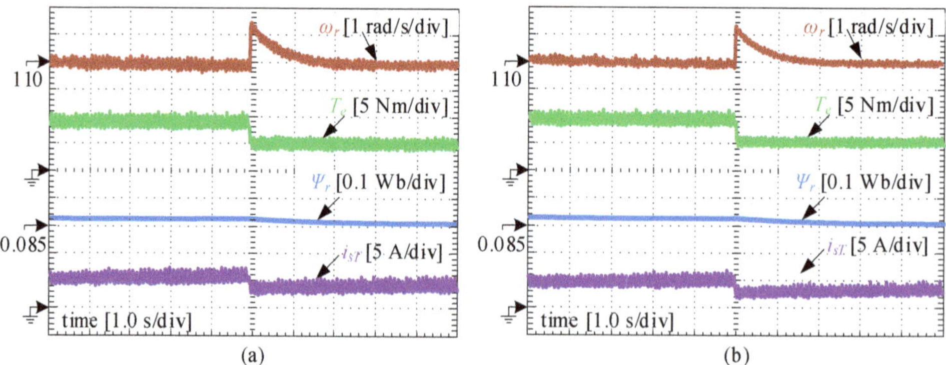

Figure 12. HIL testing results with different weighting factors. (**a**) Performance of traditional FCS-MPTC with fixed PI parameters and fixed weighting factor. (**b**) Performance of proposed FCS-MPTC with fixed PI parameters but adaptive weighting factor.

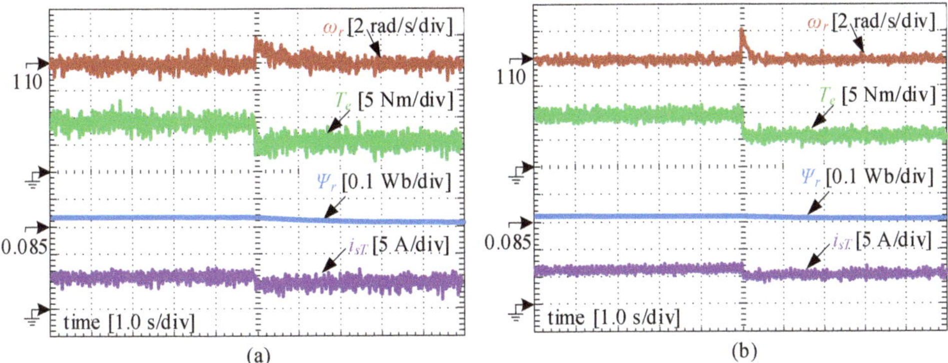

Figure 13. HIL testing results when external disturbances occur. (**a**) Performance of traditional FCS-MPTC with fixed PI parameters and fixed weighting factor. (**b**) Performance of proposed FCS-MPTC with adaptive PI parameters and adaptive weighting factor.

6. Conclusions

This paper proposes FCS-MPTC ripple attenuation techniques for IM drives (mainly working at the high-speed conditions) by using novel fuzzy adaptive PI controller and fuzzy adaptive weighting factor tuning strategy. The main contributions and novelties are as follows:

1. The PPM of the IM drives is established for achieving the FCS-MPTC algorithm. It needs to be mentioned that, in this paper, in order to calculate the rotor flux more accurately, a novel flux observer based on the numerical solution of flux differential equation is derived.
2. The reasons why the traditional fuzzy adaptive speed controller contributes less in suppressing the system ripples are analyzed theoretically. On this basis, an improved three-dimensional fuzzy adaptive PI controller was developed, with a new structure and modified input scaling factors. The proposed fuzzy adaptive PI controller is suitable for ripple reduction because the bandwidth of the system becomes narrow when the speed ripples are large, eliminating the impact of the high-frequency disturbances.
3. A novel weighting factor tuning strategy based on fuzzy adaptive theory was developed. It is achieved by balancing the ripples of flux and torque. The proposed weighting factor strategy is adjusted in real-time, avoiding the ripples caused by the inappropriate internal parameters.

Author Contributions: Writing—original draft, Z.Z.; writing—review and editing, H.W.; supervision, W.Z.; validation, J.J. All authors have read and agreed to the published version of the manuscript.

Funding: This research received no external funding.

Institutional Review Board Statement: Not applicable.

Informed Consent Statement: Not applicable.

Conflicts of Interest: The authors declare no conflict of interest.

References

1. Listwan, J.; Pieńkowski, K. Comparative Analysis of Control Methods with Model Reference Adaptive System Estimators of a Seven-Phase Induction Motor with Encoder Failure. *Energies* **2021**, *14*, 1147. [CrossRef]
2. Gunabalan, R.; Sanjeevikumar, P.; Blaabjerg, F.; Ojo, O.; Subbiah, V. Analysis and Implementation of Parallel Connected Two-Induction Motor Single-Inverter Drive by Direct Vector Control for Industrial Application. *IEEE Trans. Power Electron.* **2015**, *30*, 6472–6475. [CrossRef]
3. Xu, Q.; Cui, S.; Zhang, Q.; Song, L.; Li, X. Research on a New Accurate Thrust Control Strategy for Linear Induction Motor. *IEEE Trans. Plasma Sci.* **2015**, *43*, 1321–1325.

4. Dhar, S.; Jayakumar, A.; Lavanya, R.; Kumar, M.D. Techno-economic assessment of various motors for three-wheeler E-auto rickshaw: From Indian context. *Mater. Today Proc.* **2021**. [CrossRef]
5. Wang, Z.; Chen, J.; Cheng, M.; Chau, K.T. Field-Oriented Control and Direct Torque Control for Paralleled VSIs Fed PMSM Drives with Variable Switching Frequencies. *IEEE Trans. Power Electron.* **2016**, *31*, 2417–2428. [CrossRef]
6. Jain, A.K.; Ranganathan, V.T. Modeling and Field Oriented Control of Salient Pole Wound Field Synchronous Machine in Stator Flux Coordinates. *IEEE Trans. Ind. Electron.* **2011**, *58*, 960–970. [CrossRef]
7. Gong, C.; Hu, Y.; Chen, G.; Wen, H.; Wang, Z.; Ni, K. A DC-Bus Capacitor Discharge Strategy for PMSM Drive System with Large Inertia and Small System Safe Current in EVs. *IEEE Trans. Ind. Inform.* **2019**, *15*, 4709–4718. [CrossRef]
8. Song, S.; Xia, Z.; Fang, G.; Ma, R.; Liu, W. Phase Current Reconstruction and Control of Three-Phase Switched Reluctance Machine with Modular Power Converter Using Single DC-Link Current Sensor. *IEEE Trans. Power Electron.* **2018**, *33*, 8637–8649. [CrossRef]
9. Fu, X.; Li, S. A Novel Neural Network Vector Control Technique for Induction Motor Drive. *IEEE Trans. Energy Convers.* **2015**, *30*, 1428–1437. [CrossRef]
10. Wróbel, K.; Serkies, P.; Szabat, K. Model Predictive Base Direct Speed Control of Induction Motor Drive—Continuous and Finite Set Approaches. *Energies* **2020**, *13*, 1193. [CrossRef]
11. Ahmed, A.A.; Koh, B.K.; Lee, Y.I. A Comparison of Finite Control Set and Continuous Control Set Model Predictive Control Schemes for Speed Control of Induction Motors. *IEEE Trans. Ind. Inform.* **2018**, *14*, 1334–1346. [CrossRef]
12. Yan, L.; Dou, M.; Hua, Z.; Zhang, H.; Yang, J. Robustness Improvement of FCS-MPTC for Induction Machine Drives Using Disturbance Feedforward Compensation Technique. *IEEE Trans. Power Electron.* **2019**, *34*, 2874–2886. [CrossRef]
13. Gong, C.; Hu, Y.; Ni, K.; Liu, J.; Gao, J. SM Load Torque Observer-Based FCS-MPDSC with Single Prediction Horizon for High Dynamics of Surface-Mounted PMSM. *IEEE Tran. Power Electron.* **2020**, *35*, 20–24. [CrossRef]
14. Jun, E.S.; Park, S.Y.; Kwak, S. Model Predictive Current Control Method with Improved Performances for Three-Phase Voltage Source Inverters. *Electronics* **2019**, *8*, 625. [CrossRef]
15. Gong, C.; Hu, Y.; Gao, J.; Wang, Y.; Yan, L. An Improved Delay-Suppressed Sliding-Mode Observer for Sensorless Vector-Controlled PMSM. *IEEE Trans. Ind. Electron.* **2020**, *67*, 5913–5923. [CrossRef]
16. Zhang, Z.; Fang, H.; Gao, F.; Rodríguez, J.; Kennel, R. Multiple-Vector Model Predictive Power Control for Grid-Tied Wind Turbine System with Enhanced Steady-State Control Performance. *IEEE Trans. Ind. Electron.* **2017**, *64*, 6287–6298. [CrossRef]
17. Liu, Z.; Zhao, Y. Robust Perturbation Observer-based Finite Control Set Model Predictive Current Control for SPMSM Considering Parameter Mismatch. *Energies* **2019**, *12*, 3711. [CrossRef]
18. Zhang, Y.; Yang, H. Model-Predictive Flux Control of Induction Motor Drives with Switching Instant Optimization. *IEEE Trans. Energy Convers.* **2015**, *30*, 1113–1122. [CrossRef]
19. Mamdouh, M.; Abido, M.; Hamouz, Z. Weighting Factor Selection Techniques for Predictive Torque Control of Induction Motor Drives: A Comparison Study. *Arab. J. Sci. Eng.* **2018**, *43*, 433–445. [CrossRef]
20. Arahal, M.R.; Barrero, F.; Duran, M.J.; Ortega, M.G.; Martin, C. Trade-offs Analysis in Predictive Current Control of Multi-Phase Induction Machines. *Control Eng. Pract.* **2018**, *81*, 105–113. [CrossRef]
21. Alamir, M. A State-Dependent Updating Period for Certified Real-Time Model Predictive Control. *IEEE Tran. Automat. Contr.* **2017**, *62*, 2464–2469. [CrossRef]
22. Gao, J.; Gong, C.; Li, W.; Liu, J. Novel Compensation Strategy for Calculation Delay of Finite Control Set Model Predictive Current Control in PMSM. *IEEE Trans. Ind. Electron.* **2020**, *67*, 5816–5819. [CrossRef]
23. Li, H.; Lin, J.; Lu, Z. Three Vectors Model Predictive Torque Control without Weighting Factor Based on Electromagnetic Torque Feedback Compensation. *Energies* **2019**, *12*, 1393. [CrossRef]
24. Li, Z.; Yuan, Y.; Ke, F.; He, W.; Su, C. Robust Vision-Based Tube Model Predictive Control of Multiple Mobile Robots for Leader–Follower Formation. *IEEE Tran. Ind. Electron.* **2020**, *67*, 3096–3106. [CrossRef]
25. Yang, Y.; Wen, H.; Fan, M.; Xie, M.; Chen, R. Fast Finite-Switching-State Model Predictive Control Method without Weighting Factors for T-Type Three-Level Three-Phase Inverters. *IEEE Trans. Ind. Inform.* **2019**, *15*, 1298–1310. [CrossRef]
26. Zhang, X.; Hou, B.; Mei, Y. Deadbeat Predictive Current Control of Permanent-Magnet Synchronous Motors with Stator Current and Disturbance Observer. *IEEE Trans. Power Electron.* **2017**, *32*, 3818–3834. [CrossRef]
27. Arshad, M.H.; Abido, M.A.; Salem, A.; Elsayed, A.H. Weighting Factors Optimization of Model Predictive Torque Control of Induction Motor Using NSGA-II with TOPSIS Decision Making. *IEEE Access* **2019**, *7*, 177595–177606. [CrossRef]
28. Rojas, C.A.; Rodriguez, J.; Villarroel, F.; Espinoza, J.R.; Silva, C.A.; Trincado, M. Predictive Torque and Flux Control without Weighting Factors. *IEEE Trans. Ind. Electron.* **2013**, *60*, 681–690. [CrossRef]
29. Wang, J.; Wang, F.; Zhang, Z.; Li, S.; Rodríguez, J. Design and Implementation of Disturbance Compensation-Based Enhanced Robust Finite Control Set Predictive Torque Control for Induction Motor Systems. *IEEE Trans. Ind. Inform.* **2017**, *13*, 2645–2656. [CrossRef]
30. Rathi, M.K.; Prabha, N.R. Interval Type-2 Fuzzy Logic Controller-Based Multi-level Shunt Active Power Line Conditioner for Harmonic Mitigation. *Int. J. Fuzzy Syst.* **2019**, *21*, 104–114. [CrossRef]
31. Robles Algarín, C.; Callejas Cabarcas, J.; Polo Llanos, A. Low-Cost Fuzzy Logic Control for Greenhouse Environments with Web Monitoring. *Electronics* **2017**, *6*, 71. [CrossRef]
32. Shi, J.Z. A Fractional Order General Type-2 Fuzzy PID Controller Design Algorithm. *IEEE Access* **2020**, *8*, 52151–52172. [CrossRef]

33. Asl, R.M.; Palm, R.; Wu, H.; Handroos, H. Fuzzy-Based Parameter Optimization of Adaptive Unscented Kalman Filter: Methodology and Experimental Validation. *IEEE Access* **2020**, *8*, 54887–54904. [CrossRef]

34. Neji, B.; Hamrouni, C.; Alimi, A.M.; Nakajima, H.; Alim, A.R. Hierarchical Fuzzy-Logic-Based Electrical Power Subsystem for Pico Satellite ERPSat-1. *IEEE Syst. J.* **2015**, *9*, 474–486. [CrossRef]

35. Shaker, M.M.; Al-khashab, Y.M. Design and implementation of fuzzy logic system for DC motor speed control. In Proceedings of the 2010 1st International Conference on Energy, Power and Control (EPC-IQ), Basrah, Iraq, 30 November–2 December 2010; pp. 123–130.

36. Baranov, G.; Zolotarev, A.; Ostrovskii, V.; Karimov, T.; Voznesensky, A. Analytical Model for the Design of Axial Flux Induction Motors with Maximum Torque Density. *World Electr. Veh. J.* **2021**, *12*, 24. [CrossRef]

37. Białoń, T.; Pasko, M.; Niestrój, R. Developing Induction Motor State Observers with Increased Robustness. *Energies* **2020**, *13*, 5487. [CrossRef]

38. Rehman, H.; Xu, L. Alternative Energy Vehicles Drive System: Control, Flux and Torque Estimation, and Efficiency Optimization. *IEEE Trans. Veh. Technol.* **2011**, *60*, 3625–3634. [CrossRef]

39. Utrata, G.; Rolek, J.; Kaplon, A. The Novel Rotor Flux Estimation Scheme Based on the Induction Motor Mathematical Model Including Rotor Deep-Bar Effect. *Energies* **2019**, *12*, 2676. [CrossRef]

40. Xue, H.; Zhang, Z.; Wu, M.; Chen, P. Fuzzy Controller for Autonomous Vehicle Based on Rough Sets. *IEEE Access* **2019**, *7*, 147350–147361. [CrossRef]

41. Liang, Y.; He, Y.; Niu, Y. Microgrid Frequency Fluctuation Attenuation Using Improved Fuzzy Adaptive Damping-Based VSG Considering Dynamics and Allowable Deviation. *Energies* **2020**, *13*, 4885. [CrossRef]

42. Liu, J.; Gong, C.; Wang, X.; Yu, H.; Zhang, E. A Parameter Tuning Strategy for Fuzzy Adaptive PI Controller. China Patent CN106292285A, 11 December 2018.

43. Arahal, M.R.; Barrero, F.; Toral, S.; Duran, M.; Gregor, R. Multi-Phase Current Control Using Finite-State Model-Predictive Control. *Control. Eng. Pract.* **2009**, *17*, 579–587. [CrossRef]

44. Wang, Y.; Wang, X.; Xie, W.; Wang, F.; Dou, M.; Kennel, R.M.; Lorenz, R.D.; Gerling, D. Deadbeat Model-Predictive Torque Control with Discrete Space-Vector Modulation for PMSM Drives. *IEEE Trans. Ind. Electron.* **2017**, *64*, 3537–3547. [CrossRef]

Article

A Quality Integrated Fuzzy Inference System for the Reliability Estimating of Fluorochemical Engineering Processes

Feng Xue [1], Xintong Li [1], Kun Zhou [1], Xiaoxia Ge [2], Weiping Deng [2], Xu Chen [1] and Kai Song [1,*]

[1] School of Chemical Engineering and Technology, Tianjin University, Tianjin 300350, China; 3014207238@tju.edu.cn (F.X.); 3014207218@tju.edu.cn (X.L.); kzhou@tju.edu.cn (K.Z.); xchen@tju.edu.cn (X.C.)

[2] Health, Safety and Environmental Protection Department, Juhua Group Co., Ltd., Quzhou 324004, China; ahbgxx@juhua.com.cn (X.G.); jhdwp@juhua.com.cn (W.D.)

* Correspondence: ksong@tju.edu.cn; Tel.: +86-189-2031-7821

Abstract: Hypertoxic materials make it critical to ensure the safety of the fluorochemical engineering processes. This mainly depends on the over maintenance or the manual operations due to the lack of precise models and mechanism knowledge. To quantify the deviations of the operating variables and the product quality from their target values at the same time and to overcome the measurement delay of the product quality, a novel quality integrated fuzzy inference system (QFIS) was proposed to estimate the reliability of the operation status as well as the product quality to enhance the performance of the safety monitoring system. To this end, a novel quality-weighted multivariate inverted normal loss function was proposed to quantify the deviation of the product quality from the target value to overcome the measurement delay. Vital safety process variables were identified according to the expert knowledge. Afterward, the quality loss and the vital variables were inputs to an elaborate fuzzy inference system to estimate the process reliability of the fluorochemical engineering processes. By integrating the abundant expert knowledge and a data-driven quality prediction model to design the fuzzy rules of QFIS, not only the operation reliability but also the product quality can be monitored on-line. Its superiority in estimating system reliability has been strongly proved by the application of a real fluorochemical engineering process located in East China. Moreover, the application of the Tennessee Eastman process also confirmed its generalization performance for other complicated black-box chemical processes.

Keywords: process reliability estimating; fluorochemical engineering process; fuzzy inference system; quality prediction; prognostics and health management

Citation: Xue, F.; Li, X.; Zhou, K.; Ge, X.; Deng, W.; Chen, X.; Song, K. A Quality Integrated Fuzzy Inference System for the Reliability Estimating of Fluorochemical Engineering Processes. *Processes* **2021**, *9*, 292. https://doi.org/10.3390/pr9020292

Academic Editors: Jie Zhang and Meihong Wang

Received: 11 January 2021

Accepted: 31 January 2021

Published: 3 February 2021

Publisher's Note: MDPI stays neutral with regard to jurisdictional claims in published maps and institutional affiliations.

1. Introduction

With the diversification of products and the continuous development of application fields, the fluorochemical industry has become more and more important. However, hypertoxic materials widely exist in the fluorochemical engineering process. Even a tiny leak of these hypertoxic materials in the environment would cause huge damage to people, equipment and even public safety. Additionally, nowadays, the operation condition of chemical industrial processes is typically monitored by a large number of different types of sensors, capturing temperature, pressure, flow, vibration, solution concentration and other process variables. This not only results in very heterogeneous data at different time scales but also introduces the signals affected by measurement and transmission noise. In many cases, consequently, the sensors are partly redundant or highly related variables. Failures in such redundant sensors would not cause the same influence on the operation reliability as what a vital process variable would cause. Therefore, the requirement of the monitoring system goes far beyond fault detection and diagnosis, whose major tasks are limited to react after there are failures or faults happen. In order to avoid any possible failures or faults, and to reduce maintenance costs and equipment uptime at the highest level, proactive maintenance measures should be taken. This means that the maintenance strategy should swift from fault detection and diagnosis (FDD) to Prognostics and Health Management (PHM).

The goal of PHM is to provide methods and tools to design optimal maintenance policies for a specific process under its distinct operating and degradation conditions, achieving a high availability at minimal costs [1,2]. It is not limited to the predictions of failure times or the remaining useful life (RUL) and supports optimal maintenance and logistics decisions by considering the process operation status, the operating context and the economic consequences of different faults. System reliability estimating (SRE) plays an important role in PHM. It focuses on assessing the operational reliability based on outputs from process operation status, available resources and operational demand.

SRE is developing rapidly and there are many methods currently available [3]. Fault tree analysis (FTA), reliability graph, Monte Carlo Simulation and Bayesian Networks (BN) are commonly used methods for it [4]. FTA is a systematic way to obtain the reliability of complex systems both qualitatively and quantitatively by using exact values of root causes' occurrence probability. Fuzzy methods are often applied in FTA to make up for the shortcomings of insufficient probability values [5,6]. Monte Carlo Simulation-based tools are useful for reliability assessment of large and complex power systems [7], but they may lead to a combinatorial explosion of the number of states to model a system [8]. They are more suitable for estimating the reliability of a component or system of low complexity rather than of highly complex systems [9].

Due to the complicated mechanism, copyright protection, hypertoxicity and so on, the lack of mathematical models is one of the major reasons hindering the applications of advanced control and monitoring methods in fluorochemical engineering processes. On the other hand, plenty of background and expert knowledge has accumulated along with the continuous operation of such processes. When there are differences in expert knowledge, expert consistency prioritization can be conducted for expertise differences instead of assuming experts identical or assigning some predefined weights [10]. Currently, because of the aforementioned reasons, over maintenance strategies and manual operations are the most commonly used strategies in the safety management of the fluorochemical engineering processes [11]. It is obvious that these methods lead to big economic losses and security risks [12]. Therefore, it is very urgent to propose an appropriate SRE method for further application of the PHM system in these processes.

Unlike principal component analysis (PCA) and other multivariate statistical process monitoring (MSPM) methods [13,14], a fuzzy inference system (FIS) can integrate data-driven modelling and the priceless expert knowledge by the designation of membership functions and fuzzy rules [15]. Successful applications of it are also attributable to its superiority to manage uncertainty and computation for noisy and imprecise data. It also takes advantage of operational experience and provides suggestions on chemical processes without hard intervention [16,17]. The success of FIS is evident from its applicability and relevance in extensive research areas: control systems, engineering, medicine, chemistry, finance and business, computer networks, computational biology, fault detection and diagnosis and pattern recognition [18,19]. It holds high promise in the realization of SRE for complicated and black-box processes like the fluorochemical engineering processes.

Additionally, an important aspect of any industrial operation is conformance to standards. This relates to how closely the operational performance, process safety, as well as quality of the final products, match the design specifications. Whether the product quality matches the expected value is an important standard to estimate how healthy the process operation status is. However, quality control has not been taken into consideration when the SRE or PHM system was designed. It is very important for decision-makers to know the overall status both reflects the safety assessment and the product quality of the chemical process to make the best response. Unfortunately, for chemical engineering industries, there is always an intolerably long time-delay in the measurement of product quality. This paper, therefore, is aimed to propose a quality integrated SRE method to fill in the gap between the PHM system and quality control.

Therefore, a quality integrated FIS model (QFIS) based system reliability estimating method was proposed and applied to a process in a fluorochemical factory located in East China. A novel quality loss function was proposed to estimate the quality deviation based on a product quality regression model trained with the Partial Least Squares (PLS) algorithm to overcome the time-delay in quality measurement. Meanwhile, vital safety variables of the fluorochemical process operation were selected under the guidance of expert knowledge. Then, these vital safety variables and the quality loss value were used as inputs to the FIS model. By making good use of the expert knowledge and the operation experience, the membership functions and fuzzy rules were well-constructed to obtain the system reliability of the fluorochemical engineering process. To test the generalization ability of our proposed QFIS method, it was also used in the Tennessee Eastman process, a widely applied benchmark for advanced control and monitoring system.

The rest of the paper is organized as follows: Section 2, brief introduction of R22 refrigerant producing process and existing algorithms used in this paper; Section 3, details of the proposed quality integrated fuzzy inference system; Section 4, applications in the R22 refrigerant producing process and the Tennessee Eastman process; Section 5, conclusion.

2. Background and Methods

2.1. Brief Introduction of R22 Refrigerant Producing Process

R22, also known as HCFC-22, is one of the most widely used fluorides. It is mainly used as a kind of common propellant and refrigerant. The global use of R22 continues to increase because it is a versatile intermediate in the organic fluorine chemical industry, e.g., as a precursor to tetrafluoroethylene.

The producing process of R22 is presented in Figure 1. The main operating units include a Feeder, a Reactor, a Water Scrubber, a Separator and two Rectifying columns. R22 is prepared from the chloroform as: $HCCl_3 + 2\,HF \rightarrow HCF_2Cl + 2\,HCl$.

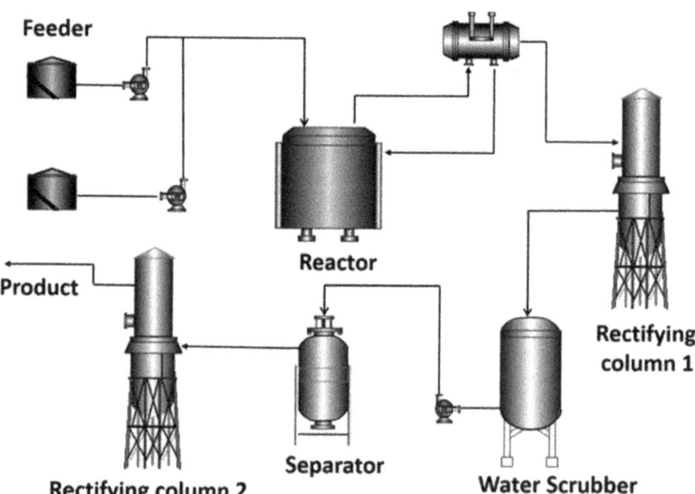

Figure 1. The producing process of R22.

All materials and byproducts like AHF, HCL and HF become intensely corrosive when meeting water in the air. Therefore, it is very vital to public safety and environmental protection to secure the safety of the R22 producing process and to improve the performance of the PHM system of it. As mentioned above, however, the complicated characteristics, the confidential agreement, and the time-varying mechanisms of it adversely hinder the performance of the traditional FDD methods. Over maintenance and manual operations are still the most commonly used strategies in the safety management of it.

The most important thing is that even a tiny amount of material or byproduct leakage into the environment can cause terrible damage to equipment and workers. PHM or at least SRE should play a bigger role than the traditional FDD, which can only react after a disaster happens, in predicting the operation stability to react and prevent a disaster from happening.

On the other hand, the sampling interval of the product quality is 180 min while the DCS process variables are 1 min. Such a long delay in product quality measurement is completely intolerable for quality control and safety management practice. It is not practical to use product quality as an input to estimate its deviation neither.

Therefore, we proposed a quality integrated fuzzy inference system to integrate the quality control and the system stability estimating at the same time for further applications of PHM in such a complicated and hypertoxic process. It also can overcome the time delay in quality measurement to evaluate the process operation through the quality control point of view.

2.2. Brief Introduction of Existing Algorithms

2.2.1. Partial Least Squares (PLS)

Partial Least Squares (PLS) is a widely used linear regression method. It aims at modeling linear relationships between the input variables $X \in \mathbf{R}_{m \times n}$ (n is the number of process variables and m is the number of observations) and output variables $Y \subset \mathbf{R}_{m \times p}$ (p is the number of output variables) [20]. Regularly, X and Y are supposed to be normalized. The PLS model structure can be described as [21]:

$$Y = XB + V \tag{1}$$

where $B \in \mathbf{R}_{n \times p}$ is the regression coefficient matrix and $V \in \mathbf{R}_{m \times p}$ is the residual matrix.

It iteratively extracts the Latent Variables (LVs) $t_i \in \mathbf{R}_m$, $u_i \in \mathbf{R}_m$ and the weight vectors $w_i \in \mathbf{R}_n$, $c_i \in \mathbf{R}_p$ from X and Y matrices in decreasing order of their corresponding singular values, where $i = 1, \ldots, v$, and v is the number of LVs, which is usually determined by cross-validation. In other words, the PLS algorithm decomposes X and Y matrices as follows [22]:

$$X^{\mathrm{T}} = \sum_{i=1}^{v} t_i p_i^{\mathrm{T}} + E = TP^{\mathrm{T}} + E \tag{2}$$

$$Y^{\mathrm{T}} = \sum_{i=1}^{v} u_i q_i^{\mathrm{T}} + F = UQ^{\mathrm{T}} + F \tag{3}$$

where E and F are the residual matrix of X and Y, respectively. Therefore, by extracting LVs, the n-dimensional original input space X is compressed into the v-dimensional LV-space. In common cases, $v << n$. By doing this, PLS can effectively remove the noise and multi-collinearity of the original data, which is especially true for the chemical process data [23].

Then the estimated regression coefficient matrix can be obtained by the following Equation:

$$\hat{B} = X^{\mathrm{T}} U \left(T^{\mathrm{T}} X X^{\mathrm{T}} U \right)^{-1} T^{\mathrm{T}} Y \tag{4}$$

The absolute value of the coefficient represents the contribution of the corresponding variable to the linear model, therefore, it can be used to quantify the importance of the corresponding variable.

Root mean square error (RMSE) is usually used as a metric on the determination of the value of v in the PLS model. The definition of RMSE is:

$$\mathrm{RMSE} = \sqrt{\sum (\hat{y}_i - y_i)^2 / n} \tag{5}$$

where y_i is the observed value of the output variable and $\hat{y}_i$ is the corresponding predicted value. PLS model performed better when the RMSE of the training data and testing data were smaller.

2.2.2. Fuzzy Inference System (FIS)

A fuzzy inference system (FIS) is a tool for modeling a complex system without a thorough mathematical explanation [24]. It is capable of modeling uncertainties commonly represented in linguistic form and extending the functionality of the engineering system. The term "fuzzy" refers to the fact that the involved logic can deal with concepts that cannot be expressed as "true" or "false" but rather as "partially true" or "partially false". The design of fuzzy rules is a delicate task, and it can be generally carried out by an expert, who, on the basis of some heuristics that s/he has developed about the system. This makes it easier to mechanize tasks that are already successfully performed by humans.

The application procedure of a FIS system consists of three steps: an input stage, a processing stage, and an output stage [15,25]:

(1)   The input stage maps inputs to the appropriate membership functions and true values. The most common shape of membership functions is triangular, although trapezoidal and bell curves are also used. The shape is generally less important than the number of curves and their placements.
(2)   The processing stage invokes each appropriate rule and generates a result, then combines the results of all rules. It is based on a collection of logic rules in the form of IF-THEN statements, where the IF part is called the "antecedent" and the THEN part is called the "consequent". Typical a fuzzy control system has dozens of rules.
(3)   The output stage converts the combined result into a specific control output value.

3. The Quality Integrated FIS Based System

PHM aims to provide an integrated framework for degradation prediction and system maintenance [2,26]. Since PHM can be considered as a holistic approach to an effective and efficient system health management, the quality of the produced products of a concerned process should be taken into serious consideration. However, most proposed methods and tools mainly focus on operation stability and maintenance cost. On the other hand, as aforementioned, for fluorochemical engineering and other chemical industries, there is always a long time-delay in the measurement of product quality. For example, in the R22 refrigerant producing process, the sampling interval of the product quality is 180 min while the DCS process variable is 1 min. It is not practical, in other words, to use product quality as an input to estimate its drift. Therefore, a novel quality integrated FIS (QFIS) based system reliability estimating method was proposed to evaluate the stability of the operation status and the quality of the product simultaneously.

3.1. A Novel Quality Weighted Multivariate Inverted Normal Loss Function (QMINLF)

Inverted normal loss function (INLF), Modified inverted normal loss function (MINLF), Inverted Beta loss function (IBLF) and Inverted Gamma loss function (IGLF) are different loss functions considering a random deviation from target values and are widely used in industrial applications [27].

For a multivariate process, to consider the deviation of a variable from its expected value as well as the importance of it to the final product, a novel quality weighted multivariate inverted normal loss function (QMINLF) was proposed as:

$$L(Q) = \frac{1}{1 - e^{-\rho^2}} \sum_{i=1}^{n} \left(1 - e^{-\frac{1}{2}\beta_i (x_i - a_i)^2}\right) \tag{6}$$

where $L(Q)$ is the estimated quality loss, ρ is a shape parameter, n is the number of process variables, x_i $(i = 1, 2, \ldots, n)$ is the observed value of the ith process variable, a_i is its expected value and β_i is the corresponding importance index. $\beta_i = \frac{|b_i|}{\sum_{i=1}^{n}|b_i|}$ and $|b_i|$ is the absolute value of the ith element value of $\hat{B}$ in Equation (4).

From the definition of $L(Q)$, we can see that $L(Q)$ is decided by the deviation values $(x_i - a_i)^2$ of all quality-related process variables. The bigger the deviation, the bigger the $L(Q)$ value. Additionally, for the same deviation value, because of being weighted by β_i, the more important the process variable x_i to the final quality, the bigger the $L(Q)$ value, which means the bigger the quality loss. Therefore, QMINLF is more sensitive to the deviations of the comparatively higher quality-related process variables.

Note: ρ should be optimized according to the operation knowledge or the performance to make sure $0 \leq L(Q) \leq 1$. It also can be optimized by a genetic algorithm or other optimization methods.

3.2. The Procedure of the Proposed QFIS Method

To make good use of operation knowledge and to consider the quality loss at the same time, a quality integrated fuzzy inference system was proposed to estimate the system's reliability. The procedure of it consists of four major steps:

(1) Identifying the vital safety variables to process stability based on operation experience and background knowledge;
(2) Quantifying the importance of variables to product quality using PLS algorithm;
(3) Estimating the quality loss according to Equation (6);
(4) Designing the membership functions and fuzzy rules for operational reliability using quality loss and vital safety variables as inputs.

The designing of the membership functions is the most time-consuming step for the QFIS method, and it is also the most important step to make sure the performance of the QFIS method. It is supposed to integrate the operation experience, background knowledge and mechanism analysis in this step. Therefore, the membership functions should be designed specifically. Strict membership functions are more preferred when the system is designed to provide warning of system reliability, so triangular and trapezoid are suggested as the membership curves. The details of membership functions and fuzzy rules construction will be given with the specific cases in Section 4.

The structure of the proposed method is presented in Figure 2 and the details are listed in Table 1.

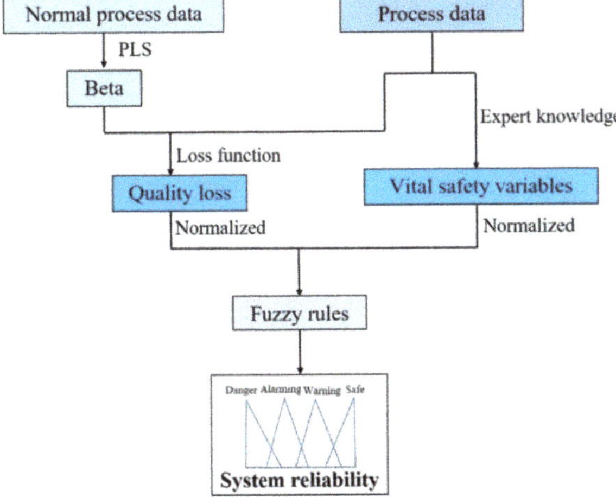

Figure 2. The structure of the quality integrated FIS.

Table 1. The procedure of the quality integrated fuzzy inference system (FIS).

The Procedure of the Quality Integrated FIS

Vital safety variables identification
  Step 1: Determining important operating unit and their process variables;
  Step 2: Selecting vital safety variables under the guidance of expert knowledge;

Quality loss estimation (after collecting a certain number of normal observations)
  Step 1: Quantifying the importance of quality-related process variables to product quality using PLS;
  Step 2: Estimating quality loss using loss function in Equation (6);

Fuzzy inference system
  Step 1: Normalizing the quality loss and vital safety variables as the inputs;
  Step 2: Constructing fuzzy rules and membership functions;
  Step 3: Obtaining the system reliability of the chemical process;

4. Application Results and Discussion

4.1. Application in R22 Refrigerant Producing Process

To test the performance of the proposed QFIS method, the observations of a part of an R22 producing process in a large-scale fluorochemical industry company located in East China were applied. The flowchart of the R22 producing process is shown in Figure 1. There are 69 process variables in total. The sampling period was from May to November 2019 (Sampling interval was one minute), which included dozens of procurement cycles of raw materials and experienced through the summer, autumn and winter of the location.

According to the operation experience, as shown in Figure 3, data in the green box were observations in normal operation status which were used as training data. Data in red boxes were observations corresponding to three types of abnormal status. Ten thousand observations for each of them were used to test the performance of the proposed method.

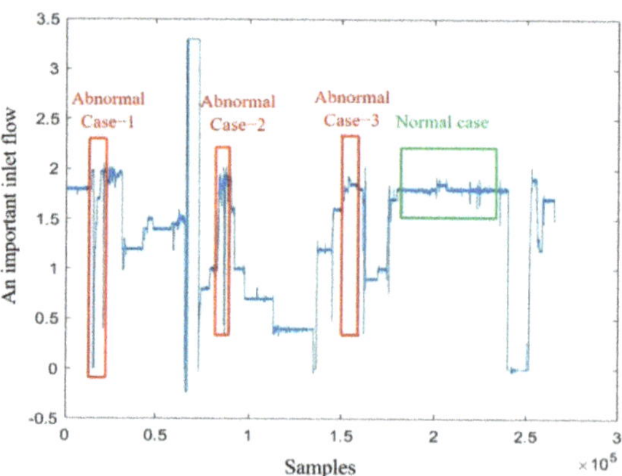

Figure 3. The selection of training and testing data for R22.

The purity of one of the major intermediate materials was used as the quality variable. Because it is sampled and measured offline per hour while the DCS process variables are 1 min, therefore, it is not practical to use product quality as an input to estimate its drift. To match with the measured product quality, only the process variables sampled at the same time were used to train the QFIS model. In this way, 849 observations of normal operation status were available. Of the observations, 699 were used as training data while the other 150 observations were used as testing data of the PLS-based quality prediction model. Three-fold cross-validation was performed in the training step to optimize the quality prediction model so that the importance index of each process variable can be evaluated comprehensively.

Besides the reactor, which is always the major unit of a process, rectifying column 1 is another major one suffering from corrosion according to the operation records and the expert knowledge. Consequently, the reactor level, reactor temperature, rectifying column 1 pressure and rectifying column 1 level were selected as vital safety variables. Then these four vital safety variables and quality loss value calculated by Equation (6) were normalized and input to the QFIS model to estimate the operation reliability.

As we mentioned above, membership functions or fuzzy rules are very important. The membership functions were determined strictly according to the characteristics of each input, and was basically divided into three levels: high, medium and low reliability. According to the expert knowledge and experimental experience, triangle and trapezoid curves were used. The membership functions for each input and output is presented in Figure 4a. For the reactor temperature, higher than the up limitation would cause much worse damage than lower than the low limit. Then the level of reactor temperature was defined as Not-high and High. For the reactor level, Rectifying column 1 pressure and Rectifying column 1 level, low pressure or level has the potential risk of leakage and high pressure or level would bring damage to the equipment. Then the levels of them were defined as Low, Medium and High. For the quality loss, obviously, normally, it is supposed to be as low as possible. Medium or high-quality loss will bring huge economic loss which is unacceptable. So the level of quality loss was defined as Low and Not-low.

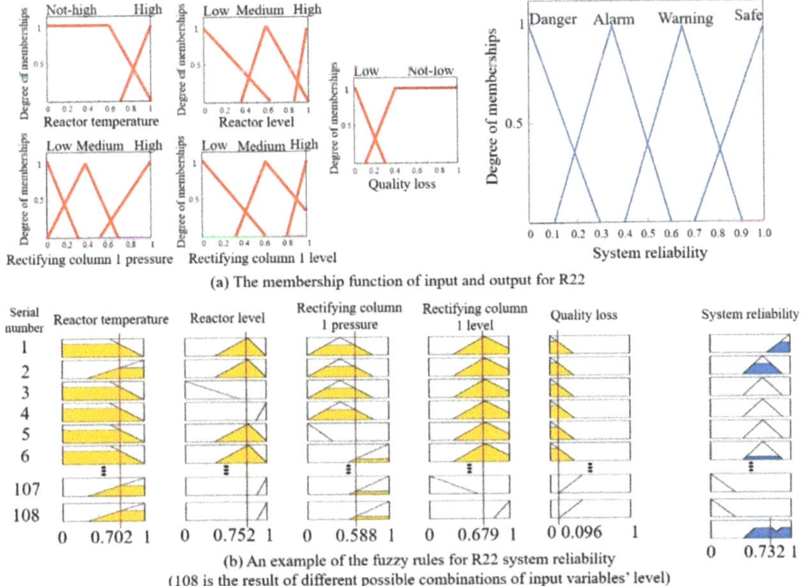

(a) The membership function of input and output for R22

(b) An example of the fuzzy rules for R22 system reliability
(108 is the result of different possible combinations of input variables' level)

Figure 4. The examples of the application of the proposed method in R22 producing process.

The shape and position of membership function curves were determined according to the range of the corresponding variables and the operation experience which plays an important role due to the particularity of the R22 producing process. For example, according to equipment condition and operating status in the field. The reactor would experience a strong decrease in reliability when the reactor temperature is over 0.7. Then the starting point of the high reactor temperature curve is set as 0.7. The vertex of the medium reactor level curve (0.63) was the average value of reactor level in normal observations.

The fuzzy rules were determined according to the level of each input. Table 2 lists the level of each input and whether it is allowed. They were formulated as follows:

(1)   If there was no input in state "No", then the system reliability was in "safe" status;

(2) If one or two inputs were in state "No", then the system reliability was in "warning" status;

(3) If three or four inputs were in state "No", then the system reliability was in "alarm" status;

(4) If there were five inputs in state "No", then the system reliability was in "danger" status.

Table 2. Constraints of inputs of R22 refrigerant producing process.

Index	Low	Medium	High
Reactor temperature		Yes	No
Reactor level	No	Yes	No
Rectifying column 1 pressure	No	Yes	No
Rectifying column 1 level	No	Yes	No
Quality loss	Yes	No	

For example, if Reactor temperature was not high, Reactor level was medium, Rectifying column 1 pressure was high, Rectifying column 1 level was low and Quality loss was low, then the system reliability was in "warning" status.

A total of 108 fuzzy rules were constructed. Parts of them are presented in Figure 4b.

Then, the estimated system reliability of three abnormal cases is shown in Figure 5 with the corresponding value of an important inlet flow (IIF), which partly indicated the operation status of the process.

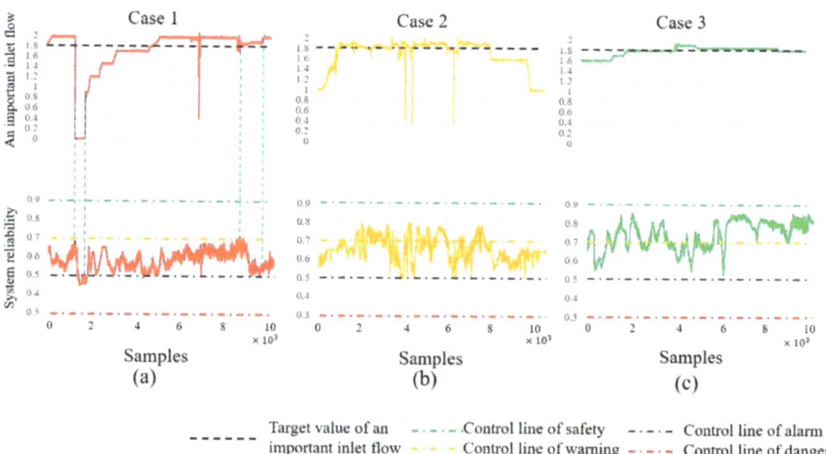

Figure 5. The important inlet flow and system reliability of three abnormal cases in R22; (**a–c**) are three abmormal cases' important inlet flow and system reliability.

In fact, the valve of the important inlet had to be shut down for a while to protect the process. Correspondingly, there was a very big deviation in the IIF value of Case 1 from its target value (1.8). As a result, there was a big fluctuation in the system reliability curve with a delay of only 40 min (see the period between the first two blue lines in Figure 5). The system reliability was continuously lower than 50% triggered a "Danger" notice. During the recovering period of the process, the overall trend of the system reliability was upward volatility because the spread of this shutdown influenced other vital variables. The system reliability was recovered to "Alarm", then "Warning". Unfortunately, another big disturbance happened at the end of the period of Case 1, and the reliability went back to "Alarm" status again. The system reliability was always under the control line of warning which represented that the system was not reliable and the "shut down" of the important inlet was necessary.

Case 2 was during the recovering period of the process after an overhaul. The system reliability was going upward at the beginning. The ups and downs of the reliability had the same trend as the IIF value, and both of them were because of the operation fluctuations caused by the DCS (distributed control system). It was obvious that the system reliability was around the control line of warning and was improved in the recovering period.

Case 3, the operation was almost recovered back to normal with only small fluctuations. The system reliability was going upward toward 90% with fluctuations too. Most of the observations were above the control line of warning. It showed that the system was reliable.

With the lowest average reliability of Case 1 and the better system reliability in Case 2 and Case 3, the same trends between the system reliability and the IIF value strongly proved the performance of our method. It can predict the system reliability with a reasonable delay using only normal observations to train the model. The proposed method comprehensively considered both safety and economic factors, and the result fully reflected the status of the system, so as to provide appropriate suggestions to the decision-makers of the R22 refrigerant producing process operation.

Due to the confidential agreement, it is not allowed to show the quality variables of the R22 producing process. The contribution of quality loss to the final estimated system reliability would be discussed in the application in the Tennessee Eastman process in Section 4.2.

4.2. Application in the Tennessee Eastman Process

To further test the proposed method, it was applied to predict the system reliability of the Tennessee Eastman chemical process (TEP). TEP is a chemical simulation process that was promoted by J. J. DOWNS and E. F. VOGEL in 1992. It consists of five major operating units namely, a reactor, a product-condenser, a vapor-liquid separator, a recycle compressor and a product stripper. Its process flow diagram is shown in Figure 6. G and H are products of the TEP. Twenty-two process variables and twelve manipulated variables are measured online. Among them, two manipulated variables are constant. Additionally, nineteen variables are measured by offline equipment. There are 28 process fault types (IDV1-IDV28) in the revised version for researchers to test their monitoring methods [28,29]. Details of them are available in Refs. [28,29].

According to the literature research, the reactor pressure, reactor level, product separator level and stripper base level were selected as the vital safety variables. Normally, the reactor temperature should be considered as a vital variable, but it will not be affected by any available IDVs provided by TEP. Therefore, it was not selected as a vital variable.

The 22 process variables and 10 non-constant processes manipulated variables were used as the input of the PLS-based quality prediction model. The ratio of the product G and H was used as the output of this quality model. A total of 960 observations, which were sampled per 3 min, were used as training data and 240 observations were used as testing data. The longest time delay in the G/H ratio was 30 min, which was ten times of the process variables. Five-fold cross-validation was taken to optimize this model. The result showed when $v = 2$, both the RMSEs of training data and testing data had the smallest values. It meant the quality model had the best regression performance when $v = 2$. So the corresponding regression coefficient vector β was used to obtain the quality loss defined in Equation (6).

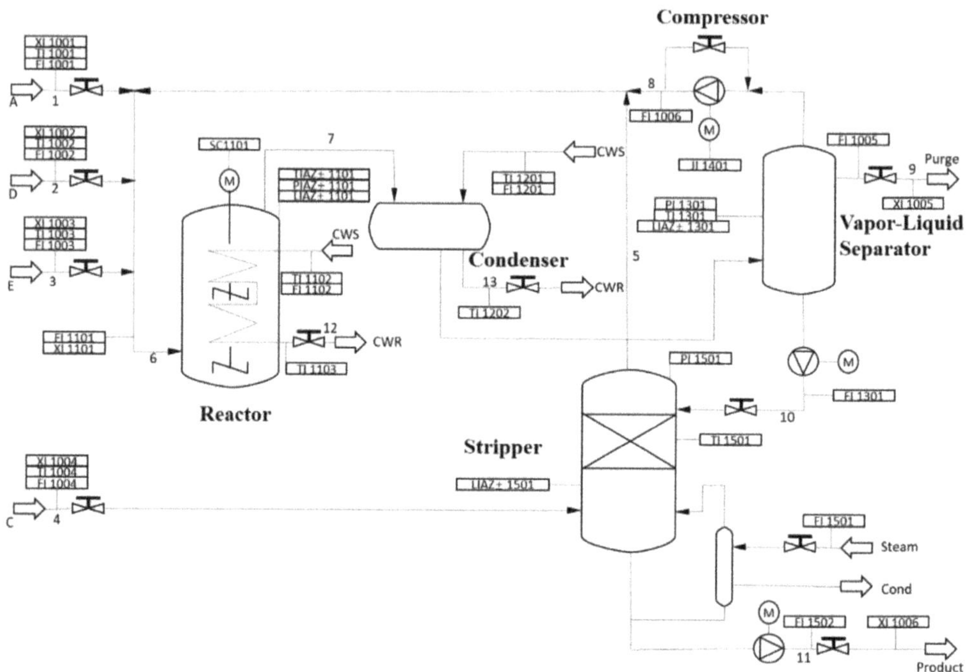

Figure 6. The producing process of Tennessee Eastman process.

The principles of designing membership functions and fuzzy rules were similar to those of the R22 refrigerant producing process. The constraints of the five inputs of QFIS are presented in Table 3. A total of 72 fuzzy rules were designed. The membership functions of input and output are presented in Figure 7a and part of the fuzzy rules viewer is presented in Figure 7b. Details of the codes for TEP is provided in the supplementary material at the end of the paper.

Table 3. Constraints of inputs of Tennessee Eastman process.

Index	Low	Medium	High
Reactor pressure		Yes	No
Reactor level	No	Yes	No
Product separator level		Yes	No
Stripper base level	No	Yes	No
Quality loss	Yes	No	

These 28 IDVs occur in a different part of the TEP with different characteristics and amplitudes, they cause different influences on the process operation. According to the system reliabilities calculated with our method, these 28 IDVs can be divided into three categories: (1) Low danger: The system reliabilities were minor impacted by some IDVs and they were always around 90%; (2) Medium danger: Some IDVs had a middle impact on the safe operation of TEP and the corresponding system reliabilities were between 60% and 90%; (3) High danger: A few IDVs impacted the operation severely and the system reliabilities were under 50%. The IDV descriptions of TEP and their danger levels were given in Table 4. We selected three typical IDVs as examples to demonstrate our QFIS method. The system reliabilities and corresponding ratio of G/H and quality loss are shown in Figure 8.

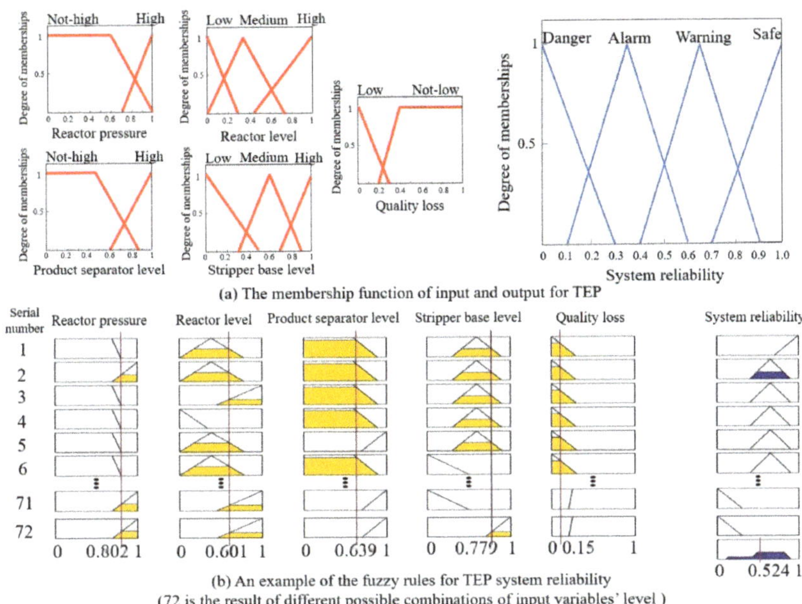

(a) The membership function of input and output for TEP

(b) An example of the fuzzy rules for TEP system reliability
(72 is the result of different possible combinations of input variables' level)

Figure 7. The example of the application of the proposed method in the Tennessee Eastman chemical process (TEP).

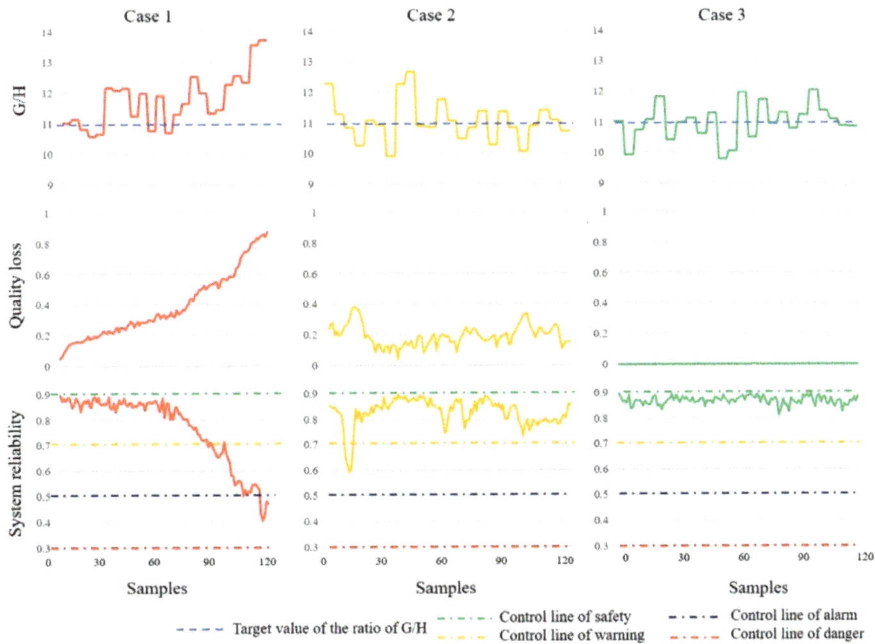

Figure 8. The ratio of G/H, quality loss and system reliability of three cases in TEP.

Table 4. The fault description of Tennessee Eastman Process.

No.	Description	Type	Danger Level
1	A/C feed ratio, B composition constant (stream 4)	Step	Medium
2	B composition. A/C ratio constant (stream 4)	Step	Low
3	D feed temperature (stream 2)	Step	Low
4	Reactor cooling water inlet temperature	Step	Low
5	Condenser cooling water inlet temperature	Step	Low
6	A feed loss (steam 1)	Step	High
7	C header pressure loss-reduced availability (steam 4)	Step	Low
8	A, B, C feed composition (stream 4)	Random	Medium
9	D feed temperature (stream 2)	Random	Low
10	C feed temperature (stream 4)	Random	Low
11	Reactor cooling water inlet temperature	Random	Low
12	Condenser cooling water inlet temperature	Random	Medium
13	Reaction kinetics	Slow drift	Medium
14	Reactor cooling water valve	Sticking	Low
15	Condenser cooling water valve	Sticking	Low
16	Unknown	Unknown	Low
17	Unknown	Unknown	Low
18	Unknown	Unknown	Medium
19	Unknown	Unknown	Medium
20	Unknown	Unknown	Low
21	A feed temperature (stream 1)	Random	Low
22	E feed temperature (stream 3)	Random	Low
23	A feed pressure (stream 1)	Random	Low
24	D feed pressure (stream 2)	Random	Medium
25	E feed pressure (stream 3)	Random	Low
26	A and C feed pressure (stream 4)	Random	Low
27	Pressure fluctuation in the cooling water re-circulating unit of the reactor	Random	Low
28	Pressure fluctuation in the cooling water re-circulating unit of the condenser	Random	Low

Case 1 was the IDV6 (A feed loss) in TEP. The loss of the main raw material of the reactor caused a severely bad influence on the product quality and the process operation. After IDV6 was introduced, the system reliability dropped very quickly to a lower 40%, which indicated that the process was in a very dangerous status. The quality loss increased quickly and the ratio of G/H deviated from its target value severely at the end of the simulation. Actually, the simulation would shut down in 6.2 h after IDV6 was introduced. This strongly proved how serious the damage was caused by IDV6. It also proved how good the performance of our system reliability estimating method.

Case 2 was the IDV12 (A random fluctuation in the condenser cooling water inlet temperature) in TEP. The condenser was not the major part of the TEP. Moreover, according to the mechanism and the flowchart of TEP, a random fluctuation in the condenser cooling water inlet temperature could not cause severe damage because of the time delay and the operation of the automatic DCS system. The product quality was only slightly affected and the simulation could still keep running. The ratio of G/H was in a reasonable fluctuation too. Consequently, the system reliability did not drop seriously. Except for several points that were lower than 70% (the control line of warning), it was around 90% most of the time.

Case 3 was IDV9 (A random fluctuation in the D feed temperature) in TEP. D was a reaction raw material in TEP. The automatic DCS system tuned the process parameters to overcome the influence caused by it. Therefore, it only caused mild fluctuations in the process operation. The quality loss was extremely small and the ratio of G/H was around the target value. The system reliability was around the control line of safety which meant the status was safe and reliable.

Therefore, the following conclusion can be summarized from the results: (1) The proposed method is sensitive to the change both in product quality and in the safety of TEP. (2) The degree of the abnormal status of TEP can be accurately estimated by our proposed method.

5. Conclusions

The proposed methodology considers both the effect of quality control and safety assessment on the reliability of the chemical process system. To estimate the quality deviation and to overcome the time delay in quality measurement, which is very common for engineering practice, a novel quality loss function was proposed by weighting the contributions of the process variables to the final product. Meanwhile, the vital safety variables of the fluorochemical process operation were selected under the guidance of expert knowledge. Finally, the system reliability was estimated with an elaborate fuzzy inference system using the quality loss and vital process variables as inputs. The membership functions and fuzzy rules were constructed by making good use of the expert knowledge and the operation experience. Applications on a practical fluorochemical engineering process in East China and on the Tennessee Eastman process strongly confirmed the superiority of QFIS in system reliability estimating of the proposed system for complicated black-box chemical processes. The most important contribution of the proposed methodology is to provide an overall system reliability assessment method on both quality control and operation status which can offer a comprehensive proposal on further PHM. However, the estimated system reliability which should serve as a reference for decision-makers can not control or regulate the chemical process directly. The result shows the overall status of the chemical process and more research needs to be carried out on identifying specific reasons leading to the decline in system reliability.

Supplementary Materials: The following are available online at https://www.mdpi.com/2227-9717/9/2/292/s1, MATLAB codes for TEP.

Author Contributions: Conceptualization, F.X. and K.S.; methodology, F.X. and K.S.; software, F.X.; validation, F.X., X.L. and X.G.; formal analysis, F.X. and X.L.; investigation, F.X.; resources, W.D.; data curation, F.X., X.L., X.G. and K.Z.; writing—original draft preparation, F.X.; writing—review and editing, K.S.; visualization, K.Z.; supervision, X.C.; project administration, K.S.; funding acquisition, K.S. All authors have read and agreed to the published version of the manuscript.

Funding: This work was supported by the National Key Research and Development Program of China (No. 2018YFC0808600), Ministry of Science and Technology of the People's Republic of China.

Conflicts of Interest: The authors declare no conflict of interest.

References

1. Lee, J.; Wu, F.; Zhao, W.; Ghaffari, M.; Liao, L.; Siegel, D. Prognostics and health management design for rotary machinery systems—Reviews, methodology and applications. *Mech. Syst. Signal Process.* **2014**, *42*, 314–334. [CrossRef]
2. Oluwasegun, A.; Jung, J.C. The application of machine learning for the prognostics and health management of control element drive system. *Nucl. Eng. Technol.* **2020**, *52*, 2262–2273. [CrossRef]
3. Rebello, S.; Yu, H.; Ma, L. An integrated approach for system functional reliability assessment using Dynamic Bayesian Network and Hidden Markov Model. *Reliab. Eng. Syst. Saf.* **2018**, *180*, 124–135. [CrossRef]
4. Jiang, Q.; Gao, D.-C.; Zhong, L.; Guo, S.; Xiao, A. Quantitative sensitivity and reliability analysis of sensor networks for well kick detection based on dynamic Bayesian networks and Markov chain. *J. Loss Prev. Process. Ind.* **2020**, *66*, 104180. [CrossRef]
5. Senol, Y.E.; Aydogdu, Y.V.; Sahin, B.; Kilic, I. Fault Tree Analysis of chemical cargo contamination by using fuzzy approach. *Expert Syst. Appl.* **2015**, *42*, 5232–5244. [CrossRef]
6. Şahin, B. Consistency control and expert consistency prioritization for FFTA by using extent analysis method of trapezoidal FAHP. *Appl. Soft Comput.* **2017**, *56*, 46–54. [CrossRef]
7. Peng, L.; Hu, B.; Xie, K.; Tai, H.-M.; Ashenayi, K. Analytical model for fast reliability evaluation of composite generation and transmission system based on sequential Monte Carlo simulation. *Int. J. Electr. Power Energy Syst.* **2019**, *109*, 548–557. [CrossRef]
8. Zhang, X.; Sun, L.; Sun, H.; Guo, Q.; Bai, X. Floating offshore wind turbine reliability analysis based on system grading and dynamic FTA. *J. Wind. Eng. Ind. Aerodyn.* **2016**, *154*, 21–33. [CrossRef]
9. Nguyen, D.Q.; Brammer, C.; Bagajewicz, M. New Tool for the Evaluation of the Scheduling of Preventive Maintenance for Chemical Process Plants. *Ind. Eng. Chem. Res.* **2008**, *47*, 1910–1924. [CrossRef]
10. Sahin, B.; Yip, T.L. Shipping technology selection for dynamic capability based on improved Gaussian fuzzy AHP model. *Ocean Eng.* **2017**, *136*, 233–242. [CrossRef]
11. Zhao, Z.; Wang, F.; Jia, M.-X.; Wang, S. Predictive maintenance policy based on process data. *Chemom. Intell. Lab. Syst.* **2010**, *103*, 137–143. [CrossRef]

12. Calabrese, F.; Regattieri, A.; Botti, L.; Galizia, F.G. Prognostic Health Management of Production Systems. New Proposed Approach and Experimental Evidences. *Procedia Manuf.* **2019**, *39*, 260–269. [CrossRef]
13. Yin, S.; Ding, S.X.; Xie, X.; Luo, H. A Review on Basic Data-Driven Approaches for Industrial Process Monitoring. *IEEE Trans. Ind. Electron.* **2014**, *61*, 6418–6428. [CrossRef]
14. Ge, Z. Review on data-driven modeling and monitoring for plant-wide industrial processes. *Chemom. Intell. Lab. Syst.* **2017**, *171*, 16–25. [CrossRef]
15. Dey, S.; Jana, D.K. Application of fuzzy inference system to polypropylene business policy in a petrochemical plant in India. *J. Clean. Prod.* **2016**, *112*, 2953–2968. [CrossRef]
16. Azadeh, A.; Ebrahimipour, V.; Bavar, P. A fuzzy inference system for pump failure diagnosis to improve maintenance process: The case of a petrochemical industry. *Expert Syst. Appl.* **2010**, *37*, 627–639. [CrossRef]
17. Hu, G.; Kaur, M.; Hewage, K.; Sadiq, R. An integrated chemical management methodology for hydraulic fracturing: A fuzzy-based indexing approach. *J. Clean. Prod.* **2018**, *187*, 63–75. [CrossRef]
18. Vafaei, L.E.; Sah, M. Predicting efficiency of flat-plate solar collector using a fuzzy inference system. *Procedia Comput. Sci.* **2017**, *120*, 221–228. [CrossRef]
19. Latysheva, O.; Rovenska, V.; Смирнова, I.I.; Nitsenko, V.; Balezentis, T.; Streimikiene, D. Management of the sustainable development of machine-building enterprises: A sustainable development space approach. *J. Enterp. Inf. Manag.* **2020**. [CrossRef]
20. Harkat, M.F.; Mansouri, M.; Nounou, M.N.; Nounou, H.N. Fault detection of uncertain chemical processes using interval partial least squares -based generalized likelihood ratio test. *Inf. Sci.* **2019**, *490*, 265–284.
21. Wold, S.; Sjöström, M.; Eriksson, L. PLS-regression: A basic tool of chemometrics. *Chemom. Intell. Lab. Syst.* **2001**, *58*, 109–130. [CrossRef]
22. Geladi, P.; Kowalski, B.R. Partial least-squares regression: A tutorial. *Anal. Chim. Acta* **1985**, *185*, 1–17. [CrossRef]
23. Zhang, Z.; Song, K.; Tong, T.P.; Wu, F. A novel nonlinear adaptive Mooney-viscosity model based on DRPLS-GP algo-rithm for rubber mixing process. *Chemom. Intell. Lab. Syst.* **2012**, *112*, 17–23. [CrossRef]
24. Ojha, V.; Abraham, A.; Snášel, V. Heuristic design of fuzzy inference systems: A review of three decades of research. *Eng. Appl. Artif. Intell.* **2019**, *85*, 845–864. [CrossRef]
25. Tiri, A.; Belkhiri, L.; Mouni, L. Evaluation of surface water quality for drinking purposes using fuzzy inference system. *Groundw. Sustain. Dev.* **2018**, *6*, 235–244. [CrossRef]
26. Xia, T.; Dong, Y.; Xiao, L.; Du, S.; Pan, E.; Xi, L. Recent advances in prognostics and health management for advanced manufacturing paradigms. *Reliab. Eng. Syst. Saf.* **2018**, *178*, 255–268. [CrossRef]
27. Khan, F.; Wang, H.; Yang, M. Application of loss functions in process economic risk assessment. *Chem. Eng. Res. Des.* **2016**, *111*, 371–386. [CrossRef]
28. Bathelt, A.; Ricker, N.L.; Jelali, M. Revision of the Tennessee Eastman Process Model. *IFAC-PapersOnLine* **2015**, *48*, 309–314. [CrossRef]
29. Downs, J.; Vogel, E. A plant-wide industrial process control problem. *Comput. Chem. Eng.* **1993**, *17*, 245–255. [CrossRef]

Article

DOA Estimation in Non-Uniform Noise Based on Subspace Maximum Likelihood Using MPSO

Jui-Chung Hung

Department of Computer Science, University of Taipei, Taipei 100, Taiwan; juichung@go.utaipei.edu.tw;
Tel.: +886-2-2311-3040

Received: 3 October 2020; Accepted: 5 November 2020; Published: 9 November 2020

Abstract: In general, the performance of a direction of arrival (DOA) estimator may decay under a non-uniform noise and low signal-to-noise ratio (SNR) environment. In this paper, a memetic particle swarm optimization (MPSO) algorithm combined with a noise variance estimator is proposed, in order to address this issue. The MPSO incorporates re-estimation of the noise variance and iterated local search algorithms into the particle swarm optimization (PSO) algorithm, resulting in higher efficiency and a reduction in non-uniform noise effects under a low SNR. The MPSO procedure is as follows: PSO is initially utilized to evaluate the signal DOA using a subspace maximum-likelihood (SML) method. Next, the best position of the swarm to estimate the noise variance is determined and the iterated local search algorithm to reduce the non-uniform noise effect is built. The proposed method uses the SML criterion to rebuild the noise variance for the iterated local search algorithm, in order to reduce non-uniform noise effects. Simulation experiments confirm that the DOA estimation methods are valid in a high SNR environment, but in a low SNR and non-uniform noise environment, the performance becomes poor because of the confusion between noise and signal sources. The proposed method incorporates the re-estimation of noise variance and an iterated local search algorithm in the PSO. This method is effectively improved by the ability to reduce estimation deviation in low SNR and non-uniform environments.

Keywords: non-uniform noise; memetic algorithms; particle swarm optimization; direction of arrival estimation; subspace maximum-likelihood

---

1. Introduction

Obtaining original signal-related information from signal sources containing interference is a very important issue [1,2]. The main sources of interference in the development of mobile communication technologies are low signal-to-noise ratio (SNR) and non-uniform noise. Array signal processing technologies have been applied to estimate the direction of arrival (DOA), using sensing elements arranged in different geometries to sample the wave field and collect spatial-related information to calculate the signal source DOA [3–6]. In wireless communications, low SNR and non-uniform noise are types of propagation phenomena, which can lead to misrecognition of the signal source and significant degradation of DOA estimation performance [7–12].

Among DOA estimation techniques, the maximum-likelihood (ML) [4,5,13] and multiple signal classification (MUSIC) [14] methods are the most representative. The ML algorithm assumes that the noise has a white Gaussian distribution and that the energy is uniform. The MUSIC method uses the autocorrelation matrix of the received signal to perform feature decomposition and decomposes its feature vector into signal and noise subspaces, utilizing the characteristics of orthogonality between the signal and noise to establish the DOA search criteria of the signal source. The performance of the ML and MUSIC methods is adversely affected by low SNR, however [12]. Therefore, Ji et al. [15] proposed the spatial MUSIC algorithm, while Zhang et al. [16] used the colony algorithm to solve computationally complex problems; however, these methods cannot deal with the DOA inaccuracy caused by excessive noise. Madurasinghe [10] suggested the power domain ML method to formulate a new objective function to solve the problem of low-energy non-uniform noise through estimation of the actual noise; however, the objective function did not propose a solution for the general low SNR and high non-uniform noise method. Wen [11] proposed the smooth space method to deal with coherent signals and non-uniform noise. This method is similar to the signal subspace projection technique under a low SNR and suffers performance loss under high non-uniform noise environments. Pesavento et al. [5] proposed a non-uniform noise and combined iterative quadratic algorithm (PIQML) to improve the estimation performance and obtain better results; however, their method shows poor performance if the non-uniform noise is too large. The judgment error will be invalid, due to early iteration. Sha et al. [17] proposed the use of projection into a subspace to establish a high-resolution estimate of the associated signal direction angle. This method can reduce the computational complexity and can handle higher resolution DOA problems, but is not suitable under low SNR conditions. This paper proposes the subspace ML (SML) method using iterated local searching by the memetic particle swarm optimization (MPSO) [8,18,19] algorithm to search the neighborhood of the signal direction, in order to build the beam-space [20]. The received data are bypassed through beamforming, which can decrease the non-uniform noise phenomenon [11,12].

The particle swarm optimization (PSO) algorithm was inspired by the social behavior of animals, such as bird flocking, swarming, and the schooling of fish. It is a branch of evolutionary algorithms, first suggested by Kennedy et al. [21]. PSO has been shown to be outstanding for the solution of DOA problems and is simple to implement [22–24]. PSO is a population-based random search optimization procedure, in which the population is called a swarm. Each swarm consists of many particles and is updated based on the influence of individual experiences, the best past experience of each individual, and the overall best experience. The swarm characteristics of parallel multi-directional search are different from the general heuristic method. The advantage of PSO is that it is simple to solve and that it has the characteristics of parallel multi-directional search, which can quickly find the optimal method but is more likely to converge to a local optimum result and does not guarantee convergence to the global optimum, especially when the objective function has a high dimension or is a complex non-linear function [25,26].

To reduce the premature convergence of PSO and to obtain an adequate solution for DOA estimation under low SNR and non-uniform noise environments, this paper proposes the combination of the iterated local search algorithm and the PSO to construct a MPSO for solving the DOA under low SNR and non-uniform noise conditions. The proposed MPSO algorithm is simple and practicable, as it adopts a first-order Taylor series expansion of the target function using the SML criterion [9,27], in order to reduce the non-uniform noise effect, therefore increasing the capacity of PSO to find the best solution. The first-order Taylor series approximates the spatial search vector and cuts it down to a direct one-dimensional optimization [20]. Simulation results show that the proposed method has a considerably improved ability to decrease the estimation bias under non-uniform noise and a low SNR environment.

The remainder of the paper is structured as follows: Section 2 describes the SML DOA estimator. Section 3 presents the SML DOA estimator using MPSO. Section 4 presents numerical simulation results, illustrating the effect of the proposed method. The final section outlines our conclusions, referring to the proposed estimator.

2. SML DOA Estimator

Assume a P narrowband signal impinges on M ($P < M$) sensors in a uniform linear array (ULA) system. The tth measured snapshot of the received signal is written as [1]:

$$\mathbf{x}(t) = \sum_{p=1}^{P} \mathbf{a}(\theta_p)s_p(t) + \mathbf{n}(t) = A(\Theta)\mathbf{s}(t) + \mathbf{n}(t), \quad t = 1, 2, \cdots, N, \tag{1}$$

where $\theta = \begin{bmatrix} \theta_1 & \theta_2 & \cdots & \theta_P \end{bmatrix}^T$ is the unknown DOA, the superscript T indicates transposition, N is the number of snapshots, $\mathbf{n}(t) = \begin{bmatrix} n_1(t) & n_2(t) & \cdots & n_M(t) \end{bmatrix}^T$ is the sensor noise, $\mathbf{x}(t) = \begin{bmatrix} x_1(t) & x_2(t) & \cdots & x_M(t) \end{bmatrix}^T$, $x_i(t)$ is the ith sensor receiving signals, $\mathbf{s}(t) = \begin{bmatrix} s_1(t) & s_2(t) & \cdots & s_P(t) \end{bmatrix}^T$ is the $P \times 1$ vector of signal amplitudes, and $\mathbf{A}(\theta)$ is the $M \times P$ composite steering matrix, expressed as

$$\mathbf{A}(\theta) = \begin{bmatrix} \mathbf{a}(\theta_1) & \mathbf{a}(\theta_2) & \cdots & \mathbf{a}(\theta_P) \end{bmatrix}^T,$$
$$\mathbf{a}(\theta_i) = \begin{bmatrix} 1 & \exp(-j2\pi d \sin\theta_i/\chi) & \cdots & \exp(-j2\pi d(M-1)\sin\theta_i/\chi) \end{bmatrix}^T, \tag{2}$$

where $\mathbf{a}(\theta_i)$ is the steering vector, χ is the wavelength, and d is the sensor spacing between two neighboring sensors. In this paper, the sensor noise, $\mathbf{n}(t)$, is considered to be non-uniform and to be a zero mean Gaussian process, such that

$$E[\mathbf{n}(t)] = 0$$
$$\mathbf{R}_n = E\left[\mathbf{n}(t)\mathbf{n}(t)^H\right] = diag\{\sigma_1^2, \sigma_2^2, \cdots, \sigma_M^2\}, \tag{3}$$

where $E[\cdot]$ is the expectation, the superscript H denotes the complex conjugate transpose, $diag\{.\}$ is a diagonal matrix composed of the bracketed elements, and σ_i^2 is the ith sensor's noise variance. In general, the sensor noise $\mathbf{n}(t)$ is uncorrelated with all signals. The array covariance corresponding to Equation (1) can be expressed as

$$\mathbf{R} = E\{\mathbf{x}(t)\mathbf{x}^H(t)\} = \mathbf{A}(\theta)\mathbf{R}_s\mathbf{A}^H(\theta) + \mathbf{R}_n, \tag{4}$$

where $\mathbf{R}_s = E[\mathbf{s}(t)\mathbf{s}^H(t)]$ is the covariance matrix of the signal amplitudes. The array covariance matrix can be estimated by the sample average, $\hat{\mathbf{R}}$:

$$\hat{\mathbf{R}} = \frac{1}{N}\sum_{t=1}^{N} \mathbf{x}(t)\mathbf{x}^H(t). \tag{5}$$

The ML estimator for non-uniform noise can be found using the weighted least-squares approach [10], using the normalized composite steering matrix and noise component. The maximum likelihood problem becomes a least-squares solution [10]:

$$L(\theta, \sigma^2) = \min_{\theta, \sigma^2} \sum_{t=1}^{N} \left| \overline{\mathbf{x}}(t) - \overline{\mathbf{A}}(\theta)\mathbf{s}(t) \right|^2, \tag{6}$$

where σ^2 is the $M \times 1$ vector of noise variance, $|.|^2$ denotes the l_2 norm, $\overline{\mathbf{x}}(t)$ is the normalized receiving signal, $\overline{\mathbf{x}}(t) = \mathbf{R}_n^{-1/2}\mathbf{x}(t)$, and $\overline{\mathbf{A}}(\theta)$ is the normalized steering matrix, $\overline{\mathbf{A}}(\theta) = \mathbf{R}_n^{-1/2}\mathbf{A}(\theta)$. The $\overline{\mathbf{A}}(\theta)\mathbf{s}(t)$ of Equation (6) is separable and, for fixed $\overline{\mathbf{A}}(\theta)$, $\mathbf{s}(t)$ can be obtained by using the pseudo-inverse [9]:

$$\mathbf{s}(t) = \left[\overline{\mathbf{A}}^H(\theta)\overline{\mathbf{A}}(\theta)\right]^{-1}\overline{\mathbf{A}}^H(\theta)\overline{\mathbf{x}}(t). \tag{7}$$

Given Equation (7), substituting the ML estimator into Equation (6) results in

$$L(\theta, \sigma^2) = \min_{\theta, \sigma^2} tr\{\mathbf{P}_{\overline{\mathbf{A}}}\hat{\mathbf{R}}\}, \tag{8}$$

where $\mathbf{P}_{\overline{\mathbf{A}}} = I - \overline{\mathbf{A}}(\theta)\left[\overline{\mathbf{A}}^H(\theta)\overline{\mathbf{A}}(\theta)\right]^{-1}\overline{\mathbf{A}}^H(\theta)$, I is the unit diagonal matrix, and $tr\{.\}$ is the trace of the matrix. Equation (8) is a multi-objective minimization problem. In general, $L(\theta, \sigma^2)$ is a very highly non-linear function of θ and σ^2, and the cost function is highly non-linear; it is impossible to represent the target function with any closed-form expression [16]. However, the ML estimator still has inferior performance when the noise is non-uniform and in a low SNR environment. It is well-known that the signal subspace projection method can weaken the noise effect of the received noisy data vector [9].

Applying eigende composition, the sample covariance matrix Equation (5) can be expressed as

$$\hat{\mathbf{R}} = \sum_{m=1}^{M} \lambda_m \mathbf{e}_m \mathbf{e}_m^H \qquad m = 1, \ldots, M,$$
(9)

where $\lambda_1 \geq \lambda_2 \geq \cdots \geq \lambda_M$ are the eigenvalues of $\hat{\mathbf{R}}$ and $\mathbf{e}_m$ denotes the eigenvector associated with λ_m for $m = 1, 2, \cdots, M$. The column spans of $\mathbf{E_s} = [\mathbf{e}_1, \ldots, \mathbf{e}_P]$ and $\mathbf{E}_n = [\mathbf{e}_{P+1}, \ldots, \mathbf{e}_M]$ are defined as the signal and noise subspaces, respectively. The covariance matrix can be expressed as the summation of two orthogonal components, $\mathbf{E_s}\mathbf{E}_s^H$ and $\mathbf{E}_n\mathbf{E}_n^H$. Hence, this paper adopts the signal subspace projection $\mathbf{E_s}\mathbf{E}_s^H\mathbf{x}(t)$, the projection of $\mathbf{x}(t)$ onto the columns of $\mathbf{E_s}$. Properly constructing the signal subspace projection-based approach to filter the non-uniform noise can effectively enhance the performance. The SML estimator using the ML estimator of Equation (8) can be expressed as:

$$L(\theta, \sigma^2) = \min_{\theta, \sigma^2} tr\{\mathbf{P}_{\overline{\mathbf{A}}}\mathbf{E_s}\mathbf{E}_s^H\hat{\mathbf{R}}\}.$$
(10)

Selection of a signal subspace under a low SNR or non-uniform noise is very difficult when using eigendecomposition. As the received signal has low SNR or non-uniform noise, $\mathbf{E}_s$ may contain a noise subspace and $\mathbf{E}_n$ may contain a signal subspace. This paper adopts the reiterated procedure of a method to reduce such confusion. First, it is assumed that the noise variance is constant for all sensors (λ_M) and that the SML estimator in Equation (10) can be expressed as

$$L(\theta) = \min_{\theta} tr\{\mathbf{P}_{\overline{\mathbf{A}}}\mathbf{E_s}\mathbf{E}_s^H\hat{\mathbf{R}}\}.$$
(11)

Next, using Equation (11) to obtain the DOA, the noise variance $\hat{\mathbf{R}}_n$ can be estimated by

$$\hat{\mathbf{R}}_n = \frac{1}{N}\sum_{t=1}^{N}\left|\mathbf{x}(t) - \mathbf{A}(\hat{\theta})\mathbf{s}(t)\right|^2,$$
$$\mathbf{s}(t) = \left[\mathbf{A}^H(\hat{\theta})\mathbf{A}(\hat{\theta})\right]^{-1}\mathbf{A}^H(\hat{\theta})\mathbf{x}(t),$$
(12)

where $\hat{\theta}$ is the estimated DOA. This procedure implies that, for a fixed $\hat{\mathbf{R}}_n$, the solution θ minimizes $L(\theta)$ in Equation (11), and vice versa. Once θ is obtained, a refined result for $\hat{\mathbf{R}}_n$ can be achieved using Equation (12). Hence, $\hat{\theta}$ and $\hat{\mathbf{R}}_n$ can be estimated in an iterative manner.

The steering matrix can be determined using a first-order Taylor series approximation expansion at the estimation DOA $\hat{\theta}$, expressed as [1]:

$$\overline{\mathbf{A}}(\theta) = \overline{\mathbf{A}}(\hat{\theta} + \eta\theta) \cong \overline{\mathbf{A}}(\hat{\theta}) + \eta\theta\,\overline{\mathbf{A}'}(\hat{\theta}),$$
(13)

where $\eta\theta$ is a small value and $\overline{\mathbf{A}'}(\hat{\theta}) = \frac{d}{d\theta}\overline{\mathbf{A}}(\theta)\big|_{\theta=\hat{\theta}}$. Substituting Equation (13) into Equation (11), the following expression can be obtained:

$$L(\theta) = \min_{\eta\theta} tr\{\mathbf{P}_{\overline{\mathbf{A}}(\hat{\theta})+\eta\theta\overline{\mathbf{A}'}(\hat{\theta})}\mathbf{E_s}\mathbf{E}_s^H\hat{\mathbf{R}}\},$$
(14)

which is a direct one-dimensional optimization problem. Then, it can easily be shown that the optimum $\eta\theta$ is given by $\dfrac{d\{tr\{\mathbf{P}_{\overline{\mathbf{A}}(\hat{\theta})+\eta\theta\overline{\mathbf{A}'}(\hat{\theta})}\mathbf{E_s}\mathbf{E}_s^H\hat{\mathbf{R}}\}\}}{d(\eta\theta)} = 0$. The value of $|\eta\theta|$ has two characteristics [8]: first, if the value of $|\eta\theta|$ is small, it can achieve a convergence result that may be local or global. However, if the value of $|\eta\theta|$ is large, the results may remain far from convergence. Using these characteristics of $|\eta\theta|$, $\hat{\theta}$ is updated by:

$$\hat{\theta} = \begin{cases} \hat{\theta} + \eta\theta, & \text{if } |\eta\theta| > \varepsilon \\ \hat{\theta}, & \text{if } |\eta\theta| \leq \varepsilon \end{cases} , \tag{15}$$

where ε is the search precision value. In this paper, Equation (15) is repeated until $|\eta\theta| < \varepsilon$. The proposed procedure is as follows:

1. Given the received signal $x(t)$, compute $\hat{R}$ in Equation (5) and the eigendecomposition in Equation (9).
2. Assume that the noise variance, which is constant for all sensors, is λ_M.
3. Estimate the DOA θ using the SML estimator in Equation (11).
4. Update the estimate of θ in Equation (15) using the $\eta\theta$ property until $|\eta\theta| \leq \varepsilon$.
5. Update the estimated non-uniform noise $\hat{R}_n$ in Equation (12).
6. Reiterate Steps 3 to 5 until $\hat{\theta}$ and $\hat{R}_n$ converge.

From the above, we propose a new method, which is a hybrid algorithm that combines PSO and the estimated DOA θ with an iterated local search algorithm for $|\eta\theta|$.

3. The Proposed Method

This paper proposed a hybrid algorithm that incorporates PSO with an iterated local search algorithm. The DOA with SML estimator criterion cannot be directly carried out under a non-uniform noise and low SNR environment, in which the closed form of the contained criteria are $\{\theta, \sigma^2\}$ at the same time. Therefore, MPSO is introduced to solve the issue.

The MPSO includes the following two components: First, the PSO-based SML estimation method searches the entire space. Second, the local search using the property achieves a more accurate search around potential solutions of the first component. The design process of the MPSO is expressed below.

3.1. The PSO-Based SML Estimation

The PSO algorithm is an optimized search method based on a group that is easy to use, as the algorithm requires few parameters. The swarm of the PSO algorithm consists of many particles. Each individual particle represents a solution; it has its own position and velocity, the initial values of which are set randomly. Then, each particle obtains a value measure from a target function; the changing of the particle position is regulated by the value of the objective function. Our objective is to minimize the value of the fitness function. We use the fitness function $L(\theta)$ presented in Equation (11). There are three kinds of influences on the movement of particles. First, movement in the previous direction; second, movement towards the position of the individual particle's optimization situation; and, finally, movement towards the position of the global optimization situation of the overall swarm [21].

Let S denote the swarm size, $\mathbf{v}_i(t) = [\begin{array}{cccc} v_{i,1}(t) & v_{i,2}(t) & \cdots & v_{i,P}(t) \end{array}]^T$ be the current velocity, and $\theta_i(t) = [\begin{array}{cccc} \theta_{i,1}(t) & \theta_{i,2}(t) & \cdots & \theta_{i,P}(t) \end{array}]^T$ be the current position. During each iteration, the update to the velocity $v_{i,j}(t+1)$ and position $\theta_i(t+1)$ of each particle is as follows:

$$v_{i,j}(t+1) = \kappa \cdot v_{i,j}(t) + c_1 \times \varphi_{1,i}(t) \times (p_{i,j} - \theta_{i,j}(t)) + c_2 \times \varphi_{2,i}(t) \times (g_j - \theta_{i,j}(t)), \\ \text{for all } i = 1, 2, \cdots, S, j = 1, 2, \cdots, P \tag{16}$$

$$\theta_i(t+1) = \theta_i(t) + \mathbf{v}_i(t+1), \tag{17}$$

where $v_{i,j}(t)$ is the velocity of the jth dimension of the ith particle, κ is the inertia weight (this value is typically set as $0 \leq \kappa < 1$), c_1 and c_2 are set near 2.0 [27], $\varphi_{1,i}(t)$ and $\varphi_{2,i}(t)$ are uniformly distributed random numbers in the range [0,1] $p_{i,j}$ is the individual best position of the jth dimension of the ith particle, and g_j is the global best position of the jth dimension. The individual best position of each particle is updated using

$$\mathbf{p}_i = \begin{cases} \mathbf{p}_{i\prime} & \text{if } L(\theta_i(t+1)) \geq L(\theta_i(t)) \\ \theta_i(t+1), & \text{if } L(\theta_i(t+1)) < L(\theta_i(t)) \end{cases}, \tag{18}$$

where $\mathbf{p}_i = \begin{bmatrix} p_{i,1} & p_{i,2} & \cdots & p_{i,P} \end{bmatrix}^T$. The overall best position can be found as follows:

$$\mathbf{g} = \min L(\mathbf{p}_i), \, for \, i = 1, 2, \cdots, S, \tag{19}$$

where $\mathbf{g} = \begin{bmatrix} g_1 & g_2 & \cdots & g_P \end{bmatrix}^T$. The value of the velocity $v_{i,j}(t)$ can be limited to the range $[-v_{\max}, v_{\max}]$, in order to reduce the number of particles escaping the search space with an uncontrollable trajectory [22]. In this paper, we use $v_{\max} = 0.1 \times \theta_{\max} = 18°$ for the ULA system [8].

3.2. The MPSO Estimator

The MPSO proposed in this paper combines the application of a local search method into the PSO algorithm, in order to solve the problem that the SML estimator criterion cannot obtain a closed form solution. Using the PSO to estimate $\hat{\theta}$ in Equation (12), the characters of $|\eta\theta|$ in the local search method and the re-estimated $\hat{\mathbf{R}}_n$ are combined. A small $|\eta\theta|$ value will generate convergent results, while a large $|\eta\theta|$ value requires a greater time to converge and may cause a value that changes the estimated deviation towards a more correct value. The pseudocode of the Algorithm 1 MPSO is as follows:

Algorithms 1. MPSO

Input: received signal $\mathbf{x}(k)$ and set of initial values for $S, c_1, c_2, \varepsilon, \theta_{\min}, \theta_{\max}$ in MPSO.
Output: DOA $\theta \in \begin{bmatrix} \theta_{\min} & \theta_{\max} \end{bmatrix}$
Step 1: Set $t = 0$.
Step 2: Evaluate $\hat{\mathbf{R}}$ by Equation (5), λ_M by Equation (9), and randomly uniformly generate $\theta_i(t)$ and $\mathbf{v}_i(t)$ for $i = 1, 2, \cdots, S$.
Step 3: Evaluate fitness value $L(\theta_i(t))$ in Equation (11).
Step 4: Set $\mathbf{p}_i \leftarrow \theta_i(t)$ and $\mathbf{g} = \min\{\mathbf{p}_i\}$.
Step 5: Update the velocities $v_{i,j}(t+1)$ using Equation (16) and positions $\theta_i(t+1)$ using Equation (17).
Step 6: Evaluate the fitness value $L(\theta_i(t+1))$ in Equation (11).
Step 7: IF $L(\theta_i(t+1)) < L(\theta_i(t))$, **then** $\mathbf{p}_i \leftarrow \theta_i(t+1)$.
Step 8: $\mathbf{g} \leftarrow \min\{\mathbf{p}_i\}$ and $\hat{\theta} \leftarrow \mathbf{g}$
Step 9: While $|\eta\theta| > \varepsilon$, **Do**
        Update $\hat{\theta} \leftarrow \hat{\theta} + \eta\theta$
        Evaluate $|\eta\theta|$
      End While.
Step 10: Evaluate $\hat{\mathbf{R}}_n$ in Equation (12)
Step 11: Set $\mathbf{g} \leftarrow \hat{\theta}$.
Step 12: Set $t = t + 1$.
Step 13: Go to Step 5 until the stopping criterion is satisfied.

Based on the above analysis, Figure 1 presents a flowchart of the proposed algorithm.

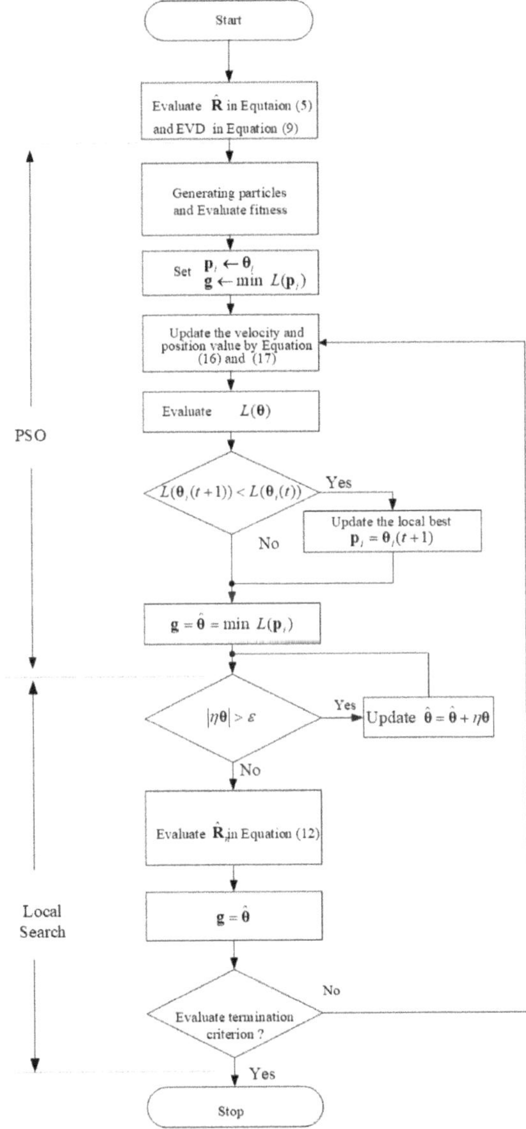

Figure 1. The flowchart of the proposed method.

4. Simulation Result

Two examples were considered in order to illustrate the practicability of using the proposed algorithm for DOA estimation in a non-uniform noise and low SNR environment. Simulation results were used to compare the performance of the proposed algorithm with the MUSIC [14], minimum variance distortionless response (MVDR) [1], and power domain ML (Power-Domain) methods [10]. The non-uniform noise (using the worst noise power ratio; WNPR) and SNR were defined using:

$$\text{WNPR} = \frac{\sigma_{\max}^2}{\sigma_{\min}^2},$$

$$\text{SNR} = \frac{s_p^2}{M} \sum_{i=1}^{M} \frac{1}{\sigma_i^2}, \ for \ p = 1, 2, \cdots, P, \tag{20}$$

where $\sigma^2_{\max}$ and $\sigma^2_{\min}$ are the maximum and minimum non-uniform noise variances, respectively. The simulated results were obtained by averaging 500 independent Monte Carlo (MC) runs. In the literature, a variety of statistical methods have been applied to compare performance. These include the mean absolute error (MAE) and the root-mean-squared error (RMSE), which are defined as

$$\mathrm{MAE} = \frac{1}{500 \times P} \sum_{j=1}^{500} \sum_{p=1}^{P} \left| \hat{\theta}_{j,p} - \theta_p \right|,$$

$$\mathrm{RMSE} = \sqrt{\frac{1}{500 \times P} \sum_{j=1}^{500} \sum_{p=1}^{P} \left(\hat{\theta}_{j,p} - \theta_p \right)^2}, \tag{21}$$

where $\hat{\theta}_{j,p}$ is the jth MC run for the θ_p estimate value. The search grid capacity for the spectrum scan of MUSIC [14] was set as $0.001°$. The initial set of parameters of the proposed estimators were defined as

$$c_1 = c_2 = 2.05, \chi = 0.99, S = 200, T = 50, \varepsilon = 0.001, \tag{22}$$

where c_1, c_2 are acceleration coefficients, χ is the inertial weight, S is the size of the swarm, T is the number of iterations, and ε is the search precision value.

The first example considered a four-element ULA with half-wavelength inter-element spacing, where the noise power was given as $\sigma^2 = \begin{bmatrix} 5 & 10 & 0.1 & 6 \end{bmatrix}^t$, WNPR = 50, and the source had the DOA $\theta_1 = 5°$. Figure 2 illustrates the RMSE values of the estimated DOA versus the number of snapshots under SNR = −15 dB. The proposed method achieved a faster convergence approach with 75 snapshots, while the other methods converged at about 150 snapshots. Figure 3 shows that the RMSE of the various estimators versus different SNRs under snapshots was 100. In Figure 3, we can see that all estimator RMSEs were near zero under high SNR, but the performance of other estimators decayed under low SNR conditions. Table 1 shows the RMSE and MAE under various SNRs. The bold text is the best estimated value under the same SNR. The proposed algorithm had RMSE and MAE values smaller than those of the other estimators, especially under low SNR environments.

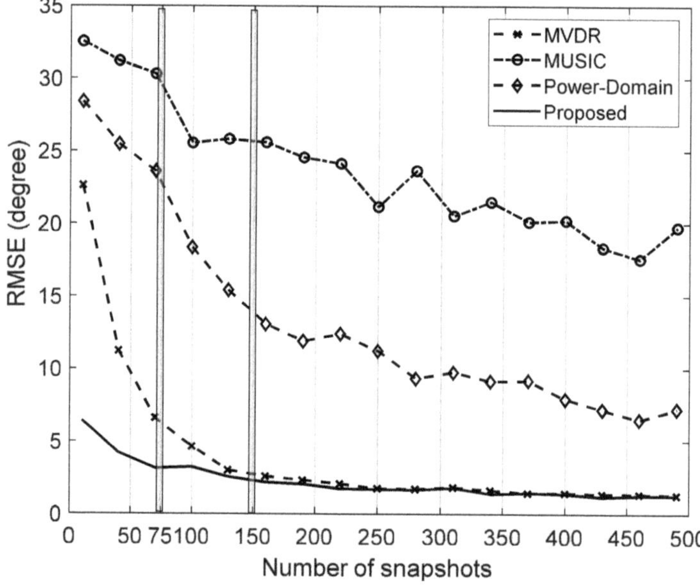

Figure 2. Root-mean-squared error (RMSE) versus the snapshot for the different estimators (MUSIC, MVDR, Power-Domain, and the proposed estimator) for Example 1.

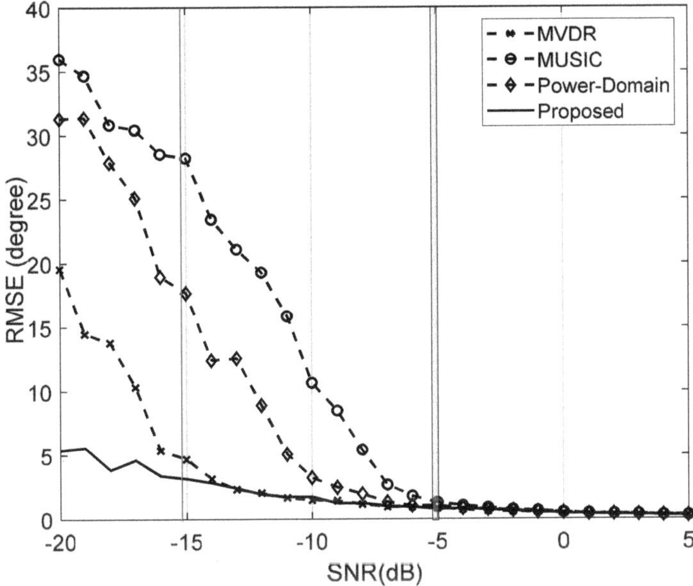

Figure 3. RMSE versus the signal-to-noise ratio (SNR) for the different estimators (MUSIC, MVDR, Power-Domain, and the proposed estimator) in Example 1.

Table 1. Direction of arrival (DOA) evaluation for the different estimators (multiple signal classification (MUSIC), minimum variance distortionless response (MVDR), power domain maximum likelihood (Power-Domain), and the proposed estimator) in Example 1.

		MVDR		MUSIC		Power-Domain		Proposed Method	
		RMSE	MAE	RMSE	MAE	RMSE	MAE	RMSE	MAE
	−20 dB	19.5392		35.9564		31.3039		5.3792	
	−15 dB	4.6780		28.2148		17.6577		3.1546	
SNR	−10 dB	1.4788		10.6452		3.2314		1.7144	
	−5 dB	0.8232		1.2765		1.0158		0.7440	
	0 dB	0.4282		1.2765		0.5046		0.4368	
	5 dB	0.2368		0.2728		0.2545		0.2392	

The second example considered an eight-element ULA and two sources with DOAs $\theta = \begin{bmatrix} -3° & 6° \end{bmatrix}^t$. The additive background noise variance was $\sigma^2 = \begin{bmatrix} 6 & 2 & 0.5 & 2.5 & 3 & 1 & 1.5 & 10 \end{bmatrix}^t$ and WNPR = 20. The other parameters were the same as in Example 1. Figure 4 indicates the RMSE of the various estimators versus different snapshots under SNR = 0 dB. Again, this figure indicates that the proposed estimator not only carried out faster convergence (at 50 snapshots) but also offered an improvement in DOA evaluation accuracy. Figure 5 indicates that the RMSE of the various estimators versus different SNRs under snapshots was 100. In Figure 5, we can see that the other estimators were not sensitive to various low SNRs (SNR < −5 dB), when comparing their performance with the proposed method. As hoped, the results indicate again that the DOA evaluation accuracy was improved by the proposed estimator. Table 2 gives the DOA estimates (RMSE, MAE) of the MUSIC, MVDR, and Power-Domain methods, along with those of the proposed estimator. The evaluation accuracy of the MUSIC, MVDR, and Power-Domain estimators worsened under a low SNR. Moreover, the accuracy of the proposed estimator was better than those of the other estimators under low SNR conditions.

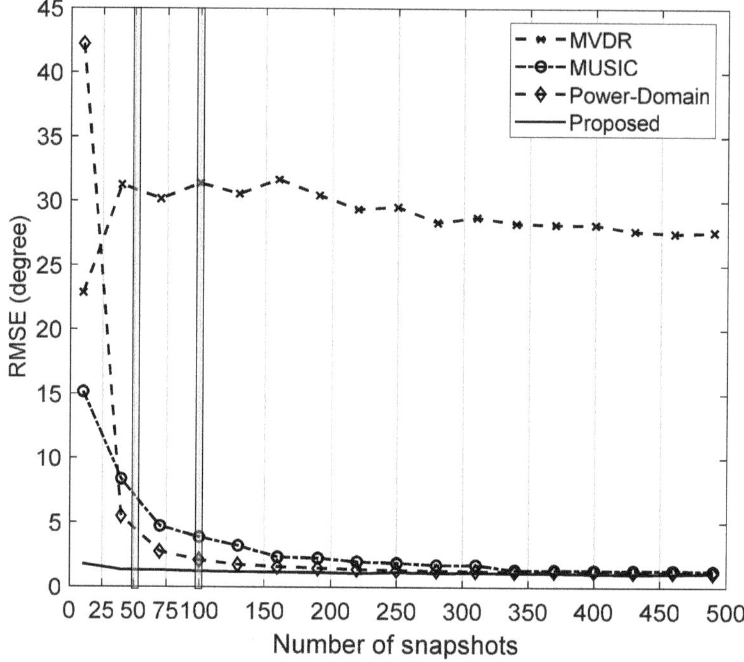

Figure 4. RMSE versus snapshot for the different estimators (MUSIC, MVDR, Power-Domain, and the proposed estimator) in Example 2.

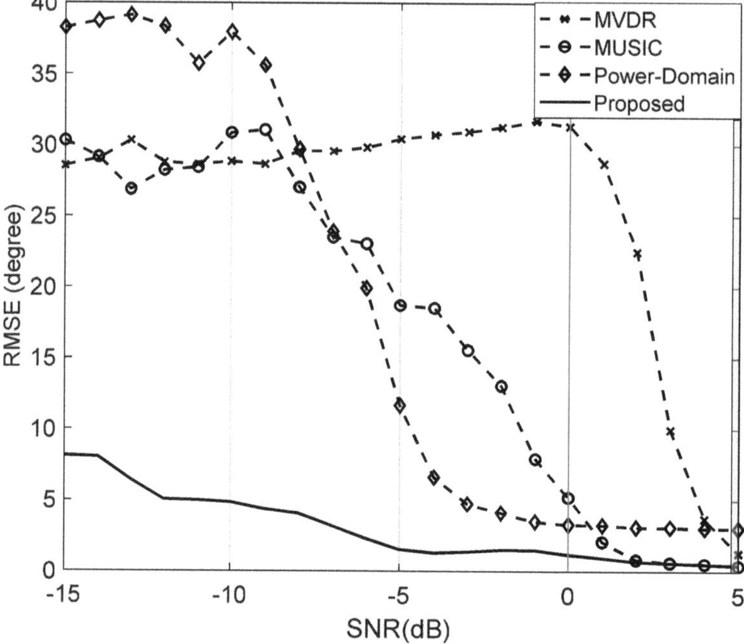

Figure 5. RMSE versus SNR for the different estimators (MUSIC, MVDR, Power-Domain, and the proposed estimator) in Example 2.

Table 2. DOA evaluation for the different estimators (MUSIC, MVDR, Power-Domain, and the proposed estimator) in Example 2.

		MVDR		MUSIC		Power-Domain		Proposed Method	
		RMSE	MAE	RMSE	MAE	RMSE	MAE	RMSE	MAE
	−15 dB	28.5391	22.2818	30.3017	25.5400	38.2552	30.3440	8.1092	7.3989
	−10 dB	28.8271	22.4764	30.8397	24.0450	37.9480	29.2230	4.8597	4.4936
SNR	−5 dB	30.4322	23.3296	18.7202	15.3910	11.6566	9.4562	1.5421	1.4312
	0 dB	31.3296	24.2212	5.2164	4.4160	3.2706	2.9772	1.1865	1.1242
	5 dB	1.2848	1.2488	0.3967	0.3590	3.0139	2.9830	0.4313	0.3950

5. Conclusions

DOA estimation cannot be directly carried out under a non-uniform noise and low SNR environment, in which the closed form of the contained criteria is $\{\theta, \sigma^2\}$. Generally, it is necessary to process the DOA (θ) and the noise variance (σ^2). In this paper, a new re-iterated process was introduced, in which the noise variance is fixed to estimate the DOA and, vice versa, the DOA is fixed to estimate the noise variance. After several iterations, the procedure converges to the nearest correct estimates of the DOA and the noise variance. The proposed solution combines the MPSO scheme, which uses the fixed noise variance to estimate the DOA through the PSO algorithm, using the best particle to estimate the noise variance. An MPSO that incorporates the re-estimation of noise variance and an iterated local search algorithm is applied in the PSO, resulting in an efficient reduction of the non-uniform noise effect under a low SNR. The iterated local search of the MPSO method exploits the characteristics of the first-order Taylor expansion $|\eta\theta|$. A small $|\eta\theta|$ value can guarantee convergent results that may be local or global while, with a large $|\eta\theta|$ value, convergence will take a longer time and may provide a value that updates the estimated deviation toward the correct value. Empirical evidence shows that the DOA estimation methods are valid in a high SNR environment, but in a low SNR and non-uniform noise environment, the performance becomes poor because the noise is confused by the source of the signal. The proposed method incorporates the re-estimation of noise variance and an iterated local search algorithm in the PSO. This method is effectively able to reduce estimation deviation and to achieve high accuracy and fast convergence in low SNR and non-uniform environments. Generally, a low SNR and non-uniform environment is caused by foul weather. This problem occurs in mountaineering, so this method provides valid DOA estimation in this environment, and can be used for positioning system issues.

Funding: This research received no external funding.

Conflicts of Interest: The author declares no conflict of interest.

References

1. Haykin, S. *Array Signal Processing*; Prentice Hall: Upper Saddle River, NJ, USA, 1985.
2. Roy, R.; Kailath, T. ESPRIT-estimation of signal parameters via rotational invariance techniques. *IEEE Trans. Acoust. Speech Signal Process.* **1989**, *37*, 984–995. [CrossRef]
3. Arikan, F.; Koroglu, O.; Fidan, S.; Arikan, O.; Guldogan, M.B. Multipath separation-direction of arrival (MS-DOA) with genetic search algorithm for HF channels. *Adv. Space Res.* **2009**, *44*, 641–652. [CrossRef]
4. Panigrahi, T.; Roula, S.; Gantayat, H. Application of comprehensive learning particle swarm optimisation algorithm for maximum likelihood DOA estimation in wireless sensor networks. *Int. J. Swarm Intell.* **2016**, *2*, 208–228. [CrossRef]
5. Pesavento, M.; Gershman, A.B. Maximum-likelihood direction-of-arrival estimation in the presence of unknown nonuniform noise. *IEEE Trans. Signal Process.* **2001**, *49*, 1310–1324. [CrossRef]

6. Thomas, J.A.; Mini, P.R.; Kumar, M.A. DOA estimation and adaptive beamforming using MATLAB and GUI. In Proceedings of the 2017 International Conference on Energy, Communication, Data Analytics and Soft Computing (ICECDS), Chennai, India, 1–2 August 2017; pp. 1890–1896.

7. Gupta, A.; Christodoulou, C.G.; Rojo-Álvarez, J.L.; Martínez-Ramón, M. Gaussian Processes for Direction-of-Arrival Estimation with Random Arrays. *IEEE Antennas Wirel. Propag. Lett.* **2019**, *18*, 2297–2300. [CrossRef]

8. Hung, J.-C. Memetic particle swarm optimization scheme for direction-of-arrival estimation in multipath environment. *J. Intell. Fuzzy Syst.* **2018**, *34*, 3955–3968. [CrossRef]

9. Liao, B.; Chan, S.-C.; Huang, L.; Guo, C. Iterative methods for subspace and DOA estimation in nonuniform noise. *IEEE Trans. Signal Process.* **2016**, *64*, 3008–3020. [CrossRef]

10. Madurasinghe, D. A new DOA estimator in nonuniform noise. *IEEE Signal Process. Lett.* **2005**, *12*, 337–339. [CrossRef]

11. Wen, J.; Liao, B.; Guo, C. Spatial smoothing based methods for direction-of-arrival estimation of coherent signals in nonuniform noise. *Digit. Signal Process.* **2017**, *67*, 116–122. [CrossRef]

12. Yang, Y.; Hou, Y.; Mao, X.; Jiang, G. Stokes Parameters and DOA Estimation for Nested Polarization Sensitive Array in Unknown Nonuniform Noise Environment. *Signal Process.* **2020**, *175*, 107630. [CrossRef]

13. Chen, C.E.; Lorenzelli, F.; Hudson, R.E.; Yao, K. Stochastic maximum-likelihood DOA estimation in the presence of unknown nonuniform noise. *IEEE Trans. Signal Process.* **2008**, *56*, 3038–3044. [CrossRef]

14. Schmidt, R. Multiple emitter location and signal parameter estimation. *IEEE Trans. Antennas Propag.* **1986**, *34*, 276–280. [CrossRef]

15. Ji, A.G.; Wang, T.; Zhu, Z.G. DOA estimation based on the multipath of the MIMO radar target. In *Applied Mechanics and Materials*; Trans Tech Publications Ltd.: Stafa-Zurich, Switzerland, 2012; pp. 2029–2032.

16. Zhang, Z.; Lin, J.; Shi, Y. Application of artificial bee colony algorithm to maximum likelihood DOA estimation. *J. Bionic Eng.* **2013**, *10*, 100–109. [CrossRef]

17. Sha, Z.-C.; Liu, Z.; Huang, Z.; Zhou, Y. Direction estimation of correlated/coherent signals by sparsely representing the signal-subspace eigenvectors. *Prog. Electromagn. Res.* **2013**, *40*, 37–52. [CrossRef]

18. Alsmadi, M.K. An efficient similarity measure for content based image retrieval using memetic algorithm. *Egypt. J. Basic Appl. Sci.* **2017**, *4*, 112–122. [CrossRef]

19. Huang, K.-W.; Wu, Z.-X.; Peng, H.-W.; Tsai, M.-C.; Hung, Y.-C.; Lu, Y.-C. Memetic particle gravitation optimization algorithm for solving clustering problems. *IEEE Access* **2019**, *7*, 80950–80968. [CrossRef]

20. Hung, J.-C. Modified particle swarm optimization structure approach to direction of arrival estimation. *Appl. Soft Comput.* **2013**, *13*, 315–320. [CrossRef]

21. Kennedy, J.; Eberhart, R. Particle swarm optimization. In Proceedings of the ICNN'95-International Conference on Neural Networks, Perth, Australia, 27 November–1 December 1995; pp. 1942–1948.

22. Chen, H.; Li, S.; Liu, J.; Liu, F.; Suzuki, M. A novel modification of PSO algorithm for SML estimation of DOA. *Sensors* **2016**, *16*, 2188. [CrossRef]

23. Sharma, A.; Mathur, S. Comparative analysis of ML-PSO DOA estimation with conventional techniques in varied multipath channel environment. *Wirel. Pers. Commun.* **2018**, *100*, 803–817. [CrossRef]

24. Wang, H.; Moon, I.; Yang, S.; Wang, D. A memetic particle swarm optimization algorithm for multimodal optimization problems. *Inf. Sci.* **2012**, *197*, 38–52. [CrossRef]

25. Clerc, M.; Kennedy, J. The particle swarm-explosion, stability, and convergence in a multidimensional complex space. *IEEE Trans. Evol. Comput.* **2002**, *6*, 58–73. [CrossRef]

26. Hajihassani, M.; Armaghani, D.J.; Kalatehjari, R. Applications of particle swarm optimization in geotechnical engineering: A comprehensive review. *Geotech. Geol. Eng.* **2018**, *36*, 705–722. [CrossRef]

27. Liu, A.; Liao, G.; Zeng, C.; Yang, Z.; Xu, Q. An eigenstructure method for estimating DOA and sensor gain-phase errors. *IEEE Trans. Signal Process.* **2011**, *59*, 5944–5956. [CrossRef]

Publisher's Note: MDPI stays neutral with regard to jurisdictional claims in published maps and institutional affiliations.

Article

A Reference-Model-Based Neural Network Control Method for Multi-Input Multi-Output Temperature Control System

Yuan Liu [1,†,‡], Song Xu [1,2,‡], Seiji Hashimoto [1,*,‡] and Takahiro Kawaguchi [1,‡]

[1] Division of Electronics and Informatics, Gunma University, 1-5-1 Tenjincho, Kiryu 376-8515, Japan; t192d005@gunma-u.ac.jp (Y.L.); xusong0922@outlook.com (S.X.); kawaguchi@gunma-u.ac.jp (T.K.)

[2] Department of Electronics and Informatics, Jiangsu University of Science and Technology N.2 Mengxi Road, Zhenjiang 212000, China

* Correspondence: hashimotos@gunma-u.ac.jp; Tel.: +81-277-3-1741

† Current address: 5-1067-5 Hishimachi, Kiryu 376-0001, Gunma, Japan.

‡ These authors contributed equally to this work.

Received: 24 September 2020; Accepted: 23 October 2020; Published: 28 October 2020

Abstract: Neural networks (NNs), which have excellent ability of self-learning and parameter adjusting, has been widely applied to solve highly nonlinear control problems in industrial processes. This paper presents a reference-model-based neural network control method for multi-input multi-output (MIMO) temperature system. In order to improve the learning efficiency of the NN control, a reference model is introduced to provide the teaching signal for the NN controller. The control inputs for the MIMO system are given by the sum of the output of the conventional integral-proportional-derivative (I-PD) controller and the outputs of the neural network controller.The proposed NN control method can not only improve the transient response of the system, but can also realize temperature uniformity in MIMO temperature systems. To verify the proposed method, simulations are carried out in MATLAB/SIMULINK environment and experiments are carried out on the DSP (Digital Signal Processor)-based experimental platform, respectively. Both results are quantitatively compared to those obtained from the conventional I-PD control systems. The effectiveness of the proposed method has been successfully verified.

Keywords: neural network control; multi-input multi-output temperature system; transient response; temperature uniformity

---

1. Introduction

To realize the precise temperature of the industrial process, temperature controllers are widely applied to manage manufacturing processes and operations. The common uses include food processing, packaging machines, and plastic extrusion etc. Their performances can seriously affect product quality, energy consumption, and production cost. In practical application, the proportional-integral-derivative (PID) controller has been widely used for its simple structure and wide applicability, especially in linear systems or first and second order systems [1]. Common methods for determining PID controller parameters are Ziegler–Nichols (ZN) and Cohen–Coon tuning rules, which are widely used in the industry due to their simplicity and ease of implementation, but may easily result in overshoot and weak response damping [2]. Meanwhile, most industrial processes are multi-variable nonlinear systems with big time constants, strong coupling effects, and large time delays. For such controlled objects, only using the conventional PID controller may not satisfy the requirements of the system performance.

In order to ensure the robustness and stability of the controller, many methods have been proposed to improve the performance of the PID controller, such as gain and phase margin [3], pole placement [4], and internal model control (IMC) [5]. For canceling the coupling interactions between each channels in multi-input multi-output (MIMO) systems, a verity of decoupling control strategies have been applied, such as inverse Nyquist array(INA) [6], feedforward decoupling control [7], and inverse based decoupling control [8].These decoupling control methods are to design the decoupler so that the MIMO control system can be divided into multiple single-input single-output (SISO) loops, which allows one to adopt well developed single loop control methods. However, the modeling error may decrease system performance as decoupling is guaranteed under the condition that the precise mathematical model is obtained. The quality of the obtained model depends on many on-site factors, such as the excitation condition and selected identification algorithm [9–11]. In addition, the decoupling control is not enough to reach uniform temperature control in transient state without adjusting the PID parameters in different channels [12].

Considering the complex control rules and controller computation grow exponentially with a number of variables in nonlinear MIMO systems, intelligent control strategies are developed and widely used. Fuzzy logic control, genetic algorithms, and neural networks (NN) are the most promising methods among them [13–15]. NN is known for its great computing power and learning ability to emulate various systems dynamics with a highly parallel structure. Over the past few decades, NN has been successfully applied in many fields such as system modeling, pattern recognition, and signal processing [16–19]. In thermal systems, the NN has been used for heat transfer data analysis, performance prediction, and dynamic modeling etc. [20–23]. It is shown that NN is well suitable to deal with complex nonlinear relationships in control systems. NN helps solve the problem in typical heating system control methods that once the parameters of the control system are designed, they cannot be adjusted while the system is in operation. For a thermal process system with strong nonlinearly, large lag, and strong coupling, an adaptive system can improve control performance in terms of the transient response and overshoot [24–26].

Our previous research has proposed the NN control method applied to the temperature control system, the proposal has successfully improved the transient response of the single-input single-output (SISO) temperature control system [27,28]. However, many controlled objects are MIMO systems in practical applications, the dynamic uniform temperature of the MIMO system is widely required. In the MIMO temperature control system, the coupling effects and delay time differences make the system much more complex than the SISO temperature control system and courses the temperature difference between each channels. Thus, different from the SISO system control, the control performance of the MIMO system should not only focus on improving transient response and overshoot, more importantly, focus on reducing temperature differences to realize temperature uniformity of different channels. In this paper, we extend the previous NN control method from SISO to MIMO systems, clearly defining the parameters selection method of the reference model for the complex multi-point control system. Moreover, the coupling effects on the system performance can been effectively suppressed by the NN learning controller without specially designing decoupling compensators.

Based on our previous research, this proposal focuses on the multi-inputs multi-outputs (MIMO) temperature control system to improve the transient response of each channel in the MIMO system and reduce the temperature difference between each channel, realizing the temperature uniformity of the MIMO temperature system. A reference-model-based neural network control method combined with the integral-proportional-derivative (I-PD) controller is proposed. The system is driven by the error signal between the reference model output and real system outputs. The error signal of each channel is used as the teaching signal for the corresponding NN controller. The output of NN controller is added to the I-PD controller output, appropriately adjusting the control input of each channel. The MIMO temperature system is expected to achieve uniform temperature and steady state quickly. The rest of this paper is organized as follows: Section 2 describes the structure of the proposed MIMO control system. The simulation results and experimental results are presented in Sections 3 and 4, respectively.

Meanwhile, the results are quantitatively compared with those of the I-PD control system. A simple conclusion is given in Section 5.

2. Configuration of the MIMO Temperature Control System

This section describes the configuration of the proposed reference-model-based NN control method in the MIMO control system. In this paper, the MIMO system is simplified as a two-input two-output (2I2O) model. The block diagram of the proposed temperature control system is shown in Figure 1.

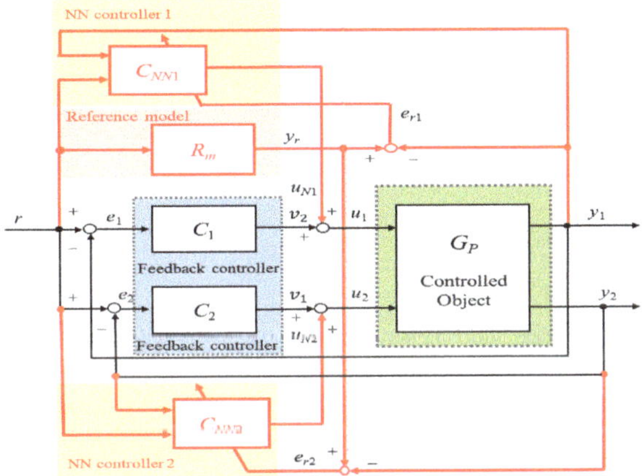

Figure 1. Block diagram of the multi-input multi-output (MIMO) temperature control system.

As shown in Figure 1, the temperature controller consists of two feedback controllers C_1 and C_2, a reference model R_m, and two neural network controllers C_{NN1} and C_{NN2}. Here, r is the reference of the system, and y_1 and y_2 indicate the actual outputs of two channels. v_1 and v_2 are the outputs of two feedback controllers C_1 and C_2, respectively. u_{N1} and u_{N2} indicate the outputs of C_{NN1} and C_{NN2}, respectively. Thus, the control inputs of the two channels can be respectively indicated by u_1 and u_2 which can be expressed as the sum of the outputs of the NN controllers and I-PD controllers. y_1 and y_2 represent the inputs and outputs of the 2I2O controlled object, respectively. Due to the control object of each channel being a plant with a time delay, the reference model R_m can be appropriately designed to provide the ideal temperature output with the same time delay which is the maximum time delay of two channels in the MIMO system. e_{r1} and e_{r2} are the errors between the outputs of the system and output of the reference model, respectively, which are the teaching signal for NN. The NN controller adjusts control inputs to compensate for the difference between the reference model output and each channel output. The explanation of the control system is divided into four main parts.

2.1. MIMO Controlled Objects with Strong Coupling Effects

The control system is designed based on the MIMO temperature system with a strong coupling influence, the schematic block diagram of the coupled system is shown in Figure 2, where u_1 and u_2 are defined as the inputs of channel ch1 and channel ch2, respectively. In addition, y_1 and y_2 indicate the output of ch1 and ch2, respectively. The coupling terms between the two channels are obtained as P_{21} and P_{12}, respectively. In this proposal, the controlled objects can be defined as a first-order plus time delay (FOPTD) system expressed as Equation (1) [29]. The step response characteristics of the controlled object is shown in Figure 3, where τ is time delay, K is the steady-state gain, and T is the time constant.

$$P(s) = \frac{K}{Ts + 1} e^{-\tau s} \tag{1}$$

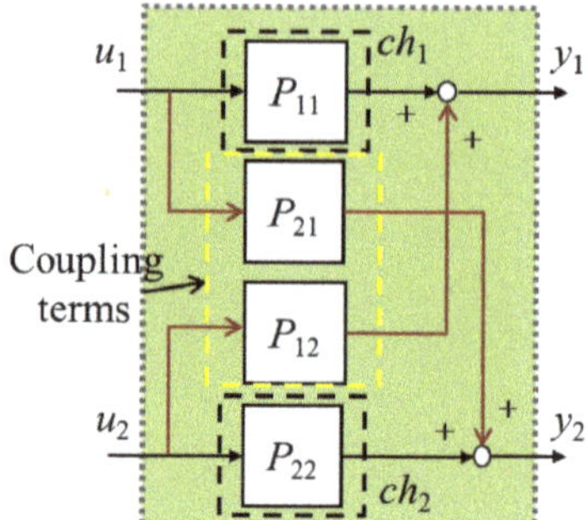

Figure 2. Block diagram of the two-input two-output (2I2O) controlled object.

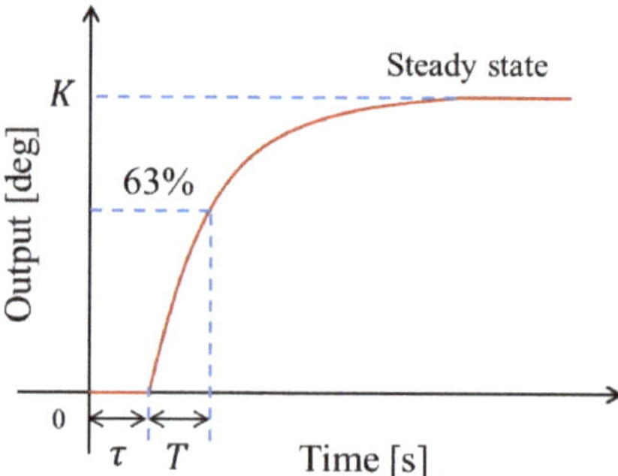

Figure 3. Step response of FOPTD plant.

2.2. Conventional I-PD Controller

Considering that the NN controller needs time to train its parameters for getting expected outputs, it will mainly act after training. Thus, the conventional I-PD controllers as the feedback controllers are designed for each channel to eliminate the proportional and derivative kick appeared during the set-point change and reduce the undesirable overshoot of the controlled variable [30]. The block diagram of the conventional I-PD controller is shown in Figure 4. For each I-PD controller parameters of the system, K_{pn} is the proportional gain and T_{in} is the integral time constant. T_{dn} represents the derivative time constant, where $n = 1, 2$. They are related to the plant parameters (K, T, and τ) as described above. In addition, μ represents the low-pass filter gain of the derivative term for reducing the high-frequency gain and noise. The feed-forward gain K_f is added to decide the system response speed ($K_f = 0$: slow; $K_f = 1$: fast).

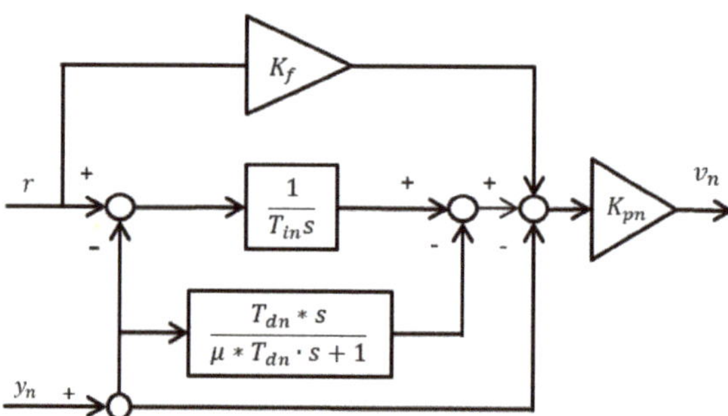

Figure 4. Block diagram of the integral-proportional-derivative (I-PD) controller.

For the industrial PID controller tuning, Ziegler–Nichols tuning rules are recognized and widely applied in actual control systems. The robustness and stability of the controllers are ensured. The parameters of two I-PD controllers C_1 and C_2 are calculated based on this tuning rule as Equations (2)–(4).

$$T_{i1} = 2\tau_{11}; \quad T_{i2} = 2\tau_{22} \tag{2}$$

$$T_{d1} = 0.5\tau_{11}; \quad T_{d2} = 0.5\tau_{22} \tag{3}$$

$$K_{p1} = \frac{1.2T_{11}}{K_{11} * \tau_{11}}; \quad K_{p2} = \frac{1.2T_{22}}{K_{22} * \tau_{22}}. \tag{4}$$

2.3. Multi-Layer NN Controller

In order to realize the uniform temperature of different channels, a multi-layer NN controller is introduced into each channel of the control system for the reference model output tracking. In the proposed system, each NN controller has one input layer with two neurons, one hidden layer with 10 neurons, and one output layer with one neuron. Thus, the structure of each multi-layer NN controller can be described as 2-10-1. Figure 5 illustrates the multi-layer neural network controllers.

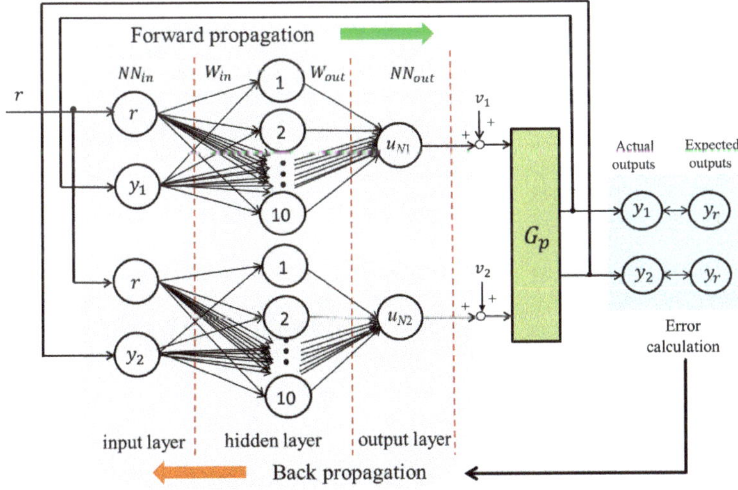

Figure 5. Structure of the multi-layer neural network controllers.

In this system, the reference value of the system r and y_1 (the output of ch1) are set as the input signals of the NN controller1, r and y_2 (the output of ch2) are set as the input signals of the NN controller2. For each controller, two inputs are transmitted in the forward direction through the network. The composition of the learning process of the network is by forward propagation and back propagation [31]. Neural network calculates and store intermediate variables from the input signals to the output signals, which is expressed as Equation (5):

$$NN_{out} = f(W_{out} * f(W_{in} * NN_{in} + b_1) + b_2). \tag{5}$$

For each NN controller, NN_{in} is the input vector and NN_{out} is the output vector. W_{in} and W_{out} are connection weights of neurons. b_1 is the bias of the hidden layer neurons, b_2 is the bias of the output layer neurons, and $f(*)$ is the activation function.

For training the network, the back propagation algorithm is used to update the weights and biases. Through this process, each NN controller constantly adjust its outputs u_{N1} and u_{N2} for achieving the minimum error between the reference model output and actual outputs of two channels in this system. Considering one of the NN controllers, a neuron in the output layer is called neuron j (here $j = 1$). The error at the output of the neuron j for nth iteration is defined by Equation (6). In the backward process, weights on the connections between all layers will be updated to minimize the error between target and output until the optimum weights are found [32]. Therefore, the total error is the sum of e_j for all neurons in the output layer, as given in Equation (7):

$$e_j(n) = y_r(n) - y_j(n) \tag{6}$$

$$\varepsilon(n) = \frac{1}{2} \sum_j e_j^2(n) \tag{7}$$

where the reference model output y_r is the expected output for neuron j and $y_j(n)$ is the actual outputs of two channels. The output for the neuron j can be expressed as Equation(8), where $k = 10$ is the number of inputs from the hidden layer. Here, W_{j0} equals the bias b_j applied to the neuron j:

$$y_j(n) = f(\sum_{i=0}^{k} W_{ji}(n)y_i(n)). \tag{8}$$

The connection weights of the neuron j are updated by the chain rule, and can be expressed in Equation (9), where $\delta_j(n)$ represents the local gradient of neuron j, given in Equation (10):

$$\frac{\partial \varepsilon(n)}{\partial W_{ji}(n)} = \delta_j(n)y_i(n) \tag{9}$$

$$\delta_j(n) = -e_j(n)f'(\sum_{i=0}^{k} W_{ji}(n)y_i(n)). \tag{10}$$

Therefore, the weight w_{ij} is updated by adopting the gradient descent, expressed as Equation (11). The correction ΔW_{ij} ensures w_{ij} changes in a way that always decreases the error, given in Equation (12), where α represents the learning rate of the back propagation:

$$W_{ji}(n+1) = W_{ji}(n) + \Delta W_{ij}(n) \tag{11}$$

$$\Delta W_{ij}(n) = -\alpha \frac{\partial \varepsilon(n)}{\partial W_{ji}(n)}. \tag{12}$$

The neuron bias connection for the neuron j is adjusted by $\delta_j(n)$ during training, as given in Equation (13), where β is the training gain of the bias:

$$b_j(n+1) = b_j(n) + \beta \delta_j(n). \tag{13}$$

In order to accelerate the training speed and solve the vanishing problem in the NN controller, the rectified linear unit (ReLU): $f(x) = max(0, x)$ is used as the activation function [33]. The derivative of the function is given in Equation (14):

$$f'(x) = \begin{cases} 1, & x \geq 0 \\ 0, & x < 0 \end{cases} \tag{14}$$

2.4. Reference Model

In the reference-model-based NN control structure, the teaching signal of each NN controller is provided by errors between the reference model output and real plant output. It also helps to prevent over learning of the NN controller.

For the controller design, time delays in the dynamic systems can be approximated by rational transfer functions. The exponential function can be defined as follows [34]:

$$e^x \approx \lim_{n \to \infty} \frac{1}{(\frac{x}{n} + 1)^n}.$$ (15)

In order to save memory and consider trade-off between accuracy and calculation burden, the time delay $e^{-\tau s}$ which is written as $1/e^{\tau s}$ can be described as the second-order rational approximation in Equation (16):

$$e^{-\tau s} \approx \frac{1}{(\frac{\tau s}{2} + 1)^2}.$$ (16)

Then the reference model with time delay can be expressed as Equation (17). Here, τ and T are delay time and the time constant of the reference model, which are set based on the real system model. It is designed to provide the ideal temperature output with the same time delay τ which is the maximum time delay of channels in the MIMO system. In addition, the time constant T of the reference model is the smaller time constant in the identified system model. The gain P_{RM} value of 0.01 is added to the plant time constant T for improving the transient response speed of the system:

$$R_m(s) = \frac{1}{T \cdot P_{RM} \cdot s + 1} * \frac{1}{(\frac{\tau s}{2} + 1)^2}.$$ (17)

3. Simulation Results

To verify the efficiency of the proposed control method, the control object is based on a real temperature control system. Figure 6 shows the experimental setup for the multi-input multi-output temperature system with strong coupling effects and large time delays.

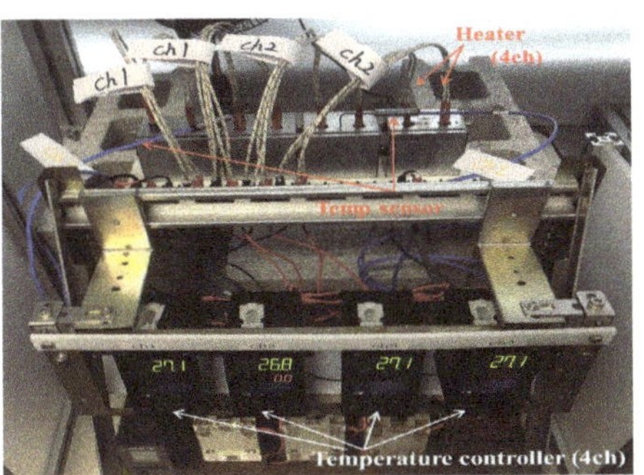

(**a**) Front view.   (**b**) Rear view.

Figure 6. Experimental setup.

3.1. Experimental Setup and System Identification

This system has four coupled aluminum blocks, the size being $60 \times 60 \times 50$ (mm). As shown in Figure 6b, they are arranged in line with the same separation distance by nuts, forming a two-input two-output system with strong coupling and large time delays. The left two blocks as a whole are channel ch1 and the right two blocks as a whole are channel ch2. Channel ch1 and channel ch2 have two heaters of 150 W (Cartridge Heater, type G2A56, WATLOW, Chiyoda, Japan) and the temperature sensor (K type thermocouple, RKC, Tokyo, Japan), respectively. The temperature sensor transforms the temperature (0–400 °C) into the output voltage (0–10 VDC). They are placed inside of holes with a depth of 30 [mm] (close to the inner center of aluminum blocks). The heaters are controlled by the solid-state relay (SSR, type G3PE-245BL, Omorn, Tokyo, Japan), which is driven by the pulse width modulation (PWM). The temperature is controlled by changing the duty ratio of the PWM signals. A digital signal processor (DSP, DS1104 R&D Controller Board, dSPACE Japan, Tokyo, Japan) is implemented as the temperature controller. The sampling time is set as 0.1 [s]. Although the SSR is a nonlinear element for the relay based on the PWM duty signal, it can be considered as the gain if the switching frequency of the relay is sufficiently large in comparison to the control bandwidth of the controlled object. In our system, the control bandwidth is $\frac{1}{2\pi * P_{RM} * T} = 0.0064$ [Hz] ($P_{RM} * T$ is the reference model time constant), while the PWM frequency is 10 [Hz]. Therefore, SSR can be handled as a linear factor.

The step signal (20% PWM duty cycle) is added to ch1 and ch2, in order. The heaters of the ch1 and heaters of ch2 are actuated, respectively. The ambient temperature during the performed identification experiments was 28 °C. The controlled object of the temperature control system is identified from the input-output(step response) measured data. The ARX (auto-regressive with eXogenous) model based on the least-squares criterion is applied to estimate the system transfer functions in MATLAB. The following Equation (18) shows the form of the ARX model, where $u(k)$ is the system inputs, $y(k)$ is the system outputs, n_k is the system delay, and $e(k)$ is the system disturbance:

$$A(q)y(k) = B(q)u(k - n_k) + e(k). \tag{18}$$

In Equation (18), $A(q)$ and $B(q)$ are given as follows:

$$A(q) = 1 + a_1 q^{-1} + \cdots + a_{n_a} q^{-n_a} \tag{19}$$

$$B(q) = b_1 q^{-1-n_k} + \cdots + b_{n_b} q^{-n_b - n_k} \tag{20}$$

where n_a and n_b are the orders of polynomial $A(q)$ and $B(q)$, respectively. The parameters of $A(q)$ and $B(q)$ are determined using the least squares method that minimizes the quadratic prediction error criteria. In MATLAB, the identified discrete-time model is transformed into the continuous-time model and model order is reduced by balanced realization, obtaining the identified system model in the form of first-order plus time delay [35,36]. The performance parameter used for validating the identified model is the percentage of fit as the following expression (21), where $y(k)$ is the actual output, $\hat{y}(k)$ is the model output, and $\bar{y}(k)$ is the mean of the actual output:

$$Fit(\%) = (1 - \frac{\sqrt{\sum_{k=1}^{N}[\hat{y}(k) - y(k)]^2}}{\sqrt{\sum_{k=1}^{N}[y(k) - \bar{y}(k)]^2}}) \times 100\%. \tag{21}$$

The identification results of the coupling terms are shown in Figure 7 and the estimation fit in results shows that the accuracy of the estimated model is above 95%.

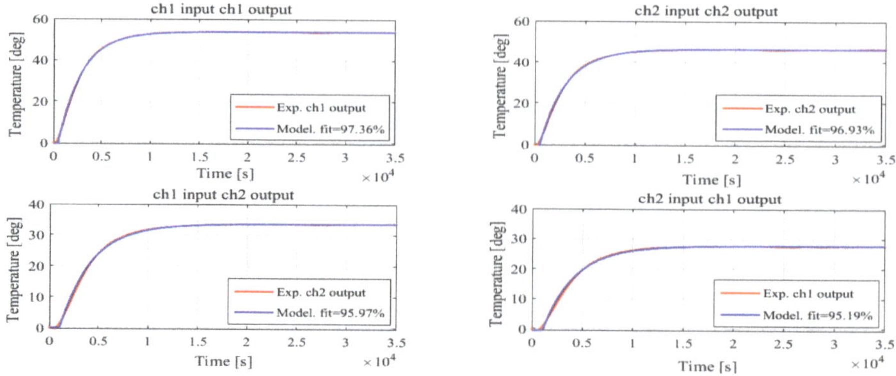

Figure 7. System identification results.

Thus, the identification results can be finalized as (22). The system controller parameter design, system simulation, and experiments are all based on the identified plant transfer functions:

$$G_p(s) = \begin{bmatrix} P_{11} & P_{12} \\ P_{21} & P_{22} \end{bmatrix} = \begin{bmatrix} \frac{2.7502}{2482.4s+1}e^{-431s} & \frac{1.4614}{3085.1s+1}e^{-1042s} \\ \frac{1.7352}{3195.9s+1}e^{-973s} & \frac{2.3937}{2588.6s+1}e^{-464s} \end{bmatrix}. \tag{22}$$

3.2. Simulation Results

According to the identified system model, the gains of two I-PD controllers are determined by the Ziegler–Nichols method, as described above. The gains of the I-PD controller (C_1) are $K_{p1} = 2.5146$, $T_{i1} = 861.5$, and $T_{d1} = 215.375$. The gains of the I-PD controller (C_2) are $K_{p2} = 2.7968$, $T_{i2} = 927.98$, and $T_{d2} = 231.995$. In addition, the reference model $R_m(s)$ is given as (23). Here, P_{RM} is set as 0.01:

$$R_m(s) \approx \frac{1}{2482.4 * 0.01 * s + 1} * \frac{1}{(\frac{464s}{2} + 1)^2}. \tag{23}$$

The hyper-parameters of both NN controllers are initialized as follows: The learning rate $\alpha = 1 \times 10^{-10}$ and the bias training gain $\beta = 1 \times 10^{-5}$. They are determined via the try and error method. The initial biases of each layer are initialized as zero. The weights of the NN controller are initialized to small random values between -1 and 1.

The simulation is divided into two phases. In the first phase, setting the reference signal to 100 °C, then the temperature of different channels increases from 0 °C to 100 °C. During this period, the neural network controller completes learning. Then, in the second phase, a repetitive step signal with an amplitude of 5 °C is added to the reference signals periodically. The offset of the reference signal is 100 °C. In one cycle, the temperature is controlled from 100 °C to 105 °C, then return to 100 °C. After multiple cycles, the first step response and last step response results in both directions can be obtained. Here, the temperature from 100 °C to 105 °C is defined as the positive direction control and the temperature from 105 °C to 100 °C is defined as the negative direction control.

Figure 8a shows the full time response of the control system and (b) shows the results of positive direction control (from 100 °C to 105 °C) and negative direction control (from 105 °C to 100 °C) of the NN control system. From the NN control system results, the first step response is almost the same as the last response, meaning the learning of the NN control ends at the beginning of the first step. Thus, the NN controllers realize the quick-learning. These results are quantitatively compared with those obtained by the conventional I-PD control with different feed-forward gains $K_f = 0$ (slow response) and $K_f = 1$ (fast response). The step-response characteristics of systems are computed from the response data in MATLAB, the rise time of a response is defined as the change in time required for the response to rise from 10% to 90% of the desired steady-state response yfinal (here, 105 °C and 100 °C, respectively). Furthermore, the settling time is defined as the time required for the error between the time response y(t) and yfinal to fall below 5% of the yfinal. Percentage overshoot is also relative to yfinal.

In the positive direction, two channels of the proposed NN control system follow the reference model as much as possible. The rise time of ch1 and ch2 in the NN control system is 753.2 s and 787.9 s, respectively. Compared with that of the I-PD ($K_f = 0$) control system, which is 931.2 s and 990.1 s, ch1 and ch2 has an improvement of 23.6% and 25.7%, respectively. Compared with that of the I-PD ($K_f = 1$) system which is 459.2 s and 482.4 s, although the transient response of the I-PD ($K_f = 1$) system is faster, the overshoots of two channels are 2.43 °C and 2.37 °C, which are about 48.6% and 47.5% of the reference value, respectively. The overshoots of two channels in the I-PD ($K_f = 0$) system are 0.51 °C and 0.4 °C, which are about 10.1% and 8% of the reference value, respectively. By contrast, both overshoots of the proposed NN system outputs are zero.

In addition, the settling time of ch1 and ch2 in the NN control system is 2225.1 s and 2178.4 s. Compared with that of the I-PD ($K_f = 1$) control system which is 3666.3 s and 3554.5 s, both channels have an improvement of about 42%. Compared with that of the I-PD ($K_f = 0$) control system which is

3634.9 s and 3144.1 s, ch1 improved by 38% and ch2 improved by 31%, respectively. The NN control system was the fastest to reach stable state without overshoots. Simulation results for the temperature from 100 °C to 105 °C are presented in Table 1. The rise time and settling time reflect the transient response and steady-state response speed of the control systems, respectively.

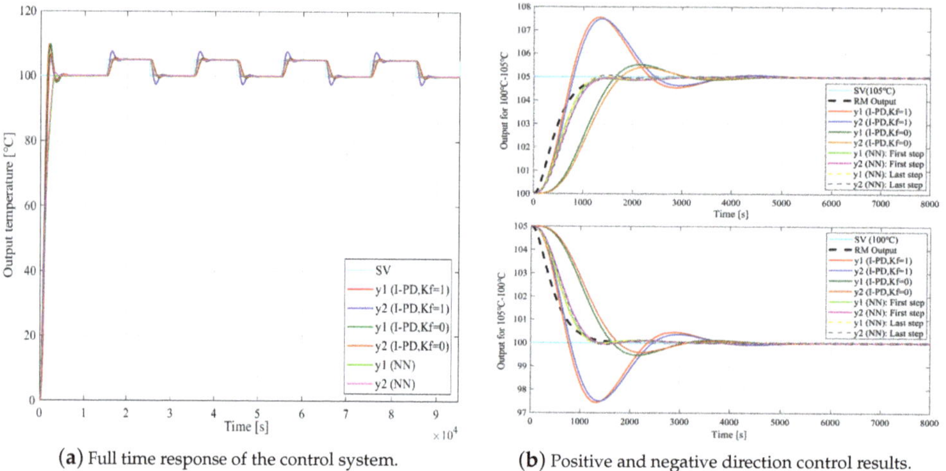

(**a**) Full time response of the control system.　　(**b**) Positive and negative direction control results.

Figure 8. Simulation results. (**a**) Full time response for the control system. (**b**) Results of positive direction control for the temperature from 100 °C to 105 °C and negative direction control for the temperature from 105 °C to 100 °C.

Table 1. Comparison of simulation results for time response characteristics (100 °C to 105 °C).

100 °C to 105 °C	Ref	ch1 ($K_f = 1$)	ch2 ($K_f = 1$)	ch1 ($K_f = 0$)	ch2 ($K_f = 0$)	ch1 (NN)	ch2 (NN)
Rise Time (s)	780.7	459.2	482.4	931.2	990.1	753.2	787.9
Settling Time (s)	1378.2	3663.0	3554.5	3634.9	3144.1	2225.1	2178.4
Overshoot (%)	0	48.6	47.5	10.1	8.0	0	0

In the negative direction, the temperature signal varies from 105 °C to 100 °C. Response characteristics of different control systems are compared and results are similar to the corresponding results in the positive direction. The rise time of ch1 and ch2 in the NN control system is 726.7 s and 772.2 s, respectively. Compared with that of the I-PD ($K_f = 0$) control system, which is 931.2 s and 990.1 s, ch1 and ch2 has an improvement of 21.9% and 22%, respectively. Compared with that of the I-PD ($K_f = 1$) system which is 459.2 s and 482.4 s, although the transient response of the I-PD ($K_f = 1$) system is faster, the overshoots of two channels are 2.43 °C and 2.37 °C, which are about 48.6% and 47.5% of the reference value, respectively. The overshoots of the two channels in the I-PD ($K_f = 0$) system are 0.51 °C and 0.4 °C, which are about 10.1% and 8% of the reference value, respectively. By contrast, both overshoots of the proposed NN system outputs are zero.

Additionally, the settling time of ch1 and ch2 in the I-PD ($K_f = 0$) control system is 3689 s and 3135.5 s. In the NN control system, it is 2241.3 s and 2164.5 s, which has an improvement of 39% and 31%, respectively. Compared with that of the I-PD ($K_f = 1$) control system which is 3669 s and 3553 s, the NN control system improved by 38.9% and 39%. From simulation results, different channels in the NN control system can quickly reach the stable state with no overshoot in contrast to the large overshoots of the I-PD control systems. Simulation results for temperature from 105 °C to 100 °C are presented in Table 2. These results show that the proposed NN control system has improved the temperature control efficiency in both directions.

Figure 9a,b, respectively show the temperature differences between ch1 and ch2 in positive and negative directions, using I-PD control with gains $K_f = 1, 0$ and the proposed NN control.

Table 2. Comparison of simulation results for time response characteristics (105 °C to 100 °C).

105 °C to 100 °C	Ref	ch1 ($K_f = 1$)	ch2 ($K_f = 1$)	ch1 ($K_f = 0$)	ch2 ($K_f = 0$)	ch1 (NN)	ch2 (NN)
Rise Time (s)	780.7	459.2	482.4	931.2	990.1	726.7	772.2
Settling Time (s)	1378.2	3669.0	3552.0	3689.0	3135.5	2241.3	2164.5
Overshoot (%)	0	48.6	47.5	10.1	7.9	0	0

In the positive direction, the maximum temperature difference of the I-PD control with gains $K_f = 1$ and $K_f = 0$ are 0.35 °C and 0.31 °C, about 7% and 6.2% of the reference temperature, respectively. The result of the proposed NN control system is 0.23 °C, about 4.6% of the reference temperature. The maximum temperature difference of the NN control system is decreased by 2.4% and 1.6%, compared to results of I-PD ($K_f = 0$) and I-PD ($K_f = 1$) control systems. Meanwhile, the temperature difference of the NN control drops to 0 °C in about 2400 s. The time has been shortened by 55% and 59% compared with the time for I-PD ($K_f = 1$) and I-PD ($K_f = 0$) control systems, which is about 5300 s and 5800 s.

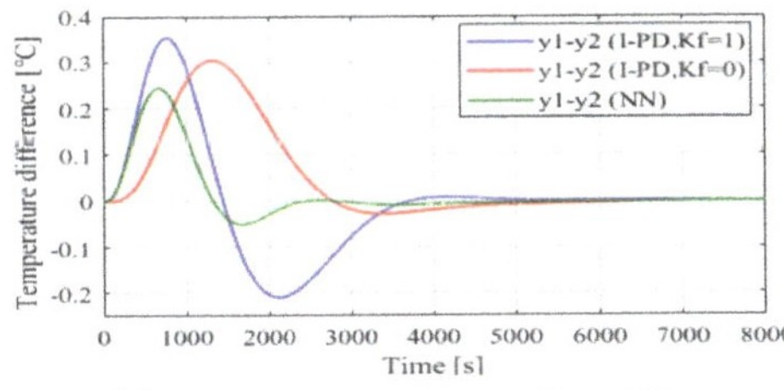

(**a**) Temperature difference (100 °C–105 °C).

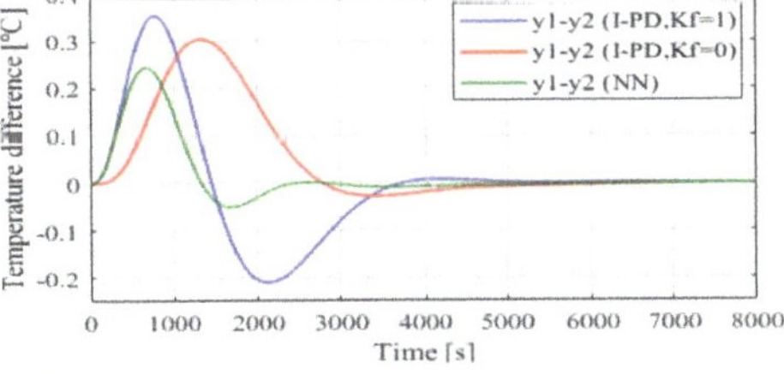

(**b**) Absolute temperature difference (105 °C–100 °C).

Figure 9. Simulation results. (**a**) Temperature differences between ch1 and ch2 from 100 °C to 105 °C. (**b**) Absolute temperature differences between ch1 and ch2 from 105 °C to 100 °C.

In the negative direction, results are similar to those in the positive direction. The maximum absolute temperature difference of the NN control system and the I-PD control system with gains $K_f = 1$ and $K_f = 0$ are about 0.24 °C, 0.34 °C and 0.31 °C, corresponding to 4.8%, 6.8%, and 6.2% of the reference value. The maximum temperature difference in the NN control system decreased by 1.4% and 1.6%, when compared to the results of I-PD ($K_f = 0$) and I-PD ($K_f = 1$) control systems. Meanwhile, the temperature difference of the NN control drops to 0 in about 2300 s. The time is shortened by 57% and 60% compared to that of I-PD ($K_f = 1$) and I-PD ($K_f = 0$) control systems, which is about 5400 s and 5700 s. From these simulation results, although the maximum temperature difference between ch1 and ch2 is not suppressed drastically, quick transient response and uniform temperature in different channels can be achieved.

In order to analyze the influences of the coupling terms between different channels, the time response results of the I-PD ($K_f = 1$) control system with coupling and without coupling effects are given in Figure 10, the reference signal is set to 100 °C. The overshoots of the system with coupling terms become bigger, about 11.2% and 8.4% of the reference signal. The overshoots of the system without coupling terms are about 9.4% and 7.8%, respectively. In addition, a slight oscillation is enhanced in both outputs for the coupling effects. From absolute temperature differences of outputs between the control system with coupling effects and without coupling effects, both output temperature differences in the I-PD system are large, with maximum values of about 11.4 °C and 16.6 °C, respectively. The maximum errors of the NN control system are about 4.2 °C and 5.1 °C in output temperatures.

The proposed NN control system can effectively weaken the coupling terms influences on the control system outputs.

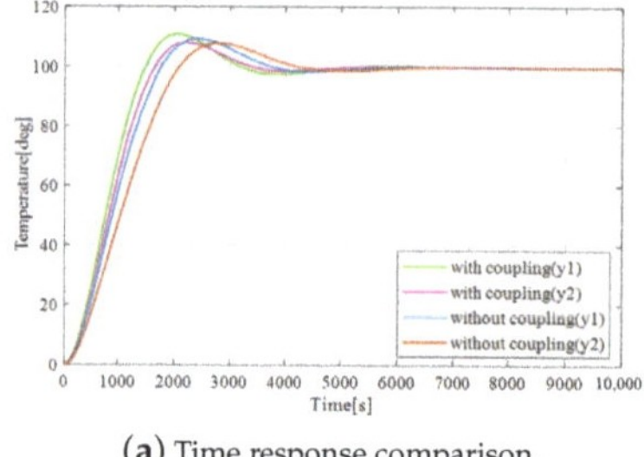

(**a**) Time response comparison.

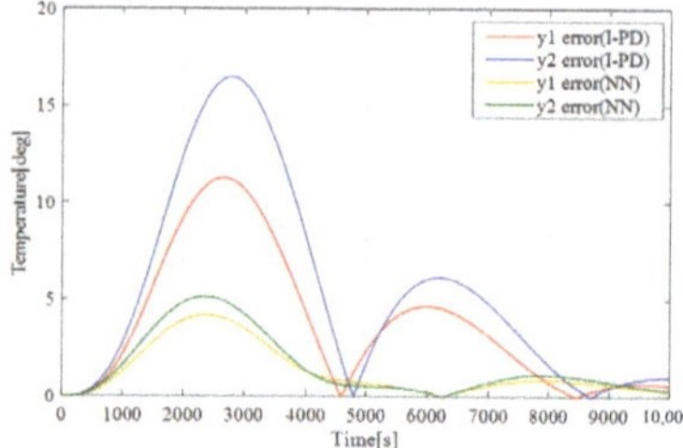

(**b**) Absolute temperature difference of outputs.

Figure 10. Analysis of coupling effects in the control system. (**a**) Compare time response results between the control system with coupling effects and without coupling effects. (**b**) Absolute temperature difference of outputs between the control system with coupling effects and without coupling effects.

4. Experimental Results

According to the identified system model introduced in the simulation, experiments with the proposed NN control method are carried out in the real temperature control system. The experimental setup is shown in Figure 6. Experimental conditions are as follows: The room temperature is set at 28 °C, sampling period is 0.1 s, controller sampling bit is 12 bits and the sensor resolution is 0.1 °C. In experiments, the reference temperature is first set to 100 °C, the temperature of two channels is controlled from the room temperature to 100 °C. Then a repetitive step signal with amplitude of 5 °C is added to the reference periodically. The temperature is controlled from 100 °C to 105 °C, then return to 100 °C. This process is repeated many times. The temperature increases from the room temperature to 100 °C, defined as the learning period of the NN controller, and the temperature from 100 °C to 105 °C and from 105 °C to 100 °C are the control results of the NN control system. For NN learning in the first step, we roughly estimated $[u_{N1}, y_1, e_{r1}, u_{N2}, y_2, e_{r2}] \times 100,000$ steps (sampling time 0.1 s) = 6.0×10^5 data is required to complete the learning. For testing the learned NN, one sequence of the step response is needed. Therefore, the same data number is required. Similarly, in the simulation, to verify the control performance of the proposed method, results are compared with those of the I-PD control with gains $K_f = 1$ (slow) and $K_f = 0$ (fast).

Figure 11a shows the results for full time response of the control system, and (b) shows the results for the temperature changes in positive and negative directions. As shown in Figure 11b, the I-PD control system with gain $K_f = 1$ has the fastest transient response speed, but is the slowest to reach the steady state in both directions. For the I-PD control system with gain $K_f = 0$, both the transient response and steady-state response are slower than the NN control system in both directions.

The time response characteristics of controlled systems are extracted from positive direction and negative direction response data in MATLAB. The response data is loaded, which is an array of response data y and corresponding time vector t. The 2nd-order butterworth digital low-pass filter with a cutoff frequency of 0.05 Hz (sampling frequency is 10 Hz) is applied to eliminate as much of the noise in the data as possible. The response characteristics were calculated from these data using the command "S = stepinfo(y, t, yfinal)" in MATLAB. Considering the noise in the data, the last value in y may not have the true steady-state response value. Therefore, the steady-state value (yfinal) should be set as 105 and 100 in the positive and negative direction control, respectively. The response characteristics of controlled systems such as rise time, settling time, and overshoot can be obtained.

In the positive direction, the rise time of ch1 and ch2 in the I-PD ($K_f = 0$) control system is 947.2 s and 973.1 s, in the NN control system, it is 812.4 s and 820.6 s, which improved by 14.2% and 15.7%. The settling time of ch1 and ch2 in the I-PD ($K_f = 0$) control system is 3414.2 s and 3217.9 s, and the NN control system had an improvement of 35.1% and 31.5%, which is 2217.3 s and 2204.8 s, respectively. In the I-PD ($K_f = 1$) control system, the settling time of ch1 and ch2 is 3824.3 s and 3718.6 s. By comparison, the NN control system improved by 42% and 40.7%. In addition, the overshoots of ch1 and ch2 in the I-PD ($K_f = 1$) control system are 2.01 °C and 1.92 °C, corresponding to 40.1% and 38.4% of the reference. The overshoots of ch1 and ch2 in the I-PD ($K_f = 0$) control system are about 0.69 °C and 0.53 °C, corresponding to 13.9% and 10.5% of the reference. In the NN control system, the overshoot of ch1 and ch2 are about 0.06 °C and 0.08 °C, corresponding to 1.3% and 1.7% of the reference, much smaller than those in the I-PD control systems.

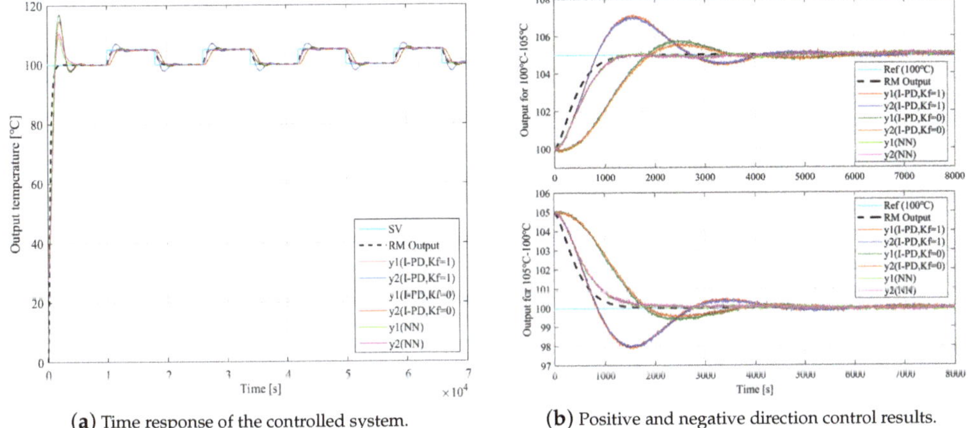

Figure 11. Experimental results. (**a**) Time response for the controlled system. (**b**) Results of positive direction control for the temperature from 100 °C to 105 °C and negative direction control for the temperature from 105 °C to 100 °C.

In the negative direction, similar results can be obtained. The rise time of ch1 and ch2 in the I-PD ($K_f = 0$) control system is 1027.3 s and 1093.4 s, in the NN control system is 877.3 s and 893.7 s, which has been improved by 14.6% and 18.3%. The settling time of ch1 and ch2 in the I-PD ($K_f = 0$) control system is 3492.7 s and 3325.1 s, the NN control system has an improvement of 38.6% and 37%, which is 2145.6 s and 2093.2 s, respectively. In the I-PD ($K_f = 1$) control system, the settling time of ch1 and ch2 is 3862.5 s and 3870.4 s. By comparison, the NN control system improved by 44.5% and 45.9%, respectively. In addition, the overshoots of ch1 and ch2 in the I-PD ($K_f = 1$) control system are about 2.14 °C and 2.07 °C, corresponding to 42.7% and 41.4% of the reference. The overshoots in the I-PD ($K_f = 0$) control system are 0.62 °C and 0.48 °C, corresponding to 12.3% and 9.7% of the reference. In the NN control system, the overshoots of ch1 and ch2 are about 0.11 °C and 0.12 °C, which are 2.1% and 2.5% of the reference. The overshoots in both directions are much smaller than the results of the I-PD control systems. Two channels can track the reference model output, realizing a quick and stable response to temperature signals. Comparing experimental results of the system response in Figure 11 with the simulation results in Figure 8, the discrepancy between the real system performance and simulation within 5% either settling time or overshoot, e.g., the errors of the NN system for the positive direction are about 2.3% and 1.7% in settling time and 1.3% and 1.7% in overshoot, respectively, the negative direction results are about 4.3% and 3.3% in settling time and 2.1% and 2.5% in overshoot. The corresponding errors of the I-PD control systems are slightly larger than 5%, e.g., the errors of the I-PD ($K_f = 0$) system for the negative direction are about 5.3% and 5.7% in settling time, respectively, the I-PD ($K_f = 1$) system errors are about 5.1% and 8.2%. Although the errors of rise time are about 10% in both directions, considering the difference between different channels is almost negligible, results are acceptable. These results show a good agreement between simulated and experimental results.

For comparison, the response characteristics of different control systems in positive and negative directions are listed in Tables 3 and 4, respectively. The percentage differences between the experimental results and simulation results are also given in both tables. From the above experimental results, it is seen that the proposed NN control gives great improvement in the performance of the temperature system.

Table 3. Comparison of experimental results for time response characteristics (100 °C to 105 °C).

Characteristics	Ref	CH1 ($K_f = 1$)	CH2 ($K_f = 1$)	CH1 ($K_f = 0$)	CH2 ($K_f = 0$)	CH1 (NN)	CH2 (NN)
Rise Time (s)	780.7	534.9 (14.1%)	557.6 (13.4%)	947.2 (1.7%)	973.1 (1.7%)	812.4 (11.6%)	820.6(4.1%)
SettlingTime 5% (s)	1378.2	3824.3 (4.2%)	3718.6 (4.4%)	3414.2 (6.1%)	3217.9 (2.3%)	2217.3 (2.3%)	2204.8 (1.7%)
Overshoot (%)	0	40.1 (4.8%)	38.4 (3.4%)	13.9 (3.9%)	10.5 (2.5%)	1.3 (1.3%)	1.7 (1.7%)

Table 4. Comparison of experimental results for time response characteristics (105 °C to 100 °C).

Characteristics	Ref	CH1 ($K_f = 1$)	CH2 ($K_f = 1$)	CH1 ($K_f = 0$)	CH2 ($K_f = 0$)	CH1 (NN)	CH2 (NN)
Rise Time (s)	780.7	519.1 (11.5%)	535.5 (9.8%)	1027.3 (9.3%)	1093.4 (9.4%)	877.3 (17.1%)	893.7 (13.6%)
SettlingTime 5% (s)	1378.2	3862.5 (5.1%)	3870.4 (8.2%)	3492.7 (5.3%)	3325.1 (5.7%)	2145.6 (4.3%)	2093.2 (3.3%)
Overshoot (%)	0	42.7 (5.9%)	41.4 (6.1%)	12.3 (2.2%)	9.7 (1.8%)	2.1 (2.1%)	2.5 (2.5%)

Figure 12a,b show the curves of temperature differences between the outputs of ch1 and ch2 in positive and negative directions, respectively. In the positive direction, the maximum temperature differences of the I-PD ($K_f = 1$), I-PD ($K_f = 0$), and NN control are 0.39 °C, 0.43 °C, and 0.15 °C, which correspond to 6.4%, 9.4%, and 3.8% of the reference value, respectively. In addition, the time for the temperature difference drops to 0 °C is about 3900 s, 3600 s, and 2300 s. The NN control system has an improvement of 41% and 36% compared to the I-PD ($K_f = 1$) and I-PD ($K_f = 0$) control system, respectively. As illustrated in Figure 12a, the temperature difference of the NN control system can be suppressed by the proposed method.

Similarly, in the negative direction, the maximum absolute temperature differences of the I-PD ($K_f = 1$), I-PD ($K_f = 0$), and NN control systems are 0.24 °C, 0.25 °C, and 0.14 °C, corresponding to 4.8%, 5%, and 2.8% of the reference value. Moreover, the time for the temperature difference drops to 0 °C is about 4400 s, 4100 s, and 3300 s, respectively. The NN control system has an improvement of 25% and 20% compared to the I-PD ($K_f = 1$) and I-PD ($K_f = 0$) control system, respectively. From these results, it was observed that the rapid uniform temperature response could be achieved in both transient state and steady state. The proposed control method improved the performance of the multi-input multi-output temperature system.

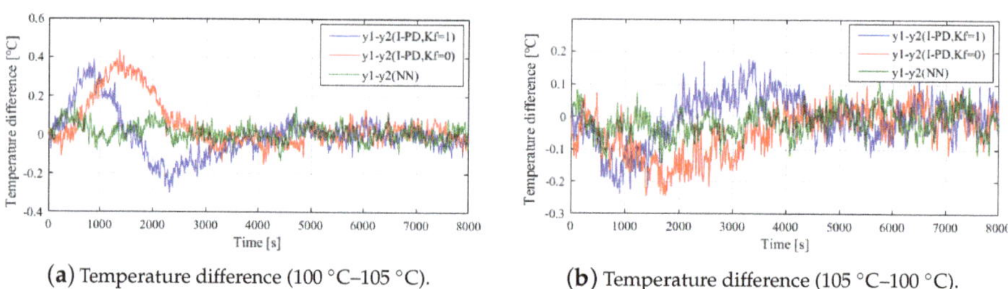

(**a**) Temperature difference (100 °C–105 °C).

(**b**) Temperature difference (105 °C–100 °C).

Figure 12. Experimental results. (**a**) Temperature differences between ch1 and ch2 from 100 °C to 105 °C. (**b**) Temperature differences between ch1 and ch2 from 105 °C to 100 °C.

5. Conclusions

In this paper, to improve transient response and realize the temperature uniformity in the multi-input multi-output temperature system with strong coupling effects, a reference-model-based neural network control method was proposed. In order to confirm the effectiveness of the proposed method, the proposed NN method was applied to a real MIMO temperature control system. Simulation and experiments were carried out, respectively. Both simulation results and experimental results were quantitatively compared with those of the I-PD control systems. The improvement of the transient response was achieved from the above experimental results, e.g., 42% and 40.7% improvements of two channels in settling time shortening compared to those of the traditional I-PD ($K_f = 1$) control system. In addition, the overshoots of different channels decreased by about 40% in both directions. The temperature uniformity of the MIMO system was achieved, e.g., in the positive direction, the temperature differences between two channels were reduced in more than half of those in the I-PD ($K_f = 1$) and I-PD ($K_f = 0$) control system. The temperature differences quickly went down to zero, with about 41% and 36% improvements in time compared to the I-PD control systems, respectively. These results show that the proposed NN control method improved the transient response and overshoot of the multi-input multi-output temperature system and realized the temperature uniformity of different channels in both transient state and steady state. The control effectiveness of the proposed method was successfully verified. For the NN control learning in the MIMO system, the selection of inputs and teaching signal is worth discussing in future study, e.g., the temperature difference between channels or derivative of difference can be added to inputs of the NN controller, thus it is possible to further improve the control performance.

Author Contributions: Conceptualization, S.H. and Y.L.; methodology, S.H.; software, S.H.; validation, Y.L., S.H. and S.X.; formal analysis, S.H. and Y.L.; investigation, S.H.; resources, S.H.; data curation, Y.L. and S.X.; writing–original draft preparation, Y.L.; writing–review and editing, S.X. and S.H.; visualization, S.H. and T.K.; supervision, S.H. and T.K.; project administration, S.H. and T.K.; funding acquisition, S.H. All authors have read and agreed to the published version of the manuscript.

Funding: This research received no external funding.

References

1. Johnson, M.A.; Moradi, M.H. *PID Control*; Springer-Verlag London Limited: London, UK, 2005.
2. Åström, K.J.; Hägglund, T. Revisiting the Ziegler–Nichols step response method for PID control. *J. Process. Control* **2004**, *14*, 635–650. [CrossRef]
3. Wang, Y.J.; Huang, S.T.; You, K.H. Calculation of robust and optimal fractional PID controllers for time delay systems with gain margin and phase margin specifications. In Proceedings of the 2017 36th Chinese Control Conference (CCC), Dalian, China, 26–28 July 2017; pp. 3077–3082.
4. Jacob, J.; Das, S.; Khaneja, N. A Concise Method of Pole Placement to Stabilize the Linear Time Invariant MIMO System. In Proceedings of the Sixth Indian Control Conference (ICC), Hyderabad, India, 18–20 December 2019; pp. 35–39.
5. Singh, J.; Chattterjee, K.; Vishwakarma, C.B. Two degree of freedom internal model control-PID design for LFC of power systems via logarithmic approximations. *ISA Trans.* **2018**, *72*, 185–196. [CrossRef] [PubMed]
6. Kurniawan, A.M.; Handogo, R.; Sutikno, J.P.; Lee, H.Y. Stability Criterion of Modified Inverse Nyquist Array on a Simple Non-square MIMO Process. *Int. J. Appl. Chem.* **2018**, *14*, 293–310.
7. Karbakhsh, F.; Gharehpetian, G.B.; Milimonfared, J.; Teymoori, A. Three phase photovoltaic grid-tied inverter based on feed-forward decoupling control using fuzzy-PI controller. In Proceedings of the 7th Power Electronics and Drive Systems Technologies Conference,Tehran, Iran, 16–18 February 2016; pp. 344–348.
8. Chen, P.Y.; Zhang, W.D. Improvement on an inverted decoupling technique for a class of stable linear multivariable processes. *ISA Trans.* **2007**, *46*, 199–210. [CrossRef] [PubMed]
9. Tuhta, S.; Günday, F.; Aydin, H.; Alalou, M. MIMO System Identification of MachineFoundation Using N4SID. *Int. J. Interdiscip. Innov. Res. Dev. (IJIIRD)* **2019**, *4*, 27–36.

10. Sands, T. Nonlinear-Adaptive Mathematical System Identification. *Computation* **2017**, *5*, 47. [CrossRef]

11. Valarmathi, R.; Guruprasath, M. System identification for a MIMO process. In Proceedings of the 2017 International Conference on Computation of Power, Energy Information and Commuincation, Melmaruvathur, India, 22–23 March 2017; pp. 435–441.

12. Suenaga, Y.; Matsunaga, N.; Kawaji, S. An Uniform Temperature Control on Thermal Conduction Surface by Gradient Control Method. In Proceedings of the 22nd SICE Kyusyu Branch Annual Conference, Fukui, Japan, 4–6 August 2003; pp. 243–246.

13. Mac Thi, T.; Copot, C.; De Keyser, R.; Tran, T.D.; Vu, T. MIMO fuzzy control for autonomous mobile robot. *J. Autom. Control Eng.* **2016**, *4*, 65–70. [CrossRef]

14. Sindhwani, N.; Bhamrah, M.S.; Garg, A.; Kumar, D. Performance analysis of particle swarm optimization and genetic algorithm in MIMO systems. In Proceedings of the 8th International Conference on Computing, Communication and Networking Technologies (ICCCNT), Delhi, India, 3–5 July 2017; pp. 1–6.

15. Slama, S.; Errachdi, A.; Benrejeb, M. Model reference adaptive control for MIMO nonlinear systems using RBF neural networks. In Proceedings of the 2018 International Conference on Advanced Systems and Electric Technologies, Hammamet, Tunisia, 22–25 March 2018; pp. 346–351.

16. Han, H.G.; Guo, Y.N.; Qiao, J.F. Nonlinear system modeling using a self-organizing recurrent radial basis function neural network. *Appl. Soft Comput.* **2018**, *71*, 1105–1116. [CrossRef]

17. Liu, W.; Wang, Z.; Liu, X.; Zeng, N. A survey of deep neural network architectures and their applications. *Neurocomputing* **2017**, *234*, 11–26. [CrossRef]

18. Baek, M.S.; Kwak, S.; Jung, J.Y.; Kim, H.M.; Choi, D.J. Implementation methodologies of deep learning-based signal detection for conventional MIMO transmitters. *IEEE Trans. Broadcast.* **2019**, *65*, 636–642. [CrossRef]

19. Sundermeyer, M.; Schlüter, R.; Ney, H. LSTM neural networks for language modeling. In Proceedings of the Thirteenth Annual Conference of the International Speech Communication Association, Portland, OR, USA, 9–13 September 2012.

20. Ryu, S.H.; Moon, H.J. Development of an occupancy prediction model using indoor environmental data based on machine learning techniques. *Build. Environ.* **2016**, *107*, 1–9. [CrossRef]

21. Esfe, M.H.; Razi, P.; Hajmohammad, M.H.; Rostamian, S.H.; Sarsam, W.S.; Arani, A.A.A.; Dahari, M. Optimization, modeling and accurate prediction of thermal conductivity and dynamic viscosity of stabilized ethylene glycol and water mixture Al2O3 nanofluids by NSGA-II using ANN. *Int. Commun. Heat Mass Transf.* **2017**, *82*, 154–160. [CrossRef]

22. Ghritlahre, H.K.; Prasad, R.K. Prediction of thermal performance of unidirectional flow porous bed solar air heater with optimal training function using artificial neural network. *Energy Procedia* **2017**, *109*, 369–376. [CrossRef]

23. Shen, C.; Song, R.; Li, J.; Zhang, X.; Tang, J.; Shi, Y.; Liu, J.; Cao, H. Temperature drift modeling of MEMS gyroscope based on genetic-Elman neural network. *Mech. Syst. Signal Process.* **2016**, *72*, 897–905.

24. Li, X.; Zhao, T.; Zhang, J.; Chen, T. Predication control for indoor temperature time-delay using Elman neural network in variable air volume system. *Energy Build.* **2017**, *154*, 545–552. [CrossRef]

25. Katic, K.; Li, R.L.; Verharrt, J.; Zeiler, W. Neural network based predictive control of personalized heating systems. *Energy Build.* **2018**, *174*, 199–213. [CrossRef]

26. Carvalho, C.B.; Carvalho, E.P.; Ravagnani, M.A. Implementation of a neural network MPC for heat exchanger network temperature control. *Braz. J. Chem. Eng.* **2020**, 1–16, doi:10.1007/s43153-020-00058-2. [CrossRef]

27. Hashimoto, S.; Xu, S.; Jiang, Y.; Nishizawa, Y. AI-Based Feedback Control Applicable to Process Control Systems. In Proceedings of the International Conference on Mechanical, Electrical and Medical Intelligent System, Kiryu, Japan, 4–6 December 2019.

28. Xu, S.; Hashimoto, S.; Jiang, W. A Novel Reference Model-Based Neural Network Approach to Temperature Control System. In Proceedings of the IOP Conference Series: Materials Science and Engineering, Wuhan, China, 10–12 October 2019.

29. Wu, S.; Zhang, R.; Lu, R.; Gao, F. Design of dynamic matrix control based PID for residual oil outlet temperature in a coke furnace. *Chemom. Intell. Lab. Syst.* **2014**, *134*, 110–117. [CrossRef]

30. Rajinikanth, V.; Latha, K. I-PD controller tuning for unstable system using bacterial foraging algorithm: A study based on various error criterion. *Appl. Comput. Intell. Soft Comput.* **2012**, *2012*. [CrossRef]

31. Hameed, A.A.; Karlik, B.; Salman, M.S. Back-propagation algorithm with variable adaptive momentum. *Knowl. Based Syst.* **2019**, *114*, 79–87. [CrossRef]

32. Ozbay, Y.; Karlik, B. A fast training back-propagation algorithm on windows. In Proceedings of the 3rd International Symposium on Mathematical and Computational Applications, Konya, Turkey, 4 September 2002; pp. 204–210.
33. Agarap, A.F. Deep learning using rectified linear units (relu). *arXiv* **2018**, arXiv:1803.08375.
34. Salas, A.H. The exponential function as a limit. *Appl. Math. Sci.* **2012**, *91*, 4519–4526.
35. Chetouani, Y. Using ARX approach for modelling and prediction of the dynamics of a reactor-exchanger. *Inst. Chem. Eng. Symp. Ser.* **2008**, *154*, 297.
36. Batlle, C.; Roqueiro, N. Balanced model order reduction for systems depending on a parameter. *arXiv* **2016**, arXiv:1604.08086.

Publisher's Note: MDPI stays neutral with regard to jurisdictional claims in published maps and institutional affiliations.

MDPI AG
Grosspeteranlage 5
4052 Basel
Switzerland
Tel.: +41 61 683 77 34

Processes Editorial Office
E-mail: processes@mdpi.com
www.mdpi.com/journal/processes